W9-BRC-990

Simplified Engineering for Architects and Builders

Simplified Engineering for Architects and Builders

The Late Harry Parker, M.S.

Formerly Professor of Architectural Construction
University of Pennsylvania

SEVENTH EDITION

prepared by

James Ambrose, M.S.

Professor of Architecture
University of Southern California

A Wiley-Interscience Publication

JOHN WILEY & SONS

New York Chichester Brisbane Toronto Singapore

Copyright © 1989 by John Wiley & Sons, Inc.

All rights reserved. Published simultaneously in Canada.

Reproduction or translation of any part of this work beyond that permitted by Section 107 or 108 of the 1976 United States Copyright Act without the permission of the copyright owner is unlawful. Requests for permission or further information should be addressed to the Permissions Department, John Wiley & Sons, Inc.

Library of Congress Cataloging-in-Publication Data:

Parker, Harry, 1887–
Simplified engineering for architects and builders.

"A Wiley-Interscience publication."
Bibliography: p.
Includes index.
1. Structural engineering. I. Ambrose, James E.
II. Title.

TA633.P3 1989 690′.21 88-37879
ISBN 0-471-61806-3

Printed in the United States of America

10 9 8 7 6 5 4 3

Preface to the Seventh Edition

With the publication of this edition of Professor Parker's book, another generation of building designers is afforded the opportunity to utilize the most enduringly popular general reference for study of the topic of design of building structures. The basic intentions and purpose of the book remain the same as those stated by Professor Parker in his preface to the first edition, which follows. As the complexity of engineering and building design steadily increases, the need for a concise presentation of basic materials for persons not thoroughly trained in engineering is, if anything, even greater now.

The body of subject material as presented in the earlier editions of the book remains essentially intact here. The principal changes in this edition consist of an expansion of the subject coverage to include topics not previously developed, including those of structural masonry, design for lateral forces, elements of dynamic behavior, plywood shear walls and diaphragms, rigid frames, and structures with internal pins. Design of roof trusses as presented in Part V in previous editions has been omitted, although discussions of trusses are included in all parts in this edition.

A major addition to this edition is the development in a new Part V of the topic of design of structural systems for buildings. This new part contains both general discussions of basic concerns for system design and examples of the design of elements of the systems for three buildings. This affords the reader an opportunity to see the applications of the materials in the preceding parts to design situations of broader scope, much as they are typically encountered by building designers.

The work presented here is intended to be understandable by persons lacking formal training in engineering. As in previous editions, the level of mathematical work has in general been kept to that involving only simple algebra, geometry, and some elementary trigonometry. Readers with more mathematical training or some engineering background may apply more rigorous analytical procedures to the investigative work, but should find the basic procedures for design work to be essentially as presented here.

This is, of course, the age of the computer, and much of the work presented here can be, and is being, done with computer-aided methods. The purpose of this book is to explain and demonstrate basic concepts and fundamentals of design, not to present sophisticated design methodologies. We have therefore set up example computations here as "hand" operations in order to involve the reader more directly and simply in the processes, which require only a pocket calculator for mathematical computations. Anyone aspiring to do professional work in building design is strongly advised to pursue familiarity with computer-aided design methods, which are sure to be more understandable after one has mastered the elementary concepts and procedures demonstrated here.

From my own experience with the use of this book and others in several years of teaching, I have developed many materials in the form of exercise problems, quizzes, and examinations. I have also recently had the experience of helping to write portions of the architectural registration examinations. In order to provide some extra help to persons who may be studying this material largely on their own, I have prepared a companion book titled *Study Manual for Simplified Engineering for Architects and Builders, Seventh Edition*. This publication contains illustrated, worked solutions for all of the exercise problems in this book that are of a computational nature. It also contains extra exercise problems with answers, lists of significant words and terms, general questions on the book subjects, a key to other sources of study, and self-tests using a multiple-choice format. This book can be obtained through most bookstores or directly from the publisher, John Wiley & Sons.

I am grateful to the American Institute of Steel Construction, American Concrete Institute, National Forest Products Association, Steel Deck Institute, Steel Joist Institute, and International Conference of Building Officials for permission to reprint and abstract materials from their publications. I am also very appreciative of the work and of the support given to me by the editors and production staff at John Wiley & Sons. Writing requires intense and extensive concentration, and writing mostly at home—as I do—requires a lot of understanding and deference by one's family. I take that for granted sometimes, but am highly appreciative of it and of the general encouragement given me by my wife, Peggy, my daughter, Julie, and my son, Jeff.

JAMES AMBROSE

Westlake Village, California
March 1989

Preface to the First Edition

To the average young architectural draftsman or builder, the problem of selecting the proper structural member for given conditions appears to be a difficult task. Most of the numerous books on engineering which are available assume that the reader has previously acquired a knowledge of fundamental principles and, thus, are almost useless to the beginner. It is true that some engineering problems are exceedingly difficult, but it is also true that many of the problems that occur so frequently are surprisingly simple in their solution. With this in mind, and with a consciousness of the seeming difficulties in solving structural problems, this book has been written.

In order to understand the discussions of engineering problems, it is essential that the student have a thorough knowledge of the various terms which are employed. In addition, basic principles of forces in equilibrium must be understood. The first section of this book, "Principles of Mechanics," is presented for those who have not had such technical preparation, as well as for those who wish a brief review of the subject. Following this section are structural problems involving the most commonly used building materials, wood, steel, reinforced concrete, and roof trusses. A major portion of the book is devoted to numerous problems and their solution, the purpose of which is to explain practical procedure in the design of structural members. Similar examples are given to be solved by the student. Although handbooks published by the manufacturers are necessities to the more advanced student, a great number of appropriate tables are presented herewith so that sufficient data are directly at hand to those using this book.

Care has been taken to avoid the use of advanced mathematics, a knowledge of arithmetic and high-school algebra being all that is required to follow the discussions presented. The usual formulas employed in the solution of structural problems are given with explanations of the terms involved and their application, but only the most elementary of these formulas are derived. These derivations are given to show how simple they are and how the underlying principle involved is used in building up a formula that has a practical application.

No attempt has been made to introduce new methods of calculation, nor have all the various methods been included. It has been the desire of the author to present to those having little or no knowledge of the subject simple solutions of everyday structural problems. Whereas thorough technical training is to be desired, it is hoped that this presentation of fundamentals will provide valuable working knowledge and, perhaps, open the doors to more advanced study.

HARRY PARKER

Philadelphia, Pennsylvania
March 1938

Contents

Simplified Engineering for Architects and Builders

Introduction

The principal purpose of this book is to develop the topic of *structural design*. However, in order to do the necessary work for design, use must be made of various methods for *structural investigation*. The work of investigation consists of the consideration of the tasks required of a structure and the evaluation of the responses of the structure in performing these tasks. Investigation may be performed in various ways, the principle ones being the use of modeling by either mathematics or the construction of physical models. For the designer, a major first step in any investigation is the visualization of the structure and the force actions to which it must respond. In this book, extensive use is made of graphic illustrations in order to encourage the reader in the development of the habit of first clearly *seeing* what is happening, before proceeding with the essentially abstract task of mathematical investigation.

Structural Mechanics

The branch of physics called *mechanics* concerns the actions of forces on material bodies. Most of engineering design and investigation is based on applications of the science of mechanics. *Stat-*

ics is the branch of mechanics that deals with bodies held motionless by the balanced nature (called *static equilibrium*) of the forces acting on them. *Dynamics* is the branch of mechanics that concerns bodies in motion or forces that are involved with time-dependent relationships.

When external forces act on a body, two things happen. First, internal forces that resist the effects of the external forces are set up in the body. These internal forces produce *stresses* in the material of the body. Second, the external forces produce *deformations,* or changes, in the shape of the body. *Strength of materials,* or mechanics of materials, is the study of the properties of material bodies that enable them to resist the actions of external forces, of the stresses within bodies, and of the deformations that result from external forces.

Taken together, the topics of applied mechanics and strength of materials are often given the overall designation of *structural mechanics* or *structural analysis*. Investigation is essentially an analytical process. Design is a refining process in which a structure is first visualized in some detail; then it is investigated for required force responses and its performance evaluated; finally—possibly after several cycles of modification and investigation—an acceptable form is derived for the structure.

Units of Measurement

At the time of preparation of this edition, the building industry in the United States is still in a state of confused transition from the use of English units (feet, pounds, etc.) to the new metric-based system referred to as the SI units (for Système International). Although a complete phase-over to SI units seems inevitable, at the time of this writing the construction-materials and products suppliers in the United States are still resisting it. Consequently, the AISC Manual and most building codes and other widely used references are still in the old units. (The old system is now more appropriately called the U.S. system because England no longer uses it!) Although it results in some degree of clumsiness in the work, we have chosen to give the data and computations in this

TABLE 1 Units of Measurement: U.S. System

Name of Unit	Abbreviation	Use
Length		
Foot	ft	large dimensions, building plans, beam spans
Inch	in.	small dimensions, size of member cross sections
Area		
Square feet	ft^2	large areas
Square inches	$in.^2$	small areas, properties of cross sections
Volume		
Cubic feet	ft^3	large volumes, quantities of materials
Cubic inches	$in.^3$	small volumes
Force, mass		
Pound	lb	specific weight, force, load
Kip	k	1000 pounds
Pounds per foot	lb/ft	linear load (as on a beam)
Kips per foot	k/ft	linear load (as on a beam)
Pounds per square foot	lb/ft^2, psf	distributed load on a surface
Kips per square foot	k/ft^2, ksf	distributed load on a surface
Pounds per cubic foot	lb/ft^3, pcf	relative density, weight
Moment		
Foot-pounds	ft-lb	rotational or bending moment
Inch-pounds	in.-lb	rotational or bending moment
Kip-feet	k-ft	rotational or bending moment
Kip-inches	k-in.	rotational or bending moment
Stress		
Pounds per square foot	lb/ft^2, psf	soil pressure
Pounds per square inch	$lb/in.^2$, psi	stresses in structures
Kips per square foot	k/ft^2, ksf	soil pressure
Kips per square inch	$k/in.^2$, ksi	stresses in structures
Temperature		
Degree Fahrenheit	°F	temperature

TABLE 2 Units of Measurement: SI System

Name of Unit	Abbreviation	Use
Length		
Meter	m	large dimensions, building plans, beam spans
Millimeter	mm	small dimensions, size of member cross sections
Area		
Square meters	m^2	large areas
Square millimeters	mm^2	small areas, properties of cross sections
Volume		
Cubic meters	m^3	large volumes
Cubic millimeters	mm^3	small volumes
Mass		
Kilogram	kg	mass of materials (equivalent to weight in U.S. system)
Kilograms per cubic meter	kg/m^3	density
Force (load on structures)		
Newton	N	force or load
Kilonewton	kN	1000 newtons
Stress		
Pascal	Pa	stress or pressure (1 pascal = 1 N/m^2)
Kilopascal	kPa	1000 pascals
Megapascal	MPa	1,000,000 pascals
Gigapascal	GPa	1,000,000,000 pascals
Temperature		
Degree Celsius	°C	temperature

book in both units as much as is practicable. The technique is generally to perform the work in U.S. units and immediately follow it with the equivalent work in SI units enclosed in brackets [thus] for separation and identity.

Table 1 lists the standard units of measurement in the U.S.

TABLE 3 Factors for Conversion of Units

To Convert from U.S. Units to SI Units Multiply by	U.S. Unit	SI Unit	To Convert from SI Units to U.S. Units Multiply by
25.4	in.	mm	0.03937
0.3048	ft	m	3.281
645.2	$in.^2$	mm^2	1.550×10^{-3}
16.39×10^3	$in.^3$	mm^3	61.02×10^{-6}
416.2×10^3	$in.^4$	mm^4	2.403×10^{-6}
0.09290	ft^2	m^2	10.76
0.02832	ft^3	m^3	35.31
0.4536	lb (mass)	kg	2.205
4.448	lb (force)	N	0.2248
4.448	kip (force)	kN	0.2248
1.356	ft-lb (moment)	N-m	0.7376
1.356	kip-ft (moment)	kN-m	0.7376
1.488	lb/ft (mass)	kg/m	0.6720
14.59	lb/ft (load)	N/m	0.06853
14.59	kips/ft (load)	kN/m	0.06853
6.895	psi (stress)	kPa	0.1450
6.895	ksi (stress)	MPa	0.1450
0.04788	psf (load or pressure)	kPa	20.93
47.88	ksf (load or pressure)	kPa	0.02093
0.566 × (°F − 32)	°F	°C	(1.8 × °C) + 32

system with the abbreviations used in this work and a description of the type of the use in structural work. In similar form, Table 2 gives the corresponding units in the SI system. The conversion units used in shifting from one system to the other are given in Table 3.

For some of the work in this book, the units of measurement are not significant. What is required in such cases is simply to find a numerical answer. The visualization of the problem, the manipulation of the mathematical processes for the solution, and the quantification of the answer are not related to the specific units—only to their relative values. In such situations we have occasionally chosen not to present the work in dual units, to provide a less confusing illustration for the reader. Although this procedure may be allowed for the learning exercises in this book, the structural designer is generally advised to develop the habit of always indicating the units for any numerical answers in structural computations.

Computations

In professional design firms, structural computations are most commonly done with computers, particularly when the work is complex or repetitive. Anyone aspiring to participation in professional design work is advised to acquire the background and experience necessary to the application of computer-aided techniques. The computational work in this book is simple and can be performed easily with a pocket calculator. The reader who has not already done so is advised to obtain one. The "scientific" type with eight-digit capacity is quite sufficient.

For the most part, structural computations can be rounded off. Accuracy beyond the third place is seldom significant, and this is the level used in this work. In some examples more accuracy is carried in early stages of the computation to ensure the desired degree in the final answer. All the work in this book, however, was performed on an eight-digit pocket calculator.

Symbols

The following "shorthand" symbols are frequently used:

Symbol	Reading
$>$	is greater than
$<$	is less than
$\geqslant$	equal to or greater than
$\leqslant$	equal to or less than
6′	six feet
6″	six inches
Σ	the sum of
ΔL	change in L

Notation

Use of standard notation in the general development of work in mechanics and strength of materials is complicated by the fact that there is some lack of consistency in the notation currently used in the field of structural design. Some of the standards used in the field are developed by individual groups (notably those relating to a single basic material, wood, steel, concrete, masonry, etc.) which each have their own particular notation. Thus the same type of stress (e.g., shear stress in a beam) or the same symbol (f_c) may have various representations in structural computations. To keep some form of consistency in this book, we use the following notation, most of which is in general agreement with that used in structural design work at present.

a	(1) Moment arm; (2) acceleration; (3) increment of an area
A	Gross (total) area of a surface or a cross section
b	Width of a beam cross section
B	Bending coefficient
c	Distance from neutral axis to edge of a beam cross section
d	Depth of a beam cross section or overall depth (height) of a truss
D	(1) Diameter; (2) deflection
e	(1) Eccentricity (dimension of the mislocation of a load resultant from the neutral axis, centroid, or simple center of the loaded object); (2) elongation
E	Modulus of elasticity (ratio of unit stress to the accompanying unit strain)
f	Computed unit stress
F	(1) Force; (2) allowable unit stress
g	Acceleration due to gravity
G	Shear modulus of elasticity
h	Height
H	Horizontal component of a force
I	Moment of inertia (second moment of an area about an axis in the plane of the area)

J	Torsional (polar) moment of inertia
K	Effective length factor for slenderness (of a column: KL/r)
M	Moment
n	Modular ratio (of the moduli of elasticity of two different materials)
N	Number of
p	(1) Percent; (2) unit pressure
P	Concentrated load (force at a point)
r	Radius of gyration of a cross section
R	Radius (of a circle, etc.)
s	(1) Center-to-center spacing of a set of objects; (2) distance of travel (displacement) of a moving object; (3) strain or unit deformation
t	(1) Thickness; (2) time
T	(1) Temperature; (2) torsional moment
V	(1) Gross (total) shear force; (2) vertical component of a force
w	(1) Width; (2) unit of a uniformly distributed load on a beam
W	(1) Gross (total) value of a uniformly distributed load on a beam; (2) gross (total) weight of an object
Δ (delta)	Change of
Σ (sigma)	Sum of
θ (theta)	Angle
μ (mu)	Coefficient of friction
ϕ (phi)	Angle

Some of the special notation that is used in individual topic areas is given at the beginning of the other parts of this book.

I

PRINCIPLES OF STRUCTURAL MECHANICS

1

Force Actions

1.1 Forces and Stresses

The idea of force is one of the fundamental concepts of mechanics and as such does not lend itself to simple, precise definition. For our purposes at this stage of study, we may define a force as that which produces, or tends to produce, motion or a change in the motion of bodies. One type of force is the effect of *gravity,* by which all bodies are attracted toward the center of the earth. The magnitude of the force of gravity is the *weight* of a body. The amount of material in a body is its *mass*. In the U.S. (old English) System the force effect of gravity is equated to the weight. In SI units a distinction is made between weight and force which results in the force unit of a *newton*. In the U.S. System the basic unit of force is the *pound,* although in engineering work a commonly used unit is the *kip* (1000 pounds or, literally, a kilopound).

Figure 1.1*a* represents a block of metal weighing 6400 lb supported on a short piece of wood having an 8 × 8 in. cross-sectional area. The wood is in turn supported on a base of masonry. The force of the metal block exerted on the wood is 6400 lb, or 6.4 kips. Note that the wood transfers a force of equal magnitude (ignoring the weight of the wood block) to the masonry base. If

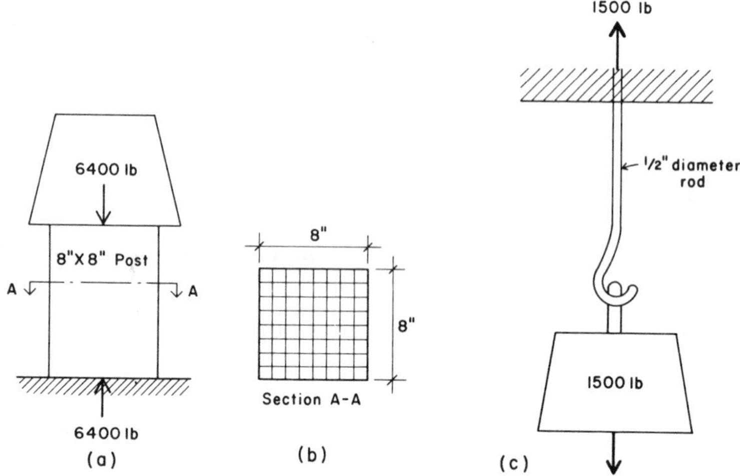

FIGURE 1.1 Direct force actions and stresses; compression and tension.

there is no motion (equilibrium), there must be an equal upward force in the base. For equilibrium, force actions must exist in opposed pairs. In this instance the magnitude of the force is 6400 lb, and the resisting force offered by the masonry and the wood block is also 6400 lb. The resisting force in the block of wood is developed by *stress,* defined as internal force per unit area of the block's cross section. For the situation shown, each square inch of the cross section must develop a stress equal to 6400/64 = 100 lb/in.^{2} (psi). (See Fig. 1.1*b*.)

External forces may result from a number of sources but essentially are distinguished only as being either static or dynamic. At present we are dealing only with static forces. Internal forces are one of three possible types, tension, compression, or shear.

When a force acts on a body in a manner that tends to shorten the body or to push the parts of the body together, the force is a compressive force and the stresses within the body are compressive stresses. The block of metal acting on the piece of wood in Fig. 1.1 represents a compressive force, and the resulting stresses in the wood are compressive stresses.

Figure 1.1*c* represents a 0.5-in.-diameter steel rod suspended from a ceiling. A weight of 1500 lb is attached to the lower end of the rod. The weight constitutes a tensile force, which is a force that tends to lengthen or pull apart the body on which it acts. In this example the rod has a cross-sectional area of πR^2, or $3.1416(0.25)^2 = 0.196$ in.^{2}. Hence the tensile unit stress in the rod is $1500/0.196 = 7653$ psi.

In this book we consider the weights given in U.S. units to be forces and make direct conversion from pounds of force to newtons of force. (See the discussion in the Introduction and the conversion factors given in Table 3.) Thus for the wood block in Fig. 1.1*a:*

$$\text{Force} = 6400 \text{ lb} = 4.448 \times 6400 = 28{,}467 \text{ N}, \quad \text{or } 28.467 \text{ kN}$$

$$\text{Stress} = 100 \text{ psi} = 6.895 \times 100 = 689.5 \text{ kPa}$$

Consider the two steel bars held together by a 0.75-in.-diameter bolt as shown in Fig. 1.2*a*. The force exerted on the bolt is 5000 lb. In addition to the tension in the bars and the bearing action of the bars on the bolt, there is a tendency for the bolt to fail by a cutting action at the plane at which the two bars are in contact. This force action is called *shear;* it results when two parallel forces having opposite sense of direction act on a body, tending to cause one part of the body to slide past an adjacent

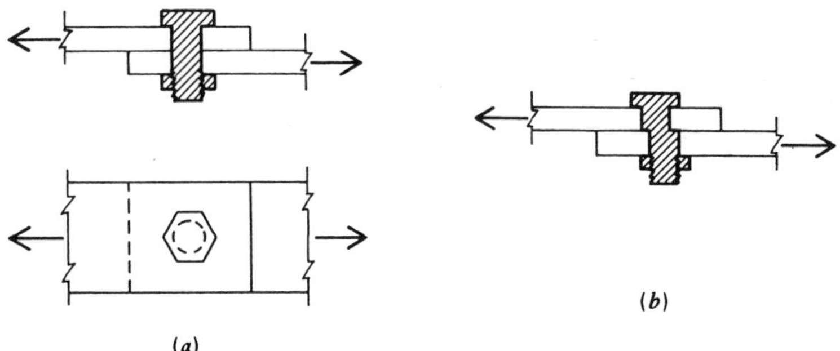

FIGURE 1.2 Direct force action and stress; shear.

part. The bolt has a cross-sectional area of $3.1416(0.75)^2/4 = 0.4418$ in.^{2} [285 mm^2], and the unit shear stress is equal to $5000/0.4418 = 11{,}317$ psi [78.03 MPa]. Note particularly that this example illustrates the computation of shear stress and that the magnitude of this stress would be the same if the forces on the bars were reversed in sense, producing compression instead of tension in the bars.

The fundamental relationship for simple direct stress may be stated as

$$f = \frac{P}{A} \quad \text{or} \quad P = fA \quad \text{or} \quad A = \frac{P}{f}$$

The first form is used for stress determinations, the second form for finding the load (total force) capacity of a member, and the third form for deriving the required area of a member for a stated load with a defined limiting stress condition.

The shear stress developed in the bolt in Fig. 1.2 is of a simple direct nature. Another situation for shear is shown in Fig. 1.3*a*, where a load is applied to a beam that is supported on walls at its ends. It is evident from the sketch in Fig. 1.3*b* that a possible

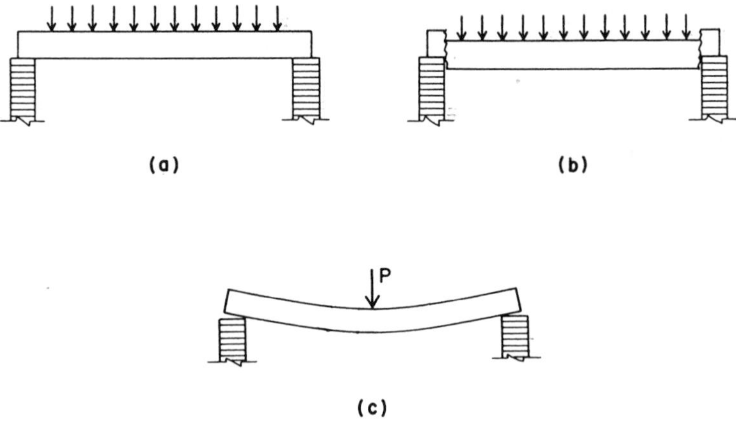

FIGURE 1.3 Shear and bending in beams.

form of failure for the beam is by dropping between the walls, resulting from a shear-type failure at the beam ends. This form of stress development is discussed in Sec. 2.11.

Problem 1.1.A. A wrought iron bar sustains a tensile force of 40 kips [177.92 kN]. If the allowable unit tensile stress is 12 ksi [82,740 kPa], what is the required cross-sectional area of the bar?

Problem 1.1.B. What axial load may be placed on a short timber post, whose actual cross-sectional dimensions are $9\frac{1}{2} \times 9\frac{1}{2}$ in. [241.3 mm], if the allowable unit compressive stress is 1100 psi [7585 kPa]?

Problem 1.1.C. What should be the diameter of the bolt shown in Fig. 1.2*a* if the shearing force is 9000 lb [40.03 kN] and the allowable unit shearing stress is 15 ksi [103,425 kPa]?

Problem 1.1.D. The allowable bearing capacity of a soil is 8000 psf [383 kPa]. What should be the length of the side of a square footing if the total load (including the weight of the footing) is 240 kips [1067.5 kN]?

Problem 1.1.E. If a steel bolt with a diameter of $1\frac{1}{4}$ in. [31.75 mm] is used for the fastener shown in Fig. 1.2*a*, find the shearing force that can be transmitted across the joint if the allowable unit shearing stress in the bolt is 15 ksi [103,425 kPa].

Problem 1.1.F. A short, hollow, cast iron column is circular in cross section, the outside diameter being 10 in. [254 mm] and the thickness of the shell $\frac{3}{4}$ in. [19.05 mm]. If the allowable unit compressive stress is 9 ksi [62,055 kPa], what load will the column support?

Problem 1.1.G. Determine the minimal cross-sectional area of a steel bar required to support a tensile force of 50 kips [222.4 kN] if the allowable unit tensile stress is 20 ksi [137,900 kPa].

Problem 1.1.H. A short, square timber post supports a load of 115 kips [511.5 kN]. If the allowable unit compressive stress is 1000 psi [6895 kPa], what nominal size square timber should be used? (See Table 3.6.)

1.2 Bending

Figure 1.3*c* illustrates a simple beam with a concentrated load *P* at the center of the span. This is an example of *bending,* or *flexure*. The fibers in the upper part of the beam are in compression, and those in the lower part are in tension. Although steel and concrete are not fibrous materials in the sense that wood is, the concept of infinitely small fibers is useful in the study of stress relationships within any material. These stresses are not uniformly distributed over the cross section of the beam and cannot be computed by the direct stress formula. The expression used to compute the value of the bending stress in either tension or compression is known as the *beam formula,* or the *flexure formula,* and is considered in Chapter 2.

1.3 Deformation

Whenever a force acts on a body, there is an accompanying change in shape or size of the body. In structural mechanics this is called *deformation*. Regardless of the magnitude of the force, some deformation is always present, although often it is so small that it is difficult to measure even with the most sensitive instruments. In the design of structures it is often necessary that we know what the deformation in certain members will be. A floor joist, for instance, may be large enough to support a given load safely but may *deflect* (the term for deformation that occurs with bending) to such an extent that the plaster ceiling below will crack, or the floor may feel excessively springy to persons walking on it. For the usual cases we can readily determine what the deformation will be. This is considered in more detail later.

1.4. Deformation and Stress: Relations and Issues

Hooke's Law. As a result of experiments with clock springs, Robert Hooke, a mathematician and physicist working in the seventeenth century, developed the theory that "deformations are directly proportional to stresses." In other words, if a force produces a certain deformation, twice the force will produce twice the amount of deformation. This law of physics is of utmost importance in structural engineering although, as we shall find, Hooke's Law holds true only up to a certain limit.

Elastic Limit and Yield Point. Suppose that we place a bar of structural steel with a cross-sectional area of 1 sq in. [645.2 mm^2] into a machine for making tension tests. We measure its length accurately and then apply a tensile force of 5000 lb [22.24 kN], which, of course, produces a unit tensile stress of 5000 psi [34,475 kPa] in the bar. We measure the length again and find that the bar has lengthened a definite amount, which we will call x inches. On applying 5000 lb more, we note that the amount of lengthening is now $2(x)$, or twice the amount noted after the first 5000 lb. If the test is continued, we will find that for each 5000-lb increment of additional load, the length of the bar will increase the same amount as noted when the initial 5000 lb was applied; that is, the deformations (length changes) are directly proportional to the stresses. So far Hooke's Law has held true, but after we reach a unit stress of about 36,000 psi [248,220 kPa], the length increases more than x for each additional 5000 lb of load. This unit stress is called the *elastic limit,* or the *yield stress,* and it varies for different grades of steel. Beyond this stress limit, Hooke's Law will no longer apply.

Another phenomenon may be noted in this connection. If we make the test again, we will discover that when any applied load which produces a unit stress *less* than the elastic limit is removed, the bar returns to its original length. If a load producing a unit stress *greater* than the elastic limit is removed, we will find that the bar has permanently increased its length. This permanent deformation is called the *permanent set.* This fact permits another way of defining the elastic limit: it is that unit stress beyond which the material does not return to its original length when the load is removed.

If our test is continued beyond the elastic limit, we quickly reach a point where the deformation increases without any increase in the load. The unit stress at which this deformation occurs is called the *yield point;* it has a value only slightly higher than the elastic limit. Since the yield point, or yield stress, as it is sometimes called, can be determined more accurately by test than the elastic limit, it is a particularly important unit stress. Nonductile materials such as wood and cast iron have poorly defined elastic limits and no yield point.

Ultimate Strength. After passing the yield point, the steel bar of the test described in the preceding section again develops resistance to the increasing load. When the load reaches a sufficient magnitude, rupture occurs. The unit stress in the bar just before it breaks is called the *ultimate strength.* For the grade of steel assumed in our test, the ultimate strength occurs at about 70,000 psi.

Structural members are designed so that stresses under normal service conditions will not exceed the elastic limit, even though there is considerable reserve strength between this value and the ultimate strength. This procedure is followed because deformations produced by stresses above the elastic limit are permanent and hence change the shape of the structure.

Factor of Safety. The degree of uncertainty that exists, with respect to both actual loading of a structure and uniformity in the quality of materials, requires that some reserve strength be built into the design. This degree of reserve strength is the *factor of safety*. Although there is no general agreement on the definition of this term, the following discussion will serve to fix the concept in mind.

Consider a structural steel that has an ultimate tensile unit stress of 58,000 psi [399,910 kPa], a yield-point stress of 36,000 psi [248,220 kPa], and an allowable stress of 22,000 psi [151,690 kPa]. If the factor of safety is defined as the ratio of the ultimate strength to the allowable stress, its value is 58,000 ÷ 22,000, or 2.64. On the other hand, if it is defined as the ratio of the yield-point stress to the allowable stress, its value is 36,000 ÷ 22,000, or 1.64. This is a considerable variation, and since failure of a structural member begins when it is stressed beyond the elastic limit, the higher value may be misleading. Consequently, the term *factor of safety* is not employed extensively today. Building codes generally specify the allowable unit stresses that are to be used in design for the grades of structural steel to be employed.

If one should be required to pass judgment on the safety of a structure, the problem resolves itself into considering each structural element, finding its actual unit stress under the existing loading conditions, and comparing this stress with the allowable stress prescribed by the local building regulations. This procedure is called *investigation*.

Modulus of Elasticity. We have seen that, within the elastic limit of a material, deformations are directly proportional to the stresses. Now we shall compute the magnitude of these deformations by use of a number (ratio), called the *modulus of elasticity,* that indicates the degree of *stiffness* of a material.

A material is said to be stiff if its deformation is relatively small when the unit stress is high. As an example, a steel rod 1 in.^{2} [645.2 mm^2] in cross-sectional area and 10 ft [3.05 m] long will elongate about 0.008 in. [0.203 mm] under a tensile load of 2000 lb [8.90 kN]. But a piece of wood of the same dimensions will stretch about 0.24 in. [6.096 mm] with the same tensile load. We say that the steel is stiffer than the wood because, for the same unit stress, the deformation is not so great.

Modulus of elasticity is defined as the unit stress divided by the unit deformation. Unit deformation refers to the percent of deformation and is usually called strain. It is dimensionless, since it is expressed as a ratio, as follows:

$$\text{strain} = s = \frac{e}{l}$$

in which s = the unit strain,
e = the actual dimensional change,
l = the original length of the member.

The modulus of elasticity is represented by the letter E, expressed in pounds per square inch, and has the same value in compression and tension for most structural materials. Letting f represent the unit stress and s the unit deformation, we have, by definition,

$$E = \frac{f}{s}$$

From Sec. 1.3 we remember that $f = P/A$. It is obvious that, if l represents the length of the member and e the total deformation, then s, the deformation per unit of length, must equal the total deformation divided by the length, or $s = e/l$. Now, if we substitute these values in the equation determined by definition,

$$E = \frac{f}{s} = \frac{P/A}{e/l} = \frac{P}{A} \times \frac{l}{e}$$

This can also be written

$$e = \frac{Pl}{AE}$$

in which e = total deformation in inches,
P = force in pounds,
l = length in inches,
A = cross-sectional area in square inches,
E = modulus of elasticity in pounds per square inch.

Note that E is expressed in the same units as f (pounds per square inch) because in the equation $E = f/s$, s is a dimensionless number. For steel E = 29,000,000 psi [200,000,000 kPa], and for wood, depending on the species and grade, it varies from something less than 1,000,000 psi [6,895,000 kPa] to about 1,900,000 psi [13,100,000 kPa]. For concrete E ranges from about 2,000,000 psi [13,790,000 kPa] to about 5,000,000 psi [34,475,000 kPa] for common structural grades. The important thing to remember is that the foregoing formula is valid only when the unit stress lies within the elastic limit of the material.

Example. A 2-in. [50.8-mm] diameter round steel rod 10 ft [3.05 m] long is subjected to a tensile force of 60 kips [266.88 kN]. How much will it elongate under the load?

Solution: (1) The area of the 2-in. rod is 3.1416 in.^{2} [2027 mm^2].

(2) Checking to determine whether the stress in the bar is within the elastic limit, we find that

$$f = \frac{P}{A} = \frac{60}{3.1416} = 19.1 \text{ ksi}$$

$$\left[f = \frac{266.88 \times 10^6}{2027} = 131{,}663 \text{ kPa}\right]$$

which is within the elastic limit of structural steel, so the formula for finding the deformation is applicable.

(3) From data, P = 60 kips, l = 120 (length in inches), A = 3.1416, and E = 29,000,000. Substituting these values, we calcu-

late the total lengthening of the rod is

$$e = \frac{Pl}{AE} = \frac{(60{,}000)(120)}{(3.1416)(29{,}000{,}000)} = 0.079 \text{ in.}$$

$$\left[e = \frac{(266.88 \times 10^6)(3050)}{(2027)(200{,}000{,}000)} = 2.0 \text{ mm}\right]$$

Problem 1.4.A. What force must be applied to a steel bar, 1 in. [25.4 mm] square and 2 ft [610 mm] long, to produce an elongation of 0.016 in. [0.4064 mm]?

Problem 1.4.B. How much will a nominal 8 × 8-in. [actually 190.5-mm] Douglas fir post, 12 ft [3.658 m] long, shorten under an axial load of 45 kips [200 kN]?

Problem 1.4.C. A routine quality control test is made on a structural steel bar 1 in. [25.4 mm] square and 16 in. [406 mm] long. The data developed during the test show that the bar elongated 0.0111 in. [0.282 mm] when subjected to a tensile force of 20.5 kips [91.184 kN]. Compute the modulus of elasticity of the steel.

Problem 1.4.D. A $\frac{1}{2}$-in. [12.7-mm] diameter steel rod 40 ft [12.19 m] long supports a load of 4 kips [17.79 kN]. How much will it elongate?

1.5 Design Use of Direct Stress

In the examples and problems dealing with the direct stress equation, we have differentiated between the unit stress developed in a member sustaining a given load ($f = P/A$) and the *allowable unit stress* used when determining the size of a member required to carry a given load ($A = P/f$). The latter form of the equation is, of course, the one used in design.

From the discussion in Secs. 1.2–1.4, we can see that the allowable unit stresses should be set within the elastic limit of the structural material being used. The procedures for establishing allowable unit stresses in tension, compression, shear, and bending are different for different materials and are prescribed in specifications promulgated by the American Society for Testing and Materials. In general, allowable stresses for structural steel are expressed as fractions of the yield stress, those for wood involve an adjustment of clear wood strength as modified by lumber grading rules and conditions of use, and allowable stresses for concrete are given as fractions of the specified compressive strength of concrete. Tables 4.1, 7.1, and 13.1 give allowable stresses for steel, wood, and reinforced concrete construction, respectively, as recommended by the industry associations concerned. These are the American Institute of Steel Construction, the National Forest Products Association, and the American Concrete Institute. When scanning these tables, you will notice that they contain several terms that have not been introduced in this book thus far. These will be identified in subsequent sections dealing with the design of members to which they apply. However, in order to provide information for convenient reference when solving the problems at the end of this chapter, selected data from the more complete allowable stress tables are presented in Table 1.1.

In actual design work, the building code governing the construction of buildings in the particular locality must be consulted for specific requirements. Many municipal codes are revised infrequently and, consequently, may not be in agreement with current editions of the industry-recommended allowable stresses. Unless otherwise noted, the allowable stresses used in this book are those given in the three tables referenced above.

Except for shear the stresses we have discussed so far have been direct or axial stresses. This, we recall, means they are assumed to be uniformly distributed over the cross section. The examples and problems presented fall under three general types: first, the design of structural members ($A = P/f$); second, the determination of safe loads ($P = fA$); third, the investigation of members for safety ($f = P/A$). The following examples will serve to fix in mind each of these types.

Example 1. Design (determine the size of) a short, square post of Southern pine, No. 1 dense SR grade, to carry an axial compressive load of 30,000 lb [133,440 N].

Solution: (1) Referring to Table 1.1, we find that the allowable unit compressive stress for this wood parallel to the grain is 925 psi [6378 kPa].

(2) The required area of the post is

$$A = \frac{P}{f} = \frac{30{,}000}{925} = 32.43 \text{ in.}^2$$

$$\left[A = \frac{133{,}440 \times 10^3}{6378} = 20{,}922 \text{ mm}^2\right]$$

TABLE 1.1 Selected Values for Common Structural Materials

Material and Property	Common Values (in pounds per sq in.)	(in kPa)
Structural steel		
Yield strength	36,000	248,220
Allowable tension	22,000	151,690
Allowable shear (on rivets)	15,000	103,425
E	29,000,000	200,000,000
Concrete		
f'_c (specified compressive strength)	3,000	20,685
Usable compression (in bearing)	900	6,206
Shear on concrete, in beams	60	414
E	3,100,000	21,374,500
Structural lumber		
1. Douglas fir, select structural grade, posts & timbers Compression parallel to grain	1,150	7,929
E	1,600,000	11,032,000
2. Southern pine, No. 1 dense SR grade, 5 in. & thicker compression parallel to grain	925	6,378
E	1,600,000	11,032,000

Source: Taken from Tables 4.1, 7.1, and 13.1.

(3) From Table 3.6 an area of 30.25 in.^{2} [19,517 mm^2] is provided by a 6 × 6-in. post with a dressed size of $5\frac{1}{2} \times 5\frac{1}{2}$ in. [139.7 mm].

Example 2. Determine the safe axial compressive load for a short, square concrete pier with a side dimension of 2 ft [0.6096 m].

Solution: (1) The area of the pier is 4 ft^2 or 576 $in.^2$ [0.3716 m^2].

(2) Table 1.1 gives the allowable unit compressive stress for concrete as 900 psi [6206 kPa].

(3) Therefore the safe load on the pier is

$$P = (f)(A) = (900)(576) = 518{,}400 \text{ lb}$$

$$[P = (6206)(0.3716) = 2306 \text{ kN}]$$

Example 3. A running track in a gymnasium is hung from the roof trusses by steel rods, each of which supports a tensile load of 11,200 lb [49,818 N]. The round rods have a diameter of $\frac{7}{8}$ in. [22.23 mm] with the ends *upset,* that is, made larger by forging. This upset is necessary if the full cross-sectional area of the rod (0.601 $in.^2$) [388 mm^2] is to be utilized; otherwise the cutting of the threads will reduce the cross section of the rod. Investigate this design to determine whether it is safe.

Solution: (1) Since the gross area of the hanger rod is effective, the unit stress developed is

$$f = \frac{P}{A} = \frac{11{,}200}{0.601} = 18{,}636 \text{ psi}$$

$$\left[f = \frac{49{,}818 \times 10^3}{388} = 128{,}397 \text{ kPa}\right]$$

(2) Table 1.1 gives the allowable unit tensile stress for steel as 22,000 psi [151,690 kPa], which is greater than that developed by the loading. Therefore the design is safe.

Shearing Stress Formula. The foregoing manipulations of the direct stress formula can, of course, be carried out also with the shearing stress formula $f_v = P/A$. However, it must be borne in mind that the shearing stress acts transversely to the cross section—not at right angles to it. Furthermore, while the shearing stress equation applies directly to the situation illustrated by Fig. 1.2*a* and *b*, it requires modification for application to beams (Fig. 1.3*a* and *b*). The latter situation will be considered in more detail later.

Problem 1.5.A. What should be the minimum cross-sectional area of a steel rod to support a tensile load of 26 kips [115.648 kN]?

Problem 1.5.B. A short, square post of Douglas fir, select structural grade, is to support an axial load of 61 kips [271.3 kN]. What should its nominal dimensions be?

Problem 1.5.C. A steel rod has a diameter of 1.25 in. [31.75 mm]. What safe tensile load will it support if its ends are upset?

Problem 1.5.D. What safe load will a short, 12 × 12-in. [actually 292.1-mm] Southern pine post support if the grade of the wood is No. 1 dense SR?

Problem 1.5.E. A short post of Douglas fir, select structural grade, with nominal dimensions of 6 × 8 in. [actually 139.7 × 190.5 mm] supports an axial load of 50 kips [222.4 kN]. Investigate this design to determine whether it is safe.

Problem 1.5.F. A short concrete pier, 1 ft 6 in. [457.2 mm] square, supports an axial load of 150 kips [667.2 kN]. Is the construction safe?

Problem 1.5.G. The shearing load on a $\frac{7}{8}$-in. [22.23-mm] diameter bolt in a lap joint (Fig. 1.2*a*) is 8.5 kips [37.808 kN]. Is this a safe condition?

1.6 Aspects of Dynamic Behavior

A good lab course in physics should provide a reasonable understanding of the basic ideas and relationships involved in dynamic behavior. A better preparation is a course in engineering dynamics that focuses on the topics in an applied fashion, dealing directly with their applications in various engineering problems. The material in this section consists of a brief summary of basic concepts in dynamics that will be useful to those with a limited background and that will serve as a refresher for those who have studied the topics before.

The general field of dynamics may be divided into the areas of *kinetics* and *kinematics*. *Kinematics* deals exclusively with motion, that is, with time–displacement relationships and the geometry of movements. *Kinetics* adds the consideration of the forces that produce or resist motion.

Kinematics. Motion can be visualized in terms of a moving point, or in terms of the motion of a related set of points that constitute a body. The motion can be qualified geometrically and quantified dimensionally. In Fig. 1.4*a* the point is seen to move along a path (its geometric character) a particular distance. The distance traveled by the point between any two separate locations on its path is called *displacement*. The idea of motion is that this displacement occurs over time, and the general mathematical expression for the time–displacement function is

$$s = f(t)$$

Velocity is defined as the rate of change of the displacement with respect to time. As an instantaneous value, the velocity is expressed as the ratio of an increment of displacement (ds) divided by the increment of time (dt) elapsed during the displacement. Using the calculus, the velocity is thus defined as

$$v = \frac{ds}{dt}$$

That is, the velocity is the first derivative of the displacement.

If the displacement occurs at a constant rate with respect to time, it is said to have *constant velocity*. In this case the velocity may be expressed more simply without the calculus as

$$v = \frac{\text{total displacement}}{\text{total elapsed time}}$$

When the velocity changes over time, its rate of change is called the *acceleration* (a). Thus, as an instantaneous change

$$a = \frac{dv}{dt} = \frac{d^2s}{dt^2}$$

That is, the acceleration is the first derivative of the velocity or the second derivative of the displacement with respect to time.

Except for the simplest cases, the derivation of the equations of motion for an object generally requires the use of the calculus in the operation of these basic relationships. Once derived, however, motion equations are generally in algebraic form and can be

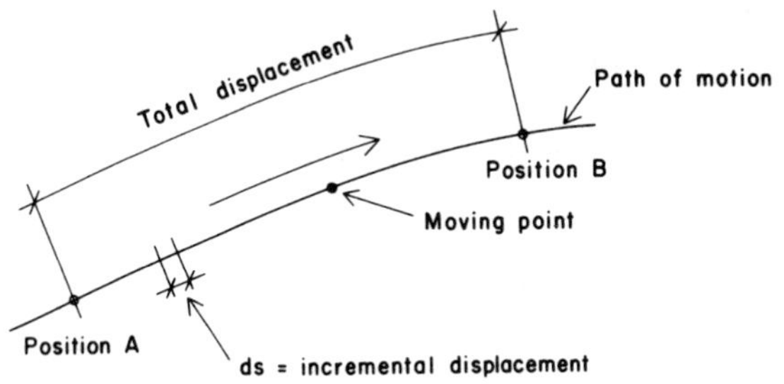

(a) Motion of a Point

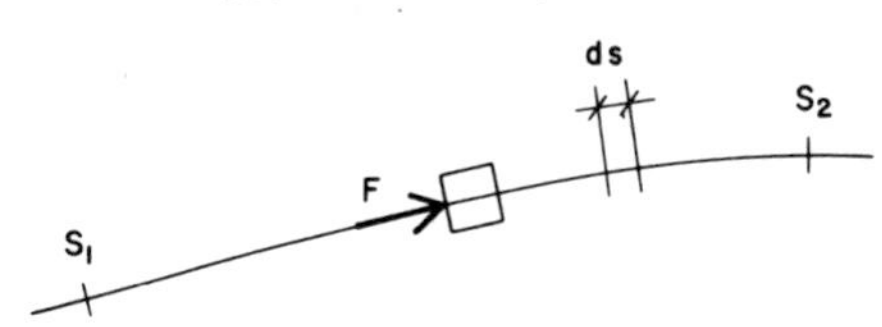

$$\text{Work} = \int_{S_1}^{S_2} F_t \, ds \quad \text{(F variable with time)}$$

$$= F(S_2 - S_1) \quad \text{(F constant with time)}$$

(b) Kinetics of a Moving Object

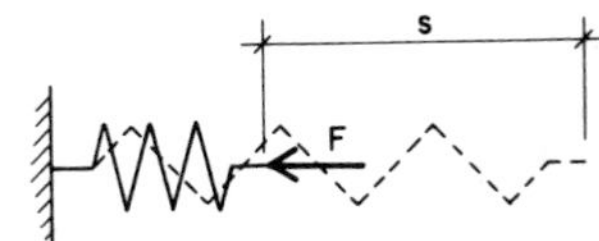

Potential (stored) energy:

$E = F \cdot k \cdot s$

k = spring constant

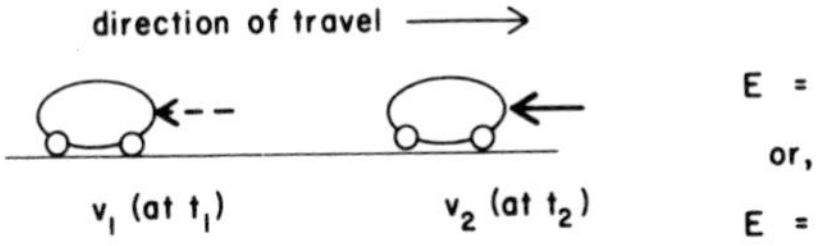

Kinetic energy:

$E = \frac{1}{2} m (v_1^2 - v_2^2)$

or, if $v_2 = 0$:

$E = \frac{1}{2} m v_1^2$

(c) Forms of Mechanical Energy

FIGURE 1.4 Aspects of dynamic effects.

used without the calculus for application to problems. An example is the set of equations that describes the motion of a free-falling object acted on by the earth's gravity field. Under idealized conditions (ignoring air friction, etc.) the distance of fall from a rest position will be

$$s = f(t) = 16.1t^2 \qquad (s \text{ in ft, } t \text{ in sec})$$

This equation indicates that the rate of fall (the velocity) is not a constant but increases with the elapsed time, so that the velocity at any instant of time may be expressed as

$$v = \frac{ds}{dt} = \frac{d(16.1t^2)}{dt} = 32.2t \qquad (v \text{ in ft/sec})$$

and the acceleration as

$$a = \frac{dv}{dt} = \frac{d(32.2t)}{dt} = 32.2 \text{ ft/sec}^2$$

which is the acceleration of gravity.

The following examples illustrate some additional displacement–velocity–acceleration relationships. We first consider the simplest case: a straight-line motion at constant velocity.

Example 1. A point moves along a straight line described by $y = \frac{3}{4}x$ at a constant velocity of 10 ft/sec. Express the relation of displacement along the line and also of displacement in the x direction with time. What is the x-direction displacement in 4 sec?

Solution: The relation of displacement along the line (Fig. 1.5) is expressed as

$$s = 10t$$

and the velocity along the line as

$$v = \frac{ds}{dt} = \frac{d(10t)}{dt} = 10 \text{ ft/sec}$$

which simply confirms the relationship, as the velocity was given.

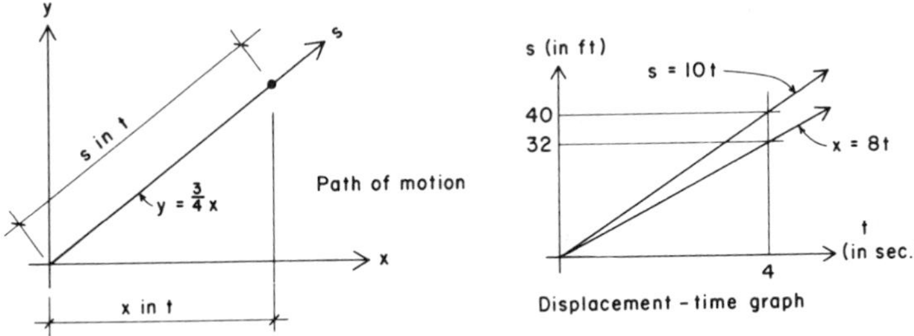

FIGURE 1.5

The relation of displacement in the x direction is found by first noting that

$$s = \sqrt{x^2 + y^2} = \sqrt{x^2 + (\tfrac{3}{4}\,x)^2} = \sqrt{\tfrac{25}{16}\,x^2} = \tfrac{5}{4}\,x$$

Then

$$s = \tfrac{5}{4}\,x = 10t$$

and

$$x = \tfrac{4}{5}\,(10t) = 8t$$

and in 4 sec

$$x = 8(4) = 32 \text{ ft}$$

We next consider an example in which the velocity is a function of time.

Example 2. Data similar to Example 1, except that the velocity varies from zero when s is zero and increases with time such that $v = 2t$.

Solution: A change in velocity implies an acceleration, thus

$$a = \frac{dv}{dt} = \frac{d(2t)}{dt} = 2 \text{ ft/sec}^2$$

For the displacement, we note that

$$v = \frac{ds}{dt} \quad \text{and thus} \quad s = \int v\,dt = \int (2t)\,dt = t^2$$

Then, as before,

$$s = \tfrac{5}{4}x = t^2, \qquad x = \tfrac{4}{5}t^2$$

and in 4 sec

$$x = \tfrac{4}{5}(4)^2 = 12.8 \text{ ft}$$

Example 3. A car accelerates from rest with a constant acceleration to a speed (velocity) of 60 mph in 30 sec ($\frac{1}{120}$ h). Derive the expressions for s, v, and a and find the distance traveled.

Solution: Using the given value for v at the time of the elapsed 30 sec, we note that

$$a = C, \qquad v = \int a\,dt = \int_0^{1/120} C\,dt = 60$$

from which

$$(1/120)C = 60, \qquad C = 7200 \text{ miles/h}^2$$

Then

$$a = 7200 \text{ miles/h}^2$$

$$v = \int_0^t a\,dt = \int_0^t 7200\,dt = 7200t$$

$$s = \int_0^t v\,dt = \int_0^t (7200t)\,dt = 3600t^2$$

and in the elapsed $\frac{1}{120}$ hour,

$$s = 3600(1/120)^2 = 0.25 \text{ mile}$$

Motion. A major aspect of consideration in dynamics is the nature of motion. While building structures are not really supposed to move (like machine parts), their responses to force actions involve consideration of motions. These motions may actually occur in the form of very small deformations, or merely be the failure response that the designer must visualize resisting. The following are some basic forms of motion.

Translation. This occurs when an object moves in simple linear displacement, with the displacement measured as simple change of distance from some reference point.

Rotation. This occurs when the motion can be measured in the form of angular displacement; that is, in the form of revolving about a fixed reference point.

Rigid-Body Motion. A rigid body is one in which no internal deformation occurs and all particles of the body remain in fixed relation to each other. Three types of motion of such a body are possible. Translation occurs when all the particles of the body move in the same direction at the same time. Rotation occurs when all points in the body describe circular paths about some common fixed line in space, called the *axis of rotation*. Plane motion occurs when all the points in the body move in planes that are parallel. Motion within the planes may be any combination of translation or rotation.

Motion of Deformable Bodies. In this case motion occurs for the body as a whole, as well as for the particles of the body with respect to each other. This is generally of more complex form than rigid-body motion, although it may be broken down into simpler component motions in many cases. This is the nature of motion of fluids and of elastic solids. The deformation of elastic structures under load is of this form, involving both the movement of elements from their original positions and changes in their shapes.

Kinetics. As stated previously, kinetics includes the additional consideration of the forces that cause motion. This means that in addition to the variables of displacement and time, we must consider the mass of the moving objects. From Newtonian physics the simple definition of mechanical force is

$$F = ma = \text{mass} \times \text{acceleration}$$

Mass is the measure of the property of inertia, which is what causes an object to resist change in its state of motion. The more common term for dealing with mass is *weight,* which is a force defined as

$$W = mg$$

where g is the constant acceleration of gravity (32.2 ft/s^2).

Weight is literally a dynamic force, although it is the standard means of measurement of force in statics, when the velocity is assumed to be zero. Thus in static analysis we express forces simply as

$$F = W$$

and in dynamic analysis, when using weight as the measure of mass, we express force as

$$F = ma = \frac{W}{g}a$$

Work, Power, Energy, and Momentum. If a force moves an object, work is done. *Work* is defined as the product of the force multiplied by the displacement (distance traveled). If the force is constant during the displacement, work may be simply expressed as

$$w = Fs = \text{force} \times \text{total distance traveled}$$

If the force varies with time, the relationship is more generally expressed with the calculus as

$$w = \int_{s_2}^{s_1} F_t ds$$

indicating that the displacement is from position s_1 to position s_2, and the force varies in some manner with respect to time.

Figure 1.4*b* illustrates these basic relationships. In dynamic analysis of structures the dynamic "load" is often translated into work units in which the distance traveled is actually the deformation of the structure.

Energy may be defined as the capacity to do work. Energy exists in various forms: heat, mechanical, chemical, and so on. For structural analysis the concern is with mechanical energy, which occurs in one of two forms. *Potential energy* is stored energy, such as that in a compressed spring or an elevated weight. Work is done when the spring is released or the weight is dropped. *Kinetic energy* is possessed by bodies in motion; work is required to change their state of motion, that is, to slow them down or speed them up (see Fig. 1.4*c*).

In structural analysis energy is considered to be indestructible, that is, it cannot be destroyed, although it can be transferred or transformed. The potential energy in the compressed spring can be transferred into kinetic energy if the spring is used to propel an object. In a steam engine the chemical energy in the fuel is transformed into heat and then into pressure of the steam and finally into mechanical energy delivered as the engine's output.

An essential idea is that of the conservation of energy, which is a statement of its indestructibility in terms of input and output. This idea can be stated in terms of work by saying that the work done on an object is totally used and that it should therefore be equal to the work accomplished plus any losses due to heat, air friction, and so on. In structural analysis we make use of this concept by using a "work equilibrium" relationship similar to the static force equilibrium relationship. Just as all the forces must be in balance for static equilibrium, so the work input must equal the work output (plus losses) for "work equilibrium."

Harmonic Motion. A special problem of major concern in structural analysis for dynamic effects is that of *harmonic motion*. The two elements generally used to illustrate this type of motion are the swinging pendulum and the bouncing spring. Both the pendulum and the spring have a neutral position where they will remain at rest in static equilibrium. If one displaces either of them from this neutral position, by pulling the pendulum sideways or compressing or stretching the spring, they will tend to move back to the neutral position. Instead of stopping at the neutral position, however, they will be carried past it by their momentum to a position of displacement in the opposite direction. This sets up a cyclic form of motion (swinging of the pendulum; bouncing of the spring) that has some basic characteristics.

Figure 1.6 illustrates the typical motion of a bouncing spring. Using the calculus and the basic motion and force equations, the displacement–time relationship may be derived as

$$s = A \cos Bt$$

The cosine function produces the basic form of the graph, as shown in Fig. 1.6*b*. The maximum displacement from the neutral position is called the *amplitude*. The time elapsed for one full cycle is called the *period*. The number of full cycles in a given unit of time is called the *frequency* (usually expressed in cycles per second) and is equal to the inverse of the period. Every object subject to harmonic motion has a fundamental period (also called natural period), which is determined by its weight, stiffness, size, and so on.

Any influence that tends to reduce the amplitude in successive cycles is called a *damping effect*. Heat loss in friction, air resistance, and so on are natural damping effects. Shock absorbers, counterbalances, cushioning materials, and other devices can also be used to damp the amplitude. Figure 1.6*c* shows the form of a damped harmonic motion, which is the normal form of most such motions, because perpetual motion is not possible without a continuous reapplication of the original displacing force.

Resonance is the effect produced when the displacing effort is itself harmonic with a cyclic nature that corresponds with the period of the impelled object. An example is someone bouncing on a diving board in rhythm with the board's fundamental period, thus causing a reinforcement, or amplification, of the board's free motion. This form of motion is illustrated in Fig. 1.6*d*. Unrestrained resonant effects can result in intolerable amplitudes, producing destruction or damage of the moving object or its supports. A balance of damping and resonant effects can sometimes

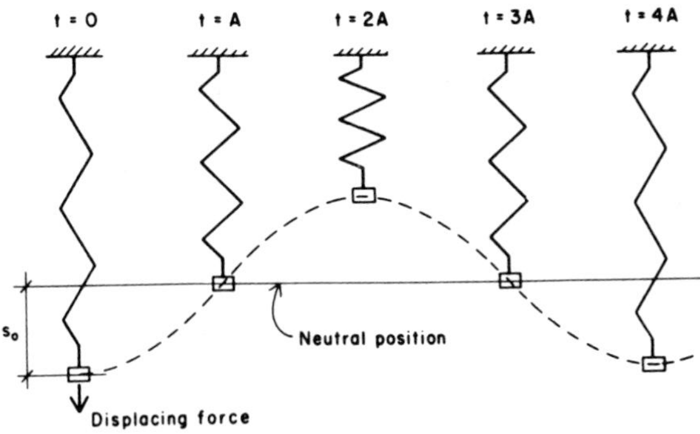

(a) The Moving Spring

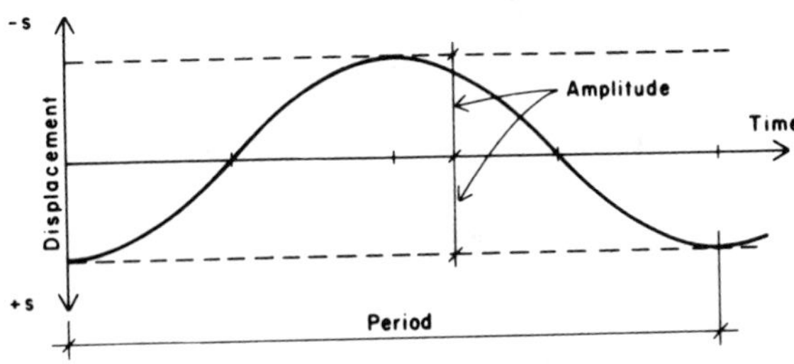

(b) Plot of the Equation: S = A cos BT

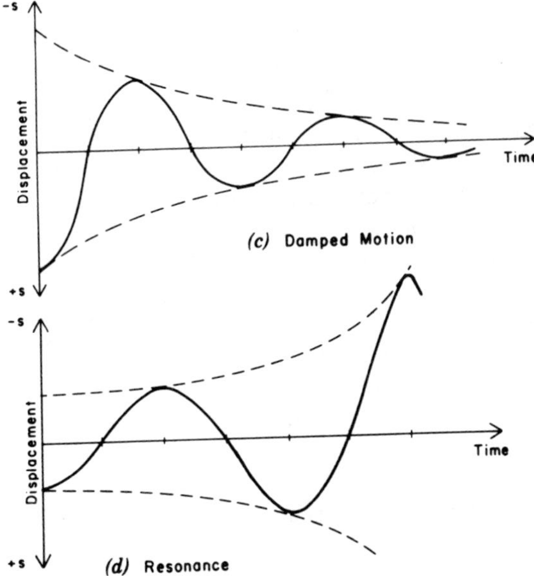

(c) Damped Motion

(d) Resonance

FIGURE 1.6 Aspects of harmonic motion.

produce a constant motion with a flat profile of the amplitude peaks.

Loaded structures tend to act like springs. Within the elastic stress range of the materials, they can be displaced from a neutral (unloaded) position and, when released, will go into a form of harmonic motion. The fundamental period of the structure as a whole, as well as the periods of its parts, are major properties that affect responses to dynamic loads.

Equivalent Static Effects. Use of equivalent static effects essentially permits simpler analysis and design by eliminating the complex procedures of dynamic analysis. To make this possible the load effects and the structure's responses must be translated into static terms.

For wind load the primary translation consists of converting the kinetic energy of the wind into an equivalent static pressure, which is then treated in a manner similar to that for a distributed gravity load. Additional considerations are made for various aerodynamic effects, such as ground surface drag, building shape, and suction, but these do not change the basic static nature of the work.

For earthquake effects the primary translation consists of establishing a hypothetical horizontal static force that is applied to the structure to simulate the effects of sideward motions during ground movements. This force is calculated as some percentage of the dead weight of the building, which is the actual source of the kinetic energy loading once the building is in motion—just as the weight of the pendulum and the spring keeps them moving after the initial displacement and release. The specific percentage used is determined by a number of factors, including some of the dynamic response characteristics of the structure.

An apparently lower safety factor is used when designing for the effects of wind and earthquake because an increase of one-third is permitted in allowable stresses. This is actually not a matter of a less-safe design but is merely a way of compensating for the fact that one is actually adding static (gravity) effects and *equivalent* static effects. The total stresses thus calculated are really quite hypothetical because in reality one is adding static strength effects to dynamic strength effects, in which case 2 + 2 does not necessarily make 4.

Regardless of the number of modifying factors and translations, there are some limits to the ability of an equivalent static analysis to account for dynamic behavior. Many effects of damping and resonance cannot be accounted for. The true energy capacity of the structure cannot be accurately measured in terms of the magnitudes of stresses and strains. There are some situations, therefore, in which a true dynamic analysis is desirable, whether it is performed by mathematics or by physical testing. These situations are actually quite rare, however. The vast majority of building designs present situations for which a great deal of experience exists. This experience permits generalizations on most occasions that the potential dynamic effects are really insignificant or that they will be adequately accounted for by design for gravity alone or with use of the equivalent static techniques.

Problem 1.6.A. A point moves along a straight line described by $y = \frac{1}{3}x$ at a constant velocity of 6 ft/sec. Express the relation of displacement along the line with time and also of displacement in the y direction with time. What is the y-direction displacement in 10 sec?

Problem 1.6.B. Data similar to Problem 1.6.A, except that the velocity varies from zero when s is zero and increases with time such that $v = 3t$.

Problem 1.6.C. A car accelerates from rest with a constant acceleration to a speed (velocity) of 60 mph in 20 sec. Derive expressions for s, v, and a and find the distance traveled.

2

Investigation of Beams and Frames

This chapter presents considerations that are made in the investigation of the behavior of beams and simple frames. For basic explanation of relationships, the units used for forces and dimensions are of less significance than their numeric values. For this reason, and for sake of brevity and simplicity, most numerical computations in the text have been done using only U.S. units. For readers who wish to use metric units, however, the exercise problems have been provided with dual units.

2.1 Moments

The term *moment of a force* is commonly used in engineering problems; it is of utmost importance that you understand exactly what the term means. It is fairly easy to visualize a length of 3 ft, an area of 16 sq in., or a force of 100 lb. A moment, however, is less readily comprehended; it is a force multiplied by a distance. *A moment is the tendency of a force to cause rotation about a given point or axis*. The magnitude of the moment of a force about a given point is the magnitude of the force (pounds, kips, etc.) multiplied by the distance (feet, inches, etc.) to the point. The point is called the *center of moments,* and the distance, which is called the *lever arm* or *moment arm,* is measured by a line drawn through the center of moments *perpendicular to the line of action* of the force. Moments are expressed in compound units such as foot-pounds and inch-pounds or kip-feet and kip-inches. In summary,

$$\text{Moment of force} = \text{magnitude of force} \times \text{moment arm}$$

Consider the horizontal force of 100 lb shown in Fig. 2.1*a*. If point *A* is the center of moments, the lever arm of the force is 5 ft 0 in. Then the moment of the 100-lb force with respect to point *A* is $100 \times 5 = 500$ ft-lb. In this illustration the force tends to cause a *clockwise* rotation (shown by the dotted arrow) about point *A* and is called a positive moment. If point *B* is the center of moments, the moment arm of the force is 3 ft 0 in. Therefore the moment of the 100-lb force about point *B* is $100 \times 3 = 300$ ft-lb. With respect to point *B,* the force tends to cause *counterclockwise* rotation; it is called a negative moment. It is important to remember that we can never consider the moment of a force without having in mind the particular point or axis about which it tends to cause rotation.

Figure 2.1*b* represents two forces acting on a bar which is supported at point *A*. The moment of force P_1 about point *A* is

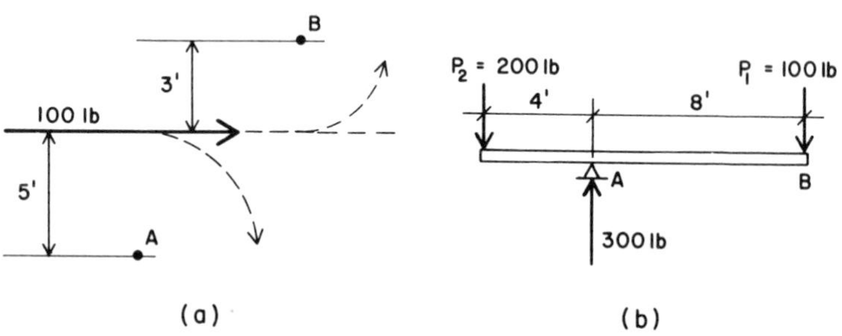

FIGURE 2.1 Development of moments.

$100 \times 8 = 800$ ft-lb, and it is clockwise or positive. The moment of force P_2 about point A is $200 \times 4 = 800$ ft-lb. The two moment values are the same, but P_2 tends to produce a counterclockwise, or negative, moment about point A. In other words, the positive and negative moments are equal in magnitude and are in equilibrium; that is, there is no motion. Another way of stating this is to say that the sum of the positive and negative moments about point A is zero, or

$$(P_1 \times 8) - (P_2 \times 4) = 0$$

Stated more generally, *if a system of forces is in equilibrium, the algebraic sum of the moments is zero.* This is one of the laws of equilibrium.

In Fig. 2.1*b* point A was taken as the center of moments, but the fundamental law holds for any point that might be selected. For example, if we take point B as the center of moments, the moment of the upward supporting force of 300 lb acting at A is clockwise (positive) and that of P_2 is counterclockwise (negative). Then

$$(300 \times 8) - (200 \times 12) = 2400 \text{ ft-lb} - 2400 \text{ ft-lb} = 0$$

Note that the moment of force P_1 about point B is $100 \times 0 = 0$; it is therefore omitted in writing the equation. The reader should be satisfied that the sum of the moments is zero also when the center of moments is taken at the left end of the bar under the point of application of P_2.

Laws of Equilibrium. When a body is acted on by a number of forces, each force tends to move the body. If the forces are of such magnitude and position that their combined effect produces no motion of the body, the forces are said to be in *equilibrium*. The three fundamental laws of equilibrium are:

1. The algebraic sum of all the vertical forces equals zero.
2. The algebraic sum of all the horizontal forces equals zero.
3. The algebraic sum of the moments of all the forces about any point equals zero.

These laws, sometimes called the conditions for equilibrium, may be expressed as follows (the symbol Σ indicates a summation, i.e., an algebraic addition of all similar terms involved in the problem):

$$\Sigma V = 0 \qquad \Sigma H = 0 \qquad \Sigma M = 0$$

The law of moments, $\Sigma M = 0$, was discussed in the preceding section.

We shall defer consideration of $\Sigma H = 0$ for the time being. Our immediate concern is with vertical loads acting on beams where the expression $\Sigma V = 0$ is another way of saying that *the sum of the downward forces equals the sum of the upward forces*. Thus the bar of Fig. 2.1*b* satisfies $\Sigma V = 0$ because the upward supporting force of 300 lb equals the sum of P_1 and P_2, the downward forces.

Moment of Forces on a Beam. Figure 2.2*a* shows two downward forces of 100 lb and 200 lb acting on a beam. The beam has a length of 8 ft between the supports; the supporting forces, which are called *reactions,* are 175 lb and 125 lb. The four forces are in equilibrium, and therefore the two laws, $\Sigma V = 0$ and $\Sigma M = 0$, apply. Let us see if this is true.

First, because the forces are in equilibrium, the sum of the downward forces must equal the sum of the upward forces. The sum of the downward forces, the loads, is 100 + 200 = 300 lb; and the sum of the upward forces, the reactions, is 175 + 125 = 300 lb. We can write 100 + 200 = 175 + 125; this is a true statement.

Second, because the forces are in equilibrium, the sum of the moments of the forces tending to cause clockwise rotation (positive moments) must equal the sum of the moments of the forces tending to produce counterclockwise rotation (negative moments) about any center of moments. Let us first write an equation of moments about point *A* at the right-hand support. The force tending to cause clockwise rotation (shown by the curved arrow) about this point is 175 lb; its moment is 175 × 8 = 1400 ft-lb. The forces tending to cause counterclockwise rotation *about the same point* are 100 lb and 200 lb, and their moments are

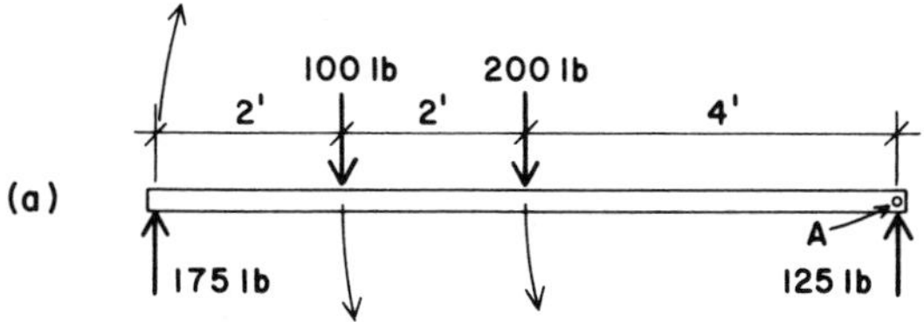

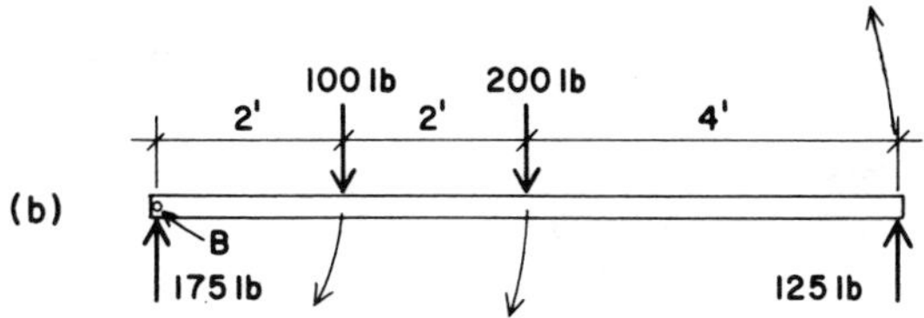

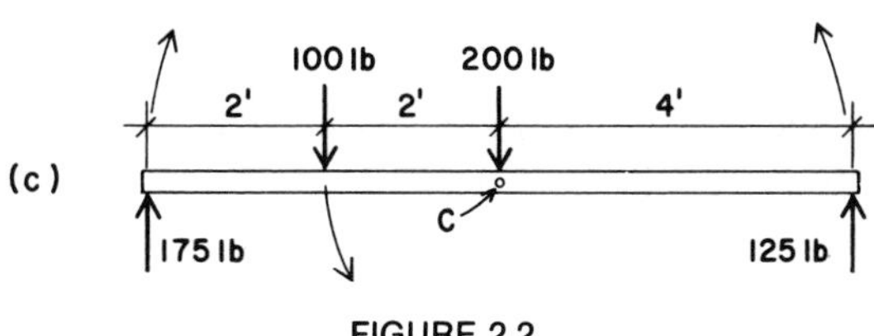

FIGURE 2.2

(100 × 6) and (200 × 4) ft-lb. Therefore we can write

$$(175 \times 8) = (100 \times 6) + (200 \times 4)$$

$$1400 = 600 + 800$$

$$1400 \text{ ft-lb} = 1400 \text{ ft-lb}$$

which is true.

The upward force of 125 lb is omitted from the above equation because its lever arm about point A is 0 ft, and consequently its moment is zero. Thus we see that a force passing through the center of moments does not cause rotation about that point.

Let us try again. This time we select point B at the left support as the center of moments (see Fig. 2.2*b*). By the same reasoning we can write

$$(100 \times 2) + (200 \times 4) = (125 \times 8)$$
$$200 + 800 = 1000$$
$$1000 \text{ ft-lb} = 1000 \text{ ft-lb}$$

Again the law holds. In this case the force 175 lb has a lever arm of 0 ft about the center of moments and its moment is zero.

Suppose we select any point, such as point C in Fig. 2.2*c*, as the center of moments; then

$$(175 \times 4) = (100 \times 2) + (125 \times 4)$$
$$700 = 200 + 500$$
$$700 \text{ ft-lb} = 700 \text{ ft-lb}$$

We have seen the law of moments holds in each case. It is of great importance that we understand this principle thoroughly before going on. Remember that the loads and reactions are usually in units of pounds or kips and that the moments are compound quantities, usually foot-pounds or kip-feet, the result of multiplying a force by a distance. When loads are given in kips, there is no intrinsic reason why moments could not be stated as

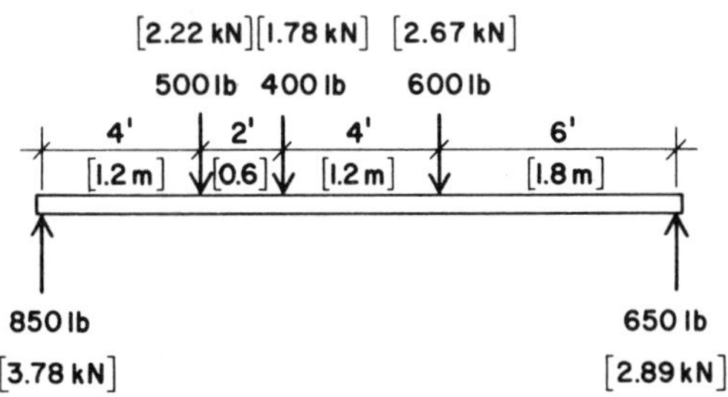

FIGURE 2.3

"foot-kips," which would be consistent with "foot-pounds." However, foot-pounds and kip-feet (or inch-pounds and kip-inches) are the terms commonly used in practice for the compound units in which moments are expressed.

Problem 2.1.A. Figure 2.3 represents a beam in equilibrium with three loads and two reactions. Select five different centers of moments and write the equation of moments for each, showing that the sum of the clockwise moments equals the sum of the counterclockwise moments.

2.2 Beam Support and Load Conditions

A beam is a structural member that resists transverse loads. The supports for beams are usually at or near the ends, and the supporting upward forces are called reactions. As noted in Sec. 1.2, the loads acting on a beam tend to *bend* it rather than shorten or lengthen it. *Girder* is the name given to a beam that supports smaller beams; all girders are beams insofar as their structural action is concerned. There are, in general, five types of beams which are identified by the number, kind, and position of the supports. Figure 2.4 shows diagrammatically the different types

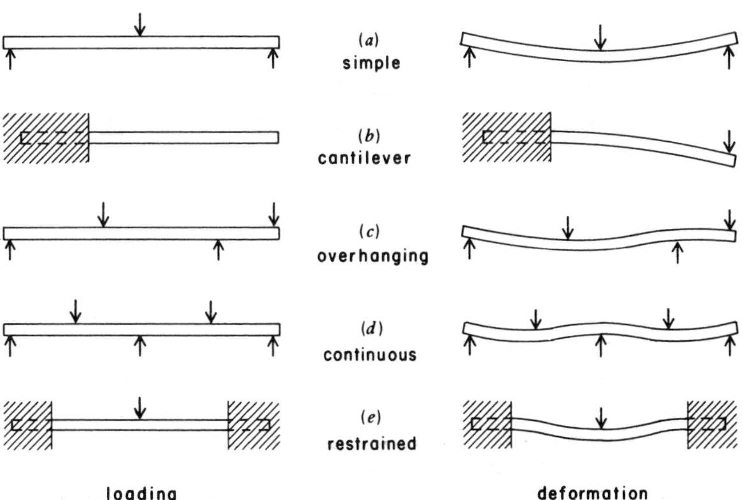

FIGURE 2.4 Types of beams.

and also the shape each beam tends to assume as it bends (deforms) under the loading. In ordinary steel or reinforced concrete beams, these deformations are not usually visible to the eye, but as noted in Sec. 1.3, some deformation is always present.

A *simple beam* rests on a support at each end, the ends of the beam being free to rotate (Fig. 2.4*a*).

A *cantilever beam* is supported at one end only. A beam embedded in a wall and projecting beyond the face of the wall is a typical example (Fig. 2.4*b*).

An *overhanging beam* is a beam whose end or ends project beyond its supports. Figure 2.4*c* indicates a beam overhanging one support only.

A *continuous beam* rests on more than two supports (Fig. 2.4*d*). Continuous beams are commonly used in reinforced concrete and welded steel construction.

A *restrained beam* has one or both ends restrained or *fixed* against rotation (Fig. 2.4*e*).

The two types of loads that commonly occur on beams are called *concentrated* and *distributed*. A concentrated load is assumed to act at a definite point, such as a column resting on a beam. A distributed load is one that acts over a considerable length of the beam. A concrete floor slab supported by a beam is an example of a distributed load. If the distributed load exerts a force of equal magnitude for each unit of length of the beam, it is known as a *uniformly distributed load*. Obviously, a distributed load need not extend over the entire length of the beam.

2.3 Beam Reactions

We have already defined reactions as the upward forces acting at the supports which hold in equilibrium the downward forces or loads. The left and right reactions are usually called R_1 and R_2, respectively.

If we have a beam 18 ft in length with a concentrated load of 9000 lb located 9 ft from the supports, it is readily seen that each upward force at the supports will be equal and will be one-half the load in magnitude, or 4500 lb. But consider, for instance, the 9000-lb load placed 10 ft from one end, as shown in Fig. 2.5. What

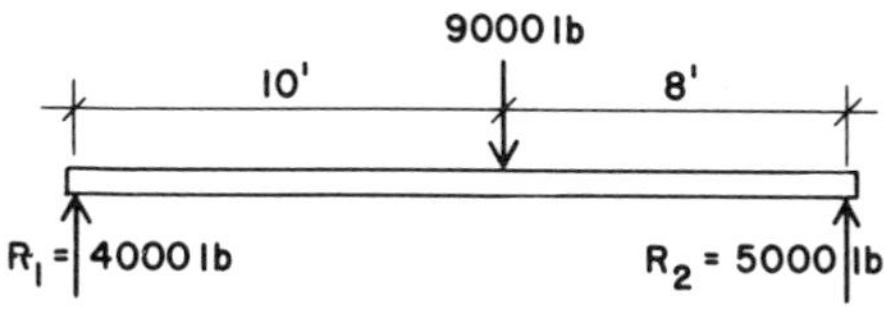

FIGURE 2.5

will the upward supporting forces be? Certainly they will not be equal.

Now this is where the principle of moments applies (see Fig. 2.5). Let us write an equation of moments, taking the center of moments about the right-hand support R_2.

$$18R_1 = 9000 \times 8$$

$$R_1 = \frac{72{,}000}{18}$$

$$R_1 = 4000 \text{ lb}$$

Because we know that the sum of the loads is equal to the sum of the reactions, we can easily compute R_2, for

$$R_1 + R_2 = 9000$$

$$4000 + R_2 = 9000$$

$$R_2 = 5000 \text{ lb}$$

To check this value of R_2, write an equation of moments about the left-hand support R_1.

$$18R_2 = 9000 \times 10$$

$$R_2 = \frac{90{,}000}{18}$$

$$R_2 = 5000 \text{ lb}$$

Example 1. A simple beam 20 ft [6.096 m] in length has three concentrated loads, as indicated in Fig. 2.6. Find the magnitudes of the reactions.

Solution: (1) With the right-hand support as the center of moments, write the equation of moments.

$$20R_1 = (2000 \times 16) + (8000 \times 10) + (4000 \times 8)$$

$$20R_1 = 32{,}000 + 80{,}000 + 32{,}000$$

$$20R_1 = 144{,}000$$

$$R_1 = \frac{144{,}000}{20} = 7200 \text{ lb}$$

(2) The sum of the reactions equals the sum of the loads.

$$R_1 + R_2 = 2000 + 8000 + 4000$$

Thus

$$7200 + R_2 = 14{,}000$$

$$R_2 = 14{,}000 - 7200 = 6800 \text{ lb}$$

(3) To check R_2, write the equation of moments about R_1.

$$20(6800) = (2000 \times 4) + (8000 \times 10) + (4000 \times 12)$$

$$136{,}000 = 8000 + 80{,}000 + 48{,}000$$

$$136{,}000 = 136{,}000$$

So far we have considered only concentrated loads in computing the magnitudes of reactions. The method of dealing with distributed loads is quite similar but there is one key point to remem-

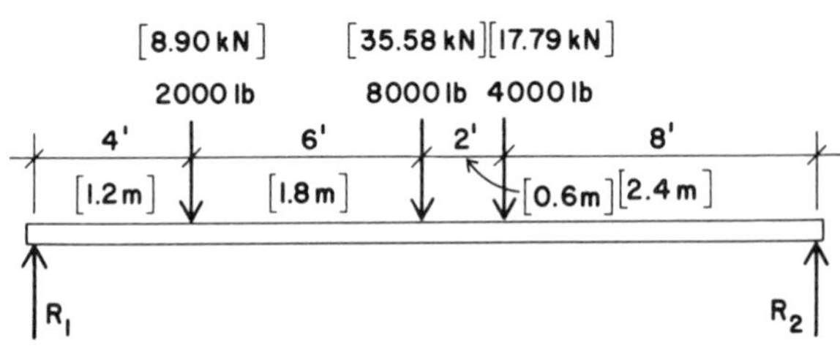

FIGURE 2.6

ber: *a distributed load on a beam produces the same reactions as a concentrated load of the same magnitude acting through the center of gravity of the distributed load.* If we bear in mind that the center of gravity of a uniformly distributed load lies at the middle of its length, the problem becomes a very simple one.

Example 2. A simple beam 16 ft long carries a concentrated load of 8000 lb and a uniformly distributed load of 14,000 lb arranged as shown in Fig. 2-7*a*. Find the reactions.

Solution: (1) Note that the uniformly distributed load extends over a length of 10 ft. Let us write an equation of moments about R_2, considering that the uniformly distributed load acts at the middle of its length, or 5 ft from R_2 (see Fig. 2.7*b*).

$$16R_1 = (8000 \times 12) + (14{,}000 \times 5)$$

$$16R_1 = 96{,}000 + 70{,}000$$

$$16R_1 = 166{,}000$$

$$R_1 = \frac{166{,}000}{16} = 10{,}375 \text{ lb}$$

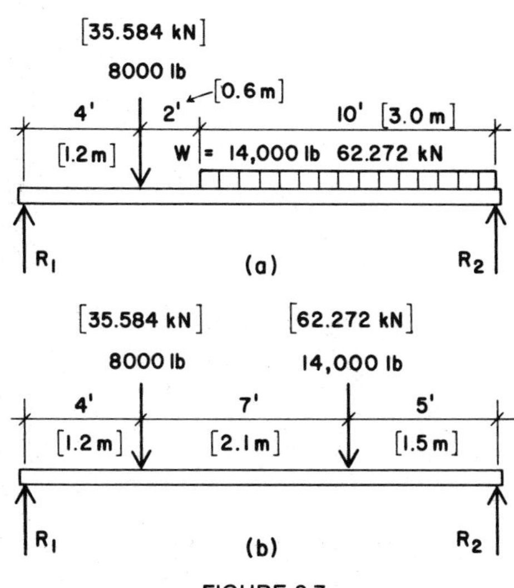

FIGURE 2.7

(2) To find R_2, take the center of moments at R_1. The lever arm of the 14,000-lb load is 11 ft because we consider that the uniformly distributed load acts at its midpoint, or 11 ft from R_1.

$$16R_2 = (8000 \times 4) + (14{,}000 \times 11)$$

$$16R_2 = 32{,}000 + 154{,}000$$

$$16R_2 = 186{,}000$$

$$R_2 = \frac{186{,}000}{16} = 11{,}625 \text{ lb}$$

(3) When we check these results, the sum of the loads should equal the sum of the reactions, or

$$8000 + 14{,}000 = 10{,}375 + 11{,}625$$

$$22{,}000 = 22{,}000$$

The method of computing reactions for overhanging beams is the same as that employed in the preceding examples. Select one of the reactions as the center of moments. On one side of the equation place the sum of the moments tending to cause clockwise rotation, and on the other side place the sum of the moments tending to cause rotation in the opposite direction. When writing a moment equation, bear two points in mind: (1) *be consistent and take the same center of moments for each force,* and (2) *consider uniformly distributed loads to act at their midpoints.*

Example 3. Find the reactions for the overhanging beam shown in Fig. 2.8*a* and check the results.

Solution: (1) Select R_1 as the center of moments. The forces tending to cause clockwise rotation about R_1 are the three loads, and the only force tending to cause counterclockwise rotation is R_2. Note the directions of the arrows. Therefore

$$13R_2 = (4000 \times 3) + (6000 \times 9) + (2000 \times 17)$$

$$13R_2 = 12{,}000 + 54{,}000 + 34{,}000 = 100{,}000$$

$$R_2 = \frac{100{,}000}{13} = 7692 \text{ lb}$$

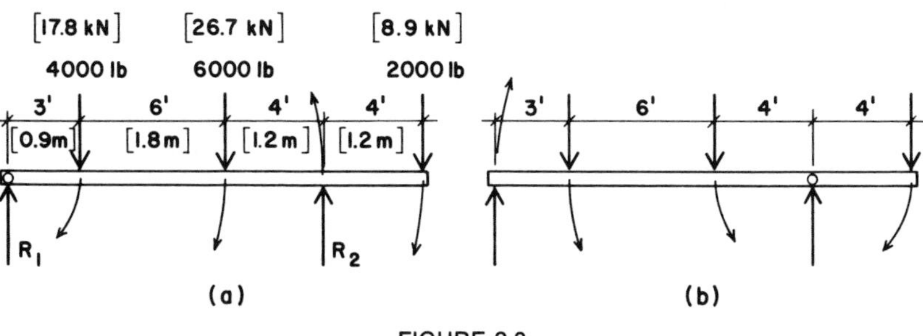

FIGURE 2.8

(2) To find R_1, take the center of moments at R_2. The forces tending to cause clockwise rotation about this point are R_1 and the 2000-lb load; those tending to cause counterclockwise rotation about the same point are the 4000-lb and 6000-lb loads. Note the directions of the arrows in Fig. 2.8*b*. Then

$$13R_1 + (2000 \times 4) = (4000 \times 10) + (6000 \times 4)$$

$$13R_1 = 40{,}000 + 24{,}000 - 8000 = 56{,}000$$

$$R_1 = \frac{56{,}000}{13} = 4308 \text{ lb}$$

(3) If these are correct results, the sum of the loads should equal the sum of the reactions; therefore

$$4000 + 6000 + 2000 = 4308 + 7692$$

$$12{,}000 = 12{,}000$$

Example 4. The overhanging beam shown in Fig. 2.9 supports a concentrated load of 4 kips and a uniformly distributed load of 6 kips, arranged as shown. Find the reactions and check the results.

Solution: (1) Note that this beam overhangs both of its reactions. The uniformly distributed load extends over a length of 10 ft, and its midpoint lies 1 ft to the right of R_2. With R_1 as the

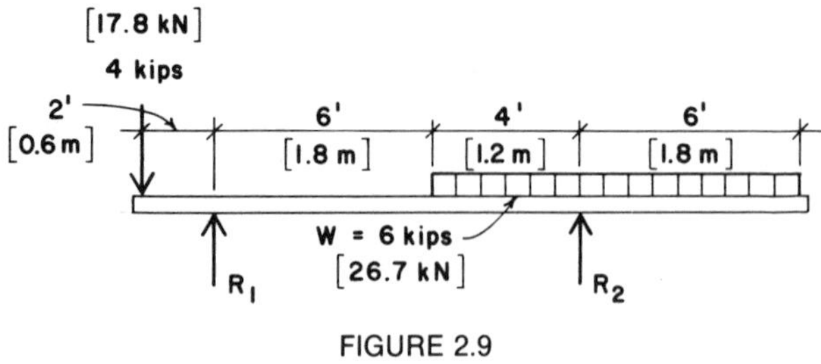

FIGURE 2.9

center of moments,

$$10R_2 + (4 \times 2) = (6 \times 11)$$

$$10R_2 = 66 - 8 = 58$$

$$R_2 = \frac{58}{10} = 5.8 \text{ kips}$$

(2) With R_2 as the center of moments,

$$10R_1 + (6 \times 1) = (4 \times 12)$$

$$10R_1 = 48 - 6 = 42$$

$$R_1 = 42 \div 10 = 4.2 \text{ kips}$$

(3) Checking the results, we find

$$4 + 6 = 5.8 + 4.2$$

$$10 = 10$$

Problems 2.3.A,B,C,D,E,F. Find the reactions for the beams shown in Fig. 2.10.

2.4 Shear in Beams

Figure 2.11*a* represents a simple beam with a uniformly distributed load *W* over its entire length. Examination of an actual beam so loaded probably would not reveal any effects of the loading on

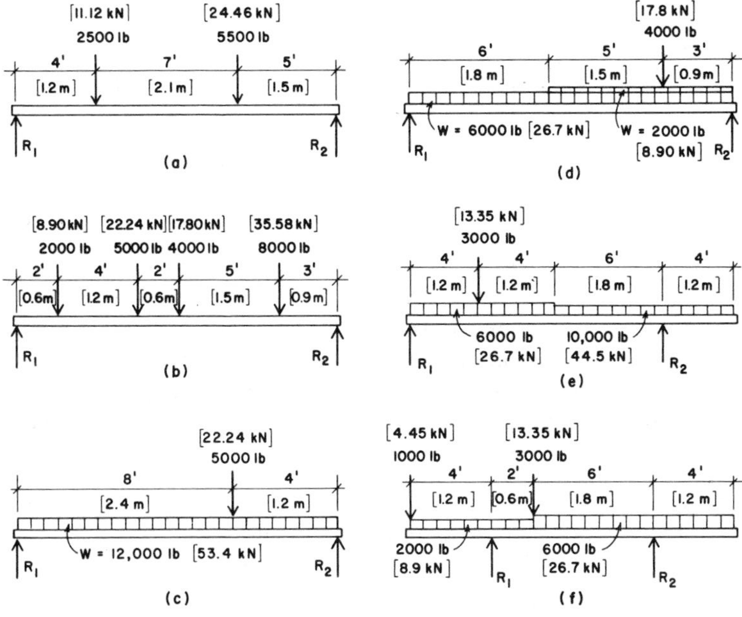

FIGURE 2.10

the beam. However, there are three distinct major tendencies for the beam to fail. Figure 2.11*b–d* illustrates the three phenomena, with deformations greatly exaggerated.

First, there is a tendency for the beam to fail by dropping between the supports (Fig. 2.11*b*). This is called *vertical shear.*

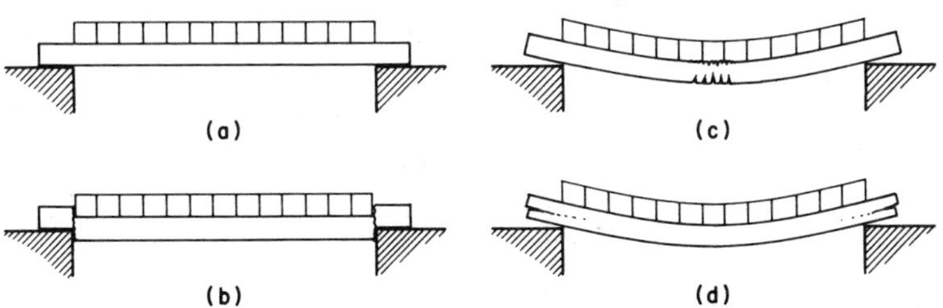

FIGURE 2.11 Development of shear in beams.

Second, the beam may fail by bending (Fig. 2.11*c*). Third, there is a tendency for the fibers of the beam to slide past each other in a horizontal direction (Fig. 2.11*d*). The name given to this action is *horizontal shear,* and it will be discussed further under steel, wood, and reinforced concrete construction. Naturally, a beam properly designed does not fail in any of the ways just mentioned, but these tendencies to fail are always present and must be considered in structural design.

The forces that prevent failure are supplied by the resisting stresses developed within the beam. Our problem in design is to select beams with dimensions that will provide adequate material to develop these resisting stresses. In this chapter we shall be concerned with methods of measuring the magnitude of shearing and bending forces caused by the beam loading or, as sometimes stated, by the external forces on the beam.

Vertical Shear. We can define vertical shear as the tendency for one part of a beam to move vertically with respect to an adjacent part. *The magnitude of the shear at any section in the length of a beam is equal to the algebraic sum of the vertical forces on either side of the section.* Vertical shear is usually represented by the letter V. In computing its values in our examples and problems, we consider the forces to the left of the section, but keep in mind that the same result will be obtained if we work with the forces on the right. We may say then that *the vertical shear at any section of a beam is equal to the reactions minus the loads to the left of the section.* Fix this definition firmly in mind. Now, if we wish to find the magnitude of the vertical shear at any section in the length of a beam, we simply repeat the foregoing statement and write an equation accordingly. It follows from this procedure that the maximum value of the shear for simple beams is equal to the greater reaction.

If the loads and reactions are in units of pounds or kips, the magnitude of the vertical shear will be in units of pounds or kips also. Form the habit of writing the denomination of the units after the numerical values; it will prevent many errors.

Example 1. Figure 2.12*a* illustrates a simple beam with two concentrated loads of 600 lb and 1000 lb. Our problem is to find the

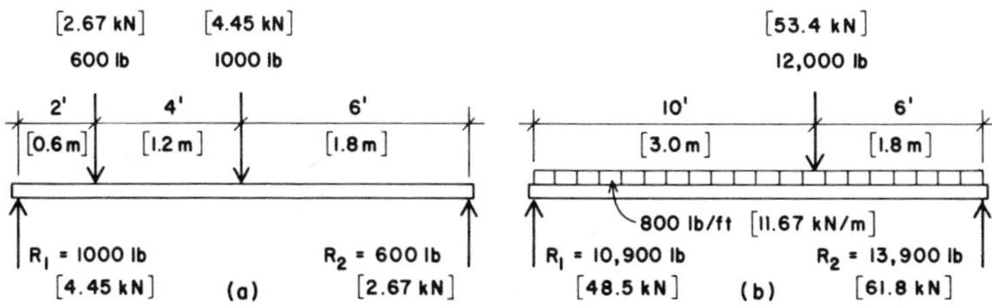

FIGURE 2.12

value of the vertical shear at various points along the length of the beam. Although the weight of the beam constitutes a uniformly distributed load, it is neglected in this example.

Solution: (1) The reactions are computed by the principle of moments previously described, and we find that they are $R_1 = 1000$ lb and $R_2 = 600$ lb.

(2) Consider first the value of the vertical shear V at an infinitely short distance to the right of R_1. Applying the rule that the shear is equal to the reactions minus the loads to the left of the section, we write $V = R_1 - 0$, or $V = 1000$ lb. The zero represents the value of the loads to the left of the section, which, of course, is zero. Now take a section 1 ft to the right of R_1; again $V_{(x=1)} = R_1 - 0$, or $V_{(x=1)} = 1000$ lb. The subscript $(x = 1)$ indicates the position of the section at which the shear is taken, the distance of the section from R_1. We find that the shear is still 1000 lb and has the same magnitude up to the 600-lb load. The next section to consider is a very short distance to the right of the load 600 lb. Then $V_{(x=2+)} = 1000 - 600 = 400$ lb. Because there are no loads intervening, the shear continues to be the same magnitude up to the load 1000 lb. At a section a short distance to the right of the 1000-lb load, $V_{(x=6+)} = 1000 - (600 + 1000) = -600$ lb. This magnitude continues up to the right-hand reaction R_2.

The preceding example dealt only with concentrated loads. Let us see if we can apply the same procedure to a beam having a uniformly distributed load in addition to a concentrated load.

Example 2. The beam shown in Fig. 2.12*b* supports a concentrated load of 12,000 lb located 6 ft from R_2 and a uniformly distributed load of 800 pounds per linear foot (lb per lin ft) over its entire length. Compute the value of the vertical shear at various sections along the span.

Solution: (1) Note that the uniform load is given in lb per lin ft (frequently abbreviated to lb/ft). The symbol for uniform load per foot is w, the capital letter W being used to represent the *total* uniformly distributed load. In this instance $w = 800$ lb and $W = 800 \times 16 = 12{,}800$ lb. The equation of moments about R_2 as the center is

$$16R_1 = (800 \times 16 \times 8) + (12{,}000 \times 6)$$

$$16R_1 = 102{,}400 + 72{,}000$$

$$R_1 = \frac{174{,}400}{16} = 10{,}900 \text{ lb}$$

In a similar manner R_2 is found to be 13,900 lb. In the quantity (800 × 16× 8), the load (800 × 16) lb has a lever arm of 8 ft, the distance of its center of gravity to the reaction.

(2) Following the rule used in the preceding example, write the value of V at various sections along the beam.

$$V_{(x=0)} = 10{,}900 - 0 = 10{,}900 \text{ lb}$$

$$V_{(x=1)} = 10{,}900 - 800 = 10{,}100 \text{ lb}$$

$$V_{(x=5)} = 10{,}900 - (800 \times 5) = 6900 \text{ lb}$$

$$V_{(x=10-)} = 10{,}900 - (800 \times 10) = 2900 \text{ lb}$$

$$V_{(x=10+)} = 10{,}900 - [(800 \times 10) + 12{,}000] = -9100 \text{ lb}$$

$$V_{(x=16)} = 10{,}900 - [(800 \times 16) + 12{,}000] = -13{,}900 \text{ lb}$$

Note that for simple span beams the value of the vertical shear at the supports has the same magnitude as the reactions.

Shear Diagrams. In the two preceding examples we computed the value of the shear at several sections along the length of the beams. In order to visualize the results we have obtained, we may make diagrams to plot these values. They are called *shear diagrams* and are constructed as explained below.

To make such a diagram, first draw the beam to scale and locate the loads. This has been done in Fig. 2.13*a* and *b* by repeating the load diagrams of Fig. 2.12*a* and *b*, respectively. Beneath the beam draw a horizontal base line representing zero shear. Above and below this line, plot at any convenient scale the values of the shear at the various sections; the positive, or plus, values are placed above the line and the negative, or minus, values below. In Fig. 2.13*a*, for instance, the value of the shear at R_1 is +1000 lb. The shear continues to have the same value up to the load of 600 lb, at which point it drops to +400 lb. The same value continues up to the next load, 1000 lb, where it drops to −600 lb and continues to the right-hand reaction. Obviously, to draw a shear diagram it is necessary to compute the values at significant points only. Having made the diagram, we may readily find the value of the shear at any section of the beam by scaling the vertical distance in the diagram. The shear diagram for the beam in Fig. 2.13*b* is made in the same manner.

There are two important facts to note concerning the vertical shear. The first is the maximum value. We see that the diagrams

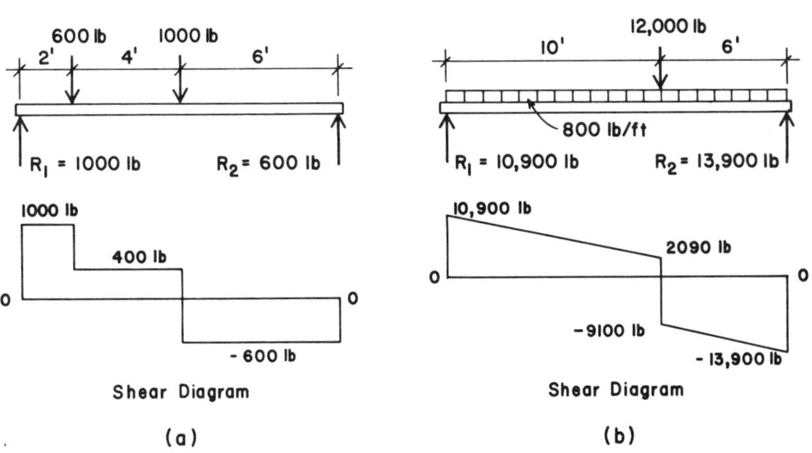

FIGURE 2.13

in each case confirm our earlier observation that the maximum shear is at the reaction having the greater value, and its magnitude is equal to that of the greater reaction. In Fig. 2.13*a* the maximum shear is 1000 lb, and in Fig. 2.13*b* it is 13,900 lb. We disregard the positive or negative signs in reading the maximum values of the shear, for the diagrams are merely conventional methods of representing the absolute numerical values.

Another important fact to note is the point at which the shear changes from a plus to a minus quantity. We call this the point at which the shear passes through zero. In Fig. 2.13*a* it is under the 1000-lb load, 6 ft from R_1. In Fig. 2.13*b* it is under the 12,000-lb load, 10 ft from R_1. A major concern for noting this point is that it indicates the location of the maximum value of bending moment in the beam, as discussed in the next section.

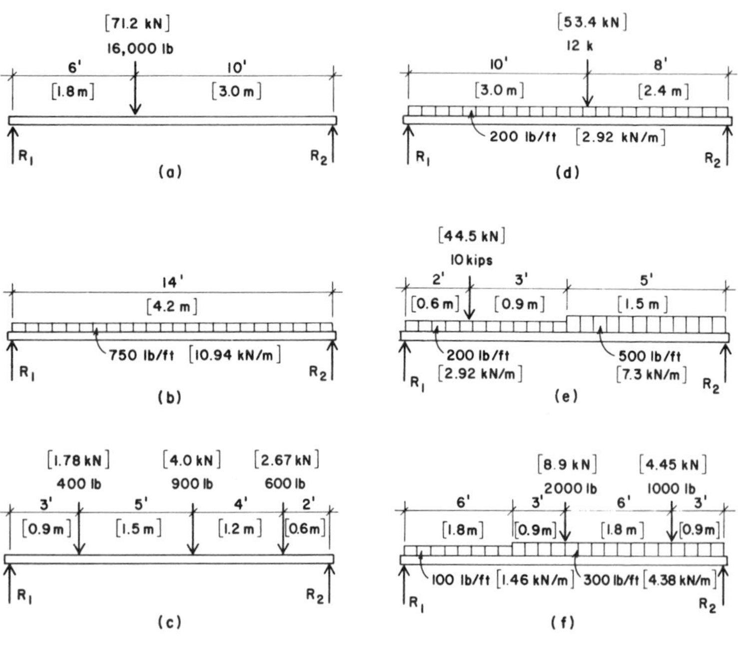

FIGURE 2.14

Problems 2.4.A,B,C,D,E,F. For the beams shown in Fig. 2.14, draw the shear diagrams and note all critical values for shear. Note particularly the maximum value for shear and the point at which the shear passes through zero.

2.5 Bending Moment in Beams

The forces that tend to cause bending in a beam are the reactions and the loads. Consider the section *X–X,* 6 ft from R_1 (Fig. 2.15). The force R_1, or 2000 lb, tends to cause a clockwise rotation about this point. Because the force is 2000 lb and the lever arm is 6 ft, the moment of the force is 2000 × 6 = 12,000 ft-lb. This same value may be found by considering the forces to the right of the section *X–X*. Let us see. There are two forces to the right of section *X–X:* R_2, which is 6000 lb, and the load 8000 lb, with lever arms of 10 and 6 ft, respectively. The moment of the reaction is 6000 × 10 = 60,000 ft-lb, and its direction is counterclockwise with respect to the section *X–X*. The moment of the force 8000 lb is 8000 × 6 = 48,000 ft-lb, and its direction is clockwise. Then 60,000 ft-lb − 48,000 ft-lb = 12,000 ft-lb, the resultant moment tending to cause counterclockwise rotation about the section *X–X*. This is the same magnitude as the moment of the forces on the left which tends to cause a clockwise rotation.

Thus it makes no difference whether we consider the forces to the right of the section or the left; the magnitude of the moment is the same. It is called the *bending moment* because it is the moment of the forces that cause bending stresses in the beam. Its magnitude varies throughout the length of the beam. For in-

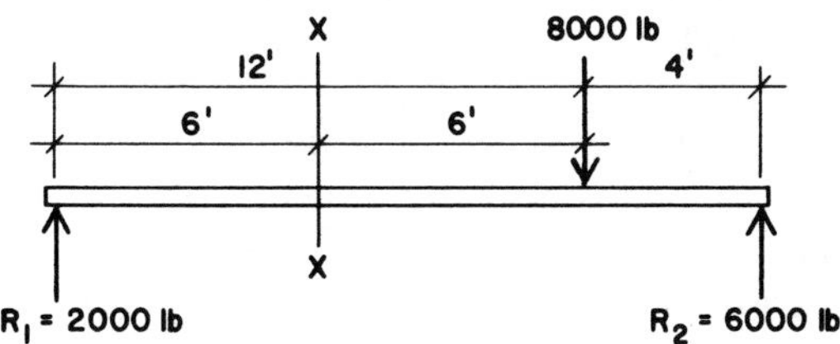

FIGURE 2.15

stance, at 4 ft from R_1 it is only 2000 × 4, or 8000 ft-lb. *The bending moment is the algebraic sum of the moments of the forces on either side of the section.* For simplicity, let us take the forces on the left; then we may say *the bending moment at any section of a beam is equal to the moments of the reactions minus the moments of the loads to the left of the section.* Because the bending moment is the result of multiplying forces by distances, the denominations are foot-pounds or kip-feet.

Almost everyone confuses shear and bending moment at first. Remember that the shear is the result of subtracting loads from reactions, the units being pounds or kips; the bending moment is the result of subtracting *moments* of loads from *moments* of reactions, with units of foot-pounds or kip-feet.

Bending Moment Diagrams. The construction of bending moment diagrams follows the procedure used for shear diagrams. The beam span is drawn to scale showing the locations of the loads. Below this, and usually below the shear diagram, a horizontal base line is drawn representing zero bending moment. Then the bending moments are computed at various sections along the beam span, and the values are plotted vertically to any convenient scale. In simple beams all bending moments are positive and therefore are plotted above the base line. In overhanging or continuous beams we shall find negative moments, and these are plotted below the base line.

Example 1. The load diagram in Fig. 2.16 shows a simple beam with two concentrated loads. Draw the shear and bending moment diagrams.

Solution: (1) R_1 and R_2 are first computed and are found to be 16,000 lb and 14,000 lb, respectively. These values are recorded on the load diagram.

(2) The shear diagram is drawn as described in Sec. 2.4. Note that in this instance it is necessary to compute the shear at only one section (between the concentrated loads) because there is no distributed load, and we know that the shear at the reactions is equal in magnitude to the reactions.

(3) Because the value of the bending moment at any section of the beam is equal to the moments of the reactions minus the

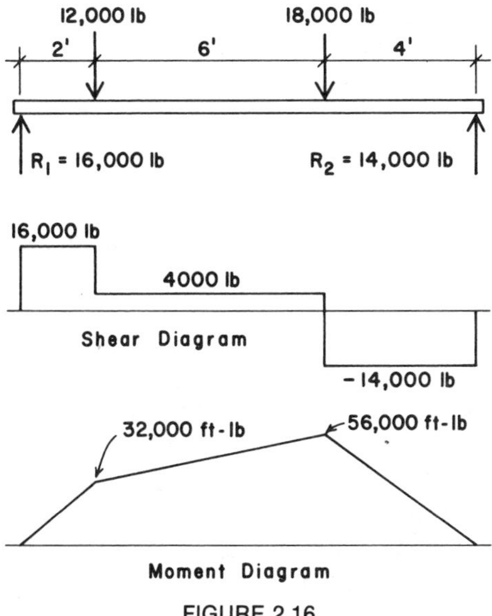

FIGURE 2.16

moments of the loads to the left of the section, the moment at R_1 must be zero, for there are no forces to the left. We might also say that R_1 has a lever arm of zero and $R_1 \times 0 = 0$. Other values in the length of the beam are computed as follows. The subscripts ($x = 1$, etc.) show the distance from R_1 at which the bending moment is computed.

$$M_{(x=1)} = (16{,}000 \times 1) = 16{,}000 \text{ ft-lb}$$

$$M_{(x=2)} = (16{,}000 \times 2) = 32{,}000 \text{ ft-lb}$$

$$M_{(x=5)} = (16{,}000 \times 5) - (12{,}000 \times 3) = 44{,}000 \text{ ft-lb}$$

$$M_{(x=8)} = (16{,}000 \times 8) - (12{,}000 \times 6) = 56{,}000 \text{ ft-lb}$$

$$M_{(x=10)} = (16{,}000 \times 10) - [(12{,}000 \times 8) + (18{,}000 \times 2)]$$
$$= 28{,}000 \text{ ft-lb}$$

$$M_{(x=12)} = (16{,}000 \times 12) - [(12{,}000 \times 10) + (18{,}000 \times 4)] = 0$$

The result of plotting these values is shown in the bending moment diagram of Fig. 2.16. More moments were computed than were actually necessary. We know that the bending moments at the supports of simple beams are zero, and in this instance only the bending moments directly under the loads were needed.

Relations Between Shear and Bending Moment. In simple beams the shear diagram passes through zero at some point between the supports. As stated earlier, an important principle in this respect is that *the bending moment has a maximum magnitude wherever the shear passes through zero*. In Fig. 2.16 the shear passes through zero under the 18,000-lb load, that is, at $x = 8$. Note that the bending moment has its greatest value at this same point, 56,000 ft-lb. In order to design beams, we must know the value of the maximum bending moment. Frequently we draw only enough of the shear diagram to find the section at which the shear passes through zero and then compute the bending moment at this point.

Example 2. Draw the shear and bending moment diagrams for the beam shown in Fig. 2.17, which carries a uniformly distributed load of 400 lb per lin ft and a concentrated load of 21,000 lb located 4 ft from R_1.

Solution: (1) Computing the reactions, we find

$$14R_1 = (21{,}000 \times 10) + (400 \times 14 \times 7)$$

$$R_1 = 17{,}800 \text{ lb}$$

$$\text{Total load} = 21{,}000 + (400 \times 14) = 26{,}600 \text{ lb}$$

Therefore

$$R_1 + R_2 = 26{,}600$$

$$17{,}800 + R_2 = 26{,}600$$

$$R_2 = 26{,}600 - 17{,}800 = 8800 \text{ lb}$$

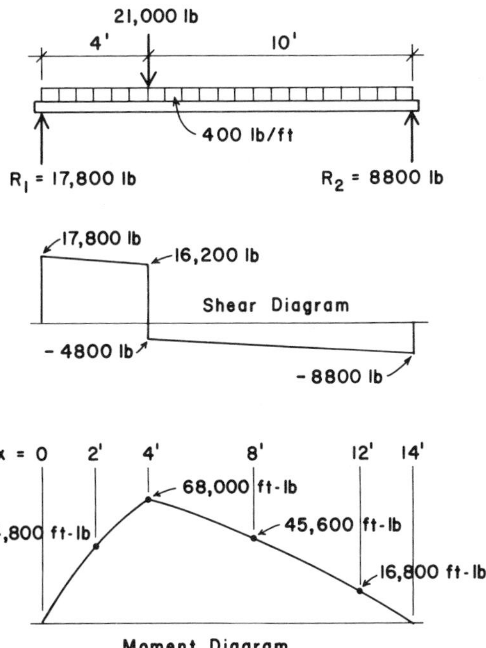

FIGURE 2.17

(2) Computing the value of the shear at essential points, we find

$$V_{(\text{at } R_1)} = 17{,}800 \text{ lb}$$

$$V_{(x=4-)} = 17{,}800 - (400 \times 4) = 16{,}200 \text{ lb}$$

$$V_{(x=4+)} = 17{,}800 - [(400 \times 4) + 21{,}000] = -4800 \text{ lb}$$

$$V_{(\text{at } R_2)} = -8800 \text{ lb}$$

Note that the shear passes through zero under the 21,000-lb load; therefore at this point we shall expect to find that the bending moment is a maximum.

(3) Computing the value of the bending moment at selected points, we get

$$M_{(x=0)} = 0$$

$$M_{(x=2)} = (17{,}800 \times 2) - (400 \times 2 \times 1) = 34{,}800 \text{ ft-lb}$$

In the foregoing equation the reaction to the left of the section is 17,800 lb and its lever arm is 2 ft. The load to the left of the section is (400 × 2) lb and its lever arm is 1 ft, the distance from the center of the load, (400 × 2) lb, to the center of moments.

$$M_{(x=4)} = (17{,}800 \times 4) - (400 \times 4 \times 2) = 68{,}000 \text{ ft-lb}$$

$$M_{(x=8)} = (17{,}800 \times 8) - [(400 \times 8 \times 4) + (21{,}000 \times 4)]$$

$$= 45{,}600 \text{ ft-lb}$$

$$M_{(x=12)} = (17{,}800 \times 12) - [(400 \times 12 \times 6) + (21{,}000 \times 8)]$$

$$= 16{,}800 \text{ ft-lb}$$

$$M_{(x=14)} = (17{,}800 \times 14) - [(400 \times 14 \times 7) + (21{,}000 \times 10)] = 0$$

From the two preceding examples (Figs. 2.16 and 2.17), it will be observed that the shear diagram for the parts of the beam on which no loads occur is represented by horizontal lines. For the parts of the beam on which a uniformly distributed load occurs, the shear diagram consists of straight inclined lines. The bending moment diagram is represented by straight inclined lines when only concentrated loads occur and by a curved line if the load is distributed.

Occasionally, when a beam has both concentrated and uniformly distributed loads, the shear does not pass through zero under one of the concentrated loads. This frequently occurs when the distributed load is relatively large compared with the concentrated loads. Since it is necessary in designing beams to find the maximum bending moment, we must know the point at which it occurs. This, of course, is the point where the shear passes through zero, and its location is readily determined by the procedure illustrated in the following example.

Example 3. The load diagram in Fig. 2.18 shows a beam with a concentrated load of 7000 lb, applied 4 ft from the left reaction, and a uniformly distributed load of 800 lb per lin ft extending over the full span. Compute the maximum bending moment on the beam.

Solution: (1) Compute the values of the reactions. These are found to be $R_1 = 10{,}600$ lb and $R_2 = 7600$ lb and are recorded on the load diagram.

(2) Constructing the shear diagram in Fig. 2.18, we see that the shear passes through zero at some point between the concentrated load of 7000 lb and the right reaction. Call this distance x ft from R_1. Now, we know that the value of the shear at this section is zero; therefore we write an expression for the shear for this point, using the terms of the reaction and loads, and equate the quantity to zero. This equation contains the distance x:

$$V_{(\text{at } x)} = 10{,}600 - [7000 + (800x)] = 0$$

$$800x = 3600$$

$$x = 4.5 \text{ ft}$$

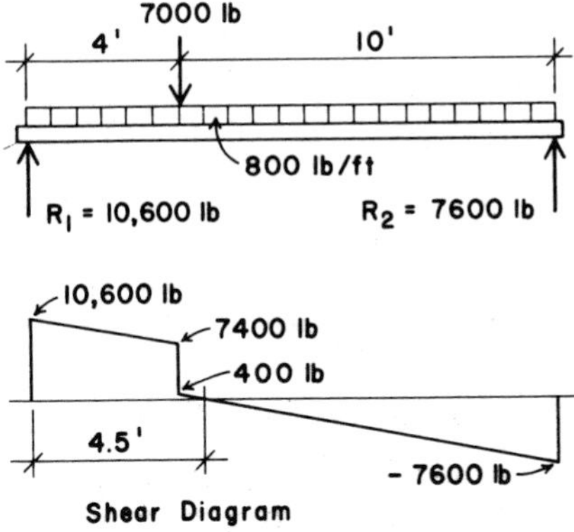

FIGURE 2.18

(3) The value of the bending moment at this section is given by the expression

$$M_{(x=4.5)} = (10{,}600 \times 4.5) - \left[(7000 \times 0.5) + \left(800 \times 4.5 \times \frac{4.5}{2}\right)\right]$$

$$M = 36{,}100 \text{ ft-lb}$$

Problems 2.5.A,B,C,D,E,F. Draw the shear and bending moment diagrams for the beams in Fig. 2.14, indicating all critical values for shear and moment and all significant dimensions. (*Note:* These are the same beams as in problems following Sec. 2.4.)

2.6 Beams With Overhanging Ends

When a simple beam bends, it has a tendency to assume the shape shown in Fig. 2.19*a*. In this case the fibers in the upper part of the beam are in compression. For this condition we say the bending moment is positive (+). Another way to describe a positive bending moment is to say that it is positive when the curve assumed by the bent beam is concave upward. When a beam projects beyond a support (Fig. 2.19*b*), this portion of the beam has tensile stresses in its upper part. The bending moment for this condition is called negative (−); the beam is bent concave downward. If we construct moment diagrams, following the method previously described, the positive and negative moments are shown graphically.

Example 1. Draw the shear and bending moment diagrams for the overhanging beam shown in Fig. 2.20.

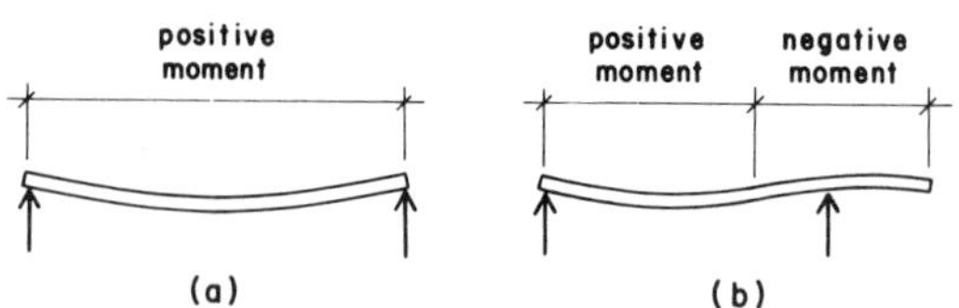

FIGURE 2.19 Positive and negative moments.

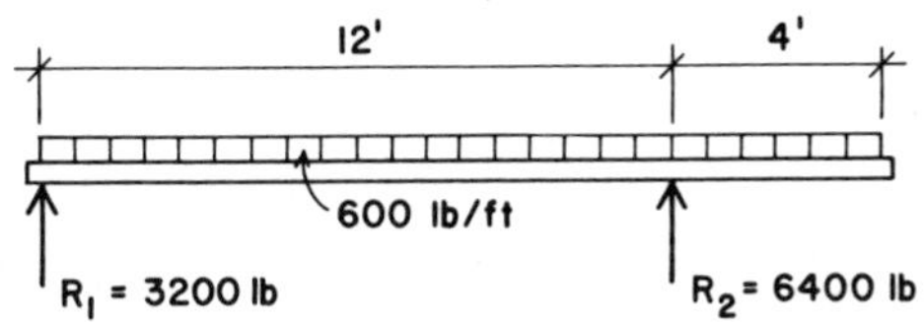

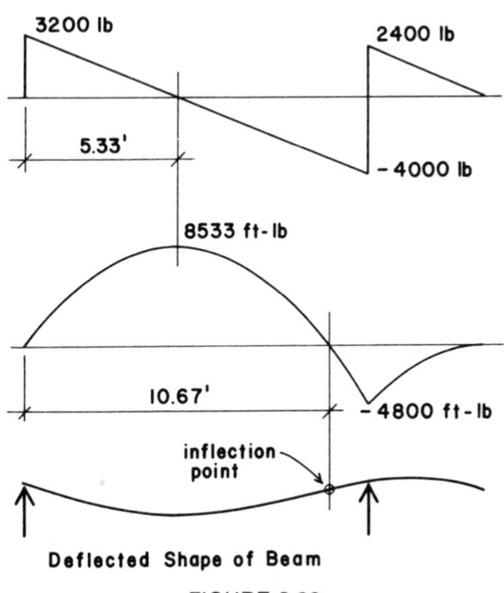

FIGURE 2.20

Solution: (1) Computing the reactions, we find

$$12R_2 = 600 \times 16 \times 8 \qquad R_2 = 6400 \text{ lb}$$

$$12R_1 = 600 \times 16 \times 4 \qquad R_1 = 3200 \text{ lb}$$

(2) Computing the values of the shear, we find

$$V_{(\text{at } R_1)} = +3200 \text{ lb}$$

$$V_{(x=12-)} = 3200 - (600 \times 12) = -4000 \text{ lb}$$

$$V_{(x=12+)} = (3200 + 6400) - (600 \times 12) = +2400 \text{ lb}$$

$$V_{(x=16)} = (3200 + 6400) - (600 \times 16) = 0$$

To find the point at which the shear passes through zero between the supports (see Example 2, Sec. 2.5), calculate

$$3200 - 600x = 0$$

$$x = 5.33 \text{ ft}$$

(3) Computing the values of the bending moment, we find

$$M_{(\text{at } R_1)} = 0$$

$$M_{(x=5.33)} = (3200 \times 5.33) - \left(600 \times 5.33 \times \frac{5.33}{2}\right) = 8533 \text{ ft-lb}$$

$$M_{(x=12)} = (3200 \times 12) - (600 \times 12 \times 6) = -4800 \text{ ft-lb}$$

$$M_{(x=16)} = 0$$

To draw the bending moment diagram accurately, we may compute the magnitudes at other points; it will be a curved line.

It is seen in plotting the shear values that there are two points at which the shear passes through zero: at $x = 5.33$ ft and at $x = 12$ ft. The bending moment diagram shows that maximum values, one for a positive moment and the other a negative moment, are found at each point. When we design beams, we are concerned only with the maximum value, regardless of whether it is positive or negative. The shear diagram does not indicate which of the two points gives the greater value of the bending moment, and often it is necessary to compute the value at each point at which the shear passes through zero, to determine which one is greater numerically. For this beam it is 8533 ft-lb.

Inflection Point. The bending moment diagram in Fig. 2.20 indicates a point between the supports at which the value of $M = 0$. This is called the *inflection point;* it is the point at which the curvature reverses as it changes from concave to convex. It is important to know the position of the inflection point in the study of reinforced concrete beams, for this is the position at which the tensile steel reinforcement is bent upward.

In this problem call x the distance from the left support to the point at which $M = 0$. Then writing an expression for the value of the bending moment and equating it to zero,

$$(3200 \times x) - \left(600 \times x \times \frac{x}{2}\right) = 0$$

$$3200x - 300x^2 = 0$$

$$x = 10.67 \text{ ft}$$

Examine the curve of the bending moment. You will see that the curve is symmetrical between the left reaction and the inflection point. Now, since the curve reaches its highest point at 5.33 ft from the left support, the inflection point occurs at 2×5.33, or 10.67 ft from R_1. This is the same result found by solving for x in the foregoing equation.

For the beam and load shown in Fig. 2.20, note that the value of the maximum vertical shear is −4000 lb. It occurs immediately to the left of the right-hand support.

Example 2. Compute the maximum bending moment for the overhanging beam shown in Fig. 2.21.

Solution: (1) Computing the reactions, we find

$$12R_1 + (200 \times 2) = (800 \times 16) + (1000 \times 10) + (4000 \times 4)$$

$$R_1 = 3200 \text{ lb}$$

$$\text{Total load} = 800 + 1000 + 4000 + 200 = 6000 \text{ lb}$$

Therefore

$$R_2 = 6000 - 3200 = 2800 \text{ lb}$$

(2) Computing the values of the shear, we find

$$V_{(x=1)} = -800 \text{ lb}$$

$$V_{(x=4+)} = 3200 - 800 = +2400 \text{ lb}$$

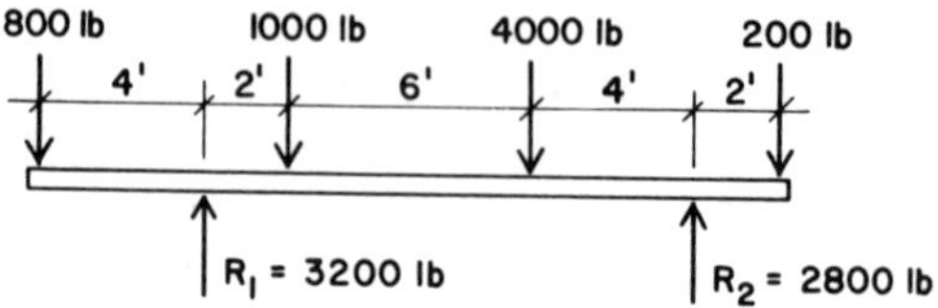

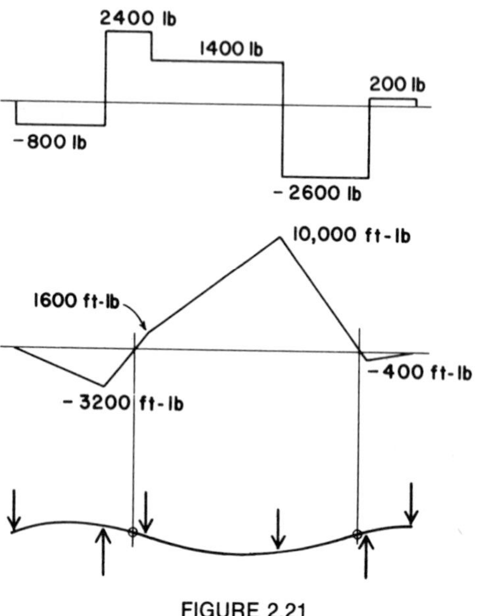

FIGURE 2.21

$$V_{(x=6+)} = 3200 - (800 + 1000) = +1400 \text{ lb}$$

$$V_{(x=12+)} = 3200 - (800 + 1000 + 4000) = -2600 \text{ lb}$$

$$V_{(x=16+)} = (3200 + 2800) - (800 + 1000 + 4000) = +200 \text{ lb}$$

With these values we now plot the shear diagram. We note that the shear passes through zero at three points, R_1, R_2, and under the 4000-lb load. We expect the bending moment to reach maximum values at these points.

(3) Computing the values of the bending moment, we find

$$M_{(x=0)} = 0$$

$$M_{(x=4)} = -(800 \times 4) = -3200 \text{ ft-lb}$$

$$M_{(x=6)} = (3200 \times 2) - (800 \times 6) = +1600 \text{ ft-lb}$$

$$M_{(x=12)} = (3200 \times 8) - [(800 \times 12) + (1000 \times 6)]$$
$$= +10{,}000 \text{ ft-lb}$$

$$M_{(x=16)} = (3200 \times 12) - [(800 \times 16) + (1000 \times 10) + (4000 \times 4)] = -400 \text{ ft-lb}$$

$$M_{(x=18)} = 0$$

The maximum value of the bending moment (10,000 ft-lb) occurs under the 4000-lb load.

The value of the maximum vertical shear is −2600 lb.

Note that the bending moment diagram changes from a plus value to a minus value at two points. There are two inflection points. If it were possible to scale the moment diagram of Fig. 2.21 with sufficient accuracy, we would find that one of these

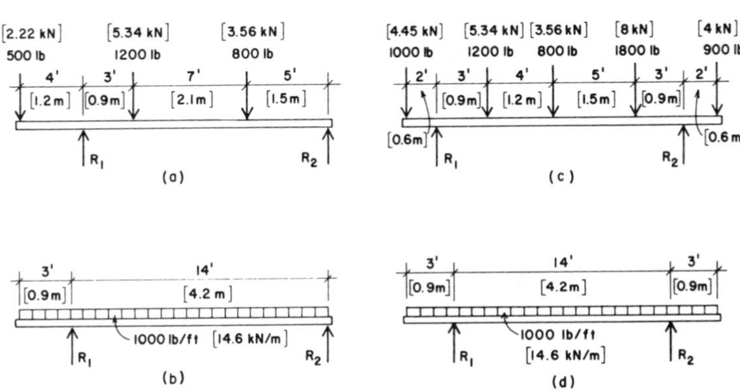

FIGURE 2.22

inflection points is located approximately 1.3 ft to the right of R_1 and the other approximately 0.15 ft to the left of R_2. The reader should check the accuracy of these locations by writing moment equations about each inflection point. If the locations are "exactly" correct, the value of the bending moment at each point will, by definition, equal zero.

Problems 2.6.A,B,C,D. Draw the shear and bending moment diagrams for the beams in Fig. 2.22, indicating all critical values for shear and moment and all significant dimensions.

2.7 Cantilever Beams

In order to keep the signs for shear and moment consistent with those for other beams, it is convenient to draw a cantilever beam with its fixed end to the right, as shown in Fig. 2.23. We then plot the values for shear and moment on the diagrams as before, proceeding from the left end.

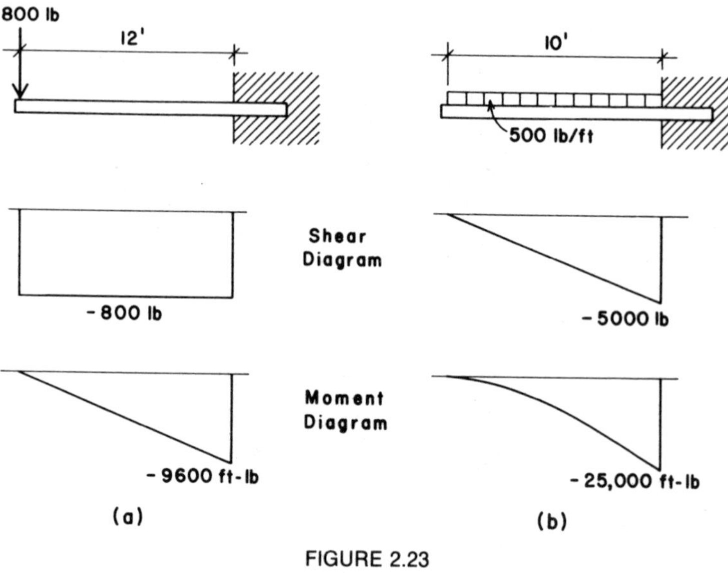

FIGURE 2.23

Example 1. The cantilever beam shown in Fig. 2.23*a* projects 12 ft from the face of the wall and has a concentrated load of 800 lb at the unsupported end. Draw the shear and moment diagrams. What are the values of the maximum shear and maximum bending moment?

Solution: (1) The value of the shear is −800 lb throughout the entire length of the beam.

(2) The bending moment is maximum at the wall; its value is −9600 ft-lb.

$$M_{(x=0)} = 0$$
$$M_{(x=1)} = -(800 \times 1) = -800 \text{ ft-lb}$$
$$M_{(x=2)} = -(800 \times 2) = -1600 \text{ ft-lb}$$
$$M_{(x=12)} = -(800 \times 12) = -9600 \text{ ft-lb}$$

Example 2. Draw the shear and bending moment diagrams for the cantilever beam, shown in Fig. 2.23*b*, which carries a uniformly distributed load of 500 lb per lin ft over its full length.

Solution: (1) Computing the values of the shear, we find

$$V_{(x=1)} = -(500 \times 1) = -500 \text{ lb}$$
$$V_{(x=2)} = -(500 \times 2) = -1000 \text{ lb}$$
$$V_{(x=10)} = -(500 \times 10) = -5000 \text{ lb, the maximum value}$$

(2) Computing the moments, we find

$$M_{(x=0)} = 0$$
$$M_{(x=2)} = -(500 \times 2 \times 1) = -1000 \text{ ft-lb}$$
$$M_{(x=4)} = -(500 \times 4 \times 2) = -4000 \text{ ft-lb}$$
$$M_{(x=10)} = -(500 \times 10 \times 5) = -25{,}000 \text{ ft-lb, the maximum value}$$

Example 3. The cantilever beam indicated in Fig. 2.24 has a concentrated load of 2000 lb and a uniformly distributed load of

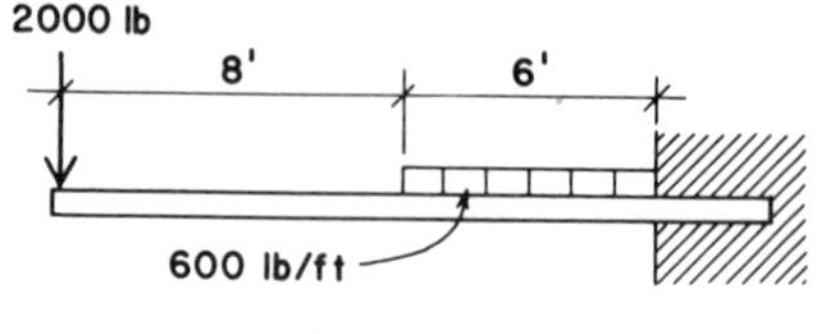

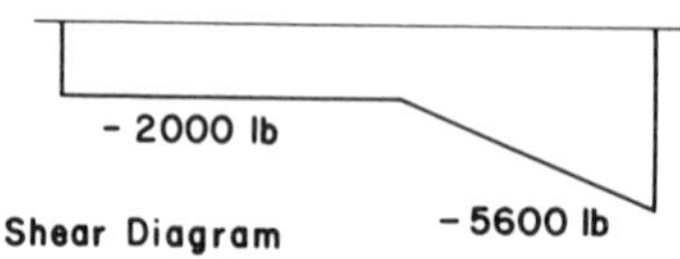

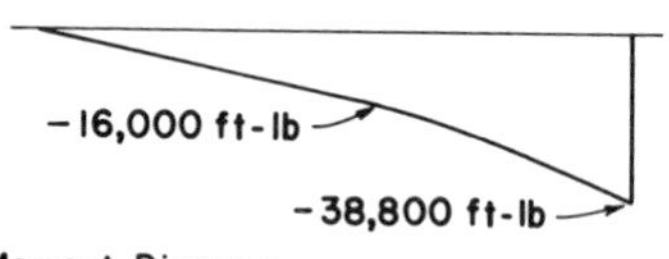

FIGURE 2.24

600 lb per lin ft at the positions shown. Draw the shear and bending moment diagrams. What are the magnitudes of the maximum shear and maximum bending moment?

Solution: (1) Computing the values of the shear, we find

$$V_{(x=1)} = -2000 \text{ lb}$$

$$V_{(x=8)} = -2000 \text{ lb}$$

$$V_{(x=10)} = -[2000 + (600 \times 2)] = -3200 \text{ lb}$$

$$V_{(x=14)} = -[2000 + (600 \times 6)] = -5600 \text{ lb}$$

(2) Computing the moments, we find

$$M_{(x=0)} = 0$$

$$M_{(x=4)} = -(2000 \times 4) = -8000 \text{ ft-lb}$$

$$M_{(x=8)} = -(2000 \times 8) = -16{,}000 \text{ ft-lb}$$

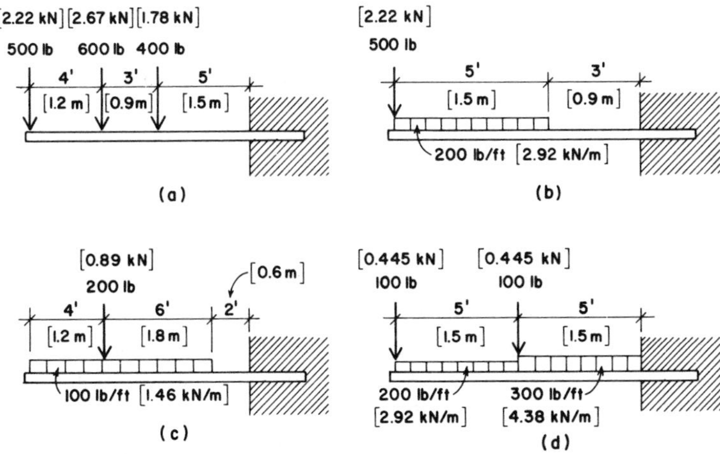

FIGURE 2.25

$$M_{(x=10)} = -[(2000 \times 10) + (600 \times 2 \times 1)] = -21{,}200 \text{ ft-lb}$$

$$M_{(x=14)} = -[(2000 \times 14) + (600 \times 6 \times 3)] = -38{,}800 \text{ ft-lb}$$

The maximum shear is 5600 lb, and the maximum bending moment is 38,800 ft-lb. In cantilever beams the maximum values of both shear and moment occur at the support; the beams are concave downward and the bending moment is negative throughout the length. The reader should be satisfied that this statement holds when the fixed end of a cantilever beam is taken to the left. It is suggested that Example 3 be reworked with Fig. 2.24 reversed, left for right. All numerical results will be the same, but the shear diagram will be positive over its full length.

Problems 2.7.A,B,C,D. Draw the shear and bending moment diagrams for the beams in Fig. 2.25, indicating all critical values for shear and moment and all significant dimensions.

2.8 Bending Moment Formulas

The method of computing bending moments presented thus far in this chapter enables us to find the maximum value under a wide variety of loading conditions. However, certain conditions occur so frequently that it is convenient to use formulas that give the maximum values directly. Structural design handbooks contain many such formulas; two of the most commonly used formulas are derived in the following examples.

Simple Beam, Concentrated Load at Center of Span. A simple beam with a concentrated load at the center of the span occurs very frequently in practice. Call the load P and the span length between supports L, as indicated in the load diagram of Fig. 2.26*a*. For this symmetrical loading each reaction is $P/2$, and it is readily apparent that the shear will pass through zero at distance $x = L/2$ from R_1. Therefore the maximum bending moment occurs at the center of the span, under the load. Now let us compute the value of the bending moment at this section.

$$M_{(x=L/2)} = \frac{P}{2} \times \frac{L}{2}$$

or

$$M = \frac{PL}{4}$$

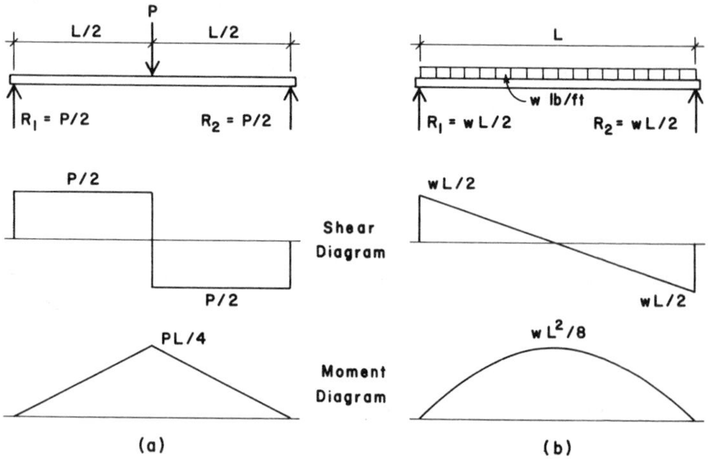

FIGURE 2.26

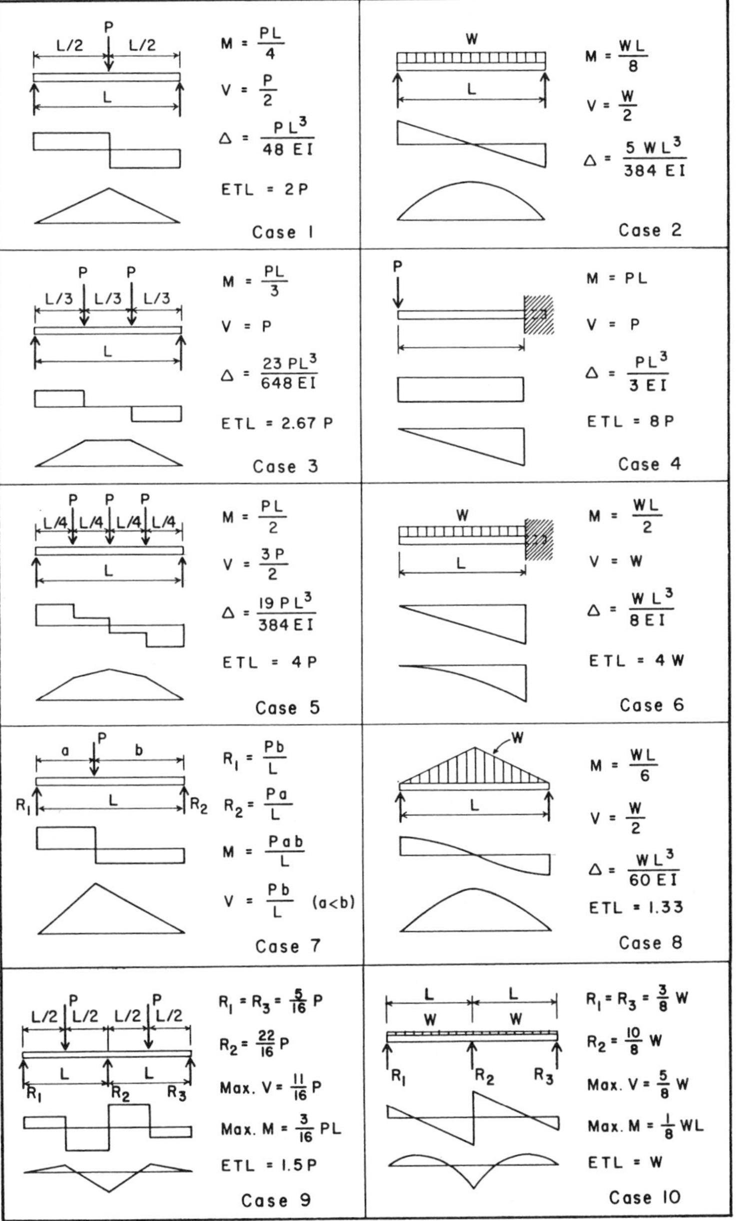

FIGURE 2.27 Values for typical beam loadings.

Note that this value is given in case 1, Fig. 2.27. This formula is well worth remembering. Observe how quickly bending moments are computed by its use.

Example 1. A simple beam 20 ft in length has a concentrated load of 8000 lb at the center of the span. Compute the maximum bending moment.

Solution: The formula giving the value of the maximum bending moment for this condition is $M = PL/4$. Therefore

$$M = \frac{8000 \times 20}{4} = 40{,}000 \text{ ft-lb}$$

If we want the results in inch-pounds, as we frequently do in design, we simply multiply by 12; that is,

$$M = 40{,}000 \text{ ft-lb} = (40{,}000 \times 12)$$

or

$$480{,}000 \text{ in-lb}$$

Simple Beam, Uniformly Distributed Load. This is probably the most common beam loading. It occurs time and again. Let us call the span L and the load w lb per lin ft, as indicated in Fig. 2.26*b*. The total load on the beam is wL; hence each reaction is $wL/2$. The maximum bending moment occurs at the center of the span at distance $L/2$ from R_1. Writing the value of M for this section, we have

$$M_{(x=L/2)} = \left(\frac{wL}{2} \times \frac{L}{2}\right) - \left(\frac{wL}{2} \times \frac{L}{4}\right)$$

$$M = \frac{wL^2}{4} - \frac{wL^2}{8}$$

$$M = \frac{wL^2}{8}$$

If, instead of being given the load per linear foot, we are given the total uniformly distributed load, we call the total load W. Because $wL = W$, the value of the maximum bending moment can be written $M = wL^2/8$, or $WL/8$. (See case 2, Fig. 2.27.) Remember this formula. You will use it many times. Its convenience is demonstrated in the practical example below.

Example 2. A simple beam 14 ft long has a uniformly distributed load of 800 lb per lin ft. Compute the maximum bending moment.

Solution: The formula that gives the maximum bending moment for a simple beam with a uniformly distributed load is $M = wL^2/8$. Substituting these values, we find

$$M = \frac{800 \times 14 \times 14}{8}$$

$$M = 19{,}600 \text{ ft-lb}$$

or

$$M = 19{,}600 \times 12 = 235{,}200 \text{ in.-lb}$$

Suppose in this problem we had been given the total load of 11,200 lb instead of 800 lb per ft. Then

$$M = \frac{WL}{8}$$

$$M = \frac{11{,}200 \times 14}{8} = 19{,}600 \text{ ft-lb}$$

$$19{,}600 \text{ ft-lb} = 235{,}200 \text{ in.-lb}$$

The result, of course, is the same.

2.9 Use of Tabulated Values for Beams

Some of the most common beam loadings are shown in Fig. 2.27. In addition to the formulas for maximum shear V and maximum bending moment M, expressions for maximum deflection D are given also. (Discussion of deflection formulas will be deferred for the time being but will be considered under beam design in subsequent sections.)

In Fig. 2.27, if the loads P and W are in pounds or kips, the vertical shear V will also be in units of pounds or kips. When the loads are given in pounds or kips and the span in feet, the bending moment M will be in units of foot-pounds or kip-feet.

An extensive series of beam diagrams and formulas is contained in Part 2 of the *Manual of Steel Construction,* 8th ed., published by the American Institute of Steel Construction (AISC).

Also given in Fig. 2.27 are values designated ETL, which stands for *equivalent tabular load.* These may be used to derive a hypothetical uniformly distributed load which when applied to the beam will produce the same magnitude of maximum bending moment as that for the given case of loading. Use of these factors is illustrated in later parts of the book.

Problem 2.9.A. A simple span beam has two concentrated loads of 4 kips [17.8 kN] each placed at the third points of the 24-ft [7.32-m] span. Find the value for the maximum bending moment in the beam.

Problem 2.9.B. A simple span beam with a span of 32 ft [9.75 m] has a concentrated load of 12 kips [53.4 kN] at 12 ft [3.66 m] from one end. Find the value for the maximum bending moment in the beam.

2.10 Development of Resisting Moment

We learned in the preceding chapter that bending moment is a measure of the tendency of the external forces on a beam to deform it by bending. We will now consider the action within the beam that resists bending and is called the *resisting moment.*

Figure 2.28*a* shows a simple beam, rectangular in cross section, supporting a single concentrated load P. Figure 2.28*b* is an enlarged sketch of the left-hand portion of the beam between the reaction and section X–X. From the preceding discussions we know that the reaction R_1 tends to cause a clockwise rotation about point A in the section under consideration; this we have defined as the bending moment at the section. In this type of beam

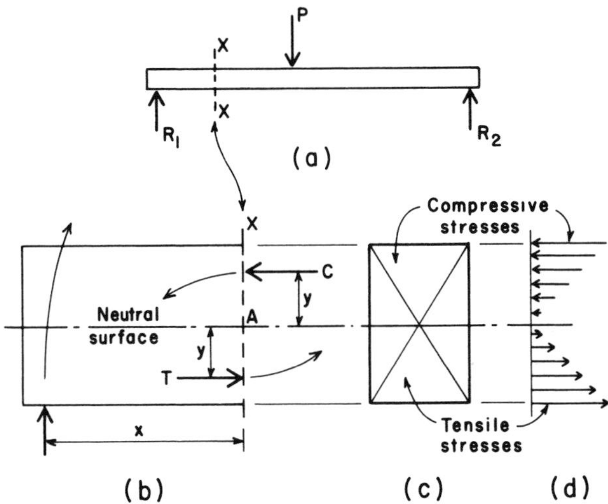

FIGURE 2.28 Development of bending stress in a beam.

the fibers in the upper part are in compression, and those in the lower part are in tension. There is a horizontal plane separating the compressive and tensile stresses; it is called the *neutral surface,* and at this plane there are neither compressive nor tensile stresses with respect to bending. The line in which the neutral surface intersects the beam cross section (Fig. 2.28*c*) is called the *neutral axis,* NA.

Call C the sum of all the compressive stresses acting on the upper part of the cross section, and call T the sum of all the tensile stresses acting on the lower part. It is the sum of the moments of these stresses at the section that holds the beam in equilibrium; this is called the *resisting moment* and is equal to the bending moment in magnitude. The bending moment about point A is $R_1 \times x$, and the resisting moment about the same point is $(C \times y) + (T \times y)$. The bending moment tends to cause a clockwise rotation, and the resisting moment tends to cause a counterclockwise rotation. If the beam is in equilibrium, these moments are equal, or

$$R_1 \times x = (C \times y) + (T \times y)$$

that is, the bending moment equals the resisting moment. This is the theory of flexure (bending) in beams. For any type of beam, we can compute the bending moment; and if we wish to design a beam to withstand this tendency to bend, we must select a member with a cross section of such shape, area, and material that it is capable of developing a resisting moment equal to the bending moment.

The Flexure Formula. The flexure formula, $M = fS$, is an expression for resisting moment that involves the size and shape of the beam cross section (represented by S in the formula) and the material of which the beam is made (represented by f). It is used in the design of all homogeneous beams, that is, beams made of one material only, such as steel or wood. You will never need to derive this formula, but you will use it many times. The following brief derivation is presented to show the principles on which the formula is based.

Figure 2.29 represents a partial side elevation and the cross section of a homogeneous beam subjected to bending stresses. The cross section shown is unsymmetrical about the neutral axis, but this discussion applies to a cross section of any shape. In Fig. 2.29*a* let c be the distance of the fiber farthest from the neutral axis, and let f be the unit stress on the fiber at distance c. If f, the extreme fiber stress, does not exceed the elastic limit of the material, the stresses in the other fibers are directly proportional to their distances from the neutral axis. That is to say, if one fiber is twice the distance from the neutral axis than another fiber, the

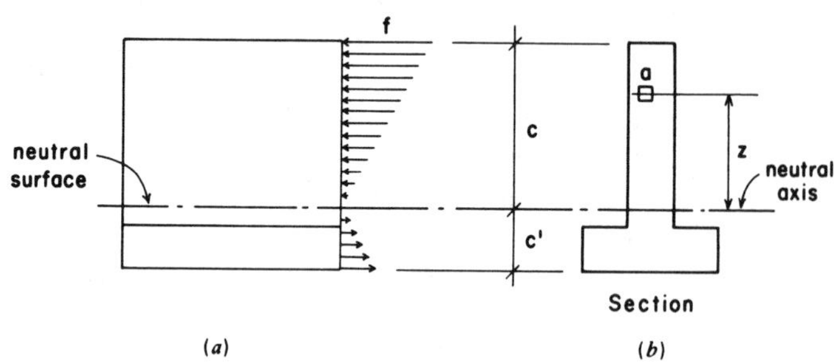

FIGURE 2.29 Distribution of bending stress on a beam section.

fiber at the greater distance will have twice the stress. The stresses are indicated in the figure by the small lines with arrows, which represent the compressive and tensile stresses acting toward and away from the section, respectively. If c is in inches, the unit stress on a fiber at 1 in. distance is f/c. Now imagine an infinitely small area a at z distance from the neutral axis. The unit stress on this fiber is $(f/c) \times z$, and because this small area contains a square inches, the total stress on fiber a is $(f/c) \times z \times a$. The *moment* of the stress on fiber a at z distance is

$$\frac{f}{c} \times z \times a \times z \qquad \text{or} \qquad \frac{f}{c} \times a \times z^2$$

We know, however, that there is an extremely large number of these minute areas, and if we use the symbol Σ to represent the sum of this very large number, we can write

$$\Sigma \frac{f}{c} \times a \times z^2$$

which means the sum of the moments of all the stresses in the cross section with respect to the neutral axis. This we know is the *resisting moment,* and it is equal to the bending moment.

Therefore

$$M = \frac{f}{c} \Sigma\, a \times z^2$$

The quantity $\Sigma\, a \times z^2$ may be read "the sum of the products of all the elementary areas times the square of their distances from the neutral axis." We call this the *moment of inertia* and represent it by the letter I. Therefore, substituting in the above, we have

$$M = \frac{f}{c} \times I \qquad \text{or} \qquad M = \frac{fI}{c}$$

This is known as the *flexure formula* or *beam formula,* and by its use we may design any beam that is composed of a single material. The expression may be simplified further by substituting S

for I/c, called the *section modulus,* a term that is described more fully in Sec. 3.5. Making this substitution, the formula becomes

$$M = fS$$

Use of the flexural formula is discussed in Sec. 4.2 for wood beams and Sec. 8.2 for steel beams.

2.11 Shear Stress in Beams

Development of shear stress in a beam tends to produce lateral deformation. This may be visualized by considering the beam to consist of a layer of loose boards. Under the beam loading on a simple span, the boards tend to slide over each other, taking the form shown in Fig. 2.30. This type of deformation also tends to occur in a solid beam, but is resisted by the development of horizontal shearing stresses.

Shear stresses in beams are not distributed evenly over cross sections of the beam as was assumed for the case of simple direct shear (see Sec. 1.1). From observations of tested beams and derivations considering the equilibrium of beam segments under combined shear and moment actions, the following expression has been obtained for shear stress in a beam:

$$f_v = \frac{VQ}{Ib}$$

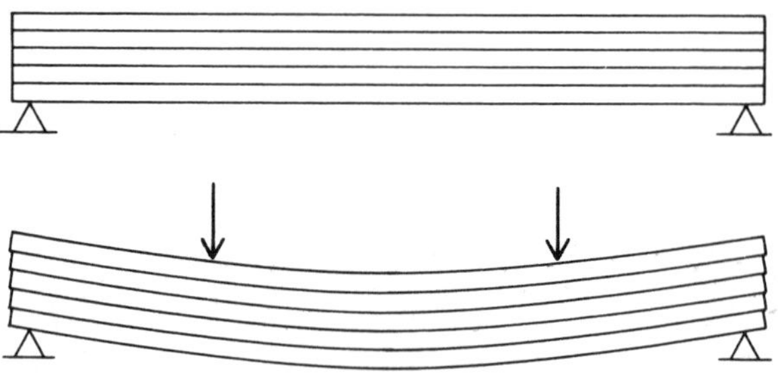

FIGURE 2.30

FIGURE 2.31 Development of shear in beams.

where V = shear force at the beam section,

Q = moment about the neutral axis of the area of the section between the point of stress and the edge of the section,

I = moment of inertia of the section with respect to the neutral (centroidal) axis,

b = width of the section at the point of stress.

It may be observed from this formula that the maximum value for Q will be obtained at the neutral axis of the section, and that the stress will be zero at the edges of the section farthest from the neutral axis. The form of shear stress distribution for various geometric shapes of beam sections is shown in Fig. 2.31.

The following examples illustrate the use of this stress relationship.

Example 1. A beam section with depth of 8 in. and width of 4 in. [200 and 100 mm] sustains a shear force of 4 kips [18 kN]. Find the maximum shear stress.

Solution: For the rectangular section the moment of inertia about the centroidal axis is

$$I = \frac{bd^3}{12} = \frac{4(8)^3}{12} = 170.7 \text{ in.}^4 \; [67 \times 10^6 \text{ mm}^4]$$

The static moment (Q) is the product of the area a' and its centroidal distance from the axis of the section ($\bar{y}$) as shown in Fig. 2.32*b*. We thus compute Q as

$$Q = a'\bar{y} = [4(4)]2 = 32 \text{ in.}^3 \; [500 \times 10^3 \text{ mm}^3]$$

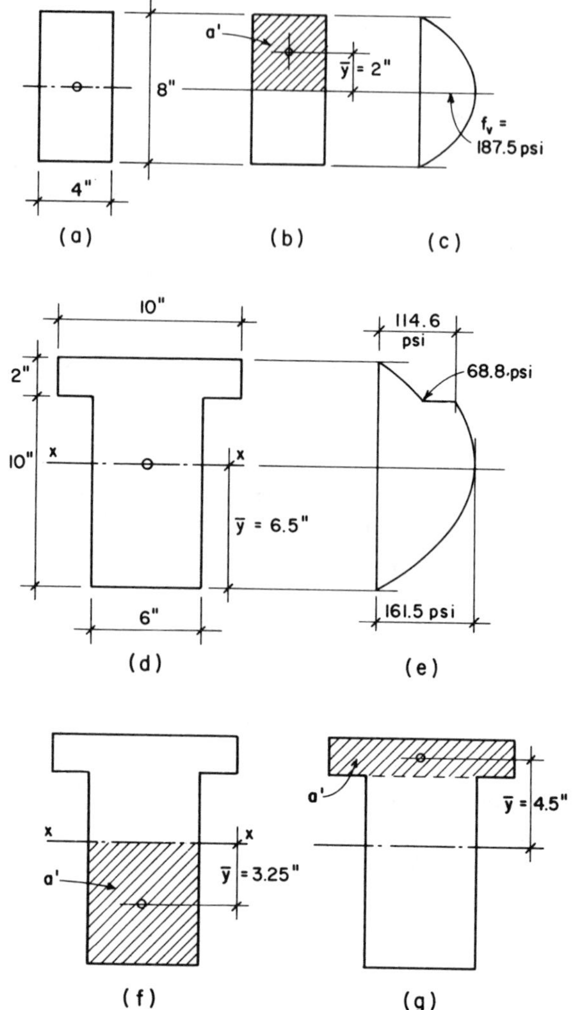

FIGURE 2.32 Form of shear stress distribution in beams with various cross sections.

The maximum shear stress at the neutral axis is thus

$$f_v = \frac{VQ}{Ib} = \frac{4000(32)}{170.7(4)} = 187.5 \text{ psi } [1.34 \text{ MPa}]$$

Example 2. A beam with the T-section shown in Fig. 2.32*d* is subjected to a shear of 8 kips [36 kN]. Find the maximum shear stress and the shear stress at the location of the juncture of the web and the flange of the T.

Solution: For this section the location of the centroid and the determination of the moment of inertia about the centroidal axis must be accomplished using processes explained in Chapter 3. This work is summarized in Table 2.1. For determination of the maximum shear stress at the neutral axis (centroidal axis x–x, as shown in Fig. 2.32*d*) we find Q using the bottom portion of the web, as shown in Fig. 2.32*f*. Thus

$$Q = a'\bar{y} = [6.5(6)]3.25 = 126.75 \text{ in.}^3 [1.98 \times 10^6 \text{ mm}^3]$$

and the maximum stress at the neutral axis is thus

$$f_v = \frac{VQ}{Ib} = \frac{8000(126.75)}{1046.7(6)} = 161.5 \text{ psi } [1160 \text{ kPa}]$$

For the stress at the juncture of the web and flange we use the area shown in Fig. 2.32*g* for Q; thus

$$Q = [2(10)]4.5 = 90 \text{ in.}^3 [1.41 \times 10^6 \text{ mm}^3]$$

TABLE 2.1 Computation of Properties for the Section in Example 2

Part	Area (in.^{2})	y from Bottom	Ay_1	I_0	Ay_2^2	I_x (in.^{4})
1	$6(12) = 72$	6	432	$\frac{6(12)^3}{12} = 864$	$72(0.5)^2 = 18$	882
2	$2(2)(2) = 8$	11	88	$2(2)(2)^3 = 2.7$	$8(4.5)^2 = 162$	164.7
Σ	80 in.^{2} [50×10^3 mm^2]		520 in.^{3} [8.125×10^6 mm^3]			1046.7 in.^{4} [4.088×10^8 mm^4]

$y_x = \frac{520}{80} = 6.5$ in. [162.5 mm] (see Fig. 2.32*d*) and the two shear stresses at this location, as displayed in Fig. 2.32*e*, are

$$f_{v1} = \frac{8000(90)}{1046.7(6)} = 114.6 \text{ psi [828 kPa]}$$

$$f_{v2} = \frac{8000(90)}{1046.7(10)} = 68.8 \text{ psi [497 kPa]}$$

In most design situations it is not necessary to use the complex form of the general expression for beam shear. In wood structures the beam sections are almost always of simple rectangular shape. For this shape we can make the following simplification.

$$I = \frac{bd^3}{12}, \qquad Q = (b)\frac{d}{2}\frac{d}{4} = \frac{bd^2}{8}$$

$$f_v = \frac{VQ}{Ib} = \frac{V(bd^2/8)}{(bd^3/12)b} = \frac{3}{2}\frac{V}{bd}$$

For steel beams—which are mostly I-shaped cross sections—the shear is taken almost entirely by the web. (See shear stress distribution for the I-shape in Fig. 2.31.) Since the stress distribution in the web is so close to uniform, it is considered adequate to use a simplified computation of the form

$$f_v = \frac{V}{dt_w}$$

in which d is the overall beam depth and t_w is the thickness of the beam web.

For beams of reinforced concrete the shear mechanisms of the composite section are so complex that it is customary to use a highly simplified form for the shear computation. Various limits for this simplified stress and its use in design situations are then established by numerous requirements. This process is explained in Part IV in discussions of the investigation and design for shear conditions of various types.

There are situations in which the general form of the shear stress formula must be used. Most of these involve the use of complex geometries for beam cross sections which commonly occur in beams of prestressed concrete and the built-up, compound sections of steel or timber plus plywood.

Problem 2.11.A. A beam has an I-shaped cross section with an overall depth of 16 in. [400 mm], web thickness of 2 in. [50 mm], and flanges that are 8 in. [200 mm] wide and 3 in. [75 mm] thick. Compute the critical shear stresses and plot the distribution of shear stress on the cross section if the beam sustains a shear force of 20 kips [89 kN].

2.12 Continuous Beams

It is beyond the scope of this book to give a detailed discussion of bending in members continuous over supports, but the material presented in this chapter will serve as an introduction to the subject. A *continuous beam* is a beam that rests on more than two supports. For most continuous beams the maximum bending moment is smaller than that found in a series of simply supported beams having the same spans and loads. Continuous beams are characteristic of reinforced concrete and welded steel floor framing.

The concepts underlying continuity and bending under restraint are illustrated in Fig. 2.33. Figure 2.33*a* represents a single

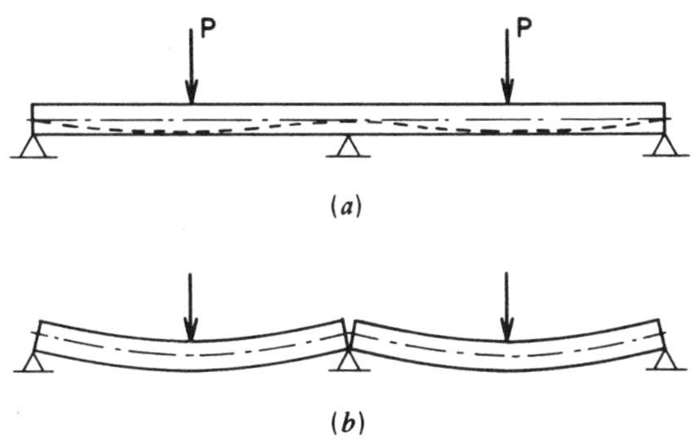

FIGURE 2.33

beam resting on three supports and carrying equal loads at the centers of the two spans. If we imagine the beam to be cut over the middle support as shown in Fig. 2.33*b*, the result will be two simple beams. Each of these simple beams will deflect as shown. However, when the beam is made continuous over the support, the deflection curve has a shape similar to that indicated by the dotted line in the first figure.

It is evident that there is no bending moment developed over the support in Fig. 2.33*b*, while there must be a moment over the support in Fig. 2.33*a*. Study of the figures shows that, in both cases, there is tension in the bottom of the beams near midspan; in addition, there is tension developed in the top of the beam of Fig. 2.33*a* over the center support. In other words, the continuous beam has a positive bending moment near the center of each span and a negative moment over the middle support. It can be shown that the positive bending moment near midspan for the continuous beam is less than that for the simple spans of Fig. 2.33*b*.

The value of the bending moments in continuous beams cannot be found by the usual equations of static equilibrium; additional equations which involve the elasticity of the material are required. Consequently a continuous beam is described as "statically indeterminate." There are several methods of analysis for statically indeterminate structures, and bending moment formulas have been developed for various typical conditions of continuous-beam loading and restraint. An extensive table of such formulas is presented under the title "Beam Diagrams and Formulas" in Part 2 of the *Manual of Steel Construction*, 8th ed., published by the American Institute of Steel Construction. These must be used with judgment, however, since the actual conditions under which the structure is built may not duplicate the theoretical ones on which the formulas are based.

Theorem of Three Moments. One method of determining reactions and constructing the shear and bending moment diagrams for continuous beams is based on the *theorem of three moments*. This theorem deals with the relation among the bending moments at any three consecutive supports. The equation of this relation is known as the *three-moment equation,* and its form varies with the type of loading. When the bending moments at all supports are established, the magnitudes of the reactions may be computed, and, by the methods previously described, the shear and moment diagrams may be drawn.

Figure 2.34*a* represents a continuous beam of two spans with uniformly distributed loads. If the beam is simply supported at both ends, the bending moment at each of these two supports must be zero, and the bending moment diagram will have the general shape shown in Fig. 2.34*b*. There will be a negative bending moment at the middle support.

The three-moment equation for a continuous beam of two spans with uniformly distributed loads and constant moment of inertia is

$$M_1L_1 + 2M_2(L_1 + L_2) + M_3L_2 = -\frac{w_1L_1^3}{4} - \frac{w_2L_2^3}{4}$$

in which the various terms are as shown in Fig. 2.34 with w expressed in pounds per linear foot. Several simple examples are presented to show how the theorem of three moments is applied.

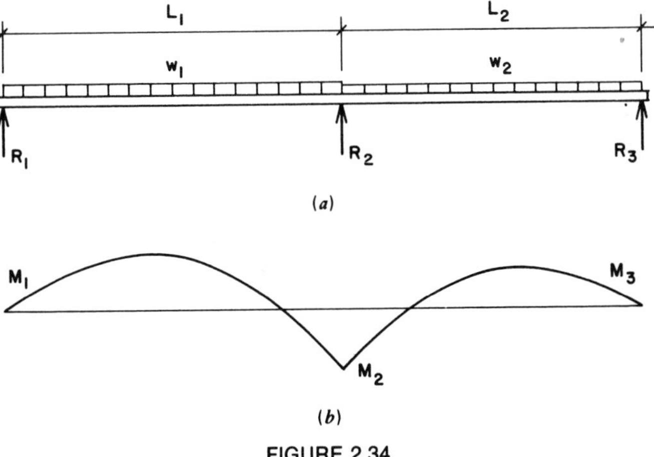

FIGURE 2.34

Continuous Beam With Two Equal Spans. We consider first a continuous beam of two equal spans with the same uniformly distributed load extending the full length of each span. In this example there is no restraint at the end supports (ends simply supported).

Example 1. A continuous beam has two spans of 10 ft each with a uniformly distributed load of 100 lb/ft extending over its entire length. Compute the magnitude of the reactions and construct the shear and bending moment diagrams.

Solution: (1) The beam and loading are shown in Fig. 2.35*a*. Bearing in mind that the moments M_1 and M_3 are each zero,

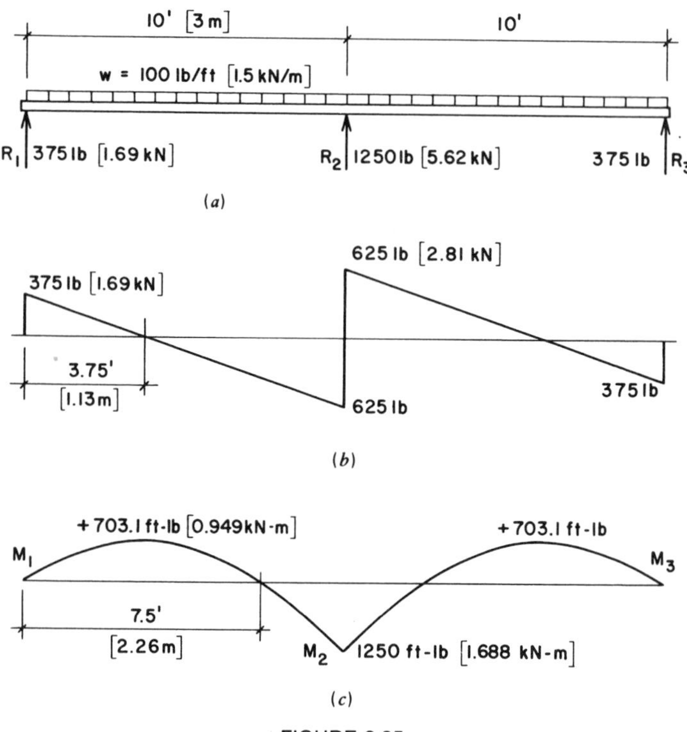

FIGURE 2.35

substitute the known values in the three-moment equation given before. Then

$$(0 \times 10) + [2\ M_2(10 + 10)] + (0 \times 10) = -\frac{100 \times 1000}{4} \times 2$$

$$40\ M_2 = -50{,}000$$

$$M_2 = -1250 \text{ ft-lb}$$

which is the negative bending moment at the center support.

(2) Next we write an expression for the bending moment at 10 ft from the left support and equate it to the value we have just found (−1250 ft-lb). Then

$$M_{(x=10)} = (R_1 \times 10) - (100 \times 10 \times 5) = -1250$$

$$10R_1 = 3750$$

$$R_1 = 375 \text{ lb}$$

Since R_1 and R_3 are equal, R_3 also equals 375 lb.

(3) To find R_2 we note that the total load on the entire length of the beam is $2 \times 100 \times 10 = 2000$ lb, hence

$$R_1 + R_3 + R_2 = 2000$$

$$375 + 375 + R_2 = 2000$$

$$R_2 = 1250 \text{ lb}$$

(4) Now that the magnitudes of the reactions have been established, we can construct the shear diagram as shown in Fig. 2.35*b*. Let x be the distance in feet from R_1 to the point between R_1 and R_2 at which the shear diagram passes through zero. Then, writing an expression for the shear at this point and equating it to zero,

$$375 - (100 \times x) = 0 \quad \text{and} \quad x = 3.75 \text{ ft } [1.125 \text{ m}]$$

(5) The maximum positive bending moment is

$$M_{(x=3.75)} = (375 \times 3.75) - \left(3.75 \times 100 \times \frac{3.75}{2}\right)$$

$$= 703.1 \text{ ft-lb}$$

Since this beam is symmetrical, the positive bending moment between R_2 and R_3 will likewise have a magnitude of 703.1 ft-lb. Now that both the maximum positive and negative moments have been computed, we may construct the bending moment diagram, Fig. 2.35*c*.

Continuous Beam With Unequal Spans. The preceding example dealt with a continuous beam having two equal spans. The following example applies the same form of the three-moment equation to the case of unequal spans.

Example 2. A continuous beam with no restraint at the end supports has spans of 14 ft and 10 ft. A uniformly distributed load of 1000 lb/ft extends over its entire length, as shown in Fig. 2.36*a*. Construct the shear and bending moment diagrams.

Solution: (1) By data we know that $L_1 = 14$ ft, $L_2 = 10$ ft, w_1 and w_2 each equal 1000 lb per lin ft, and $M_1 = M_3 = 0$. Therefore the first step is to determine M_2 by substituting the known values in the three-moment equation

$$M_1L_1 + 2M_2(L_1 + L_2) + M_3L_2 = -\frac{w_1L_1^3}{4} - \frac{w_2L_2^3}{4}$$

Thus

$$0 + 2M_2(14 + 10) + 0 = -\frac{1000 \times 14^3}{4} - \frac{1000 \times 10^3}{4}$$

$$48\ M_2 = -936{,}000$$

$$M_2 = -19{,}500 \text{ ft-lb}$$

(2) Now let us write an expression for the bending moment at $x = 14$ (at R_2), considering the forces to the left. This expression

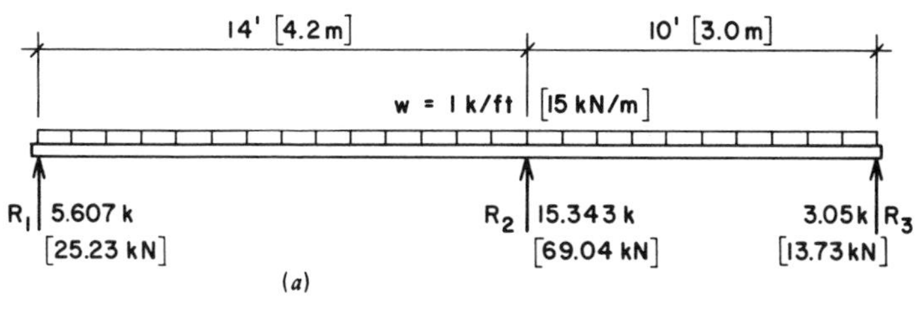

(a)

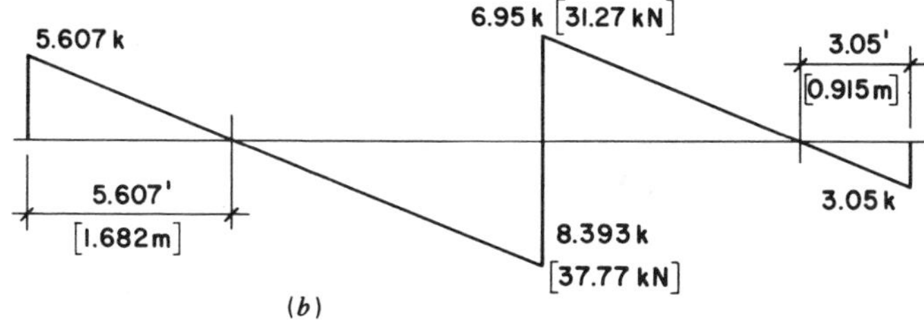

(b)

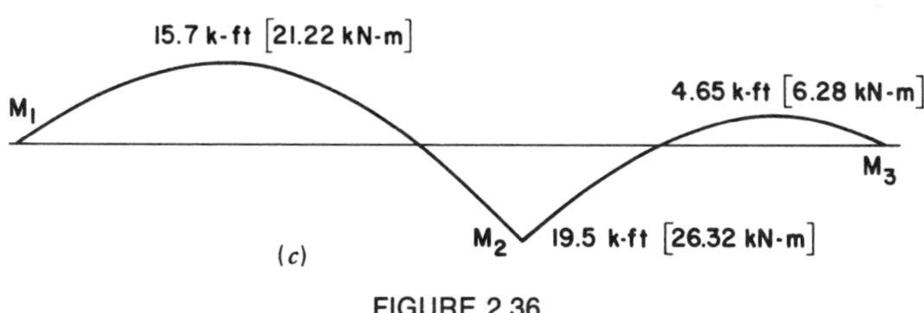

(c)

FIGURE 2.36

can be equated to −19,500 ft-lb since we have just established its magnitude. Thus

$$14R_1 - (14 \times 1000 \times 7) = -19{,}500 \quad \text{and}$$

$$R_1 = 5607 \text{ lb}$$

Similarly, writing an equation for the bending moment at R_2, considering the forces to the *right,*

$$10R_3 - (10 \times 1000 \times 5) = -19{,}500 \quad \text{and}$$

$$R_3 = 3050 \text{ lb}$$

(3) Computing R_2,

$$R_1 + R_2 + R_3 = (14 \times 1000) + (10 \times 1000)$$

$$5607 + 3050 + R_2 = 24{,}000 \text{ lb}$$

$$R_2 = 15{,}343 \text{ lb}$$

(4) Now that the values of R_1, R_2, and R_3 have been determined, we can draw the shear diagram, Fig. 2.36*b*.

Let x = the distance from R_1 to the point at which the shear is zero. Then

$$5607 - (1000 \times x) = 0 \quad \text{and} \quad x = 5.607 \text{ ft}$$

Similarly, we find that the shear is zero in the right-hand span at 3.05 ft from R_3.

(5) The maximum positive bending moment between R_1 and R_2 is

$$M_{(x=5.607)} = (5607 \times 5.607) - \left(5.607 \times 1000 \times \frac{5.607}{2}\right)$$

$$= 15{,}700 \text{ ft-lb}$$

and the maximum positive moment between R_2 and R_3 is

$$(3050 \times 3.05) - \left(3.05 \times 1000 \times \frac{3.05}{2}\right)$$

$$= 4651.25 \text{ ft-lb}$$

The bending moment diagram is shown in Fig. 2.36*c*.

Continuous Beam With Concentrated Loads. In the previous examples the loads on the continuous beams were uniformly distributed. Figure 2.37*a* shows a continuous beam of two spans with a concentrated load on each span. The general shape of the bending moment diagram for this beam is shown in Fig. 2.37*b*. For these conditions, the form of the three-moment equa-

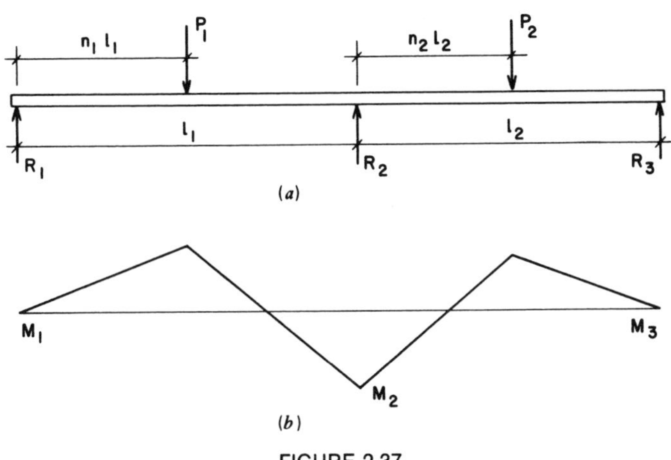

FIGURE 2.37

tion is

$$M_1L_1 + 2M_2(L_1 + L_2) + M_3L_2 = -P_1L_1^2[n_1(1 - n_1)(1 + n_1)] \\ -P_2L_2^2[n_2(1 - n_2)(2 - n_2)]$$

in which the various terms are as shown in Fig. 2.37.

Example 3. A two-span continuous beam has equal spans of 20 ft each. There is a concentrated load of 4000 lb at the center of each span. The ends of the beam at R_1 and R_2 are simply supported. Compute the magnitude of the reactions and construct the shear and bending moment diagrams (see Fig. 2.38*a*).

Solution: (1) For the given spans and loading, note that $L_1 = L_2 = 20$ ft, P_1 and $P_2 = 4000$ lb each, and both n_1 and $n_2 = 0.5$. Substituting the known values in the form of the three-moment equation stated previously,

$$(M_1 \times 20) + 2M_2(20 + 20) + (M_3 \times 20) \\ = -4000 \times 20^2(0.5 \times 0.5 \times 1.5) \\ - 4000 \times 20^2(0.5 \times 0.5 \times 1.5)$$

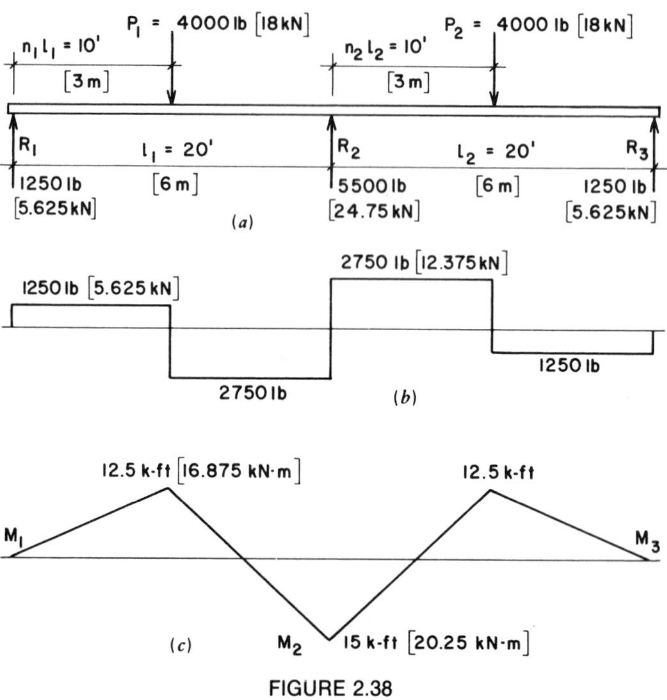

FIGURE 2.38

Because there is no restraint at R_1 and R_3, the values of M_1 and M_3 each equal zero. Thus

$$0 + 80M_2 + 0 = -(4000 \times 400 \times 0.375) - (4000 \times 400 \times 0.375)$$

or

$$80M_2 = -1{,}200{,}000$$

and

$$M_2 = -15{,}000 \text{ ft-lb}$$

which is the value of the negative bending moment at R_2, the center support.

(2) Next, write an expression for the bending moment at R_2 and equate it to $-15{,}000$ ft-lb, the value we have just found. Thus

$$M_{(x=20)} = (R_1 \times 20) - (4000 \times 10) = -15{,}000$$

$$20R_1 = 25{,}000 \quad \text{and} \quad R_1 = 1250 \text{ lb}$$

R_1 and R_3 will have equal magnitudes, hence $R_1 = R_3 = 1250$ lb and $R_2 = (4000 + 4000) - (1250 + 1250) = 5500$ lb. Now that all the vertical forces are known, the shear diagram can be constructed. It is shown in Fig. 2.38*b*.

(3) Since the magnitude of R_1 has been established, we can now compute the value of the bending moment under the first concentrated load. Thus $M_{(x=10)} = 1250 \times 10 = 12{,}500$ ft-lb. This is also the magnitude of the bending moment under the second concentrated load. We know that the bending moment has a magnitude of $-15{,}000$ ft-lb at R_2. Therefore, we have sufficient data to construct the bending moment diagram for this continuous beam; it is shown in Fig. 2.38*c*.

Continuous Beam With Three Spans. The preceding examples demonstrate that the key operation in design and investigation of continuous beams is the determination of negative bending moment values at the supports. Once these values are established, we can compute the reactions and the positive bending moments. When a continuous beam has more than two spans, the three-moment equation is applied to successive pairs of spans. For the three-span beam shown in Fig. 2.39*a*, the first pair would consist of the span to the left and the middle span; the second, the middle span and the span to the right. This process can be extended to continuous beams with any number of spans.

Figure 2.39*c* and *d* represent the shear and bending moment diagrams for the continuous beam shown in Fig. 2.39*a*. The uniformly distributed load is w lb/lin ft and extends the full length of the three equal spans. Successive application of the three-moment equation yields the values recorded in the figure. The values

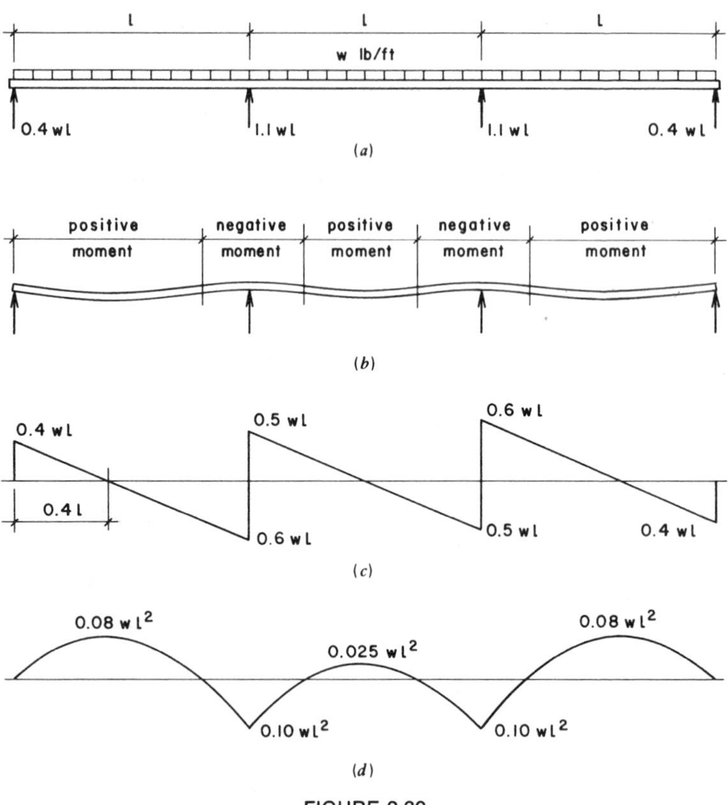

FIGURE 2.39

for reactions and shear are coefficients of wL; for the moments the values are coefficients of wL^2. Similar coefficients for continuous beams of various spans and types of loading may be found in engineering reference books. Such tabulations enable one to determine quickly the bending moments at critical sections.

Example 4. A continuous beam has three spans of 20 ft each and a uniformly distributed load of 800 lb/ft extending over the entire length of the beam. The ends of the beam are simply supported. Compute the maximum bending moment and the maximum shear.

Solution: (1) Referring to Fig. 2.39*d*, we note that the maximum positive bending moment (0.08 wL^2) occurs near the middle of each end span, and the maximum negative moment (0.10 wL^2) occurs over each of the interior supports. Using the larger value, the maximum bending moment on the beam is

$$M = -0.10\ wL^2 = -(0.10 \times 800 \times 20 \times 20)$$
$$= -32{,}000 \text{ ft-lb}$$

(2) Figure 2.39*c* shows that the maximum shear occurs at the face of the first interior support (working from either left or right) and is

$$V = 0.6\ wL = 0.6 \times 800 \times 20 = 9600 \text{ lb}$$

Problem 2.2.A. A beam is continuous through two spans and sustains a uniformly distributed load of 2 kips/ft [29.2 kN/m], including its own weight. The span lengths are 12 ft [3.66 m] and 16 ft [4.88 m]. Find the values of the three reactions and draw the complete shear and moment diagrams.

Problem 2.2.B. A beam is continuous through two spans and sustains a uniformly distributed load of 1 kip/ft [14.6 kN/m], including its own weight, plus concentrated loads of 6 kips [26.7 kN] at the centers of the two 24-ft [7.32-m] spans. Find the values of the three reactions and draw the complete shear and moment diagrams.

2.13 Restrained Beams

A simple beam was defined in Sec. 2.2 as a beam that rests on a support at each end, there being no restraint against bending at the supports; the ends are *simply supported*. The shape a simple beam tends to assume under load is shown in Fig. 2.40*a*. Figure 2.40*b* shows a beam built into a wall at the left end and simply supported at the right end. The left end is *restrained* or *fixed;* the tangent to the elastic curve is horizontal at the face of the wall, at which point there is a negative bending moment. For such a beam, symmetrically loaded, the reactions are not equal in magnitude as they are in simple beams. Figure 2.40*c* illustrates a beam fixed at both ends. In this beam negative bending moments occur

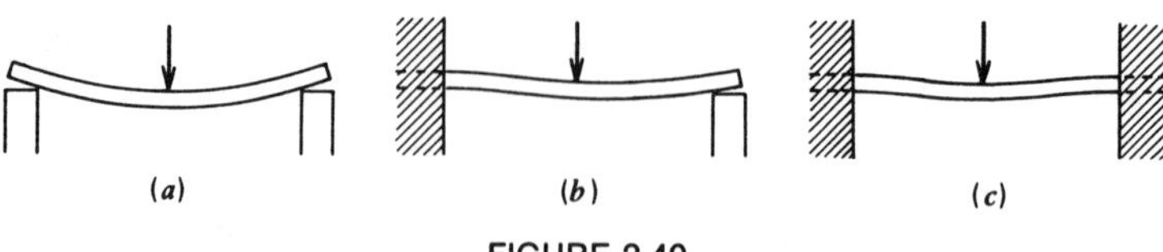

FIGURE 2.40

at the supports, and positive moment occurs in a portion of the beam in the central part of the span. A negative bending moment always occurs at a fixed support.

The reactions for beams with fixed ends cannot be computed by the principle of moments, as in simple beams, and the necessary mathematics involved is beyond the scope of this book. Fig-

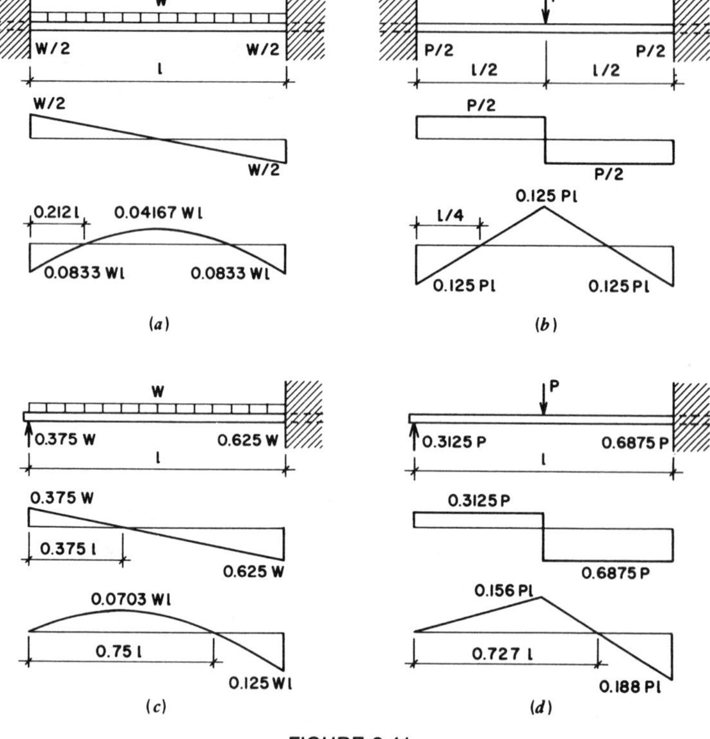

FIGURE 2.41

ure 2.41*a* and *b* show beams fixed at both ends, and Fig. 2.41*c* and *d* indicate the condition where one end is fixed and the other is simply supported. For each of these two cases there are two types of loading—uniformly distributed and concentrated at the center of the span. For each beam, the magnitude of the maximum shear and bending moment is given in the shear and moment diagrams.

Example 1. Figure 2.42*a* represents a beam with both ends fixed and having a span of 20 ft. It supports a total load of 8000 lb uniformly distributed over its entire length. Draw the shear and bending moment diagrams.

Solution: (1) Referring to Fig. 2.41*a,* we find that the shear at each end, and consequently each reaction, is $W/2$ or $8000 \div 2 =$ 4000 lb. Since the load is uniformly distributed, the shear diagram takes the shape shown in Fig. 2.42*b*.

(2) Consulting Fig. 2.41 for moment coefficients, we find that the negative bending moment at each support is

$$M = -\frac{WL}{12} = -\frac{8000 \times 20}{12} = -13{,}330 \text{ ft-lb}$$

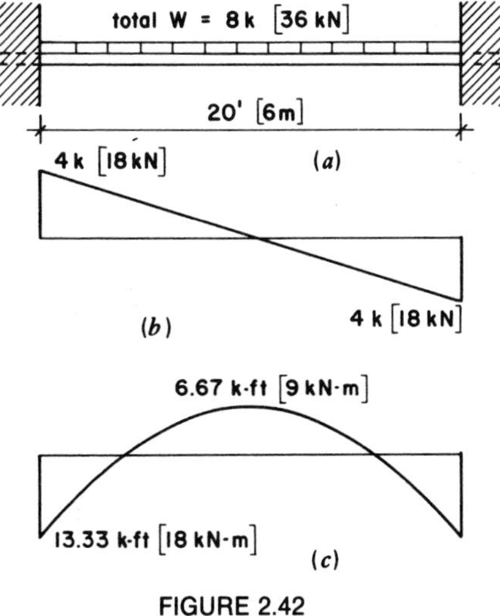

FIGURE 2.42

and the positive moment at the center of the span is

$$M = \frac{WL}{24} = \frac{8000 \times 20}{24} = 6667 \text{ ft-lb}$$

The bending moment diagram is shown in Fig. 2.42*c*.

Example 2. A beam having a span of 16 ft is fixed at one end and simply supported at the other (Fig. 2.43*a*). There is a load of 9600 lb placed at the center of the span. Draw the shear and bending moment diagrams.

Solution: (1) Shear and moment coefficients for this beam and loading are given in Fig. 2.41*d*. The shear at the simply supported end is

$$V = \tfrac{5}{16} \times P = \tfrac{5}{16} \times 9600 = 3000 \text{ lb}$$

and the shear at the fixed end is

$$V = -(\tfrac{11}{16} \times P) = -(\tfrac{11}{16} \times 9600) = -6600 \text{ lb}$$

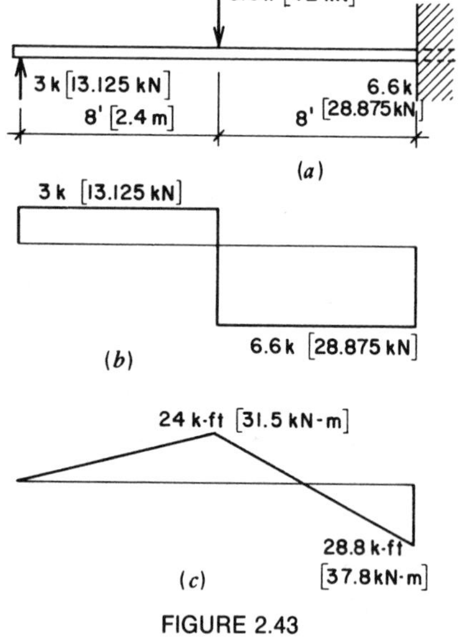

FIGURE 2.43

These two values enable us to draw the shear diagram shown in Fig. 2.43*b*.

(2) The maximum positive bending moment is

$$M = \tfrac{5}{32} \times Pl = \tfrac{5}{32} \times 9600 \times 16 = 24{,}000 \text{ ft-lb}$$

and the maximum negative moment at the fixed end is

$$\begin{aligned} M &= -(\tfrac{6}{32} \times Pl) = -(\tfrac{6}{32} \times 9600 \times 16) \\ &= -28{,}800 \text{ ft-lb} \end{aligned}$$

The bending moment diagram is shown in Fig. 2.43*c*.

Problem 2.13.A. A beam having a span of 22 ft [6.71 m] is fixed at both ends and has a concentrated load of 16 kips [71.2 kN] at the center of the span. Compute the maximum shear and bending moment, locate the inflection points, and draw the shear and moment diagrams.

Problem 2.13.B. A beam fixed at one end and simply supported at the other has a span of 20 ft [6 m]. There is a total load of 8 kips [36 kN] uniformly distributed over the full span. Compute the maximum shear and bending moment, locate the inflection point, and draw the shear and moment diagrams.

2.14 Structures With Internal Pins

In many structures qualifying conditions exist at supports or within the structure that modify the behavior of the structure, often eliminating some potential components of force actions. Qualification of supports as fixed or pinned has been a situation in most of the structures presented in this work. We now consider some qualification of conditions *within* the structure that modify its behavior.

Internal Pins. Consider the structure shown in Fig. 2.44*a*. It may be observed that there are four potential components of the reaction forces: A_x, A_y, B_x, and B_y. These are all required for the stability of the structure for the loading shown, and there are thus four unknowns in the investigation of the external forces. Since the loads and reactions constitute a general planar force system, there are three conditions of equilibrium (see Sec. 2.1; for example: $\Sigma F_x = 0$, $\Sigma F_y = 0$, $\Sigma M_p = 0$). As shown in Fig. 2.44*a*,

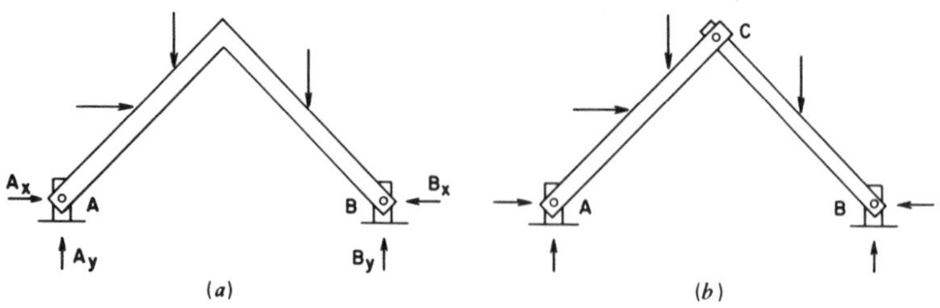

FIGURE 2.44

therefore, the structure is statically indeterminate, not yielding to complete investigation by use of static equilibrium conditions alone.

If the two members of the structure in Fig. 2.44 are connected to each other by a pinned joint, as shown in *b,* the number of reaction components is not reduced, and the structure is still stable. However, the internal pin establishes a fourth condition which may be added to the three equilibrium conditions. There are then four conditions that may be used to find the four reaction components. The method of solution for the reactions of this type of structure is illustrated in the following example.

Example 1. Find the components of the reactions for the structure shown in Fig. 2.45*a*.

Solution: It is possible to write four equilibrium equations and to solve them simultaneously for the four unknown forces. How-

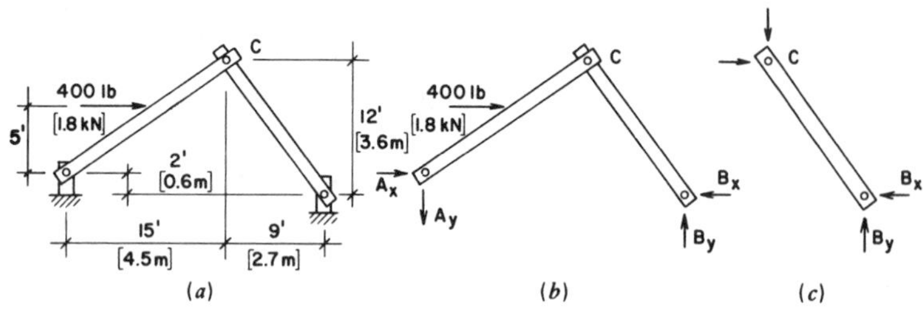

FIGURE 2.45

ever, it is always easier to solve these problems if a few tricks are used to simplify the equations. One trick is to write moment equations about points that eliminate some of the unknowns, thus reducing the number of unknowns in a single equation. This was used in the finding of beam reactions in Sec. 2.3. Consider the free body of the entire structure, as shown in Fig. 2.45*b*.

$$\Sigma M_A = 0 = +(400 \times 5) + (B_x \times 2) - (B_y \times 24)$$

$$24B_y - 2B_x = 2000, \quad \text{or,} \quad 12B_y - B_x = 1000 \qquad (1)$$

Now consider the free-body diagram of the right member, as shown in Fig. 2.45*c*.

$$\Sigma M_C = 0 = +(B_x \times 12) - (B_y \times 9)$$

thus

$$B_x = \frac{9}{12} B_y = 0.75 B_y \qquad (2)$$

Substituting equation (2) in equation (1),

$$12B_y - 0.75B_y = 1000, \qquad B_y = \frac{1000}{11.25} = 88.89 \text{ lb}$$

Then, from equation (2),

$$B_x = 0.75B_y = 0.75 \times 88.89 = 66.67$$

Referring again to Fig. 2.45*b*:

$$\Sigma F_x = \quad 0 = A_x + B_x - 400$$

$$0 = A_x + 66.67 - 400$$

$$A_x = 333.33 \text{ lb}$$

$$\Sigma F_y = \quad 0 = A_y + B_y = A_y + 88.89, \qquad A_y = 88.89 \text{ lb}$$

Note that the condition stated in equation (2) is true in this case because the right member behaves as a two-force member. This is not the case if load is directly applied to the member, and the solution of simultaneous equations would be necessary.

Continuous Beams With Internal Pins. The actions of continuous beams are discussed in Sec. 2.12. It is observed that a beam such as that shown in Fig. 2.46*a* is statically indeterminate, having a number of reaction components (3) in excess of the conditions of equilibrium for a parallel force system (2). The continuity of such a beam results in the deflected shape and variation of moment as shown beneath the beam in Fig. 2.46*a*. If the beam is made discontinuous at the middle support, as shown in Fig. 2.46*b*, the two spans each behave independently as simple beams, with the deflected shapes and moment as shown.

If a multiple-span beam is made internally discontinuous at some point, its behavior may emulate that of a truly continuous beam. For the beam shown in Fig. 2.46*c* the internal pin is located at the point where the continuous beam inflects. Inflection of the deflected shape is an indication of zero moment, and thus the pin does not actually change the continuous nature of the structure. The deflected shape and moment variation for the beam in Fig. 2.46*c* is therefore the same as for the beam in Fig. 2.46*a*. This is true, of course, only for the single loading pattern that results in the inflection point at the same location as the internal pin.

In the first of the following examples the internal pin is deliber-

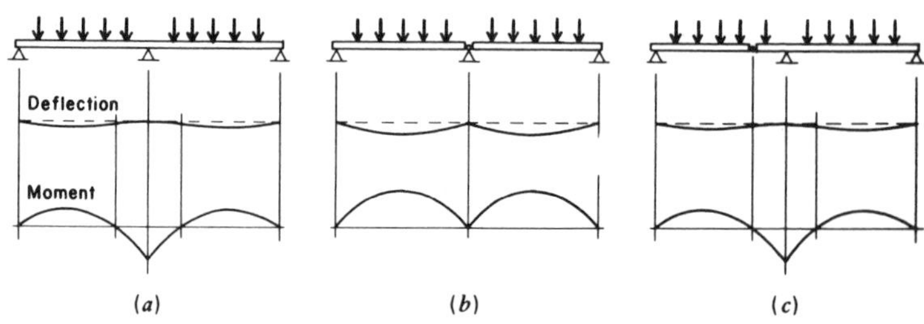

FIGURE 2.46

ately placed at the point where the beam would inflect if it was continuous. In the second example the pins are placed slightly closer to the support than the location of the natural inflection points. The modification in the second example results in slightly increasing the positive moment in the outer spans, while reducing the negative moments at the supports; thus the values of maximum moment are made closer. If it is desired to use a single-size beam for the entire length, the modification in Example 3 permits design selection of a slightly smaller size.

Example 2. Investigate the beam shown in Fig. 2.47*a*. Find the reactions, draw the shear and moment diagrams, and sketch the deflected shape.

Solution: Because of the internal pin, the first 12 ft of the left-hand span acts as a simple beam. Its two reactions are therefore equal, being one-half the total load, and its shear, moment, and deflected shape diagrams are those for a simple beam with a uniformly distributed load. (See case 2, Fig. 2.27.) As shown in *b* and *c* in Fig. 2.47, the simple beam reaction at the right end of the 12-ft portion of the left span becomes a 6-kip concentrated load at the left end of the remainder of the beam. This beam (Fig. 2.47*c*) is then investigated as a beam with one overhanging end, carrying a single concentrated load at the cantilevered end and the total distributed load of 20 kips. (Note that on the diagram we indicate the total uniformly distributed load in the form of a single force, representing its resultant.) The second portion of the beam is statically determinate, and we can proceed to find its reactions.

With the reactions known, the shear diagram can be completed. We note the relation between the point of zero shear in the span and the location of maximum positive moment. For this loading the positive moment curve is symmetrical, and thus the location of the zero moment (and beam inflection) is at twice the distance from the end as the point of zero shear. As noted previously, the pin in this example is located exactly at the inflection point of the continuous beam. (For comparison, see Sec. 2.11, Example 1.)

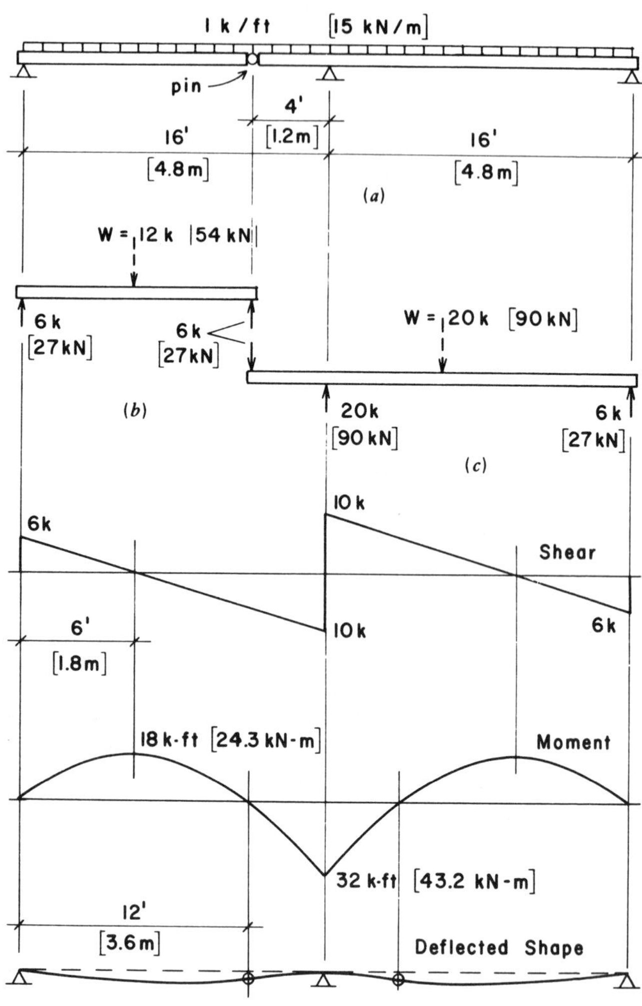

1 k/ft [15 kN/m]
pin
4' [1.2m]
16' [4.8m]
16' [4.8m]
(a)
W = 12 k [54 kN]
6 k [27kN]
6 k [27kN]
W = 20 k [90 kN]
(b)
20k [90 kN]
6 k [27kN]
(c)
10 k
6k
Shear
10 k
6k
6' [1.8m]
18 k-ft [24.3 kN-m]
Moment
32 k-ft [43.2 kN-m]
12' [3.6m]
Deflected Shape

FIGURE 2.47

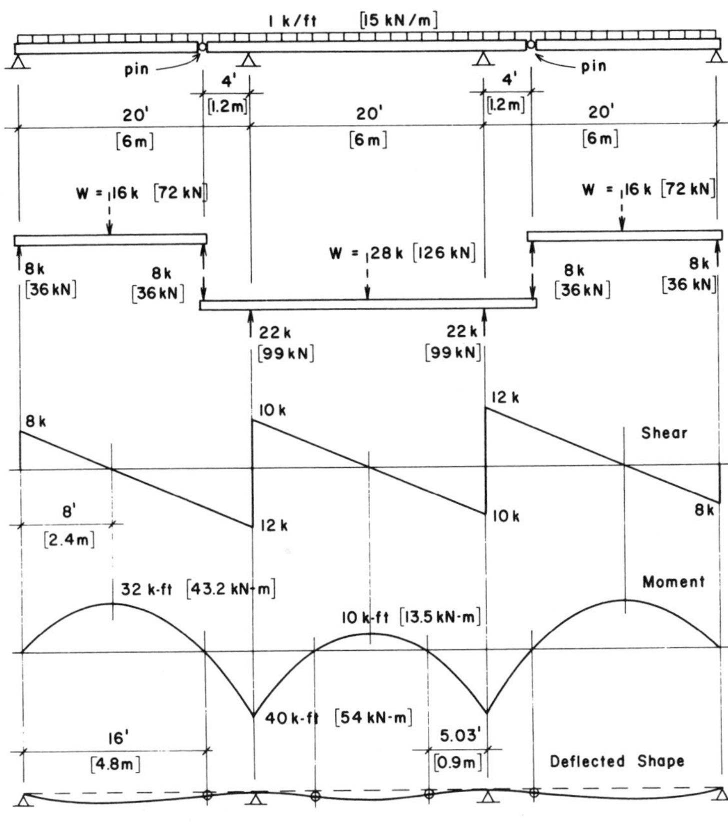

FIGURE 2.48

Example 3. Investigate the beam shown in Fig. 2.48.

Solution: The procedure is essentially the same as for Example 2. Note that this beam with four supports requires two internal pins to become statically determinate. As before, the investigation begins with the consideration of the end portion acting as a simple beam. The second step is to consider the center portion as a beam with two overhanging ends.

Problems 2.14.A,B. Find the components of the reactions for the structures shown in Fig. 2.49*a* and *b*.

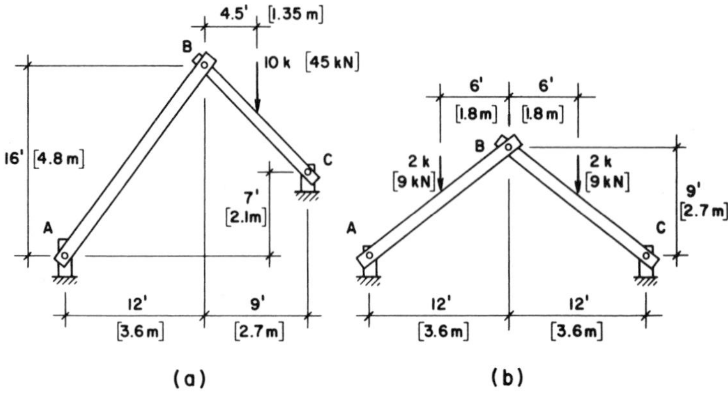

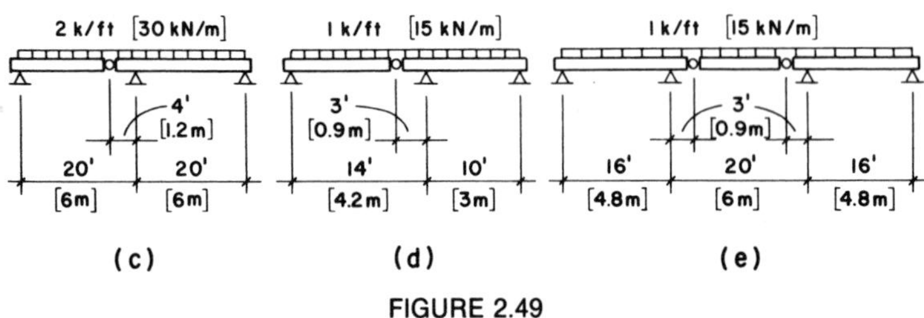

FIGURE 2.49

Problems 2.14.C,D,E. Investigate the beams shown in Fig. 2.49*c–e*. Find the reactions and draw the shear and moment diagrams, indicating all critical values. Sketch the deflected shape and determine the locations of any inflection points not related to the internal pins. (*Note:* Problem 2.14.E has the same spans and loading as Example 2 in Sec. 2.11.)

2.15 Rigid Frames

Frames in which two or more of the members are attached to each other with connections that are capable of transmitting bending between the ends of the members are called *rigid frames*. The connections used to achieve such a frame are called *moment connections*. Most rigid frame structures are statically indeterminate and do not yield to investigation by consideration of static equilibrium alone. The examples presented in this chapter are all rigid frames that have conditions that make them statically determinate and thus capable of being fully investigated by methods developed in this book.

Cantilever Frames. Consider the frame shown in Fig. 2.50*a*, consisting of two members rigidly joined at their intersection. The vertical member is fixed at its base, providing the necessary support condition for stability of the frame. The horizontal member is loaded with a uniformly distributed loading and functions as a simple cantilever beam. (See Sec. 2.7.) The frame is described as a cantilever frame because of the single fixed support. The five

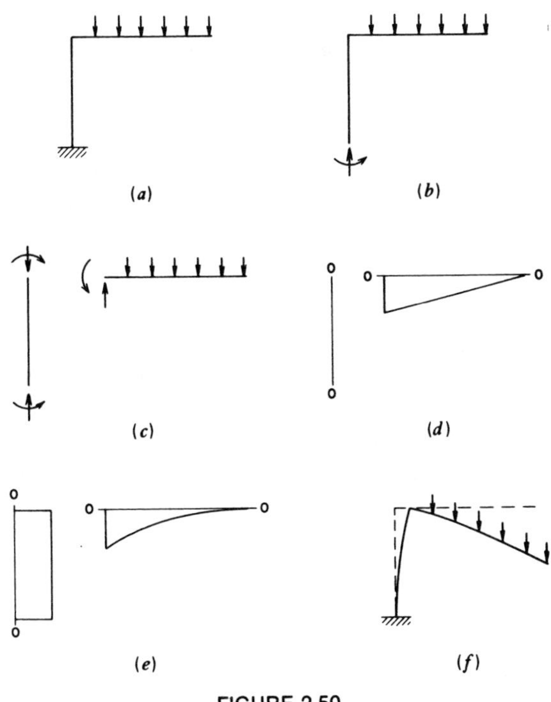

FIGURE 2.50

sets of figures shown in Fig. 2.50*b–f* are useful elements for the investigation of the behavior of the frame:

1. The free-body diagram of the entire frame, showing the loads and the components of the reactions (Fig. 2.50*b*). Study of this figure will help in establishing the nature of the reactions and in the determination of the conditions necessary for stability of the frame.
2. The free-body diagrams of the individual elements (Fig. 2.50*c*). These are of great value in visualizing the interaction of the parts of the frame. They are also useful in the computations for the internal forces in the frame.
3. The shear diagrams of the individual elements (Fig. 2.50*d*). These are sometimes useful for visualizing, or for actually computing, the variations of moment in the individual elements. No particular sign convention is necessary unless in conformity with the sign used for moment. (See discussion of relation of shear and moment in Sec. 2.5.) Although good as exercises in visualization, the shear diagrams have limited value in the investigation in most cases.
4. The moment diagrams for the individual elements (Fig. 2.50*e*). These are very useful, especially in determination of the deformation of the frame. The sign convention used is that of plotting the moment on the compression side of the element.
5. The deformed shape of the loaded frame (Fig. 2.50*f*). This is the exaggerated profile of the bent frame, usually superimposed on an outline of the unloaded frame for reference. This is very useful for the general visualization of the frame behavior. It is particularly useful for determination of the character of the external reactions and the form of interaction between the parts of the frame.

When performing investigations, these elements are not usually produced in the sequence just described. In fact, it is generally recommended that the deformed shape be sketched first so that its correlation with other factors in the investigation may be used as a check on the work. The following examples illustrate the process of investigation for simple cantilever frames.

Example 1. Find the components of the reactions and draw the free-body diagrams, shear and moment diagrams, and the deformed shape of the frame shown in Fig. 2.51*a*.

Solution: The first step is the determination of the reactions. Considering the free-body diagram of the whole frame (Fig. 2.51*b*), we compute the reactions as follows:

$$\Sigma F = 0 = +8 - R_V, \qquad R_V = 8 \text{ k (up)}$$

and, with respect to the support, labeled 0

$$\Sigma M_0 = 0 = M_R - (8 \times 4), \qquad M_R = 32 \text{ k-ft (counterclockwise)}$$

Note that the sense, or sign, of the reaction components is visualized from the logical development of the free-body diagram.

Consideration of the free-body diagrams of the individual members will yield the actions required to be transmitted by the moment connection. These may be computed by application of the conditions for equilibrium for either of the members of the frame. Note that the sense of the force and moment is opposite for the two members, simply indicating that what one does to the other is the opposite of what is done to it.

In this example there is no shear in the vertical member. As a result, there is no variation in the moment from the top to the bottom of the member. The free-body diagram of the member, the shear and moment diagrams, and the deformed shape should all corroborate this fact.

The shear and moment diagrams for the horizontal member are simply those for a cantilever beam. (See case 6, Fig. 2.27.)

It is possible with this example, as with many simple frames, to visualize the nature of the deformed shape without recourse to any mathematical computations. It is advisable to do so, and to check continually during the work that individual computations are logical with regard to the nature of the deformed structure.

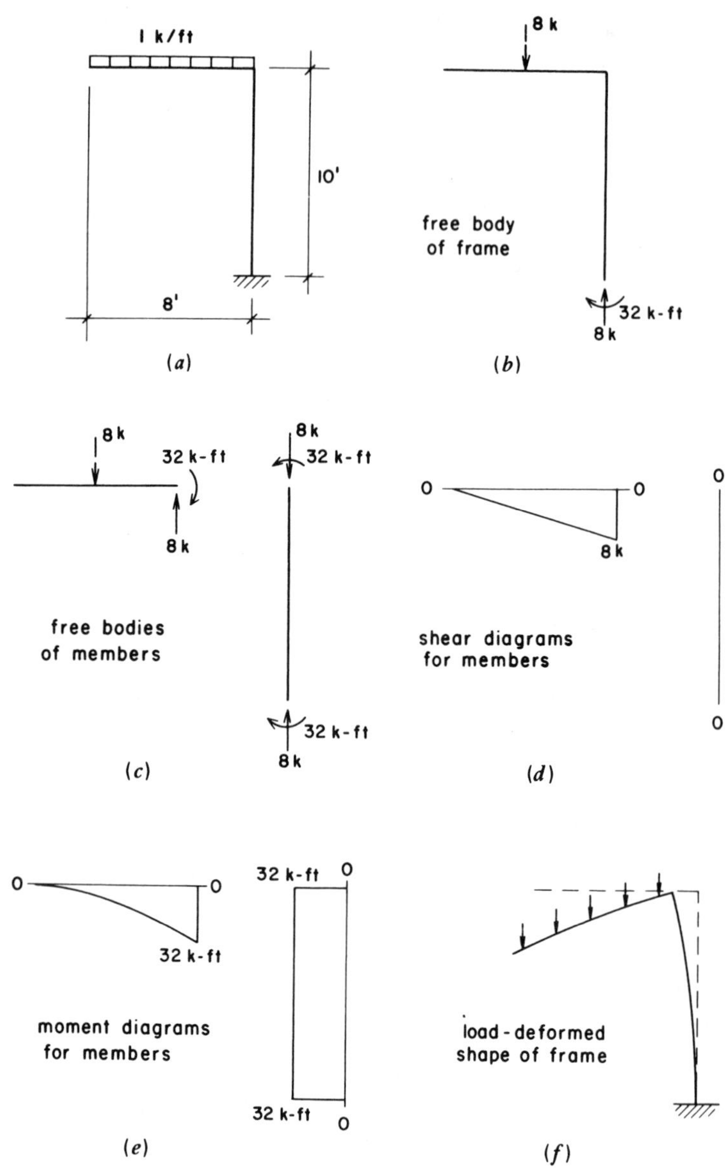

1 k/ft
10'
8'
(a)
8 k
free body
of frame
32 k-ft
8 k
(b)
8 k
32 k-ft
8 k
free bodies
of members
8 k
32 k-ft
32 k-ft
8 k
(c)
0
0
8 k
shear diagrams
for members
0
0
(d)
0
0
32 k-ft
moment diagrams
for members
32 k-ft
0
32 k-ft
0
(e)
load-deformed
shape of frame
(f)

FIGURE 2.51

Example 2. Find the components of the reactions and draw the shear and moment diagrams and the deformed shape of the frame in Fig. 2.52*a*.

Solution: In this frame there are three reaction components required for stability, since the loads and reactions constitute a

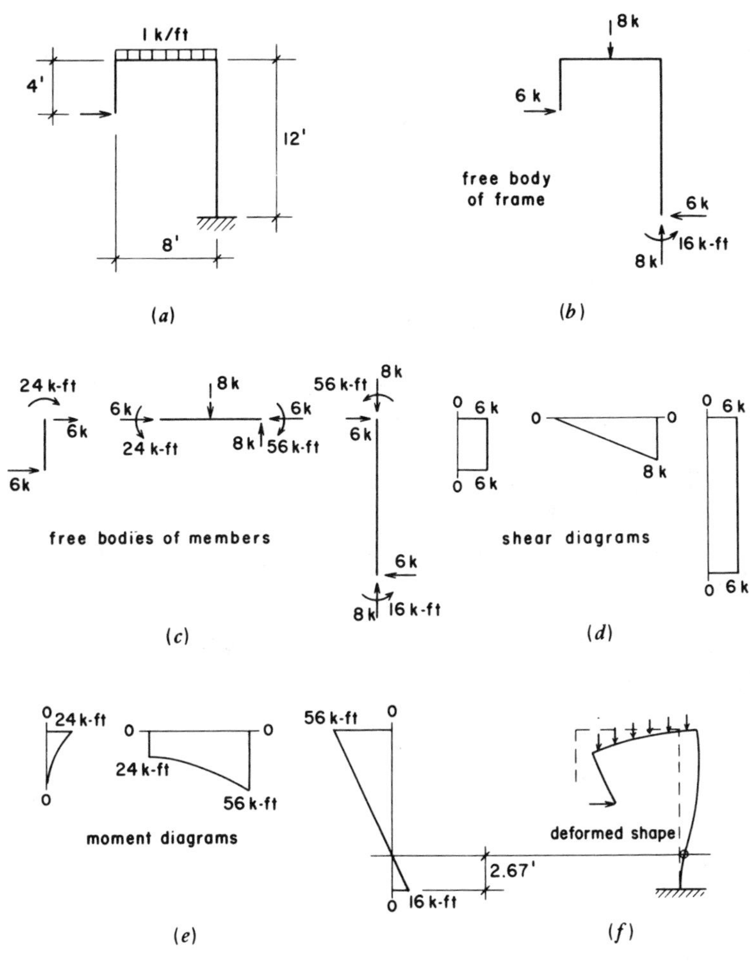

FIGURE 2.52

general coplanar force system. Using the free-body diagram of the whole frame (Fig. 2.52*b*), the three conditions for equilibrium for a coplanar system are used to find the horizontal and vertical reaction components and the moment component. If necessary, the reaction force components could be combined into a single-force vector, although this is seldom required for design purposes.

Note that the inflection occurs in the larger vertical member because the moment of the horizontal load about the support is greater than that of the vertical load. In this case, this computation must be done before the deformed shape can be accurately drawn.

The reader should verify that the free-body diagrams of the individual members are truly in equilibrium and that there is the required correlation between all the diagrams.

Single-Bent Frames. Figure 2.53 shows two possibilities for a single-bent rigid frame. In *a* the frame has pinned bases for the columns, resulting in the form of deformation under loading as shown in Fig. 2.53*c*, and the reaction components as shown in the free-body diagram for the whole frame in Fig. 2.53*e*. The second frame (Fig. 2.53*b*) has fixed bases for the columns, resulting in the slightly modified behavior indicated. These are the two most common forms of single-bent frames, the choice of the column base condition depending on a number of design factors for each case.

Both of the frames shown in Fig. 2.53 are statically indeterminate, and their investigation is beyond the scope of work in this book. The following examples of single-bent frames consist of frames with combinations of support and internal conditions that make the frames statically determinate. These conditions are technically achievable, but a bit on the weird side for practical use. We offer them here simply as exercises within the scope of our readers so that some experience in investigation may be gained.

Example 3. Investigate the frame shown in Fig. 2.54 for the reactions and internal conditions.

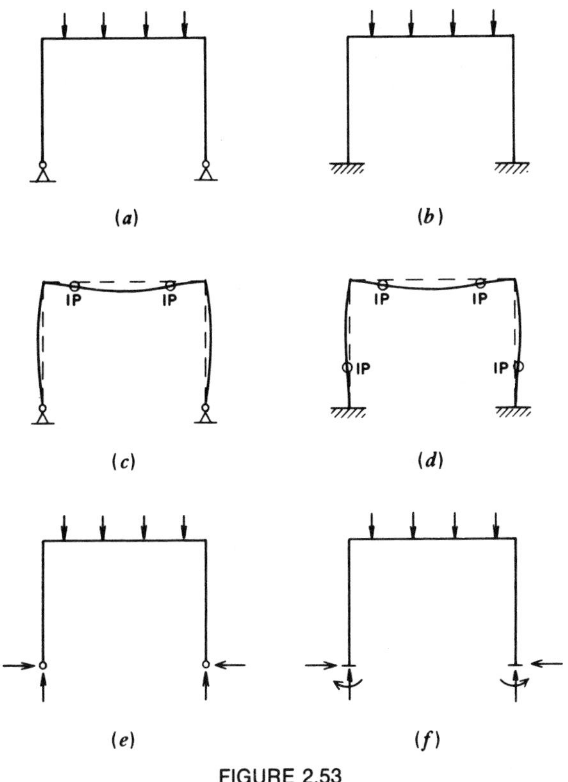

FIGURE 2.53

Solution: The typical elements of investigation, as illustrated for the preceding examples, are shown in the figure. The suggested procedure for the work is as follows:

1. Sketch the deflected shape. (A little tricky in this case, but a good exercise.)
2. Consider the equilibrium of the free-body diagram for the whole frame to find the reactions.
3. Consider the equilibrium of the left-hand vertical member to find the internal actions at its top.

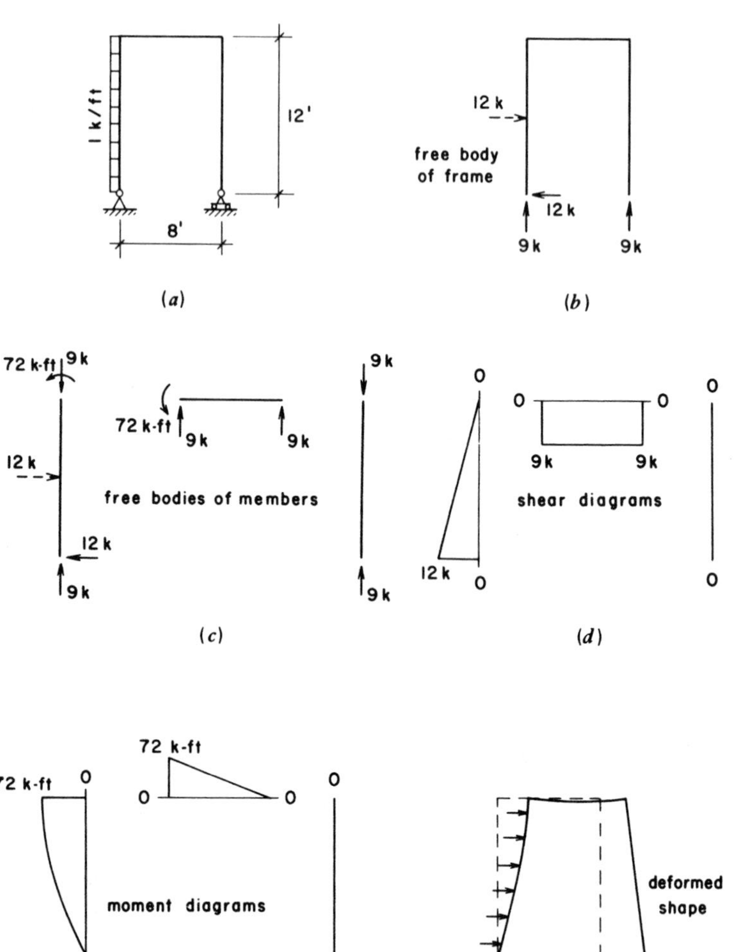
1 k/ft
12'
8'
(a)
12 k
free body
of frame
12 k
9k
9k
(b)
72 k-ft
9k
12 k
12 k
9k
72 k-ft
9k
9k
free bodies of members
9k
9k
(c)
0
0
12 k
0
0
0
9k
9k
shear diagrams
0
0
(d)
72 k-ft
0
0
72 k-ft
0
0
moment diagrams
0
0
(e)
deformed
shape
(f)

FIGURE 2.54

4. Proceed to the equilibrium of the horizontal member.
5. Finally, consider the equilibrium of the right-hand vertical member.
6. Draw the shear and moment diagrams and check for correlation of all the work.

Note that the right-hand support allows for an upward vertical reaction only, whereas the left-hand support allows for both vertical and horizontal components. Neither support provides moment resistance.

Before attempting the exercise problems, the reader is advised to attempt to produce the results shown in Fig. 2.54 independently.

Problems 2.15.A,B,C. For the frames shown in Fig. 2.55*a–c*, find the components of the reactions, draw the free-body diagrams of the whole frame and the individual members, draw the shear and moment diagrams for the individual members, and sketch the deformed shape of the loaded structure.

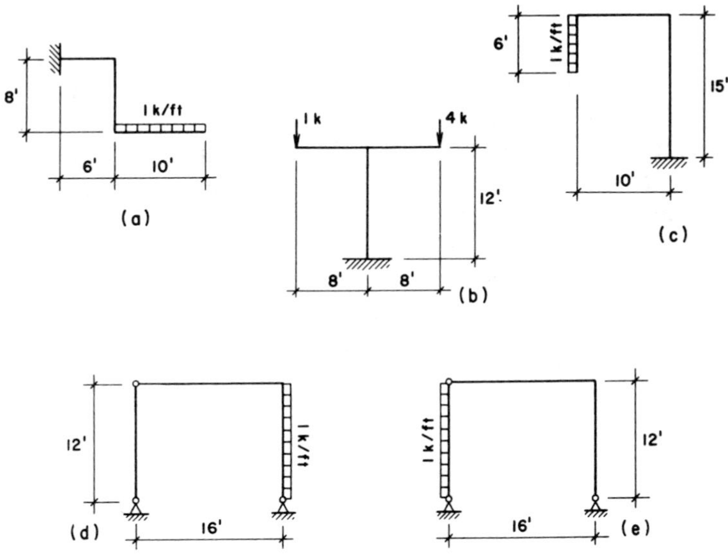

FIGURE 2.55

Problems 2.15.D,E. Investigate the frames shown in Fig. 2.55*d* and *e* for reactions and internal conditions, using the procedure shown for the preceding examples.

2.16 Approximate Analysis of Indeterminate Structures

Analysis of the behavior of indeterminate structures requires use of some conditions in addition to those provided by consideration of static equilibrium. They are thus imaged as having a negative character: being *not* capable of something—analysis by statics alone. Actually, there is nothing wrong or bad about a structure that is statically indeterminate. In fact, there are several potential advantages, including reduction of the maximum bending moments, reduction of deformations, and a general redundancy of stability and internal resisting forces that enhances safety. In some cases the structure is deliberately made redundant to gain one or more of these advantages. In other cases the structure is inherently indeterminate due to the nature of the construction—such as with poured-in-place concrete structures with multiple spans.

In some cases additional behavior conditions can be established on the basis of observations of behavior that is controlled, such as fixed supports where the rotation remains zero, any support where the deflection remains zero, internal pins where the moment must be zero, and so on. It is also possible to use the fact that the beam assumes a smooth curve as its deflected shape. In most cases, however, these conditions only allow for shortcuts or for the implementation of other, more fundamental techniques, such as the classic slope-deflection method.

We will not attempt to develop the general problem of investigation of statically indeterminate structures in this book. This is an exhaustive subject, well developed in many standard reference texts. We must, however, consider some aspects of behavior of indeterminate structures so that the reader can gain some appreciation for the issues involved. In current practice really complex indeterminate structures are routinely investigated using computer-aided techniques, the software for which is readily available (for a price) to the design professions.

The rigid-frame structure occurs quite frequently as a multiple-level, multiple-span bent, constituting part of the structure for a multistory building. In most cases such a bent is used as a lateral bracing element, although once it is formed as a moment-resistive framework it will respond characteristically for all types of loads.

The multistory rigid bent is quite indeterminate, and its investigation is complex—requiring considerations of several different loading combinations. When loaded or formed unsymmetrically, it will experience sideways movements that further complicate the analysis for internal forces. Except for very early design approximations, the analysis is now sure to be done with a computer-aided system. The software for such a system is quite readily available.

For preliminary design purposes, it is sometimes possible to use approximate analysis methods to obtain member sizes of reasonable accuracy. Actually, many of the older high-rise buildings still standing were completely designed with these techniques—a reasonable testimonial to their effectiveness. Demonstrations of these approximate methods are given in Chapter 22.

3

Properties of Sections

This chapter deals with various geometric properties of areas that are utilized in structural investigations. The areas referred to are cross-sectional areas of structural members. The geometric properties are used in the analysis for stress and deformation conditions in the members.

3.1 Centroids

The *center of gravity* of a solid is an imaginary point at which all its weight may be considered to be concentrated or the point through which the resultant weight passes. Since an area has no weight, it has no center of gravity. *The point of a plane area that corresponds to the center of gravity of a very thin homogeneous plate of the same area and shape is called the centroid of the area.*

When a simple beam is subjected to forces that cause it to bend, the fibers above a certain plane in the beam are in compression and those below the plane are in tension. This plane is called the *neutral surface*. For a cross section of the beam the line corresponding to the neutral surface is called the *neutral axis*.

The neutral axis passes through the centroid of the section; thus it is important that we know the exact position of the centroid.

The position of the centroid for symmetrical shapes is readily determined. If an area possesses a line of symmetry, the centroid will obviously be on that line; if there are two lines of symmetry, the centroid will be at their point of intersection. For instance, a rectangular area as shown in Fig. 3.1*a* has its centroid at its geometrical center, the point of intersection of the diagonals. The centroid of a circular area is its center (Fig. 3.1*b*). With respect to a triangular area (Fig. 3.1*c* and *d*), it is convenient to remember that the centroid is at a distance equal to one-third of the perpendicular distance measured from any side to the opposite vertex. The intersection of the lines drawn from the vertices to the midpoints of the opposite sides is another method of locating the centroid of a triangle. This is shown in Fig. 3.1*c*.

(*Note:* Tables 3.1–3.6, referred to in the following discussion, are presented at the end of this chapter.)

For symmetrical structural steel shapes, such as wide flange beams, the centroid is on the vertical axis through the web at a point midway between the upper and lower surfaces of the flanges at the intersection of the *X–X* and *Y–Y* axes shown in the diagram of Table 3.1. Similar tables of properties are available for unsymmetrical structural steel sections. Table 3.2, for example, gives the properties of channel sections. Consider a 10-in. channel having a weight of 15.3 lb per lin ft (designated C 10 × 15.3). In this table we find that the vertical axis is 0.634 in. from the back of the

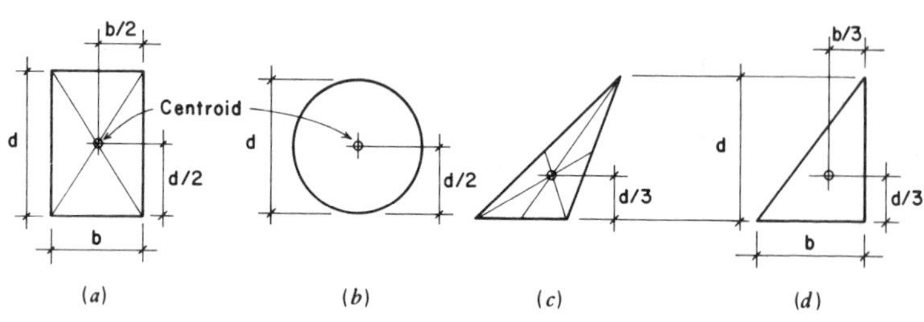

FIGURE 3.1

web, the distance $\bar{x}$ (read x bar). The centroid is on this axis at a point 5 in. between the top and bottom surfaces of the flanges.

It frequently happens, however, that we must determine the position of the centroid, and this is accomplished most readily by mathematics. The *statical moment* of a plane area with respect to a given axis is the area multiplied by the normal distance of the centroid of the area to the axis. *If an area is divided into a number of parts, the sum of the statical moments of the parts is equal to the statical moment of the entire area.* This is the principle by means of which the position of the centroid is found; its application is remarkably simple.

Example. Figure 3.2 is a beam cross section, unsymmetrical with respect to the horizontal axis. Find the value of c, the distance of the neutral axis from the most remote fiber.

Solution: (1) It is not always possible to tell by observation whether the centroid is nearer the top or bottom edge of the area. In this instance let us write an equation of moments about an axis through the uppermost edge. First divide the total area into some number of simple shapes, in this case the three rectangles shown by the diagonals. The area of the upper part is 9 in.^{2} [5625 mm^2],

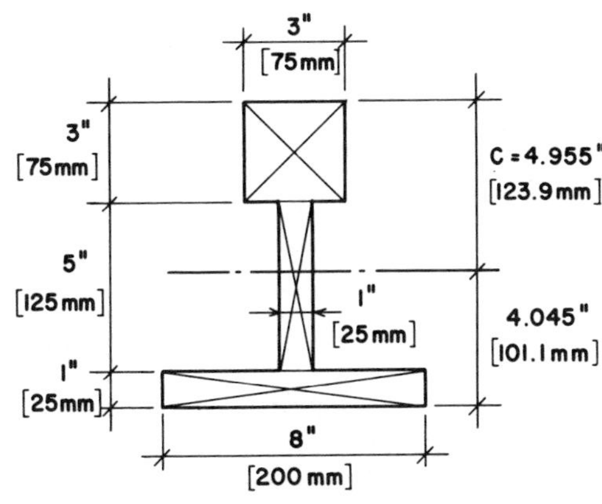

FIGURE 3.2

and its centroid is 1.5 in. [37.5 mm] from the reference axis. The center part has an area of 5 in.^{2} [3125 mm^2], and its centroid is 5.5 in. [137.5 mm] from the axis. The bottom part contains 8 in.^{2} [5000 mm^2], and its centroid is 8.5 in. [212.5 mm] from the axis. Considering these individual areas as forces, their combined moment about the reference axis is

$$M = (9 \times 1.5) + (5 \times 5.5) + (8 \times 8.5)$$
$$= 13.5 + 27.5 + 68 = 109 \text{ in.}^3$$

Since this is the moment of the entire area about the axis, it may be equated to the product of the entire area (22 in.^{2}) times the single centroid distance c, as shown in the figure. Thus

$$M = 109 = 22 \times c$$
$$c = 109 \div 22 = 4.955 \text{ in. } [123.9 \text{ mm}]$$

which is the distance of the centroid from the reference axis.

(2) The depth of the section is 9 in [225 mm]; hence the distance of the centroid from the bottom edge is

$$9 - 4.955 = 4.045 \text{ in. } [101.1 \text{ mm}]$$

Call this distance c_1. The value of c_1 may be checked by writing an equation of moments about a reference axis through the bottom edge; thus

$$(9 \times 7.5) + (5 \times 3.5) + (8 \times 0.5) = (22 \times c_1)$$

from which

$$c_1 = \frac{89}{22} = 4.045 \text{ in.}$$

Remember that the centroid of the section is the point through which the neutral axis (zero bending stress point) passes. The position of the centroid for structural steel angles may be found

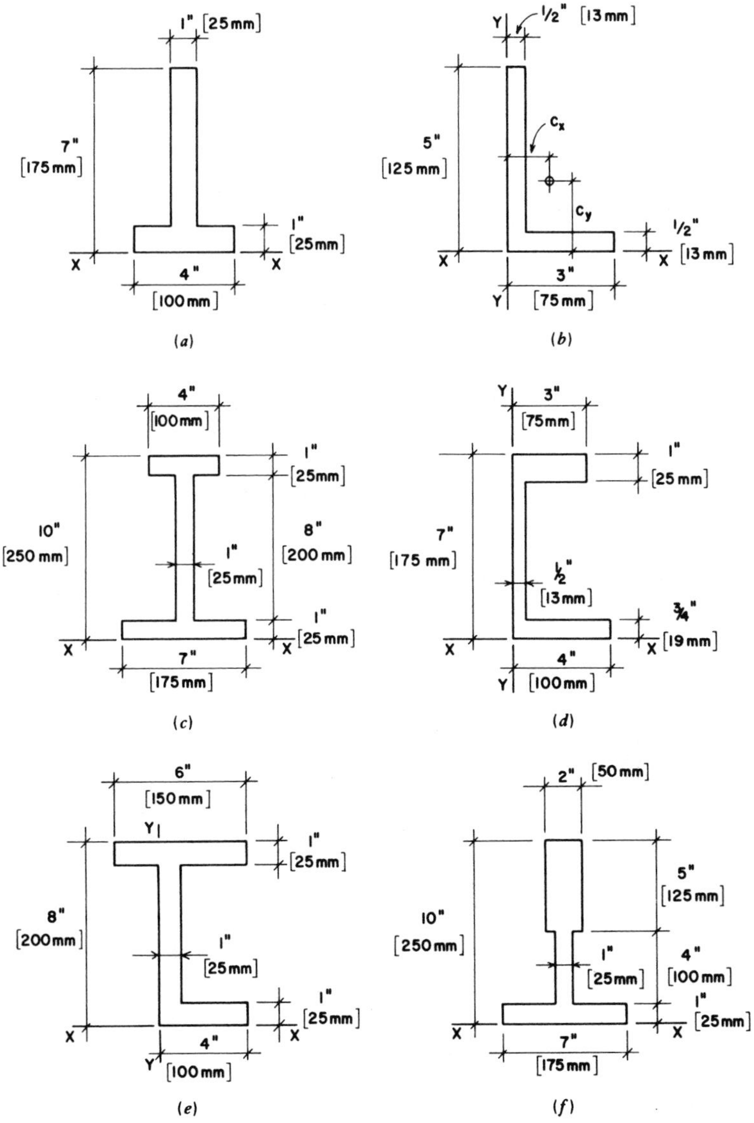
1" [25mm]
7"
[175mm]
1"
[25mm]
X
X
4"
[100mm]
(a)
Y
1/2" [13mm]
c_x
5"
[125mm]
c_y
1/2"
[13mm]
X
X
3"
Y
[75mm]
(b)
4"
[100mm]
1"
[25mm]
10"
[250mm]
8"
[200mm]
1"
[25mm]
1"
[25mm]
X
X
7"
[175mm]
(c)
Y
3"
[75mm]
1"
[25mm]
7"
[175mm]
1/2"
[13mm]
3/4"
[19mm]
X
X
4"
Y
[100mm]
(d)
6"
[150mm]
Y
1"
[25mm]
8"
[200mm]
1"
[25mm]
1"
[25mm]
X
X
4"
Y
[100mm]
(e)
2" [50mm]
5"
[125mm]
10"
[250mm]
1"
[25mm]
4"
[100mm]
1"
[25mm]
X
X
7"
[175mm]
(f)

FIGURE 3.3

directly from Table 3.3. The locating dimensions from the backs of the angle legs are given as y and x distances.

Problems 3.1.A,B,C,D,E,F. Find the location of the centroid for the cross-sectional areas shown in Fig. 3.3. Use the references and indicate the distances as c_x and c_y, as shown in Fig. 3.3*b*.

3.2 Moment of Inertia

Consider the area enclosed by the irregular line in Fig. 3.4*a*, area A. In this area an infinitely small area a is indicated at z distance from an axis marked X–X. If we multiply this tiny area by the square of its distance to the axis, we have the quantity $(a \times z^2)$. The entire area is made up of an infinite number of these small elementary areas at various distances from the X–X axis. Now, if we use the Greek letter Σ to represent the sum of an infinite number, we may write Σaz^2, which indicates the sum of all the infinitely small areas (of which the area A is composed) multiplied by the square of their distances from the X–X axis. This quantity is called the *moment of inertia* of the area and is represented by the letter I. Then $I_{X-X} = \Sigma az^2$, meaning that Σaz^2 is the moment of inertia of the area with respect to the axis marked X–X.

We may define the moment of inertia of an area as *the sum of the products of all the elementary areas multiplied by the square of their distances from an axis*. In the United States the linear dimensions of cross sections of structural members are invariably in units of inches, and, since the moment of inertia involves an area multiplied by the square of a distance, the moment of inertia

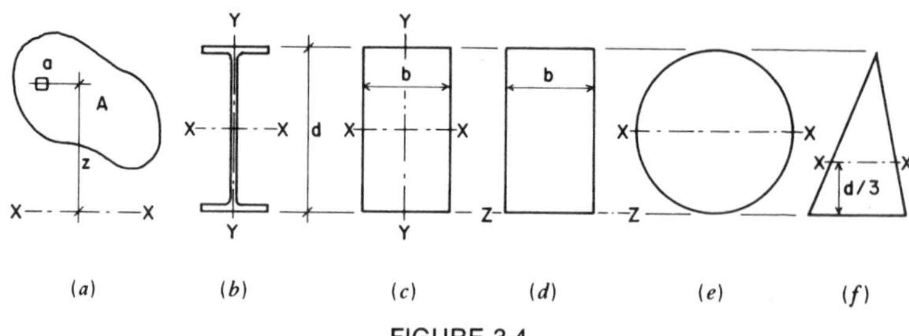

FIGURE 3.4

is expressed in inches to the fourth power and is written in.[4], sometimes called biquadratic inches. We cannot conceive such units as, for example, a linear dimension; it is a mathematical concept, expressed as Σaz^2. Thus the moment of inertia of a cross section depends not only on the number of units in its area but also on the arrangement of the area with respect to the axis under consideration. The moment of inertia of cross sections is of particular importance in the design of beams and columns. It is important to realize that the moment of inertia of cross sections of beams is taken about an axis that passes through the centroid of the area.

The moment of inertia of rolled steel sections used in building construction is given in tables of properties published by the American Institute of Steel Construction. Tables 3.1–3.5, given at the end of this chapter, have been compiled from more extensive tables in the AISC *Manual of Steel Construction,* 8th ed.

3.3 Moment of Inertia of Geometric Figures

It is frequently necessary to compute the moment of inertia for built-up sections or for sections not given in tables. In engineering problems the shapes most commonly involved are the rectangle, the circle, and the triangle. The derivations of the formulas to be used for computing moments of inertia are most readily accomplished by the use of calculus. It is beyond the scope of this book to derive such formulas, but their values are given here and the computations involved in computing I for the sections most frequently used present no difficulties.

Rectangles. Consider the rectangle shown in Fig. 3.4c. Its width is b and its depth is d. The two major axes are X–X and Y–Y, both passing through the centroid of the area. It can be shown that *the moment of inertia of a rectangular area about an axis passing through the centroid parallel to the base is* $I_{X-X} = bd^3/12$. With respect to the vertical axis, $I_{Y-Y} = db^3/12$.

Example 1. Find the value of the moment of inertia for a 6 × 12-in. wood beam about an axis through its centroid and parallel to the base of the section.

Solution: Referring to Table 3.6, we find that a nominal 6 × 12-in. section has standard dressed dimensions of 5.5 × 11.5 in. [139.7 × 292.1 mm]. Then

$$I = \frac{bd^3}{12} = \frac{(5.5)(11.5)^3}{12} = 697.07 \text{ in.}^4 \; [290.1 \times 10^6 \text{ mm}^4]$$

which is in agreement with the value of I listed in the table.

Circles. Figure 3.4*e* shows a circular area with diameter d, and axis X–X passing through its center. *The moment of inertia of a circular area about an axis passing through its centroid is* $I_{X-X} = \pi d^4/64$.

Example 2. Compute the moment of inertia of a circular cross section, 10 in. in diameter, about an axis through its centroid.

Solution: Since I_{X-X} in Fig. 3.4*e* may be any axis through the centroid of the circle, we may use the symbol I_0. Then

$$I_o = \frac{\pi d^4}{64} = \frac{3.1416 \times 10^4}{64} = 490.9 \text{ in.}^4 \; [204.3 \times 10^6 \text{ mm}^4]$$

Triangles. The triangle in Fig. 3.4*f* has a height d and a base b. The axis X–X passes through the centroid of the area. *The moment of inertia of a triangular cross section about an axis through the centroid parallel to the base is* $I_{X-X} = bd^3/36$.

Example 3. Assuming that the base of the triangular area shown in Fig. 3.4*f* is 12 in. and the height 10 in., compute the moment of inertia about axis X–X which passes through the centroid.

Solution: For this triangle $b = 12$ in. and $d = 10$ in. Then

$$I_0 = \frac{bd^3}{36} = \frac{12 \times 10^3}{36} = 333.33 \text{ in.}^4 \; [138.7 \times 10^6 \text{ mm}^4]$$

Hollow Sections and I-Shapes. The moment of inertia of some geometric figures may be found by employing the principle that states *if the moments of inertia of the component parts of an* *area are taken about the same axis, they may be added or subtracted to find the moment of inertia of the area.*

Example 4. Compute the moment of inertia of the hollow rectangular section shown in Fig. 3.5*a* about a horizontal axis through the centroid parallel to the 6-in. side.

Solution: The moment of inertia of the 6 × 10-in. rectangle is

$$I = \frac{bd^3}{12} = \frac{6 \times 10^3}{12} = 500 \text{ in.}^4 \ [208.2 \times 10^6 \text{ mm}^4]$$

and that of the 4 × 8-in. rectangle is

$$I = \frac{bd^3}{12} = \frac{4 \times 8^3}{12} = 170.7 \text{ in.}^4 \ [71.04 \times 10^6 \text{ mm}^4]$$

Since both of these moments of inertia are taken *about the same axis,* we may subtract that of the smaller rectangle from that of the larger, and the difference will be the moment of inertia of the hollow rectangular section. Thus $500 - 170.7 = 329.3$ in.^{4} [137.08×10^6 mm^4].

Example 5. Compute the moment of inertia about an axis through the centroid of the pipe cross section shown in Fig. 3.5*b*. The thickness of the pipe shell is 1 in.

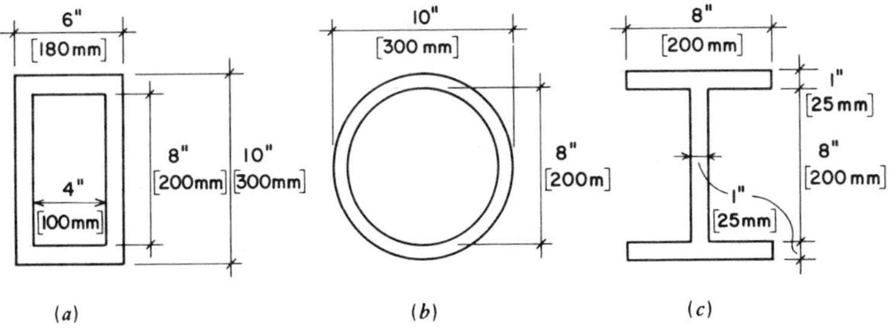

FIGURE 3.5

Solution: To find I_0 of the ring section, we will compute I for the area bounded by the outer circle and then subtract from this value the I for the inner circular area. Using the 10-in. outside diameter of the pipe,

$$I = \frac{\pi d^4}{64} = \frac{3.1416 \times 10^4}{64} = 491 \text{ in.}^4 \; [204.3 \times 10^6 \text{ mm}^4]$$

and working with the 8-in. inside diameter,

$$I = \frac{\pi d^4}{64} = \frac{3.1416 \times 8^4}{64} = 201 \text{ in.}^4 \; [83.7 \times 10^6 \text{ mm}^4]$$

Therefore the moment of inertia of the pipe section is

$$I_0 = 491 - 201 = 290 \text{ in.}^4 \; [120.6 \times 10^6 \text{ mm}^4]$$

Example 6. Referring to Fig. 3.5*c*, compute the moment of inertia of the I-section about a horizontal axis through the centroid of the section parallel to the flange.

Solution: For the full rectangular area with the width of 8 in. and depth of 10 in.,

$$I = \frac{bd^3}{12} = \frac{8 \times 10^3}{12} = 667 \text{ in.}^4 \; [277.5 \times 10.6 \text{ mm}^4]$$

This moment of inertia includes that of the two open spaces at the sides of the vertical web. Taken together, these open spaces are equivalent to a rectangle 7 in. wide and 8 in. deep. For this area

$$I = \frac{bd^3}{12} = \frac{7 \times 8^3}{12} = 299 \text{ in.}^4 \; [124.3 \times 10.6 \text{ mm}^4]$$

Subtracting I for the open spaces from I for the large rectangle, the moment of inertia of the I-section is

$$I = 667 - 299 = 368 \text{ in.}^4 \; [153.2 \times 10^6 \text{ mm}^4]$$

Formulas for moment of inertia of rectangles, circles, and triangles are given in Fig. 3.9, together with formulas for other properties considered in Sec. 3-5 and 3-6.

3.4 Transferring Moments of Inertia

When shapes are combined to form built-up structural sections, it is necessary to determine the moment of inertia of the built-up section about its neutral axis. This requires transferring the moments of inertia of some of the individual parts from one axis to another and is accomplished by means of the transfer-of-axis equation, which may be stated as follows:

The moment of inertia of a cross section about any axis parallel to an axis through its own centroid is equal to the moment of inertia of the cross section about its own centroidal axis, plus its area times the square of the distance between the two axes. Expressed mathematically,

$$I = I_0 + Az^2$$

In this formula,

I = moment of inertia of the cross section about the required axis,
I_0 = moment of inertia of the cross section about its own centroidal, axis parallel to the required axis,
A = area of the cross section,
z = distance between the two parallel axes.

These relationships are indicated in Fig. 3.6*a*, where X–X is the centroidal axis of the angle (passing through its centroid) and Y–Y is the axis about which the moment of inertia is to be found.

To illustrate the use of the equation, we may prove that the value of I for a rectangle about an axis through its base is $bd^3/3$. Since I for a rectangle about its centroidal axis is known to be

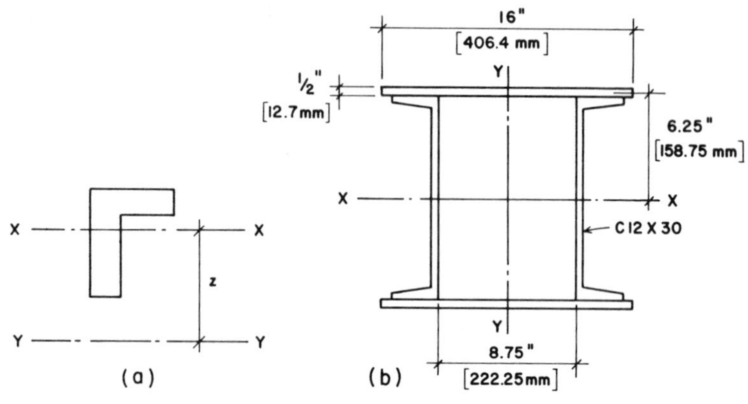

FIGURE 3.6

$bd^3/12$, and z in this instance is $d/2$, we may write

$$I = I_0 + Az^2$$

$$I = \frac{bd^3}{12} + \left[bd \times \left(\frac{d}{2}\right)^2\right]$$

$$I = \frac{bd^3}{12} + \frac{bd^3}{4} = \frac{bd^3}{3}$$

Example 1. Find the centroidal moment of inertia parallel to the flanges for the I-shaped section in Fig. 3.5*c*. (This problem was also worked as Example 6 in Sec. 3.3.)

Solution: Considering the section to be a composite of the 8-in. high web and the two flanges,

$$I = \frac{1(8)^3}{12} + 2\left[\frac{8(1)^3}{12} + (8)(4.5)^2\right]$$

$$= 42.67 + 2(0.67 + 162)$$

$$= 368 \text{ in.}^4$$

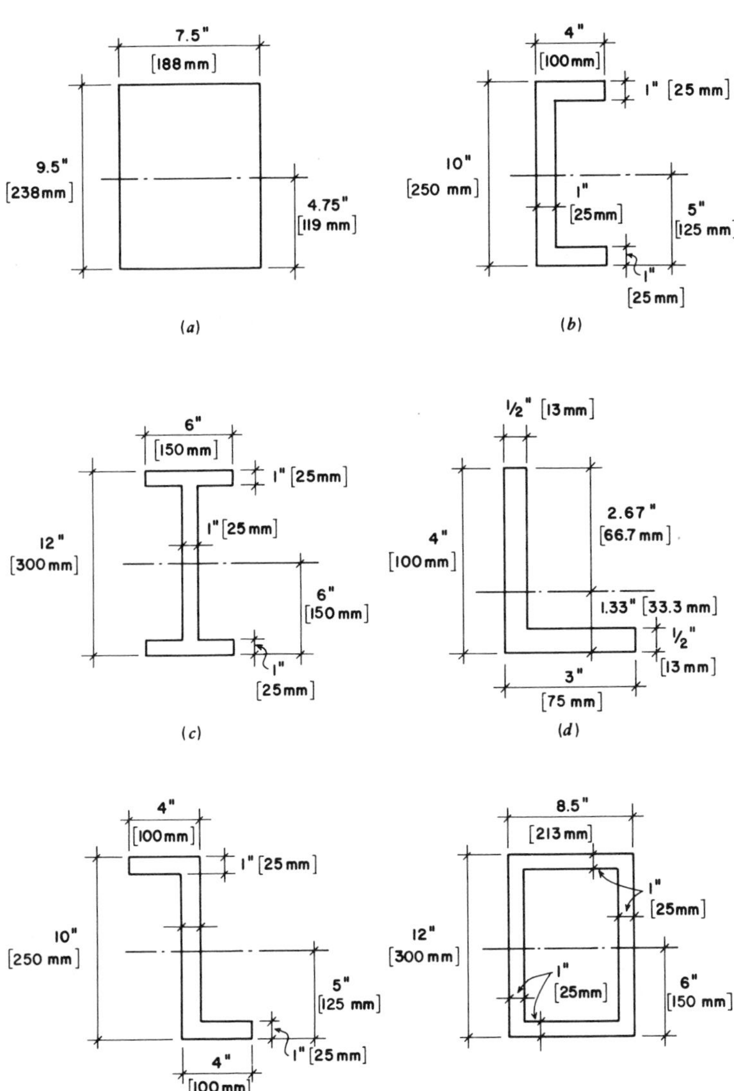
7.5"
[188 mm]
9.5"
[238mm]
4.75"
[119 mm]
(a)
4"
[100mm]
1" [25 mm]
10"
[250 mm]
1"
[25mm]
5"
[125 mm]
1"
[25 mm]
(b)
6"
[150 mm]
1" [25mm]
1" [25 mm]
12"
[300mm]
6"
[150 mm]
1"
[25mm]
(c)
½" [13 mm]
4"
[100mm]
2.67"
[66.7 mm]
1.33" [33.3 mm]
½"
[13 mm]
3"
[75 mm]
(d)
4"
[100mm]
1" [25 mm]
10"
[250 mm]
5"
[125 mm]
1" [25 mm]
4"
[100 mm]
(e)
8.5"
[213 mm]
1"
[25mm]
12"
[300 mm]
1"
[25mm]
6"
[150 mm]
(f)

FIGURE 3.7

The application of the transfer formula to the steel built-up section shown in Fig. 3.6*b* is illustrated in the following example.

Example 2. Compute the moment of inertia about the X–X axis of a built-up section composed of two C 12 × 30 channels and two 16 × 0.5-in. [406.4 × 12.7-mm] plates. (See Fig. 3.6*b*.)

Solution: (1) From Table 3.2 we find that I_x for the channel is 162 in.^{4} [67.42 × 10^6 mm^4], so the value for the two channels is twice this, or 324 in.^{4} [134.84 × 10^6 mm^4].

(2) For one plate the moment of inertia about an axis through its own centroid is

$$I_0 = \frac{bd^3}{12} = \frac{16 \times 0.5^3}{12} = 0.1667 \text{ in.}^4 \; [0.0694 \times 10^6 \text{mm}^4]$$

the distance between the centroid of the plate and the X–X axis is 6.25 in. [158.75 mm], and the area of one plate is 8 in.^{2} [5161.3 mm^2]. Therefore the I of one plate about the X–X axis of the combined section is

$$I_0 + Az^2 = 0.1667 + (8)(6.25)^2 = 312.7 \text{ in.}^4$$

$$[0.0694 \times 10^6 + (5161.3)(158.75)^2 = 130.14 \times 10^6 \text{ mm}^4]$$

and the value for two plates is twice this, or 625 in.^{4} [260.3 × 10^6 mm^4].

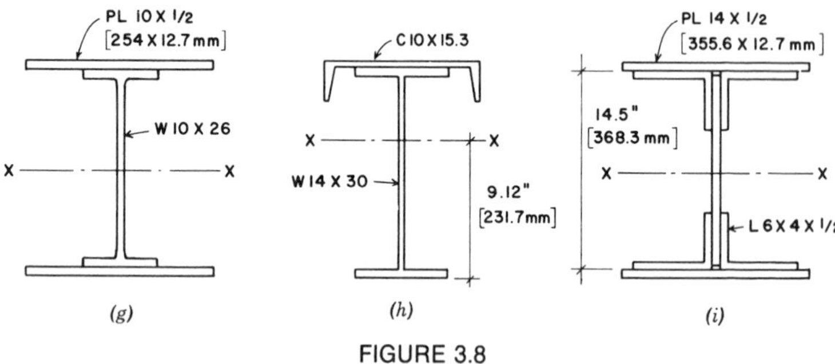

FIGURE 3.8

(3) Adding the moments of inertia of the channels and plates, we obtain the I for the entire cross section as $324 + 625 = 949$ in.^{4} [$(134.84 + 260.3) \times 10^6 = 395.14 \times 10^6$ mm^4].

Problems 3.4.A,B,C,D,E,F. Compute the moment of inertia about the neutral axes of the beam sections shown in Fig. 3.7.

Problems 3.4.G,H,I. Compute I with respect to the X–X axes for the built-up sections shown in Fig. 3.8. Make use of any appropriate data given in Tables 3.1–3.5.

3.5 Section Modulus

As noted in Sec. 2.10, the term I/c in the flexure formula is called the *section modulus*. It is defined as the moment of inertia divided by the distance of the most remote fiber from the neutral axis and is denoted by the symbol S. Since I and c always have the same values for any given cross section, values of S may be computed and tabulated for structural shapes. With I expressed in inches to the fourth power and c a linear dimension in inches, S is in units of inches to the third power, written *in.*^{3}. Section moduli are among the properties tabulated for structural steel shapes in Tables 3.1–3.5 and for structural lumber cross sections in Table 3.6.

For a rectangular cross section of breadth b and d we know that the moment of inertia about the X–X axis is $bd^3/12$ and that $c = d/2$. Therefore

$$\frac{I}{c} \text{ or } S = \frac{bd^3/12}{d/2} = \frac{bd^3}{12} \times \frac{2}{d}$$

or $S = bd^2/6$. It is often convenient to use this formula directly. It applies, of course, only to rectangular cross sections.

Example 1. Verify the tabulated value of the section modulus of a 6×12-in. wood beam.

Solution: From Table 3.6 we find that the true dimensions of the beam are 5.5 in. $\times$ 11.5 in. [139.7 $\times$ 292.1 mm]. Then

$$S = \frac{bd^2}{6} = \frac{5.5 \times 11.5 \times 11.5}{6} = 121.23 \text{ in.}^3 \ [1987 \times 10^3 \text{ mm}^3]$$

Compare this with the value of S in Table 3.6.

Example 2. Verify the tabulated value of S_x for a W 18 × 46.
Solution: From Table 3.1 we find that $I_x = 712$ in.^{4} [296.33 × 10^6 mm^4] and the actual depth is 18.06 in. [458.72 mm]. For the symmetrical section, $c = d \div 2 = 9.03$ in. Then

$$S = \frac{I}{c} = \frac{712}{9.03} = 78.848 \text{ in.}^3 \left[\frac{296.33 \times 10^6}{229.36} = 1292 \times 10^3 \text{ mm}^3\right]$$

which checks reasonably with the value in the table.

Problems 3.5.A,B,C,D. Verify the tabulated section modulus values for the following elements:

A A 6 × 8-in. timber beam, actually 5.5 × 7.5 in. [139.7 × 190.5 mm]
B S_x for a W 24 × 84 rolled shape
C S_x for an L 5 × 3.5 × 0.5 rolled steel angle
D S_y for an L 4 × 4 × 0.5 steel angle

3.6 Radius of Gyration

This property of a cross section is related to the design of compression members rather than beams and is discussed in more detail under the design of columns in subsequent sections of the book. Radius of gyration is considered briefly here, however, because it is a property listed in Tables 3.1–3.5.

Just as the section modulus is a measure of the resistance of a beam section to bending, the radius of gyration (which is also related to the size and shape of the cross section) is an index of the stiffness of a structural section when used as a column or other compression member. The radius of gyration is found from the formula

$$r = \sqrt{\frac{I}{A}}$$

and is expressed in inches, since the moment of inertia is in inches to the fourth power and the cross-sectional area is in square inches.

If a section is symmetrical about both major axes, the moment of inertia, and consequently the radius of gyration, is the same for each axis. But most common sections, particularly steel columns, are not symmetrical about the two major axes, and *in the design of columns the least moment of inertia, and therefore the least radius of gyration, is the one used in computations.* By *least* we mean the smallest in magnitude. Note in Table 3.3 that the least radius of gyration of angle sections occurs about the *Z–Z axes.*

Example. Verify the tabulated values for radii of gyration for a W 12 × 58, as given in Table 3.1.

Solution: The table shows the area of this section to be 17.0 in.^{2} [10968 mm^2] and I with respect to the X–X axis to be 475 in.^{4} [197.7 × 10^6 mm^4]. Then

$$r_x = \sqrt{\frac{I}{A}} = \sqrt{\frac{475}{17}} = \sqrt{27.94} = 5.29 \text{ in. } [134.3 \text{ mm}]$$

The table value for I with respect to the Y–Y axis is 107 in.^{4} [44.53 × 10^6 mm^4]. Therefore

$$r_y = \sqrt{\frac{I}{A}} = \sqrt{\frac{107}{17}} = \sqrt{6.29} = 2.51 \text{ in. } [63.7 \text{ mm}]$$

Compare these with the values given in Table 3.1. It may be noted that there is a minor discrepancy in the value for r_x, having to do with the manner of rounding off for the last digit.

Problem 3.6.A. Verify the value of r_y for a W 12 × 40.

Problem 3.6.B. Verify the value of r_x for an L 4 × 3 × $\frac{5}{8}$.

Problem 3.6.C. Compute the radius of gyration with respect to the X–X axis for the built-up sections shown in Fig. 3.6*b*.

3.7 Tables of Properties

The preceding sections in this chapter have pertained to the various properties of sections used in the design of structural members. Since the rectangle, circle, and triangle are the areas most commonly involved in engineering problems, the properties of these shapes are given in Fig. 3.9. This figure will serve as a

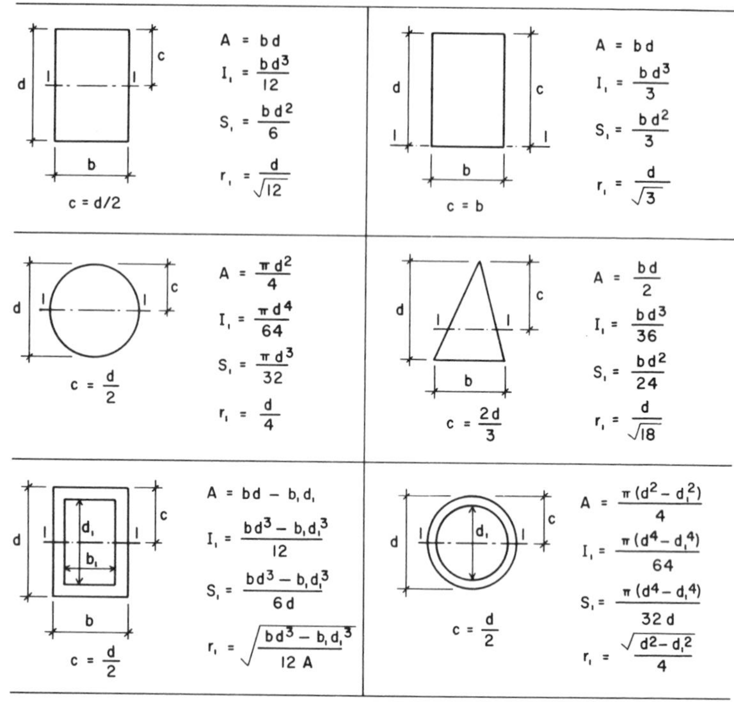

FIGURE 3.9 Properties of common geometric shapes.

reference. Although the properties of commonly used structural sections may be found directly in tables, a thorough knowledge of their significance will be of great value to the designer. Thus far nothing has been said concerning the use of these properties or their application in the design of members that resist forces. This is explained later, particularly in connection with members in bending and in the design of columns. It should be noted here, however, that all of the properties discussed relate to the size and shape of the sections considered—they are independent of the material of which the structural member is made.

For use in design, the properties of selected structural steel shapes are presented in Tables 3.1–3.5. Attention is called to the two major axes, $X–X$ and $Y–Y$. I, S, and r are given for both axes; the values to be used in design depend on the position in which the member is placed. If an I-section is used as a beam, the web is placed in a vertical position, and therefore the $X–X$ or horizontal axis determines the section modulus to be used.

Table 3.6 gives the properties of standard dressed sizes of structural lumber. Note the difference between the nominal and dressed dimensions. The properties A, I, and S are, of course, based on actual dressed sizes.

TABLE 3.1 Properties of Wide-Flange (W) Shapes

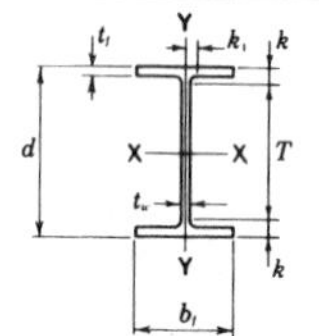

	Area	Depth	Web Thickness	Flange			Elastic Properties					
				Width	Thickness		Axis X-X			Axis Y-Y		
	A	d	t_w	b_f	t_f	k	I	S	r	I	S	r
Designation	in²	in.	in.	in.	in.	in.	in⁴	in³	in.	in⁴	in³	in.
W 36 × 300	88.3	36.74	0.945	16.655	1.680	2.81	20,300	1110	15.2	1300	156	3.83
× 260	76.5	36.26	0.840	16.550	1.440	2.56	17,300	953	15.0	1090	132	3.78
× 230	67.6	35.90	0.760	16.470	1.260	2.38	15,000	837	14.9	940	114	3.73
× 194	57.0	36.49	0.765	12.115	1.260	2.19	12,100	664	14.6	375	61.9	2.56
× 170	50.0	36.17	0.680	12.030	1.100	2.00	10,500	580	14.5	320	53.2	2.53
× 150	44.2	35.85	0.625	11.975	0.940	1.88	9,040	504	14.3	270	45.1	2.47
× 135	39.7	35.55	0.600	11.950	0.790	1.69	7,800	439	14.0	225	37.7	2.38
W 33 × 241	70.9	34.18	0.830	15.860	1.400	2.19	14,200	829	14.1	932	118	3.63
× 201	59.1	33.68	0.715	15.745	1.150	1.94	11,500	684	14.0	749	95.2	3.56
× 152	44.7	33.49	0.635	11.565	1.055	1.88	8,160	487	13.5	273	47.2	2.47
× 130	38.3	33.09	0.580	11.510	0.855	1.69	6,710	406	13.2	218	37.9	2.39
× 118	34.7	32.86	0.550	11.480	0.740	1.56	5,900	359	13.0	187	32.6	2.32
W 30 × 211	62.0	30.94	0.775	15.105	1.315	2.13	10,300	663	12.9	757	100	3.49
× 173	50.8	30.44	0.655	14.985	1.065	1.88	8,200	539	12.7	598	79.8	3.43
× 124	36.5	30.17	0.585	10.515	0.930	1.69	5,360	355	12.1	181	34.4	2.23
× 108	31.7	29.83	0.545	10.475	0.760	1.56	4,470	299	11.9	146	27.9	2.15
× 99	29.1	29.65	0.520	10.450	0.670	1.44	3,990	269	11.7	128	24.5	2.10
W 27 × 178	52.3	27.81	0.725	14.085	1.190	1.88	6,990	502	11.6	555	78.8	3.26
× 146	42.9	27.38	0.605	13.965	0.975	1.69	5,630	411	11.4	443	63.5	3.21
× 102	30.0	27.09	0.515	10.015	0.830	1.56	3,620	267	11.0	139	27.8	2.15
× 84	24.8	26.71	0.460	9.960	0.640	1.38	2,850	213	10.7	106	21.2	2.07
W 24 × 162	47.7	25.00	0.705	12.955	1.220	2.00	5,170	414	10.4	443	68.4	3.05
× 131	38.5	24.48	0.605	12.855	0.960	1.75	4,020	329	10.2	340	53.0	2.97
× 104	30.6	24.06	0.500	12.750	0.750	1.50	3,100	258	10.1	259	40.7	2.91
× 84	24.7	24.10	0.470	9.020	0.770	1.56	2,370	196	9.79	94.4	20.9	1.95
× 69	20.1	23.73	0.415	8.965	0.585	1.38	1,830	154	9.55	70.4	15.7	1.87
× 55	16.2	23.57	0.395	7.005	0.505	1.94	1,350	114	9.11	29.1	8.30	1.34
W 21 × 147	43.2	22.06	0.720	12.510	1.150	1.88	3,630	329	9.17	376	60.1	2.95
× 122	35.9	21.68	0.600	12.390	0.960	1.69	2,960	273	9.09	305	49.2	2.92
× 101	29.8	21.36	0.500	12.290	0.800	1.56	2,420	227	9.02	248	40.3	2.89
× 83	24.3	21.43	0.515	8.355	0.835	1.56	1,830	171	8.67	81.4	19.5	1.83
× 68	20.0	21.13	0.430	8.270	0.685	1.44	1,480	140	8.60	64.7	15.7	1.80
× 57	16.7	21.06	0.405	6.555	0.650	1.38	1,170	111	8.36	30.6	9.35	1.35

TABLE 3.1 (*Continued*)

				Flange			Elastic Properties					
			Web	Width	Thickness		Axis X-X			Axis Y-Y		
	Area A	Depth d	Thickness t_w	b_f	t_f	k	I	S	r	I	S	r
Designation	in²	in.	in.	in.	in.	in.	in⁴	in³	in.	in⁴	in³	in.
× 44	13.0	20.66	0.350	6.500	0.450	1.19	843	81.6	8.06	20.7	6.36	1.26
W 18 × 119	35.1	18.97	0.655	11.265	1.060	1.75	2,190	231	7.90	253	44.9	2.69
× 97	28.5	18.59	0.535	11.145	0.870	1.56	1,750	188	7.82	201	36.1	2.65
× 76	22.3	18.21	0.425	11.035	0.680	1.38	1,330	146	7.73	152	27.6	2.61
× 65	19.1	18.35	0.450	7.590	0.750	1.44	1,070	117	7.49	54.8	14.4	1.69
× 55	16.2	18.11	0.390	7.530	0.630	1.31	890	98.3	7.41	44.9	11.9	1.67
× 46	13.5	18.06	0.360	6.060	0.605	1.25	712	78.8	7.25	22.5	7.43	1.29
× 35	10.3	17.70	0.300	6.000	0.425	1.13	510	57.6	7.04	15.3	5.12	1.22
W 16 × 100	29.4	16.97	0.585	10.425	0.985	1.69	1,490	175	7.10	186	35.7	2.51
× 77	22.6	16.52	0.455	10.295	0.760	1.44	1,110	134	7.00	138	26.9	2.47
× 57	16.8	16.43	0.430	7.120	0.715	1.38	758	92.2	6.72	43.1	12.1	1.60
× 45	13.3	16.13	0.345	7.035	0.565	1.25	586	72.7	6.65	32.8	9.34	1.57
× 36	10.6	15.86	0.295	6.985	0.430	1.13	448	56.5	6.51	24.5	7.00	1.52
× 26	7.68	15.69	0.250	5.500	0.345	1.06	301	38.4	6.26	9.59	3.49	1.12
W 14 × 730	215.0	22.42	3.070	17.890	4.910	5.56	14,300	1280	8.17	4720	527	4.69
× 605	178.0	20.92	2.595	17.415	4.160	4.81	10,800	1040	7.80	3680	423	4.55
× 455	134.0	19.02	2.015	16.835	3.210	3.88	7,190	756	7.33	2560	304	4.38
× 370	109.0	17.92	1.655	16.475	2.660	3.31	5,440	607	7.07	1990	241	4.27
× 283	83.3	16.74	1.290	16.110	2.070	2.75	3,840	459	6.79	1440	179	4.17
× 211	62.0	15.72	0.980	15.800	1.560	2.25	2,660	338	6.55	1030	130	4.07
× 159	46.7	14.98	0.745	15.565	1.190	1.88	1,900	254	6.38	748	96.2	4.00
× 120	35.3	14.48	0.590	14.670	0.940	1.63	1,380	190	6.24	495	67.5	3.74
× 90	26.5	14.02	0.440	14.520	0.710	1.38	999	143	6.14	362	49.9	3.70
× 68	20.0	14.04	0.415	10.035	0.720	1.50	723	103	6.01	121	24.2	2.46
× 53	15.6	13.92	0.370	8.060	0.660	1.44	541	77.8	5.89	57.7	14.3	1.92
× 43	12.6	13.66	0.305	7.995	0.530	1.31	428	62.7	5.82	45.2	11.3	1.89
× 34	10.0	13.98	0.285	6.745	0.455	1.00	340	48.6	5.83	23.3	6.91	1.53
× 30	8.85	13.84	0.270	6.730	0.385	0.94	291	42.0	5.73	19.6	5.82	1.49
× 26	7.69	13.91	0.255	5.025	0.420	0.94	245	35.3	5.65	8.91	3.54	1.08
× 22	6.49	13.74	0.230	5.000	0.335	0.88	199	29.0	5.54	7.00	2.80	1.04
W 12 × 336	98.8	16.82	1.775	13.385	2.955	3.69	4,060	483	6.41	1190	177	3.47
× 279	81.9	15.85	1.530	13.140	2.470	3.19	3,110	393	6.16	937	143	3.38
× 210	61.8	14.71	1.180	12.790	1.900	2.63	2,140	292	5.89	664	104	3.28
× 152	44.7	13.71	0.870	12.480	1.400	2.13	1,430	209	5.66	454	72.8	3.19
× 106	31.2	12.89	0.610	12.220	0.990	1.69	933	145	5.47	301	49.3	3.11
× 79	23.2	12.38	0.470	12.080	0.735	1.44	662	107	5.34	216	35.8	3.05
× 58	17.0	12.19	0.360	10.010	0.640	1.38	475	78.0	5.28	107	21.4	2.51

TABLE 3.1 *(Continued)*

Designation	Area A	Depth d	Web Thickness t_w	Flange Width b_f	Flange Thickness t_f	k	Axis X-X I	Axis X-X S	Axis X-X r	Axis Y-Y I	Axis Y-Y S	Axis Y-Y r
	in²	in.	in.	in.	in.	in.	in⁴	in³	in.	in⁴	in³	in.
× 50	14.7	12.19	0.370	8.080	0.640	1.38	394	64.7	5.18	56.3	13.9	1.96
× 40	11.8	11.94	0.295	8.005	0.515	1.25	310	51.9	5.13	44.1	11.0	1.93
× 30	8.79	12.34	0.260	6.520	0.440	0.94	238	38.6	5.21	20.3	6.24	1.52
× 22	6.48	12.31	0.260	4.030	0.425	0.88	156	25.4	4.91	4.66	2.31	0.847
× 19	5.57	12.16	0.235	4.005	0.350	0.81	130	21.3	4.82	3.76	1.88	0.822
× 16	4.71	11.99	0.220	3.990	0.265	0.75	103	17.1	4.67	2.82	1.41	0.773
× 14	4.16	11.91	0.200	3.970	0.225	0.69	88.6	14.9	4.62	2.36	1.19	0.753
W 10 × 112	32.9	11.36	0.755	10.415	1.250	1.88	716	126	4.66	236	45.3	2.68
× 88	25.9	10.84	0.605	10.265	0.990	1.63	534	98.5	4.54	179	34.8	2.63
× 60	17.6	10.22	0.420	10.080	0.680	1.31	341	66.7	4.39	116	23.0	2.57
× 45	13.3	10.10	0.350	8.020	0.620	1.25	248	49.1	4.32	53.4	13.3	2.01
× 33	9.71	9.73	0.290	7.960	0.435	1.06	170	35.0	4.19	36.6	9.20	1.94
× 26	7.61	10.33	0.260	5.770	0.440	0.88	144	27.9	4.35	14.1	4.89	1.36
× 19	5.62	10.24	0.250	4.020	0.395	0.81	96.3	18.8	4.14	4.29	2.14	0.874
× 15	4.41	9.99	0.230	4.000	0.270	0.69	68.9	13.8	3.95	2.89	1.45	0.810
× 12	3.54	9.87	0.190	3.960	0.210	0.63	53.8	10.9	3.90	2.18	1.10	0.785
W 8 × 67	19.7	9.00	0.570	8.280	0.935	1.44	272	60.4	3.72	88.6	21.4	2.12
× 48	14.1	8.50	0.400	8.110	0.685	1.19	184	43.3	3.61	60.9	15.0	2.08
× 40	11.7	8.25	0.360	8.070	0.560	1.06	146	35.5	3.53	49.1	12.2	2.04
× 35	10.3	8.12	0.310	8.020	0.495	1.00	127	31.2	3.51	42.6	10.6	2.03
× 31	9.13	8.00	0.285	7.995	0.435	0.94	110	27.5	3.47	37.1	9.27	2.02
× 28	8.25	8.06	0.285	6.535	0.465	0.94	98.0	24.3	3.45	21.7	6.63	1.62
× 24	7.08	7.93	0.245	6.495	0.400	0.88	82.8	20.9	3.42	18.3	5.63	1.61
× 21	6.16	8.28	0.250	5.270	0.400	0.81	75.3	18.2	3.49	9.77	3.71	1.26
× 18	5.26	8.14	0.230	5.250	0.330	0.75	61.9	15.2	3.43	7.97	3.04	1.23
× 15	4.44	8.11	0.245	4.015	0.315	0.75	48.0	11.8	3.29	3.41	1.70	0.876
× 13	3.84	7.99	0.230	4.000	0.255	0.69	39.6	9.91	3.21	2.73	1.37	0.843
× 10	2.96	7.89	0.170	3.940	0.205	0.63	30.8	7.81	3.22	2.09	1.06	0.841
W 6 × 25	7.34	6.38	0.320	6.080	0.455	0.81	53.4	16.7	2.70	17.1	5.61	1.52
× 20	5.87	6.20	0.260	6.020	0.365	0.75	41.4	13.4	2.66	13.3	4.41	1.50
× 15	4.43	5.99	0.230	5.990	0.260	0.63	29.1	9.72	2.56	9.32	3.11	1.46
× 16	4.74	6.28	0.260	4.030	0.405	0.75	32.1	10.2	2.60	4.43	2.20	0.966
× 12	3.55	6.03	0.230	4.000	0.280	0.63	22.1	7.31	2.49	2.99	1.50	0.918
× 9	2.68	5.90	0.170	3.940	0.215	0.56	16.4	5.56	2.47	2.19	1.11	0.905
W 5 × 19	5.54	5.15	0.270	5.030	0.430	0.81	26.2	10.2	2.17	9.13	3.63	1.28
× 16	4.68	5.01	0.240	5.000	0.360	0.75	21.3	8.51	2.13	7.51	3.00	1.27
W 4 × 13	3.83	4.16	0.280	4.060	0.345	0.69	11.3	5.46	1.72	3.86	1.90	1.00

Source: Adapted from data in the *Manual of Steel Construction*, 8th ed., with permission of the publishers, American Institute of Steel Construction.

TABLE 3.2 Properties of American Standard Channel (C) Shapes

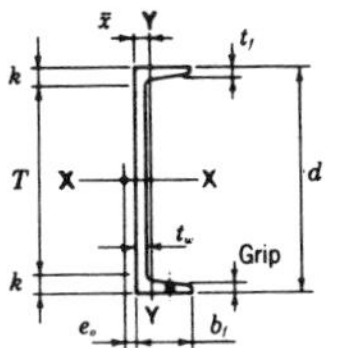

				Flange			Elastic Properties						
				Width	Thickness		Axis X-X			Axis Y-Y			
	Area A	Depth d	Web Thickness t_w	t_f	b_f	k	I	S	r	I	S	r	$\bar{x}$
Designation	in²	in.	in.	in.	in.	in.	in⁴	in³	in.	in⁴	in³	in.	in.
C 15 × 50	14.7	15.00	0.716	3.716	0.650	1.44	404	53.8	5.24	11.0	3.78	0.867	0.798
× 40	11.8	15.00	0.520	3.520	0.650	1.44	349	46.5	5.44	9.23	3.37	0.886	0.777
× 33.9	9.96	15.00	0.400	3.400	0.650	1.44	315	42.0	5.62	8.13	3.11	0.904	0.787
C 12 × 30	8.82	12.00	0.510	3.170	0.501	1.13	162	27.0	4.29	5.14	2.06	0.763	0.674
× 25	7.35	12.00	0.387	3.047	0.501	1.13	144	24.1	4.43	4.47	1.88	0.780	0.674
× 20.7	6.09	12.00	0.282	2.942	0.501	1.13	129	21.5	4.61	3.88	1.73	0.799	0.698
C 10 × 30	8.82	10.00	0.673	3.033	0.436	1.00	103	20.7	3.42	3.94	1.65	0.669	0.649
× 25	7.35	10.00	0.526	2.886	0.436	1.00	91.2	18.2	3.52	3.36	1.48	0.676	0.617
× 20	5.88	10.00	0.379	2.739	0.436	1.00	78.9	15.8	3.66	2.81	1.32	0.692	0.606
× 15.3	4.49	10.00	0.240	2.600	0.436	1.00	67.4	13.5	3.87	2.28	1.16	0.713	0.634
C 9 × 20	5.88	9.00	0.448	2.648	0.413	0.94	60.9	13.5	3.22	2.42	1.17	0.642	0.583
× 15	4.41	9.00	0.285	2.485	0.413	0.94	51.0	11.3	3.40	1.93	1.01	0.661	0.586
× 13.4	3.94	9.00	0.233	2.433	0.413	0.94	47.9	10.6	3.48	1.76	0.962	0.669	0.601
C 8 × 18.75	5.51	8.00	0.487	2.527	0.390	0.94	44.0	11.0	2.82	1.98	1.01	0.599	0.565
× 13.75	4.04	8.00	0.303	2.343	0.390	0.94	36.1	9.03	2.99	1.53	0.854	0.615	0.553
× 11.5	3.38	8.00	0.220	2.260	0.390	0.94	32.6	8.14	3.11	1.32	0.781	0.625	0.571
C 7 × 14.75	4.33	7.00	0.419	2.299	0.366	0.88	27.2	7.78	2.51	1.38	0.779	0.564	0.532
× 12.25	3.60	7.00	0.314	2.194	0.366	0.88	24.2	6.93	2.60	1.17	0.703	0.571	0.525
× 9.8	2.87	7.00	0.210	2.090	0.366	0.88	21.3	6.08	2.72	0.968	0.625	0.581	0.540
C 6 × 13	3.83	6.00	0.437	2.157	0.343	0.81	17.4	5.80	2.13	1.05	0.642	0.525	0.514
× 10.5	3.09	6.00	0.314	2.034	0.343	0.81	15.2	5.06	2.22	0.866	0.564	0.529	0.499
× 8.2	2.40	6.00	0.200	1.920	0.343	0.81	13.1	4.38	2.34	0.693	0.492	0.537	0.511
C 5 × 9	2.64	5.00	0.325	1.885	0.320	0.75	8.90	3.56	1.83	0.632	0.450	0.489	0.478
× 6.7	1.97	5.00	0.190	1.750	0.320	0.75	7.49	3.00	1.95	0.479	0.378	0.493	0.484
C 4 × 7.25	2.13	4.00	0.321	1.721	0.296	0.69	4.59	2.29	1.47	0.433	0.343	0.450	0.459
× 5.4	1.59	4.00	0.184	1.584	0.296	0.69	3.85	1.93	1.56	0.319	0.283	0.449	0.457
C 3 × 6	1.76	3.00	0.356	1.596	0.273	0.69	2.07	1.38	1.08	0.305	0.268	0.416	0.455
× 5	1.47	3.00	0.258	1.498	0.273	0.69	1.85	1.24	1.12	0.247	0.233	0.410	0.438
× 4.1	1.21	3.00	0.170	1.410	0.273	0.69	1.66	1.10	1.17	0.197	0.202	0.404	0.436

Source: Adapted from data in the *Manual of Steel Construction*, 8th ed., with permission of the publishers, American Institute of Steel Construction.

TABLE 3.3 Properties of Angles

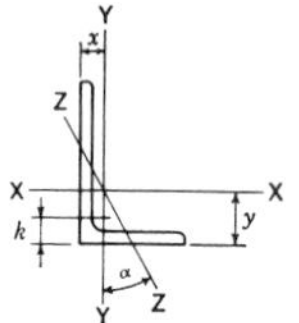

Size and Thickness in.	k in.	Weight lb	Area in²	Axis X-X: I in⁴	Axis X-X: S in³	Axis X-X: r in.	Axis X-X: y in.	Axis X-X: I in⁴	Axis X-X: S in³	Axis X-X: r in.	Axis X-X: x in.	Axis Z-Z: r in.	Axis Z-Z: Tan α
8 × 8 × 1 1/8	1.75	56.9	16.7	98.0	17.5	2.42	2.41	98.0	17.5	2.42	2.41	1.56	1.000
1	1.625	51.0	15.0	89.0	15.8	2.44	2.37	89.0	15.8	2.44	2.37	1.56	1.000
7/8	1.50	45.0	13.2	79.6	14.0	2.45	2.32	79.6	14.0	2.45	2.32	1.57	1.000
3/4	1.375	38.9	11.4	69.7	12.2	2.47	2.28	69.7	12.2	2.47	2.28	1.58	1.000
5/8	1.25	32.7	9.61	59.4	10.3	2.49	2.23	59.4	10.3	2.49	2.23	1.58	1.000
1/2	1.125	26.4	7.75	48.6	8.36	2.50	2.19	48.6	8.36	2.50	2.19	1.59	1.000
8 × 6 × 1	1.50	44.2	13.0	80.8	15.1	2.49	2.65	38.8	8.92	1.73	1.65	1.28	0.543
3/4	1.25	33.8	9.94	63.4	11.7	2.53	2.56	30.7	6.92	1.76	1.56	1.29	0.551
1/2	1.0	23.0	6.75	44.3	8.02	2.56	2.47	21.7	4.79	1.79	1.47	1.30	0.558
8 × 4 × 1	1.50	37.4	11.0	69.6	14.1	2.52	3.05	11.6	3.94	1.03	1.05	0.846	0.247
3/4	1.25	28.7	8.44	54.9	10.9	2.55	2.95	9.36	3.07	1.05	0.953	0.852	0.258
1/2	1.0	19.6	5.75	38.5	7.49	2.59	2.86	6.74	2.15	1.08	0.859	0.865	0.267
7 × 4 × 3/4	1.25	26.2	7.69	37.8	8.42	2.22	2.51	9.05	3.03	1.09	1.01	0.860	0.324
1/2	1.0	17.9	5.25	26.7	5.81	2.25	2.42	6.53	2.12	1.11	0.917	0.872	0.335
3/8	0.875	13.6	3.98	20.6	4.44	2.27	2.37	5.10	1.63	1.13	0.870	0.880	0.340
6 × 6 × 1	1.50	37.4	11.0	35.5	8.57	1.80	1.86	35.5	8.57	1.80	1.86	1.17	1.000
7/8	1.375	33.1	9.73	31.9	7.63	1.81	1.82	31.9	7.63	1.81	1.82	1.17	1.000
3/4	1.25	28.7	8.44	28.2	6.66	1.83	1.78	28.2	6.66	1.83	1.78	1.17	1.000
5/8	1.125	24.2	7.11	24.2	5.66	1.84	1.73	24.2	5.66	1.84	1.73	1.18	1.000
1/2	1.0	19.6	5.75	19.9	4.61	1.86	1.68	19.9	4.61	1.86	1.68	1.18	1.000
3/8	0.875	14.9	4.36	15.4	3.53	1.88	1.64	15.4	3.53	1.88	1.64	1.19	1.000
6 × 4 × 3/4	1.25	23.6	6.94	24.5	6.25	1.88	2.08	8.68	2.97	1.12	1.08	0.860	0.428
5/8	1.125	20.0	5.86	21.1	5.31	1.90	2.03	7.52	2.54	1.13	1.03	0.864	0.435
1/2	1.00	16.2	4.75	17.4	4.33	1.91	1.99	6.27	2.08	1.15	0.987	0.870	0.440
3/8	0.875	12.3	3.61	13.5	3.32	1.93	1.94	4.90	1.60	1.17	0.941	0.877	0.446
6 × 3 1/2 × 3/8	0.875	11.7	3.42	12.9	3.24	1.94	2.04	3.34	1.23	0.988	0.787	0.767	0.350
5/16	0.8125	9.8	2.87	10.9	2.73	1.95	2.01	2.85	1.04	0.996	0.763	0.772	0.352
5 × 5 × 7/8	1.375	27.2	7.98	17.8	5.17	1.49	1.57	17.8	5.17	1.49	1.57	0.973	1.000
3/4	1.25	23.6	6.94	15.7	4.53	1.51	1.52	15.7	4.53	1.51	1.52	0.975	1.000
1/2	1.0	16.2	4.75	11.3	3.16	1.54	1.43	11.3	3.16	1.54	1.43	0.983	1.000
5 × 5 × 3/8	0.875	12.3	3.61	8.74	2.42	1.56	1.39	8.74	2.42	1.56	1.39	0.990	1.000
5/16	0.8125	10.3	3.03	7.42	2.04	1.57	1.37	7.42	2.04	1.57	1.37	0.994	1.000
5 × 3 1/2 × 3/4	1.25	19.8	5.81	13.9	4.28	1.55	1.75	5.55	2.22	0.977	0.996	0.748	0.464
1/2	1.0	13.6	4.00	9.99	2.99	1.58	1.66	4.05	1.56	1.01	0.906	0.755	0.479
3/8	0.875	10.4	3.05	7.78	2.29	1.60	1.61	3.18	1.21	1.02	0.861	0.762	0.486
5/16	0.8125	8.7	2.56	6.60	1.94	1.61	1.59	2.72	1.02	1.03	0.838	0.766	0.489
5 × 3 × 1/2	1.0	12.8	3.75	9.45	2.91	1.59	1.75	2.58	1.15	0.829	0.750	0.648	0.357
3/8	0.875	9.8	2.86	7.37	2.24	1.61	1.70	2.04	0.888	0.845	0.704	0.654	0.364
5/16	0.8125	8.2	2.40	6.26	1.89	1.61	1.68	1.75	0.753	0.853	0.681	0.658	0.368
1/4	0.75	6.6	1.94	5.11	1.53	1.62	1.66	1.44	0.614	0.861	0.657	0.663	0.371

TABLE 3.3 *(Continued)*

Size and Thickness in.	k in.	Weight lb	Area in²	Axis X-X I in⁴	S in³	r in.	y in.	Axis Y-Y I in⁴	S in³	r in.	x in.	Axis Z-Z r in.	Tan α
4 × 4 × 3/4	1.125	18.5	5.44	7.67	2.81	1.19	1.27	7.67	2.81	1.19	1.27	0.778	1.000
5/8	1.0	15.7	4.61	6.66	2.40	1.20	1.23	6.66	2.40	1.20	1.23	0.779	1.000
1/2	0.875	12.8	3.75	5.56	1.97	1.22	1.18	5.56	1.97	1.22	1.18	0.782	1.000
3/8	0.75	9.8	2.86	4.36	1.52	1.23	1.14	4.36	1.52	1.23	1.14	0.788	1.000
5/16	0.6875	8.2	2.40	3.71	1.29	1.24	1.12	3.71	1.29	1.24	1.12	0.791	1.000
1/4	0.625	6.6	1.94	3.04	1.05	1.25	1.09	3.04	1.05	1.25	1.09	0.795	1.000
4 × 3 1/2 × 1/2	0.9375	11.9	3.50	5.32	1.94	1.23	1.25	3.79	1.52	1.04	1.00	0.722	0.750
3/8	0.8125	9.1	2.67	4.18	1.49	1.25	1.21	2.95	1.17	1.06	0.955	0.727	0.755
5/16	0.75	7.7	2.25	3.56	1.26	1.26	1.18	2.55	0.994	1.07	0.932	0.730	0.757
1/4	0.6875	6.2	1.81	2.91	1.03	1.27	1.16	2.09	0.808	1.07	0.909	0.734	0.759
4 × 3 × 1/2	0.9375	11.1	3.25	5.05	1.89	1.25	1.33	2.42	1.12	0.864	0.827	0.639	0.543
3/8	0.8125	8.5	2.48	3.96	1.46	1.26	1.28	1.92	0.866	0.879	0.782	0.644	0.551
5/16	0.75	7.2	2.09	3.38	1.23	1.27	1.26	1.65	0.734	0.887	0.759	0.647	0.554
1/4	0.6875	5.8	1.69	2.77	1.00	1.28	1.24	1.36	0.599	0.896	0.736	0.651	0.558
3 1/2 × 3 1/2 × 3/8	0.75	8.5	2.48	2.87	1.15	1.07	1.01	2.87	1.15	1.07	1.01	0.687	1.000
5/16	0.6875	7.2	2.09	2.45	0.976	1.08	0.990	2.45	0.976	1.08	0.990	0.690	1.000
1/4	0.625	5.8	1.69	2.01	0.794	1.09	0.968	2.01	0.794	1.09	0.968	0.694	1.000
3 1/2 × 3 × 3/8	0.8125	7.9	2.30	2.72	1.13	1.09	1.08	1.85	0.851	0.897	0.830	0.625	0.721
5/16	0.75	6.6	1.93	2.33	0.954	1.10	1.06	1.58	0.722	0.905	0.808	0.627	0.724
1/4	0.6875	5.4	1.56	1.91	0.776	1.11	1.04	1.30	0.589	0.914	0.785	0.631	0.727
3 1/2 × 2 1/2 × 3/8	0.8125	7.2	2.11	2.56	1.09	1.10	1.16	1.09	0.592	0.719	0.660	0.537	0.496
5/16	0.75	6.1	1.78	2.19	0.927	1.11	1.14	0.939	0.504	0.727	0.637	0.540	0.501
1/4	0.6875	4.9	1.44	1.80	0.755	1.12	1.11	0.777	0.412	0.735	0.614	0.544	0.506
3 × 3 × 1/2	0.8125	9.4	2.75	2.22	1.07	0.898	0.932	2.22	1.07	0.898	0.932	0.584	1.000
3/8	0.6875	7.2	2.11	1.76	0.833	0.913	0.888	1.76	0.833	0.913	0.888	0.587	1.000
5/16	0.625	6.1	1.78	1.51	0.707	0.922	0.865	1.51	0.707	0.922	0.865	0.589	1.000
1/4	0.5625	4.9	1.44	1.24	0.577	0.930	0.842	1.24	0.577	0.930	0.842	0.592	1.000
3/16	0.5000	3.71	1.09	0.962	0.441	0.939	0.820	0.962	0.441	0.939	0.820	0.596	1.000
3 × 2 1/2 × 3/8	0.75	6.6	1.92	1.66	0.810	0.928	0.956	1.04	0.581	0.736	0.706	0.522	0.676
1/4	0.625	4.5	1.31	1.17	0.561	0.945	0.911	0.743	0.404	0.753	0.661	0.528	0.684
3/16	0.5625	3.39	0.996	0.907	0.430	0.954	0.888	0.577	0.310	0.761	0.638	0.533	0.688
3 × 2 × 3/8	0.6875	5.9	1.73	1.53	0.781	0.940	1.04	0.543	0.371	0.559	0.539	0.430	0.428
5/16	0.625	5.0	1.46	1.32	0.664	0.948	1.02	0.470	0.317	0.567	0.516	0.432	0.435
1/4	0.5625	4.1	1.19	1.09	0.542	0.957	0.993	0.392	0.260	0.574	0.493	0.435	0.440
3/16	0.5000	3.07	0.902	0.842	0.415	0.966	0.970	0.307	0.200	0.583	0.470	0.439	0.446
2 1/2 × 2 1/2 × 3/8	0.6875	5.9	1.73	0.984	0.566	0.753	0.762	0.984	0.566	0.753	0.762	0.487	1.000
5/16	0.625	5.0	1.46	0.849	0.482	0.761	0.740	0.849	0.482	0.761	0.740	0.489	1.000
1/4	0.5625	4.1	1.19	0.703	0.394	0.769	0.717	0.703	0.394	0.769	0.717	0.491	1.000
3/16	0.5000	3.07	0.902	0.547	0.303	0.778	0.694	0.547	0.303	0.778	0.694	0.495	1.000
2 1/2 × 2 × 3/8	0.6875	5.3	1.55	0.912	0.547	0.768	0.831	0.514	0.363	0.577	0.581	0.420	0.614
5/16	0.625	4.5	1.31	0.788	0.466	0.776	0.809	0.446	0.310	0.584	0.559	0.422	0.620
1/4	0.5625	3.62	1.06	0.654	0.381	0.784	0.787	0.372	0.254	0.592	0.537	0.424	0.626
3/16	0.5000	2.75	0.809	0.509	0.293	0.793	0.764	0.291	0.196	0.600	0.514	0.427	0.631
2 × 2 × 3/8	0.6875	4.7	1.36	0.479	0.351	0.594	0.636	0.479	0.351	0.594	0.636	0.389	1.000
5/16	0.625	3.92	1.15	0.416	0.300	0.601	0.614	0.416	0.300	0.601	0.614	0.390	1.000
1/4	0.5625	3.19	0.938	0.348	0.247	0.609	0.592	0.348	0.247	0.609	0.592	0.391	1.000
3/16	0.5000	2.44	0.715	0.272	0.190	0.617	0.569	0.272	0.190	0.617	0.569	0.394	1.000

Source: Abstracted from the *Manual of Steel Construction*, 8th ed., with permission of the publishers, American Institute of Steel Construction.

TABLE 3.4 Properties of Round Steel Pipe

Dimensions				Weight per Foot Lbs. Plain Ends	Properties			
Nominal Diameter In.	Outside Diameter In.	Inside Diameter In.	Wall Thickness In.		A In.^{2}	I In.^{4}	S In.^{3}	r In.
Standard Weight								
½	.840	.622	.109	.85	.250	.017	.041	.261
¾	1.050	.824	.113	1.13	.333	.037	.071	.334
1	1.315	1.049	.133	1.68	.494	.087	.133	.421
1¼	1.660	1.380	.140	2.27	.669	.195	.235	.540
1½	1.900	1.610	.145	2.72	.799	.310	.326	.623
2	2.375	2.067	.154	3.65	1.07	.666	.561	.787
2½	2.875	2.469	.203	5.79	1.70	1.53	1.06	.947
3	3.500	3.068	.216	7.58	2.23	3.02	1.72	1.16
3½	4.000	3.548	.226	9.11	2.68	4.79	2.39	1.34
4	4.500	4.026	.237	10.79	3.17	7.23	3.21	1.51
5	5.563	5.047	.258	14.62	4.30	15.2	5.45	1.88
6	6.625	6.065	.280	18.97	5.58	28.1	8.50	2.25
8	8.625	7.981	.322	28.55	8.40	72.5	16.8	2.94
10	10.750	10.020	.365	40.48	11.9	161	29.9	3.67
12	12.750	12.000	.375	49.56	14.6	279	43.8	4.38
Extra Strong								
½	.840	.546	.147	1.09	.320	.020	.048	.250
¾	1.050	.742	.154	1.47	.433	.045	.085	.321
1	1.315	.957	.179	2.17	.639	.106	.161	.407
1¼	1.660	1.278	.191	3.00	.881	.242	.291	.524
1½	1.900	1.500	.200	3.63	1.07	.391	.412	.605
2	2.375	1.939	.218	5.02	1.48	.868	.731	.766
2½	2.875	2.323	.276	7.66	2.25	1.92	1.34	.924
3	3.500	2.900	.300	10.25	3.02	3.89	2.23	1.14
3½	4.000	3.364	.318	12.50	3.68	6.28	3.14	1.31
4	4.500	3.826	.337	14.98	4.41	9.61	4.27	1.48
5	5.563	4.813	.375	20.78	6.11	20.7	7.43	1.84
6	6.625	5.761	.432	28.57	8.40	40.5	12.2	2.19
8	8.625	7.625	.500	43.39	12.8	106	24.5	2.88
10	10.750	9.750	.500	54.74	16.1	212	39.4	3.63
12	12.750	11.750	.500	65.42	19.2	362	56.7	4.33
Double-Extra Strong								
2	2.375	1.503	.436	9.03	2.66	1.31	1.10	.703
2½	2.875	1.771	.552	13.69	4.03	2.87	2.00	.844
3	3.500	2.300	.600	18.58	5.47	5.99	3.42	1.05
4	4.500	3.152	.674	27.54	8.10	15.3	6.79	1.37
5	5.563	4.063	.750	38.55	11.3	33.6	12.1	1.72
6	6.625	4.897	.864	53.16	15.6	66.3	20.0	2.06
8	8.625	6.875	.875	72.42	21.3	162	37.6	2.76

The listed sections are available in conformance with ASTM Specification A53 Grade B or A501. Other sections are made to these specifications. Consult with pipe manufacturers or distributors for availability.

Source: Reprinted from the *Manual of Steel Construction*, 8th ed. (Ref. 2), with permission of the publishers, American Institute of Steel Construction.

TABLE 3.5 Properties of Structural Steel Tubing (TS)

DIMENSIONS				PROPERTIES**			
Nominal* Size	Wall Thickness		Weight per Foot	Area	I	S	r
In.	In.		Lb.	In.²	In.⁴	In.³	In.
16 x 16	.5000	1/2	103.30	30.4	1200	150	6.29
	.3750	3/8	78.52	23.1	931	116	6.35
	.3125	5/16	65.87	19.4	789	98.6	6.38
14 x 14	.5000	1/2	89.68	26.4	791	113	5.48
	.3750	3/8	68.31	20.1	615	87.9	5.54
	.3125	5/16	57.36	16.9	522	74.6	5.57
12 x 12	.5000	1/2	76.07	22.4	485	80.9	4.66
	.3750	3/8	58.10	17.1	380	63.4	4.72
	.3125	5/16	48.86	14.4	324	54.0	4.75
	.2500	1/4	39.43	11.6	265	44.1	4.78
10 x 10	.6250	5/8	76.33	22.4	321	64.2	3.78
	.5000	1/2	62.46	18.4	271	54.2	3.84
	.3750	3/8	47.90	14.1	214	42.9	3.90
	.3125	5/16	40.35	11.9	183	36.7	3.93
	.2500	1/4	32.63	9.59	151	30.1	3.96
8 x 8	.6250	5/8	59.32	17.4	153	38.3	2.96
	.5000	1/2	48.85	14.4	131	32.9	3.03
	.3750	3/8	37.69	11.1	106	26.4	3.09
	.3125	5/16	31.84	9.36	90.9	22.7	3.12
	.2500	1/4	25.82	7.59	75.1	18.8	3.15
	.1875	3/16	19.63	5.77	58.2	14.6	3.18
7 x 7	.5000	1/2	42.05	12.4	84.6	24.2	2.62
	.3750	3/8	32.58	9.58	68.7	19.6	2.68
	.3125	5/16	27.59	8.11	59.5	17.0	2.71
	.2500	1/4	22.42	6.59	49.4	14.1	2.74
	.1875	3/16	17.08	5.02	38.5	11.0	2.77
6 x 6	.5000	1/2	35.24	10.4	50.5	16.8	2.21
	.3750	3/8	27.48	8.08	41.6	13.9	2.27
	.3125	5/16	23.34	6.86	36.3	12.1	2.30
	.2500	1/4	19.02	5.59	30.3	10.1	2.33
	.1875	3/16	14.53	4.27	23.8	7.93	2.36
5 x 5	.5000	1/2	28.43	8.36	27.0	10.8	1.80
	.3750	3/8	22.37	6.58	22.8	9.11	1.86
	.3125	5/16	19.08	5.61	20.1	8.02	1.89
	.2500	1/4	15.62	4.59	16.9	6.78	1.92
	.1875	3/16	11.97	3.52	13.4	5.36	1.95
4 x 4	.5000	1/2	21.63	6.36	12.3	6.13	1.39
	.3750	3/8	17.27	5.08	10.7	5.35	1.45
	.3125	5/16	14.83	4.36	9.58	4.79	1.48
	.2500	1/4	12.21	3.59	8.22	4.11	1.51
	.1875	3/16	9.42	2.77	6.59	3.30	1.54
3.5 x 3.5	.3125	5/16	12.70	3.73	6.09	3.48	1.28
	.2500	1/4	10.51	3.09	5.29	3.02	1.31
	.1875	3/16	8.15	2.39	4.29	2.45	1.34
3 x 3	.3125	5/16	10.58	3.11	3.58	2.39	1.07
	.2500	1/4	8.81	2.59	3.16	2.10	1.10
	.1875	3/16	6.87	2.02	2.60	1.73	1.13
2.5 x 2.5	.2500	1/4	7.11	2.09	1.69	1.35	.899
	.1875	3/16	5.59	1.64	1.42	1.14	.930
2 x 2	.2500	1/4	5.41	1.59	.766	.766	.694
	.1875	3/16	4.32	1.27	.668	.668	.726

* Outside dimensions across flat sides.

** Properties are based upon a nominal outside corner radius equal to two times the wall thickness.

Source: Reprinted from the *Manual of Steel Construction,* 8th ed. (Ref. 2), with permission of the publishers, American Institute of Steel Construction.

TABLE 3.6 Properties of Structural Lumber

Dimensions (in.)		Area	Section Modulus	Moment of Inertia	
Nominal b h	Actual b h	A (in.^{2})	S (in.^{3})	I (in.^{4})	Weight[a] (lb/ft)
2 × 3	1.5 × 2.5	3.75	1.563	1.953	0.9
2 × 4	1.5 × 3.5	5.25	3.063	5.359	1.3
2 × 6	1.5 × 5.5	8.25	7.563	20.797	2.0
2 × 8	1.5 × 7.25	10.875	13.141	47.635	2.6
2 × 10	1.5 × 9.25	13.875	21.391	98.932	3.4
2 × 12	1.5 × 11.25	16.875	31.641	177.979	4.1
2 × 14	1.5 × 13.25	19.875	43.891	290.775	4.8
3 × 2	2.5 × 1.5	3.75	0.938	0.703	0.9
3 × 4	2.5 × 3.5	8.75	5.104	8.932	2.1
3 × 6	2.5 × 5.5	13.75	12.604	34.661	3.3
3 × 8	2.5 × 7.25	18.125	21.901	79.391	4.4
3 × 10	2.5 × 9.25	23.125	35.651	164.886	5.6
3 × 12	2.5 × 11.25	28.125	52.734	296.631	6.8
3 × 14	2.5 × 13.25	33.125	73.151	484.625	8.1
3 × 16	2.5 × 15.25	38.125	96.901	738.870	9.3
4 × 2	3.5 × 1.5	5.25	1.313	0.984	1.3
4 × 3	3.5 × 2.5	8.75	3.646	4.557	2.1
4 × 4	3.5 × 3.5	12.25	7.146	12.505	3.0
4 × 6	3.5 × 5.5	19.25	17.646	48.526	4.7
4 × 8	3.5 × 7.25	23.375	30.661	111.148	6.2
4 × 10	3.5 × 9.25	32.375	49.911	230.840	7.9
4 × 12	3.5 × 11.25	39.375	73.828	415.283	9.6
4 × 14	3.5 × 13.25	46.375	102.411	678.475	11.3
4 × 16	3.5 × 15.25	53.375	135.661	1034.418	13.0
6 × 2	5.5 × 1.5	8.25	2.063	1.547	2.0
6 × 3	5.5 × 2.5	13.75	5.729	7.161	3.3
6 × 4	5.5 × 3.5	19.25	11.229	19.651	4.7
6 × 6	5.5 × 5.5	30.25	27.729	76.255	7.4
6 × 8	5.5 × 7.5	41.25	51.563	193.359	10.0
6 × 10	5.5 × 9.5	52.25	82.729	392.963	12.7
6 × 12	5.5 × 11.5	63.25	121.229	697.068	15.4
6 × 14	5.5 × 13.5	74.25	167.063	1127.672	18.0

TABLE 3.6 (*Continued*)

Dimensions (in.) Nominal b h	Actual b h	Area A (in.^{2})	Section Modulus S (in.^{3})	Moment of Inertia I (in.^{4})	Weight[a] (lb/ft)
6 × 16	5.5 × 15.5	85.25	220.229	1706.776	20.7
8 × 2	7.25 × 1.5	10.875	2.719	2.039	2.6
8 × 3	7.25 × 2.5	18.125	7.552	9.440	4.4
8 × 4	7.25 × 3.5	25.375	14.802	25.904	6.2
8 × 6	7.5 × 5.5	41.25	37.813	103.984	10.0
8 × 8	7.5 × 7.5	56.25	70.313	263.672	13.7
8 × 10	7.5 × 9.5	71.25	112.813	535.859	17.3
8 × 12	7.5 × 11.5	86.25	165.313	950.547	21.0
8 × 14	7.5 × 13.5	101.25	227.813	1537.734	24.6
8 × 16	7.5 × 15.5	116.25	300.313	2327.422	28.3
8 × 18	7.5 × 17.5	131.25	382.813	3349.609	31.9
8 × 20	7.5 × 19.5	146.25	475.313	4634.297	35.5
10 × 10	9.5 × 9.5	90.25	142.896	678.755	21.9
10 × 12	9.5 × 11.5	109.25	209.396	1204.026	26.6
10 × 14	9.5 × 13.5	128.25	288.563	1947.797	31.2
10 × 16	9.5 × 15.5	147.25	380.396	2948.068	35.8
10 × 18	9.5 × 17.5	166.25	484.896	4242.836	40.4
10 × 20	9.5 × 19.5	185.25	602.063	5870.109	45.0
12 × 12	11.5 × 11.5	132.25	253.479	1457.505	32.1
12 × 14	11.5 × 13.5	155.25	349.313	2357.859	37.7
12 × 16	11.5 × 15.5	178.25	460.479	3568.713	43.3
12 × 18	11.5 × 17.5	201.25	586.979	5136.066	48.9
12 × 20	11.5 × 19.5	224.25	728.813	7105.922	54.5
12 × 22	11.5 × 21.5	247.25	885.979	9524.273	60.1
12 × 24	11.5 × 23.5	270.25	1058.479	12437.129	65.7
14 × 14	13.5 × 13.5	182.25	410.063	2767.922	44.3
16 × 16	15.5 × 15.5	240.25	620.646	4810.004	58.4

Source: Compiled from data in the *National Design Specification for Wood Construction*, 1986 ed. (Ref. 4), with permission of the publishers, National Forest Products Association.

[a] Based on an assumed average weight of 35 lb/ft^3.

WOOD CONSTRUCTION

Wood is the structural material of choice in the United States whenever conditions favor its use. For small-sized buildings—whenever fire codes permit—it is almost exclusively used. This has to do partly with the ready availability of good trees and the long-established timber and lumber industries, but mostly with the long habits of usage and the easy, low-skilled nature of the construction work. Industrialization and product developments have steadily simplified and standardized the materials and products used for construction and the means for building erection to the point where the systems, as utilized most of the time, are highly refined and ultimately routine.

4

Wood Spanning Elements

Horizontal spanning systems used for roofs and floors commonly employ a variety of wood products. The solid wood material is used directly for structural lumber in the form of standard-sized elements, such as the all-purpose 2 by 4 (here designated 2 × 4). For large or specially formed members, solid wood pieces may be glued together to form glued laminated elements. The plywood sheet in panel form (most commonly in a 48 × 96-in. size) is still widely used for roof and floor decking and wall sheathing. More recently, various processed materials have used the wood as a raw material, reducing it to a basic fibrous form and reconstituting it in various forms—as particleboard or cemented-fiber panels. This chapter deals with basic considerations in the use of common products for wood spanning structures for roofs and floors.

4.1 Structural Lumber

Unlike the metals, wood is not a processed material but an organic material generally used in its natural state. Aside from the natural properties of the species, the most important factors that influence its strength are density, natural defects (knots, checks, slope of grain, etc.), and moisture content. Because the effects of natural defects on the strength of lumber vary with the type of loading to which an individual piece is subjected, structural lumber is classified according to its *size* and *use*. The three major classifications are:

1. *Joists and Planks.* Rectangular cross sections with nominal dimensions 2–4 in. thick and 4 or more in. wide, graded primarily for strength in bending edgewise or flatwise.
2. *Beams and Stringers.* Rectangular cross sections with nominal dimensions 5 × 8 in. and larger, graded for strength in bending when loaded on the narrow face.
3. *Posts and Timbers.* Square or nearly square cross sections with nominal dimensions 5 × 5 in. and larger, graded primarily for use as posts or columns but adapted to other uses where bending strength is not especially important.

The two groups of trees used for building purposes are the *softwoods* and the *hardwoods*. Softwoods such as the pines or cypress are coniferous or cone bearing, whereas hardwoods have broad leaves, as exemplified by the oaks and maples. The terms softwood and hardwood are not accurate indications of the degree of hardness of the various species of trees. Certain softwoods are as hard as the medium density hardwoods, while some species of hardwoods have softer wood than some of the softwood species. The two species of trees used most extensively in the United States for structural members are Douglas fir and Southern pine, both of which are classified among the softwoods.

Nominal and Dressed Sizes. *Note:* For sake of brevity, SI units have been omitted from the text discussion and most tabular data. However, example computations and exercise problems are presented with data in both U.S. and SI units.

An individual piece of structural lumber is designated by its *nominal* cross-sectional dimensions; the size is indicated by the breadth and depth of the cross section in inches. As an example, we speak of a "6 by 12" (written 6 × 12), and by this we mean a timber with a nominal breadth of 6 in. and depth of 12 in.; the length is variable. However, after being dressed or surfaced on four sides (S4S), the actual dimensions of this piece are $5\frac{1}{2} \times 11\frac{1}{2}$ in. Since lumber used in structural design is almost exclusively dressed lumber, the sectional properties (A, I, and S) given in Table 3.6 are for standard dressed sizes conforming to those established by the industry.

Allowable Stresses for Structural Lumber. Many factors are taken into account in determining the allowable unit stresses for structural lumber. Numerous tests by the Forest Products Laboratory of the U.S. Department of Agriculture made on material free from defects have resulted in a tabulation known as *clear wood strength values*. To obtain allowable design stresses, we must reduce the clear wood values by factors that take into consideration the loss of strength from defects, size and position of knots, size of member, degree of density, and condition of seasoning. These adjustments are made in accordance with various industry grading standards. Grading is necessary to identify lumber quality. Individual grades are given a commercial designation, such as No. 1, No. 2, select structural, dense No. 2, etc., to which a schedule of allowable unit stresses is assigned.

Table 4.1, which has been compiled from more extensive data given in the 1986 edition of *Design Values for Wood Construction,* lists some of the most commonly used stress-grade woods and their allowable working stresses. The working stresses tabulated therein are for normal loading conditions and are applicable to lumber that will be used under continuously dry conditions, such as exist in most covered structures. Where wet conditions exist, and in situations where a member is fully stressed to the maximum allowable unit stress for many years (full load permanently applied), the allowable stress values in Table 4.1 are subject to adjustments. Methods for making such adjustments are given in the reference cited above. The stresses given are: extreme fiber in bending F_b; tension parallel to grain F_t; horizontal shear F_v; compression perpendicular to grain $F_{c\perp}$; compression parallel to grain F_c; and modulus of elasticity E. In addition, there is a column on size classification, which relates to the three major classifications described at the beginning of this section.

TABLE 4.1 Design Values for Structural Lumber of Douglas Fir–Larch[a]

Species and Commercial Grade	Size Classification	Design Values (psi): Extreme Fiber in Bending, F_b: Single Member Uses	Extreme Fiber in Bending, F_b: Repetitive Member Uses	Tension Parallel to Grain, F_t	Horizontal Shear, F_v	Compression Perpendicular to Grain, $F_{c\perp}$	Compression Parallel to Grain, F_c	Modulus of Elasticity, E
Dense select structural	2 to 4 in. thick,	2450	2800	1400	95	730	1850	1,900,000
Select structural	2 to 4 in. wide	2100	2400	1200	95	625	1600	1,800,000
Dense No. 1		2050	2400	1200	95	730	1450	1,900,000
No. 1		1750	2050	1050	95	625	1250	1,800,000
Dense No. 2		1700	1950	1000	95	730	1150	1,700,000
No. 2		1450	1650	850	95	625	1000	1,700,000
No. 3		800	925	475	95	625	600	1,500,000
Appearance		1750	2050	1050	95	625	1500	1,800,000
Stud		800	925	475	95	625	600	1,500,000
Construction	2 to 4 in. thick,	1050	1200	625	95	625	1150	1,500,000
Standard	4 in. wide	600	675	350	95	625	925	1,500,000
Utility		275	325	175	95	625	600	1,500,000

Dense select structural	2 to 4 in. thick, 5 in. and wider	2100	2400	1400	95	730	1650	1,900,000
Select structural		1800	2050	1200	95	625	1400	1,800,000
Dense No. 1		1800	2050	1200	95	730	1450	1,900,000
No. 1		1500	1750	1000	95	625	1250	1,800,000
Dense No. 2		1450	1700	775	95	730	1250	1,700,000
No. 2		1250	1450	650	95	625	1050	1,700,000
No. 3		725	850	375	95	625	675	1,500,000
Appearance		1500	1750	1000	95	625	1500	1,800,000
Stud		725	850	375	95	625	675	1,500,000
Dense select structural	Beams and stringers	1900	—	1100	85	730	1300	1,700,000
Select structural		1600	—	950	85	625	1100	1,600,000
Dense No. 1		1550	—	775	85	730	1100	1,700,000
No. 1		1300	—	675	85	625	925	1,600,000
Dense select structural	Posts and timbers	1750	—	1150	85	730	1350	1,700,000
Select structural		1500	—	1000	85	625	1150	1,600,000
Dense No. 1		1400	—	950	85	730	1200	1,700,000
No. 1		1200	—	825	85	625	1000	1,600,000
Select Dex	Decking	1750	2000	—	—	625	—	1,800,000
Commercial Dex		1450	1650	—	—	625	—	1,700,000

Source: Data adapted from *National Design Specification for Wood Construction*, 1986 ed. (Ref. 4), with permission of the publishers, National Forest Products Association. The table in the reference document lists values for 29 other wood species and has 19 footnotes.

[a] Values listed are for normal duration loading with wood that is surfaced dry or green and used at 19% maximum moisture content.

It will be noted that two sets of values are given in the table for F_b, the extreme fiber stress in bending. The values listed for single-member uses apply where an individual beam or girder carries its full design load; the values given for repetitive-member uses are intended for design of members in bending, such as joists, rafters, or similar members, that are spaced not more than 24 in., are three or more in number, and are joined by floor, roof, and other load-distributing construction adequate to support the design load.

The allowable unit stresses to be used in actual design practice must, of course, conform to the requirements of the local building code. As noted earlier, many municipal codes are revised only infrequently and, consequently, may not be in agreement with current editions of industry-recommended allowable stresses. However, the allowable stresses for wood construction used throughout this book are those given in the National Design Specification and recommended by the National Forest Products Association.

Modified Design Values. Basic allowable values as given in Table 4.1 may be used without modification in some instances. However, in many situations there are additional considerations that require some adjustment of these values. Some adjustments are described in the footnotes to the table in the reference document (Ref. 4) from which Table 4.1 is adapted. Others are described in the body of the standard of which the table for design values is a part. Some of the considerations are the following:

1. *Moisture.* Stress values may be increased in some cases if the wood is in a better-cured condition than that assumed in the table. On the other hand, stresses may be reduced for some other usage conditions, notably that of full exterior exposure to the weather.
2. *Repetitive Use.* This increase in the allowable bending stress is permitted when the individual member is one of a closely spaced set, as in the case of rafters, joists, and studs. The qualification is limited to members at least three in number and not spaced more than 24 in. center to center.

TABLE 4.2 Modification Factors for Design Values for Structural Lumber

Duration of Load and General Use	Multiply Design Values by:
10 years or more at the full load limit of a member (as for members carrying only dead load, such as headers in walls)	0.90
2 months' duration (as for snow)	1.15
7 days' duration (as for roof load where no snow pack is incurred)	1.25
Maximum force of wind or earthquake	1.33
Impact (e.g., wheel bumps, braking of moving equipment, or slamming of heavy doors)	2.00

Source: Adapted from specifications in *National Design Specification for Wood Construction,* 1986 ed. (Ref. 4), with permission of the publishers, National Forest Products Association.

3. *Duration of Load.* These adjustments are made on the basis of the time character of the load, as described in Table 4.2. Although they are actually load adjustments, and are sometimes used in the form of direct adjustment of loads in the strength design method (sometimes called the factored load method), the usual technique with wood is to modify the design stress values.
4. *Other Modifications.* Modification is also made for wood that is specially treated with preservatives or fire retardants, for slenderness effects in columns and thin joists or beams, and for size of large beams.

For real design situations, designers must be aware of the usage conditions and the various modifications that are applicable to specific structural members. Some examples of modifications are incorporated in the design examples in Part V of this book.

4.2 Design for Bending

The design of a wood beam for strength in bending is accomplished by use of the flexure formula (Sec. 2.10). The form of this equation used in design is

$$S = \frac{M}{F_b}$$

in which M = maximum bending moment,
F_b = allowable extreme fiber (bending) stress,
S = required section modulus.

Although section moduli for standard rectangular wood beam sizes are given in Table 3.6, it is sometimes convenient to use the formula $S = bd^2/6$, which was developed in Sec. 3.5.

To determine the dimensions of a wood beam as governed by bending, first compute the maximum bending moment. Next, refer to a table such as Table 4.1, select the species and grade of lumber that is to be used, and note the corresponding allowable extreme fiber stress F_b. Substitute these values in the flexure formula, and solve for the required section modulus. The proper beam size may be determined by referring to Table 3.6, which lists S for standard dressed sizes of structural lumber. Obviously, a number of different sections may be acceptable.

Example 1. A simple beam has a span of 16 ft [4.88 m] and supports a load, including its own weight, of 6500 lb [28.9 kN]. If the wood to be used is Douglas fir–larch, select structural grade, determine the size of the beam with the least cross-sectional area on the basis of limiting bending stress.

Solution: The maximum bending moment for the simple beam, case 2 of Fig. 2.27, is

$$M = \frac{WL}{8} = \frac{6500 \times 16}{8} = 13{,}000 \text{ ft-lb } [17.63 \text{ kN-m}]$$

Referring to Table 4.1, we find under Douglas fir–larch, beams and stringers, select structural grade, that the limiting bending stress is 1600 psi [11.03 MPa]. Then, substituting in the beam formula and converting the bending moment to inch-pounds, we can calculate the required section modulus

$$S = \frac{M}{F_b} = \frac{13{,}000 \times 12}{1600} = 97.5 \text{ in.}^3 \ [1.60 \times 10^6 \text{ mm}^3]$$

From Table 3.6 we find the section with the least area to be a 4 × 14 with $S = 102.411$ in.^{3} [1.68×10^6 mm^3]. Actually this size section falls into the higher stress category, for sections 2–4 in. thick and 5 in. and wider. Thus the allowable stress is 1800 psi [12.41 MPa], the required S drops to 87.67 in.^{3} [1.42×10^6 mm^3], and a 3 × 16 is the lightest choice.

A complete design would also require consideration of shear and deflection, as explained in Secs. 4.3 and 4.4.

Beams greater than 12 in. in depth have reduced values for the maximum flexural stress. This is done by using a size factor defined as

$$C_f = \left(\frac{12}{d}\right)^{1/9}$$

Values for this factor for standard lumber sizes are given in Table 4.3. The following example illustrates the application of this requirement.

TABLE 4.3 Size Factors for Wood Beams[a]

Beam Depth (in.)	Form Factor[b], C_F
13.5	0.987
15.5	0.972
17.5	0.959
19.5	0.947
21.5	0.937
23.5	0.928

[a] Reduce allowable bending stress or moment capacity by the factor C_F.
[b] $C_F = (12/d)^{1/9}$.

Example 2. A beam of Douglas fir–larch, No. 1 grade, is to be used on a span of 24 ft to support a load of 600 lb/ft. Select the lightest member.

Solution: The maximum bending moment is determined as

$$M = \frac{wL^2}{8} = \frac{600(24)^2}{8} = 43{,}200 \text{ ft-lb}$$

From Table 4.1: $F_b = 1300$ psi,

$$S = \frac{M}{F_b} = \frac{43{,}200 \times 12}{1300} = 399 \text{ in.}^3$$

From Table 3.6, the lightest member is a 8 × 20 with $S = 475.313$ in.^{3}. As this depth exceeds 12 in., the moment resistance must be reduced. This may be determined by multiplying the C_F factor by the section modulus. Thus

$$\text{Effective } S = C_F(S) = 0.947(475.313) = 450 \text{ in.}^3$$

As this is still in excess of the requirement, the 8 × 20 is adequate.

Example 3. Rafters of Douglas fir–larch, No. 2 grade, are to be used at 16 in. spacing for a span of 20 ft. Live load without snow is 20 psf and the total dead load, including the rafters, is 15 psf. Find the minimum size for the rafters, based only on bending stress.

Solution. At this spacing the rafters qualify for the increased bending stress described as "repetitive-member uses" in Table 4.1. Thus, for the No. 2 grade rafters, $F_b = 1450$ psi. In addition, the loading condition as described qualifies the situation with regard to load duration for an allowable stress increase factor of 1.25. (See Table 4.2.) For the rafters at 16-in. spacing, the maximum bending moment is determined as

$$M = \frac{wL^2}{8} = \left(\frac{16}{12}\right)\frac{(20 + 15)(20)^2}{8} = 2333 \text{ ft-lb}$$

and the required section modulus as

$$S = \frac{M}{F_b} = \frac{2333(12)}{1.25(1450)} = 15.45 \text{ in.}^3$$

From Table 3.6, the smallest section with this property is a 2 × 10, with an S of 21.391 in.^{3}.

As in other cases, it is also necessary to consider additional problems for a real design situation, but this is the answer to the problem as stated.

Design specifications provide for the adjustment of bending capacity or allowable bending stress when a member is vulnerable to a compression buckling failure. When beams are incorporated in a framing system, they are often provided with bracing that is adequate to resist these effects, allowing use of the full value for flexural stress. Requirements for the type of bracing required to prevent both lateral (sideways) and torsional (roll-over) buckling are given in the National Design Specification (NDS) and are summarized in Table 4.4. If details of the construction cannot provide adequate bracing, rules given in the specifications must be used to reduce bending capacity.

TABLE 4.4 Lateral Support Requirements for Wood Beams

Radio of Depth to Thickness[a]	Required Conditions
2 : 1 or less	No support required
3 : 1, 4 : 1	Ends held in position to resist rotation
5 : 1	One edge held in position for entire span
6 : 1	Bridging or blocking at maximum spacing of 8 ft; or both edges held in position for entire span; or one edge held in position for entire span (compression edge) and ends held against rotation
7 : 1	Both edges held in position for entire span

Source: Adapted from data in Sec. 4.4.1 of *National Design Specification for Wood Construction,* 1986 ed. (Ref. 4), with permission of the publishers, National Forest Products Association.

[a] Ratio of nominal dimensions for standard elements of structural lumber.

Common forms of bracing for rafters and joists include bridging and blocking. Bridging consists of crisscrossed wood or light steel members in rows; the spacing of rows is determined by the stability conditions and general code requirements. Blocking consists of solid, short pieces of lumber the same size as the framing members; these are fit between the framing members in tight rows. Bridging may also serve other purposes, such as providing for edge nailing of plywood panels. End blocking of framing is generally desired in all situations and specifically required by the code in some cases.

Problem 4.2.A. The No. 1 grade of Douglas fir–larch is to be used for a series of floor beams 6 ft [1.83 m] on center, spanning 14 ft [4.27 m]. If the total uniformly distributed load on each beam, including its own weight, is 3200 lb [14.23 kN], select the section with the least cross-sectional area based on bending stress.

Problem 4.2.B. A simple beam of Douglas fir–larch, select structural grade, has a span of 18 ft [5.49 m] with two concentrated loads of 4 kips [13.34 kN] each placed at the third points of the span. Neglecting its own weight, determine the size of the beam with the least cross-sectional area based on bending stress.

Problem 4.2.C. Rafters are to be used on 24-in. centers for a roof span of 16 ft. Live load is 20 psf and dead load is 15 psf, including the weight of the rafters. Find the rafter size required for Douglas fir–larch of (a) No. 1 grade and (b) No. 2 grade, based on bending stress.

4.3 Horizontal Shear

As discussed in Sec. 2.11 and illustrated in Fig. 2.30, a beam has a tendency to fail in shear by the fibers sliding past each other both vertically and horizontally. Also, at any point in a beam, the intensity of the vertical and horizontal shearing stresses is equal. The vertical shear strength of wood beams is seldom of concern because the shear resistance of wood *across* the grain is much larger than it is *parallel* to the grain, where the horizontal shear forces develop.

The horizontal shearing stresses are not uniformly distributed over the cross section of a beam but are greatest at the neutral surface. The maximum horizontal unit shearing stress for rectangular sections is $\frac{3}{2}$ times the average vertical unit shearing stress.

This is expressed by the formula

$$v = \frac{3}{2} \times \frac{V}{bd}$$

in which v = maximum unit horizontal shearing stress in psi,
V = total vertical shear in pounds,
b = width of cross section in inches,
d = depth of cross section in inches.

It should be noted that the depth of the cross section is called h in Table 3.6. Notation usage is not entirely consistent in wood structural design, but the variations are of minor consequence only.

This formula applies only to rectangular cross sections. Timber is relatively weak in resistance to horizontal shear, and short spans with large loads should always be tested for this shearing tendency. Frequently a beam large enough to resist bending stresses must be made larger in order to resist horizontal shear. Table 4.1 gives allowable horizontal unit shearing stresses for several stress grades of lumber under the column headed "Horizontal Shear F_v."

Example. A 6 × 10 beam of Douglas fir–larch, No. 2 dense grade, has a total uniformly distributed load of 6000 lb [26.7 kN]. Investigate the beam for structural shear.

Solution: Since the beam is symmetrically loaded, $R_1 = R_2 = 6000/2 = 3000$ lb [13.35 kN]; this is also the value of the maximum shear force. Reference to Table 3.6 shows that the dressed dimensions of a 6 × 10 are 5.5 × 9.5 in. [140 × 240 mm]. Then

$$v = \frac{3}{2} \times \frac{V}{bd} = \frac{3}{2} \times \frac{3000}{5.5 \times 9.5} = 86.1 \text{ psi } [0.594 \text{ MPa}]$$

Referring to Table 4.1, under the classification of "beams and stringers," we find that the allowable shear stress $F_v = 85$ psi [0.586 MPa]. The beam is therefore not acceptable on the basis of consideration of shear stress.

For beams that are supported by end bearing, the code permits a computation for critical shear stress at a distance of the beam depth from the support. When the computed shear stress at the support is just slightly in excess of the allowable, this more precise computation may determine that the beam is acceptable. However, for simplicity in the procedure, we will ordinarily use the maximum shear at the support and make the computation directed by the code only when this value is not acceptable.

Note: In solving the following problems, use Tables 3.6 and 4.1 and neglect the beam weight.

Problem 4.3.A. A 10 × 10 beam of Douglas fir–larch, select structural grade, supports a single concentrated load of 10 kips [44.5 kN] at the center of the span. Investigate the beam for shear stress.

Problem 4.3.B. A 10 × 14 beam of Douglas fir–larch, dense select structural grade, is loaded symmetrically with three concentrated loads of 4300 lb [19.13 kN] each, placed at the quarter points of the span. Is the beam safe in shear?

Problem 4.3.C. A 10 × 12 beam of Douglas fir–larch, No. 2 dense grade, is 8 ft [2.44 m] long and has a concentrated load of 8 kips [35.58 kN] located 3 ft [0.914 m] from one end. Investigate the beam for shear.

Problem 4.3.D. What should be the nominal cross-sectional dimensions for the beam of least weight that supports a total uniformly distributed load of 12 kips [53.4 kN] on a simple span and consists of Douglas fir–larch, No. 1 grade? Consider only the condition of limiting shear stress.

4.4 Bearing

For most beams, bearing consists of compression stress perpendicular to the grain. The allowable stress for this is that given in Table 4.1, and it is used simply by multiplying it by the bearing contact area. This is done for beam end bearing or for bearing under concentrated loads where the length of bearing along the beam is 6 in. or more. For shorter lengths an increase in allowable stress is permitted if the bearing does not occur closer than 3 in. from the end of the beam. The increase consists of multiplying the table value by a factor equal to

$$\frac{(\text{Bearing length}) + 0.375 \text{ in.}}{\text{bearing length}}$$

When a beam is not horizontal, as in the case of some roof members, there may be a bearing that occurs at some angle to the grain other than parallel or perpendicular. Allowable stress in this case is adjusted using the Hankinson formula. The same type of adjustment is made for bolted joints when the load is at an angle to the grain, and the use of the graph that is a plot of the Hankinson formula is discussed in Sec. 6.1. For bearing at intermediate angles the stress ranges between the two limiting values in Table 4.1.

Example 1. An 8 × 14 Douglas fir–larch beam, No. 1 grade, has a bearing length of 6 in. [152 mm] at its supports. If the end reaction is 7400 lb [32.9 kN], is the beam adequate for bearing?

Solution: The developed bearing stress is equal to the end reaction divided by the product of the bearing length and the beam width; 7400/(7.5 × 6) = 164 psi [1.13 MPa]. This is compared to the allowable stress from Table 4.1, 625 psi [4.31 MPa]. The beam is therefore adequate for bearing.

Example 2. A 2 × 10 roof joist cantilevers over and is supported by the top plate of a 2 × 4 stud wall. The load from the joist is 800 lb [3.56 kN]. If both the joist and the plate are Douglas fir–larch, No. 2 grade, is the situation adequate for bearing?

Solution: The developed bearing area is the product of the width of the joist (1.5 in. [38 mm]) times the width of the flat 2 × 4 plate (3.5 in. [89 mm]). The bearing stress is thus

$$F = \frac{800}{1.5 \times 3.5} = 152 \text{ psi } [1.05 \text{ MPa}]$$

This is considerably less than the allowable stress from Table 4.1: 625 psi [4.31 MPa]. Bearing stress is therefore not critical.

Example 3. A two-span 3 × 12 beam of Douglas fir–larch, No. 1 grade, bears on a 3 × 14 supporting beam at its center support. If the reaction force at the center support is 4200 lb [18.7 kN], is the situation critical for bearing?

Solution: If we assume the beam to bear at right angles to its supporting member, the developed bearing area is the product of the beam width times the support beam width: 2.5 × 2.5 = 6.25 in^2 [4032 mm^2]. The developed bearing stress is thus

$$F = \frac{4200}{6.25} = 672 \text{ psi [4.63 MPa]}$$

This exceeds the allowable stress of 625 psi [4.31 MPa] from Table 4.1. However, the situation qualifies for the increase factor for bearing lengths less than 6 in. The modified allowable stress is thus

$$F_{c\perp} = \frac{L_b + 0.375}{L_b} \times 625 = \frac{2.875}{2.5} \times 625 = 719 \text{ psi [4.96 MPa]}$$

and the condition is not critical.

Problem 4.4.A. A 6 × 12 beam of Douglas fir–larch, No. 1 grade, has 3 in. of end bearing to develop an end reaction force of 5000 lb [22.2 kN]. Is the situation adequate for bearing?

Problem 4.4.B. A 3 × 16 roof joist cantilevers over a 3 × 14 support beam. If both beams are Douglas fir–larch, No. 1 grade, is the situation adequate for bearing? The joist load on the support beam is 3000 lb [13.3 kN].

4.5 Deflection

Deflections in wood structures tend to be most critical for rafters and joists, where span-to-depth ratios are often pushed to the limit. Maximum permitted spans for specific rafter or joist arrangements are usually limited by deflection. Since rafters and joists are usually of a simple span form with uniformly distributed loading, the deflection takes the form of the equation

$$\Delta = \frac{5WL^3}{384EI}$$

Substitutions of relations between W, M, and flexural stress in this equation can result in the form

$$\Delta = \frac{5L^2 f_b}{24Ed}$$

Using average values of f_b = 1500 psi and E = 1500 ksi, the expression reduces to

$$\Delta = \frac{0.03L^2}{d}$$

where Δ = deflection, inches,
L = span, feet,
d = beam depth, inches.

Figure 4.1 is a plot of this expression with curves for standard lumber dimensions. The curves are labeled by the nominal dimensions, but the computations were done with the actual dimensions, as given in Table 3.6. For reference the lines on the graph corresponding to ratios of deflection of $L/240$ and $L/360$ are shown. These are commonly used design limitations for total load and live load deflections, respectively. Also shown for reference is the limiting ratio of beam span to depth of 25 to 1, which is an approximate practical limit, even if deflection is not especially important. For beams with values of f_b and E other than those used for the graphs, actual deflections can be obtained as follows

$$\text{true } \Delta = \frac{\text{true } f_b}{1500} \times \frac{\text{true } E \text{ (psi)}}{1{,}500{,}000} \times \Delta \text{ from graph}$$

The following examples illustrate typical problems involving computation of deflections.

Example 1. An 8 × 12 wood beam with E = 1,600,000 psi is used to carry a total uniformly distributed load of 1 kip on a simple span of 16 ft. Find the maximum deflection of the beam.

Solution: Using the deflection formula for this loading (case 2, Fig. 2.27) and the value of I = 950 in.^{4} for the 8 × 12 section (obtained from Table 3.6), we compute

$$\Delta = \frac{5WL^3}{384\,EI} = \frac{5(10{,}000)(16 \times 12)^3}{384(1{,}600{,}000)(950)} = 0.61 \text{ in.}$$

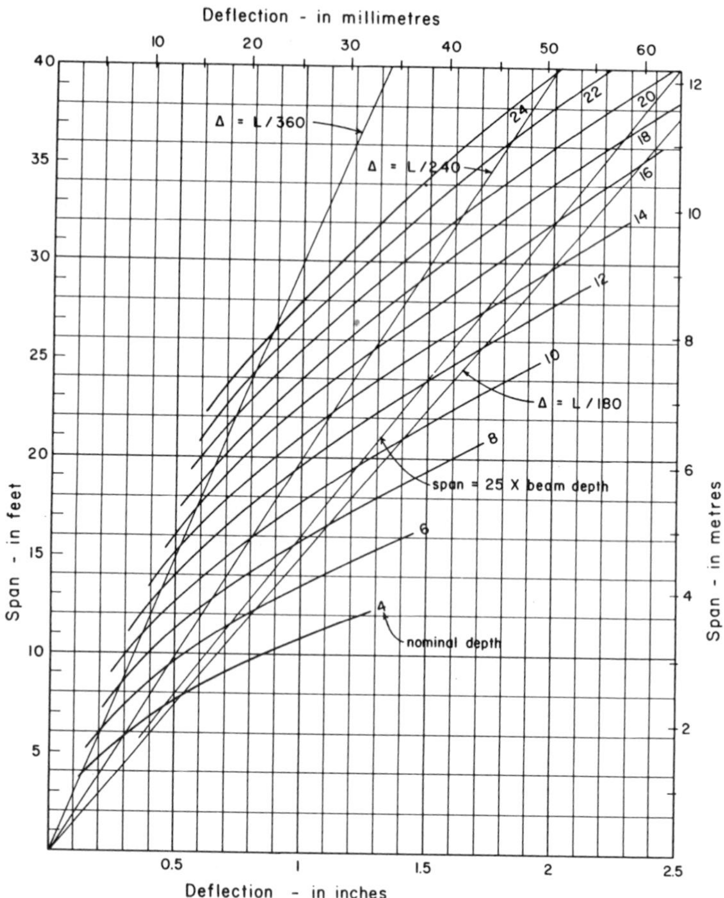

FIGURE 4.1 Deflection of wood beams. Assumed conditions: maximum bending stress = 1500 psi; E = 1,500,000 psi.

Or, using the graph in Fig. 4.1,

$$M = \frac{WL}{8} = \frac{10{,}000(16)}{8} = 20{,}000 \text{ ft-lb}$$

$$f_b = \frac{M}{S} = \frac{20{,}000(12)}{165} = 1455 \text{ psi}$$

From Fig. 4.1, Δ = approximately 0.66 in. Then

$$\text{True } \Delta = \frac{1455}{1500}\left(\frac{1{,}500{,}000}{1{,}600{,}000}\right) 0.66 = 0.60 \text{ in.}$$

which shows reasonable accuracy in comparison with the computed value.

Example 2. A beam consisting of a 6 × 10 section with E = 1,400,000 psi spans 18 ft and carries two concentrated loads. One load is 1800 lb, placed at 3 ft from one end of the beam, and the other load is 1200 lb, placed at 6 ft from the opposite end of the beam. Find the maximum deflection due only to the concentrated loads.

Solution: For an approximate computation, we use the equivalent uniform load method, consisting of finding the hypothetical total uniform load that will produce a moment equal to the actual maximum moment in the beam. For this loading the maximum moment is 6600 ft-lb (the reader should verify this by the usual procedures), and the equivalent uniform load is thus

$$W = \frac{8M}{L} = \frac{8(6600)}{18} = 2933 \text{ lb}$$

and the approximate deflection is

$$\Delta = \frac{5WL^3}{384EI} = \frac{5(2933)[(18)(12)]^3}{384(1{,}400{,}000)(393)} = 0.70 \text{ in.}$$

As in the previous example, the deflection could also be found by computing the maximum bending stress and using Fig. 4.1.

Note: In solving the following problems, use Tables 3.6 and 4.1 and Fig. 2.27. Neglect the beam weight and consider the deflection to be limited to $\frac{1}{240}$ of the span.

Problem 4.5.A. A 6 × 14 Douglas fir–larch beam, No. 1 grade, is 16 ft [4.88 m] long and supports a total uniformly distributed load of 6000 lb [26.7 kN]. Investigate the deflection.

Problem 4.5.B. An 8 × 12 beam of Douglas fir–larch, dense No. 1 grade, is 12 ft [3.66 m] in length and has a concentrated load of 5 kips [22.2 kN] at the center of the span. Investigate the deflection.

Problem 4.5.C. Two concentrated loads of 3500 lb [15.6 kN] each are located at the third points of a 15-ft [4.57-m] beam. The 10 × 14 beam is Douglas fir–larch, select structural grade. Investigate the deflection.

Problem 4.5.D. An 8 × 14 beam of Douglas fir–larch, select structural grade, has a span of 16 ft [4.88 m] and a total uniformly distributed load of 8 kips [35.6 kN]. Investigate the deflection.

Problem 4.5.E. Find the least weight nominal section that can be used for a simple beam with an 18-ft [5.49-m] span carrying a total uniformly distributed load of 10 kips [44.5 kN] based on maximum deflection. The wood is Douglas fir–larch, No. 1 grade.

4.6 Joists and Rafters

Joists are comparatively small, closely spaced beams. The sizes most commonly used are those of 2-in. nominal thickness, from 2 × 4 to 2 × 12. Joists of 3-in. nominal thickness are used when conditions require greater width for nailing decks, greater depth (up to 16-in. nominal), or simply a larger cross section for stress conditions.

Spacing of joists is usually determined by the choice of decking or—when ceilings exist—choice of ceiling construction. Most commonly used are spacings of 12, 16, and 24 in., based on the use of plywood decking or ceiling paneling in 4 × 8-ft sheets.

Cross-bridging or blocking is often required for joists as lateral bracing for the thin beams. Bridging and blocking also serve to provide for distribution of loads to adjacent joists. Such load sharing is the basis for permitting the use of higher bending stress values—those associated with the category labeled "repetitive-member uses" in Table 4.1 and discussed in Sec. 4.1. Bridging consists of rows of crisscrossed members of wood or metal. Blocking consists of rows of short pieces of the same-sized lumber as used for the joists; these are tightly fitted and nailed between the joists. Blocking is usually used where provision must be made for nailing the edges of plywood sheets that are perpendicular to the joists.

The design of joists consists of determining the load to be supported and then applying the procedures for beam design as explained in Sec. 4.2. However, to facilitate the selection of joists carrying uniformly distributed loads (by far the most common

TABLE 4.5 Allowable Spans in Feet and Inches for Floor Joists[a]

Joist Size (in)	Spacing (in)	Modulus of Elasticity, E, in 1,000,000 psi													
		0.8	0.9	1.0	1.1	1.2	1.3	1.4	1.5	1.6	1.7	1.8	1.9	2.0	2.2
2x6	12.0	8-6 720	8-10 780	9-2 830	9-6 890	9-9 940	10-0 990	10-3 1040	10-6 1090	10-9 1140	10-11 1190	11-2 1230	11-4 1280	11-7 1320	11-11 1410
	16.0	7-9 790	8-0 860	8-4 920	8-7 980	8-10 1040	9-1 1090	9-4 1150	9-6 1200	9-9 1250	9-11 1310	10-2 1360	10-4 1410	10-6 1460	10-10 1550
	24.0	6-9 900	7-0 980	7-3 1050	7-6 1120	7-9 1190	7-11 1250	8-2 1310	8-4 1380	8-6 1440	8-8 1500	8-10 1550	9-0 1610	9-2 1670	9-6 1780
2x8	12.0	11-3 720	11-8 780	12-1 830	12-6 890	12-10 940	13-2 990	13-6 1040	13-10 1090	14-2 1140	14-5 1190	14-8 1230	15-0 1280	15-3 1320	15-9 1410
	16.0	10-2 790	10-7 850	11-0 920	11-4 980	11-8 1040	12-0 1090	12-3 1150	12-7 1200	12-10 1250	13-1 1310	13-4 1360	13-7 1410	13-10 1460	14-3 1550
	24.0	8-11 900	9-3 980	9-7 1050	9-11 1120	10-2 1190	10-6 1250	10-9 1310	11-0 1380	11-3 1440	11-5 1500	11-8 1550	11-11 1610	12-1 1670	12-6 1780
2x10	12.0	14-4 720	14-11 780	15-5 830	15-11 890	16-5 940	16-10 990	17-3 1040	17-8 1090	18-0 1140	18-5 1190	18-9 1230	19-1 1280	19-5 1320	20-1 1410
	16.0	13-0 790	13-6 850	14-0 920	14-6 980	14-11 1040	15-3 1090	15-8 1150	16-0 1200	16-5 1250	16-9 1310	17-0 1360	17-4 1410	17-8 1460	18-3 1550
	24.0	11-4 900	11-10 980	12-3 1050	12-8 1120	13-0 1190	13-4 1250	13-8 1310	14-0 1380	14-4 1440	14-7 1500	14-11 1550	15-2 1610	15-5 1670	15-11 1780
2x12	12.0	17-5 720	18-1 780	18-9 830	19-4 890	19-11 940	20-6 990	21-0 1040	21-6 1090	21-11 1140	22-5 1190	22-10 1230	23-3 1280	23-7 1320	24-5 1410
	16.0	15-10 790	16-5 860	17-0 920	17-7 980	18-1 1040	18-7 1090	19-1 1150	19-6 1200	19-11 1250	20-4 1310	20-9 1360	21-1 1410	21-6 1460	22-2 1550
	24.0	13-10 900	14-4 980	14-11 1050	15-4 1120	15-10 1190	16-3 1250	16-8 1310	17-0 1380	17-5 1440	17-9 1500	18-1 1550	18-5 1610	18-9 1670	19-4 1780

Source: Reproduced from the *Uniform Building Code,* 1988 ed. (Ref. 1), with permission of the publishers, International Conference of Building Officials.

[a] *Notes:* Criteria: 40 psf live load; 10 psf dead load; live load deflection limited to $\frac{1}{360}$ of the span. Number indicated below each span is the required allowable bending stress in psi.

loading), many tables have been prepared that give maximum safe spans for joists of various sizes and spacings under different loadings per square foot. Table 4.5 is representative of such tables and has been reproduced from the 1988 edition of the *Uniform Building Code*. Examining the table, we note that spans are computed on the basis of modulus of elasticity, with the required bending stress F_b listed below each span; that is, both stiffness (deflection) and bending strength have been taken into account. Maximum safe clear spans in feet and inches are tabulated for three joist spacings and selected values of E. As stated in the table notes, deflection is limited to $\frac{1}{360}$ of the span due to live load only, whereas both live load and dead load allowance are used in determining the required bending stress.

The use of Table 4.2 is illustrated in the following example.

Example 1. Using Table 4.5, select joists to carry a live load of 40 psf on a span of 15 ft 6 in. if the spacing is 16 in. on centers.

Solution: Referring to Table 4.5, we find that 2 × 10 joists with an E of 1,400,000 psi and F_b of 1150 psi may be used on a span of 15 ft 8 in.

Turning to Table 4.1, the reader should observe that among the grades listed, the following selection would be satisfactory: Douglas fir–larch No. 2. Remember that the values for F_b in the column headed "Repetitive-Member Uses" generally applies to joists.

As with all safe load tabulations, caution must be exercised with respect to the criteria used in preparation of the table data. The 10 psf allowance provided in Table 4.5 is adequate to cover the weight of the joists, wood deck and flooring, and a gypsum drywall ceiling. However, if a heavier type of flooring or a plastered ceiling is used, this additional weight must be accounted for. This is readily accomplished, of course, if the usual beam design procedure is followed, with the table used only for a preliminary estimate of the required joist.

Rafters are the comparatively small, closely spaced beams used to support the load on roofs. As with floor joists, the most common sizes are the available range of 2- and 3-in. nominal thickness lumber, and the most used spacings are 12, 16, and 24 in. The terms *rafter* and *roof joist* are frequently used interchangeably, the former mostly for sloping roof surfaces and the latter mostly for horizontal roof surfaces.

For sloping rafters it is common practice to consider the span to be the horizontal projection, as indicated in Figure. 4.2. This applies only to consideration for gravity loads. Wind forces are considered as applied perpendicular to the roof surface, and the span must thus be the actual rafter length.

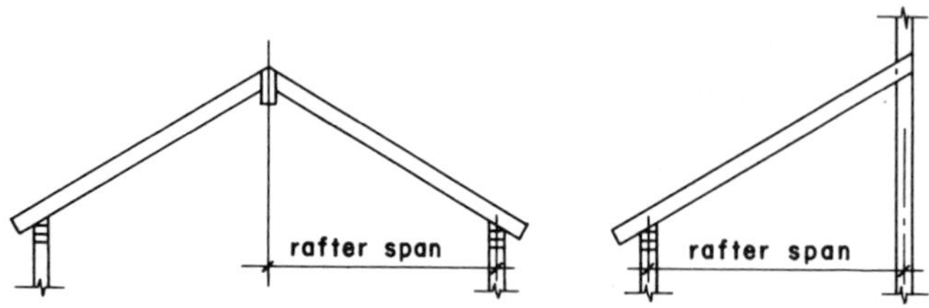

FIGURE 4.2 Determination of rafter span.

TABLE 4.6 Allowable Spans in Feet and Inches for Low- or High-Slope Rafters[a]

RAFTER SIZE (IN)	SPACING (IN)	Allowable Extreme Fiber Stress in Bending F_b (psi).														
		500	600	700	800	900	1000	1100	1200	1300	1400	1500	1600	1700	1800	1900
2x6	12.0	8-6 0.26	9-4 0.35	10-0 0.44	10-9 0.54	11-5 0.64	12-0 0.75	12-7 0.86	13-2 0.98	13-8 1.11	14-2 1.24	14-8 1.37	15-2 1.51	15-8 1.66	16-1 1.81	16-7 1.96
	16.0	7-4 0.23	8-1 0.30	8-8 0.38	9-4 0.46	9-10 0.55	10-5 0.65	10-11 0.75	11-5 0.85	11-10 0.97	12-4 1.07	12-9 1.19	13-2 1.31	13-7 1.44	13-11 1.56	14-4 1.70
	24.0	6-0 0.19	6-7 0.25	7-1 0.31	7-7 0.38	8-1 0.45	8-6 0.53	8-11 0.61	9-4 0.70	9-8 0.78	10-0 0.88	10-5 0.97	10-9 1.07	11-1 1.17	11-5 1.28	11-8 1.39
2x8	12.0	11-2 0.26	12-3 0.35	13-3 0.44	14-2 0.54	15-0 0.64	15-10 0.75	16-7 0.86	17-4 0.98	18-0 1.11	18-9 1.24	19-5 1.37	20-0 1.51	20-8 1.66	21-3 1.81	21-10 1.96
	16.0	9-8 0.23	10-7 0.30	11-6 0.38	12-3 0.46	13-0 0.55	13-8 0.65	14-4 0.75	15-0 0.85	15-7 0.96	16-3 1.07	16-9 1.19	17-4 1.31	17-10 1.44	18-5 1.56	18-11 1.70
	24.0	7-11 0.19	8-8 0.25	9-4 0.31	10-0 0.38	10-7 0.45	11-2 0.53	11-9 0.61	12-3 0.70	12-9 0.78	13-3 0.88	13-8 0.97	14-2 1.07	14-7 1.17	15-0 1.28	15-5 1.39
2x10	12.0	14-3 0.26	15-8 0.35	16-11 0.44	18-1 0.54	19-2 0.64	20-2 0.75	21-2 0.86	22-1 0.98	23-0 1.11	23-11 1.24	24-9 1.37	25-6 1.51	26-4 1.66	27-1 1.81	27-10 1.96
	16.0	12-4 0.23	13-6 0.30	14-8 0.38	15-8 0.46	16-7 0.55	17-6 0.65	18-4 0.75	19-2 0.85	19-11 0.96	20-8 1.07	21-5 1.19	22-1 1.31	22-10 1.44	23-5 1.56	24-1 1.70
	24.0	10-1 0.19	11-1 0.25	11-11 0.31	12-9 0.38	13-6 0.45	14-3 0.53	15-0 0.61	15-8 0.70	16-3 0.78	16-11 0.88	17-6 0.97	18-1 1.07	18-7 1.17	19-2 1.28	19-8 1.39
2x12	12.0	17-4 0.26	19-0 0.35	20-6 0.44	21-11 0.54	23-3 0.64	24-7 0.75	25-9 0.86	26-11 0.98	28-0 1.11	29-1 1.24	30-1 1.37	31-1 1.51	32-0 1.66	32-11 1.81	33-10 1.96
	16.0	15-0 0.23	16-6 0.30	17-9 0.38	19-0 0.46	20-2 0.55	21-3 0.65	22-4 0.75	23-3 0.85	24-3 0.97	25-2 1.07	26-0 1.19	26-11 1.31	27-9 1.44	28-6 1.56	29-4 1.70
	24.0	12-3 0.19	13-5 0.25	14-6 0.31	15-6 0.38	16-6 0.45	17-4 0.53	18-2 0.61	19-0 0.70	19-10 0.78	20-6 0.88	21-3 0.97	21-11 1.07	22-8 1.17	23-3 1.28	23-11 1.39

Source: Reproduced from the *Uniform Building Code*, 1988 ed. (Ref. 1), with permission of the publishers, International Conference of Building Officials.

[a] *Notes:* Criteria: 20 psf live load; 15 psf dead load; live load deflection limited to $\frac{1}{240}$ of the span. Number indicated below each span is the required minimum value for the modulus of elasticity E; E in psi equals the listed value times 1,000,000.

Design of rafters is generally accomplished by the use of safe-load tables. Table 4.6 is representative of such tables and has been reproduced from the 1988 edition of the *Uniform Building Code*. The table gives maximum safe spans for rafters in feet and inches for three spacings and selected values of the allowable bending stress F_b. The modulus of elasticity E, required to maintain the stated deflection limit, is listed below each span.

The live load value provided for in Table 4.6 is 20 psf. This is the usual minimum value required by codes; where snow accumulation is possible, a higher value is usually specified. For high-slope rafters, codes usually provide for some reduction of the live load, based on the unlikelihood of accumulation of anything on the sloping surface. Magnitudes of wind loads and the manner of their application to roof surfaces are a regional matter, and the prevailing code must be used. Except for very steep slopes or exceptionally light construction, wind loads are usually critical only in areas with histories of frequent windstorm conditions.

The following example illustrates the use of the data in Table 4.6.

Example 2. Rafters are to be used on 24-in. centers for a roof span of 16 ft. Live load is 20 psf; total dead load is 15 psf; live load deflection is limited to $\frac{1}{240}$ of the span. Find the rafter size required for Douglas fir–larch of (1) No. 1 grade and (2) No. 2 grade.

Solution: From Table 4.1 we find the design values for No. 1 grade to be E = 1,800,000 psi and F_b = 1750 psi. Although Table 4.6 does not have a column for F_b = 1750 psi, it is apparent that the size choice is for a 2 × 10 rafter. This observation is made by comparing the listed data for F_b = 1700 and 1800 for both 2 × 8 and 2 × 10 rafters. Thus a 2 × 8 will span between 14 ft 7 in. and 15 ft, while a 2 × 10 will span between 18 ft 7 in. and 19 ft 2 in. It should also be apparent that E will not be critical.

From Table 4.1 we find design values of E = 1,700,000 psi and F_b = 1450 psi for No. 2 grade. Observing the values listed in Table 4.6 for F_b = 1400 and 1500 psi, it should be apparent that the 2 × 10 is again the choice and that E is not critical.

Joists and rafters are frequently quite narrow in width with respect to their depth, thus falling within the categories described in Table 4.4 that require some consideration for bracing to prevent compression buckling effects. Forms of bracing are discussed in the examples in Part V.

Problems 4.6.A,B,C,D. Using Douglas fir–larch, No. 2 grade, pick the joist size required from Table 4.5 for the stated conditions. Live load is 40 psf; dead load is 10 psf; deflection is limited to $\frac{1}{360}$ of the span under live load only.

	Joist Spacing (in.)	Joist Span (ft)
A	16	14
B	12	14
C	16	18
D	12	22

Problems 4.6.E,F,G,H. Select the smallest rafter from Table 4.6 for the conditions stated. Wood is Southern pine, No. 2 grade; live load is 20 psf; dead load is 15 psf; live load deflection is limited to $\frac{1}{240}$ of the span.

	Rafter Spacing (in.)	Rafter Span (ft)
A	16	12
B	24	12
C	16	18
D	24	18

4.7 Decking for Roofs and Floors

Board and Plank Decks. Before plywood established its dominance as a decking material, most roof and floor decks were made with $\frac{3}{4}$-in. (nominal 1-in.) boards with interlocking edges of the type shown in Fig. 4.3—called tongue-and-groove joints. Today this type of deck is mostly used only in regions where labor cost is relatively low and the boards are locally competitive in availability and cost in comparison to plywood.

When installed in a position with the boards perpendicular to the joists, board decks produce rather poor horizontal diaphragms. It is common, therefore, when significant diaphragm action for lateral loads is required, to install the deck at an angle of 45° to the joists, creating a trussed structure.

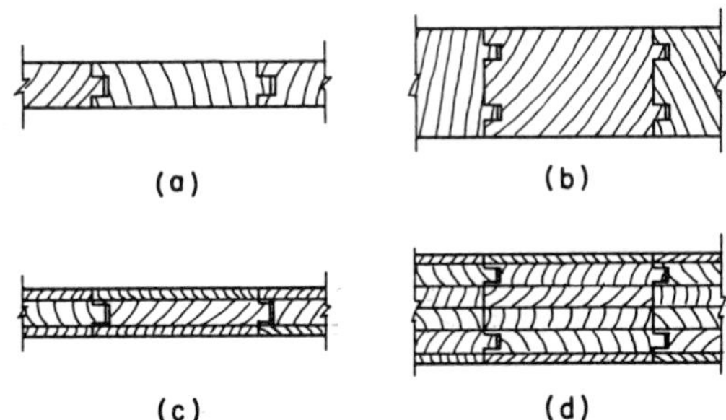

FIGURE 4.3 Typical forms of plank or timber decking.

The $\frac{3}{4}$-in.-thick board deck is usually adequate for the spanning tasks of roof and floor decks where spacing of joists or rafters does not exceed 24 in. The type of roofing or the type of finish floor to be used must be considered, however. Roofing of all types must be anchored to the deck with nails of some kind, for which the board deck is quite adequate, possibly better than the thinner plywoods. Membrane roofing for flat roofs usually requires a minimum of $\frac{1}{2}$-in. plywood, making the board deck more competitive than in the case of sloping roofs with shingles which might be achieved with thinner plywood.

For floors it is common to use some additional material on top of the structural deck, such as a layer of concrete fill or particleboard sheets. These will add considerable stiffness to the floor, so when they are *not* to be used, some conservative judgment should be exercised in choice of deck thickness and support spacings.

If a deck of the board type is thicker than $\frac{3}{4}$ in., it is generally referred to as planking or plank deck. The most widely used form is that made with 2-in.-nominal-thickness units, approximately 1.5 in. in actual dimension. There are usually special reasons for using such a deck, including one or all of the following.

1. The deck is to be exposed on the underside, and the appearance of the plank deck is hands down more handsome than board or plywood decks.
2. Exposed or not, the deck may require a fire rating, which is highly limited for other decks.
3. It may be desired to have supporting members with spacing exceeding that feasible for board or plywood decks.
4. Concentrated loadings from vehicles or equipment may be too high for the thinner decks.

Nominal 2-in. plank deck may be of the same form as board deck (Fig. 4.3*a*), but is also often made with laminated units as shown in Fig. 4.3*c*. Plank deck is also available in thicknesses greater than 2 in. nominal. When the thickness exceeds 2.5 in. or so, the units usually have a double tongue and groove on each face, as shown in Fig. 4.3*b* for a solid-sawn unit and in Fig. 4.3*d* for a laminated unit. The thicker plank deck units are capable of achieving considerable span distances and may be used in structures that are essentially without the usual rafters or joists, having clear deck spans from wall to wall or beam to beam.

One problem with plank decks, as with board decks, is the low diaphragm capacity achieved when units are in a perpendicular orientation to supports. Diagonal placement, while common with 1-in. board decks, is not so common with the plank decks. When diaphragm capacity of some significant level is required, the usual solution is to use plywood nailed to the top of the deck units. This is quite commonly done and is frankly the main reason that diaphragm capacities are rated for decks of plywood as thin as $\frac{5}{16}$ in., a thickness not usually usable for a structural spanning deck.

Plank deck units are essentially fabricated products and information about them should be obtained from the manufacturers or suppliers of the products. Type of units, finishes of exposed units, installation specifications, and structural properties vary greatly and the type of unit available regionally should be used for design.

Plywood Decks. Structural plywood consists primarily of that made with all plies of Douglas fir. Many different kinds of panels are produced; the principal distinctions other than the panel thickness are the following:

1. *Glue Type.* Panels are identified for exterior (exposed to the weather) or interior use, based on the type of glue used. Exterior type should also be used for any interior conditions involving high moisture.
2. *Grade of Plies.* Individual plies are rated—generally as A, B, C, or D, with A best—on the basis of the presence of knots, splits, patches, or plugs. The most critical ratings are those of the face plies; thus a panel is typically designated by the ratings of the front and back plies. For example, C-C indicates both faces are of C grade; C-D indicates the front is C grade and the back is D grade.
3. *Structural Classification.* In some cases panels are identified as structural I or structural II. This is mostly of concern when the panels are used for shear walls or horizontal roof

or floor diaphragms. For this rating the grade of the interior plies is also a concern.

4. *Special Faces.* Plywood with special facing, usually only on one side, is produced for a variety of uses. Special surfaces for use as exposed siding are one such example. These are usually produced as some particular manufacturer's special product and any structural properties or other usage considerations should be obtained from the supplier.
5. *Panel Identification Index.* Structural grades of plywood usually have a designation called the *Identification Index,* which is stamped on the panel back as part of the grade trademark. This index is a measure of the strength and stiffness of the panel and consists of two numbers separated by a slash (/). The first number indicates the maximum center-to-center spacing of supports for a roof deck under average loading conditions and the second number indicates the maximum spacing for a floor deck for average residential loading. There are various conditions on the use of these numbers, but they generally permit the selection of panels for a specific ordinary situation without the need for further structural computations.

Some ratings and classifications are industry-wide and some are local variations due to the use of a particular building code or the general use of particular products. Designers should be aware of the general industry standards but also of what products are generally available and widely used in a given locality.

Plywood decks quite frequently serve the dual purposes of spanning support surfaces for gravity roof or floor loads and shear-resisting diaphragms for lateral forces on the building. Choice of the plywood quality and thickness must relate to both of these concerns, as well as to any additional functional problems, such as attachment of roofing materials. Nailing of the plywood sheets to supports is done minimally for gravity loads, usually with nails at 6-in. centers at all edges and at 12-in. centers in the middle of the sheets, called field or intermediate nailing. Nailing for diaphragm action is the basic means of making the separate sheets act together as a single large unit and transfer loads to chords, shear walls, and collectors. Diaphragm nailing is discussed in Sec. 20.3.

For gravity-load spanning functions, plywood is strongest when the face grain is perpendicular to the supports. However, for various reasons, it is sometimes desired to turn the sheets the other way, which results in considerably less spanning capacity in thinner sheets that usually have only three plies, but has decreasing influence as the number of plies and overall thickness increases. Thus $\frac{3}{8}$-in.-thick sheets are strongly affected by orientation, while $\frac{3}{8}$-in.-thick sheets are only minimally affected.

One reason for the orientation with face grain parallel to supports is that it reduces the amount of total unsupported edge length in the deck. This refers to the edges of the sheets that do not fall over a supporting rafter or joist. If left substantially unsupported (and not much nailed down), these edges do not contribute to the deck diaphragm capacity and offer some problems for membrane roofing of flat roofs and some types of floor finishes. If support is considered to be necessary, it is usually achieved by one of the following means:

1. Using tongue-and-groove joints between sheets. This is generally limited to sheets of $\frac{3}{4}$ in. or greater thickness.
2. Using a clipping device—usually a short metal H-shaped element between the sheet edges.
3. Providing nailable supports at the edges not having support by the structure beneath the deck. This is typically provided by blocking when the supports are solid-sawn rafters or joists. The blocks consist of short pieces of the same-sized members as the rafters or joists, fitted between and nailed to the structural members.

Tongue-and-groove or clipped edges do not change the deck with regard to diaphragm action; thus the deck in these cases must usually need to function adequately as a so-called *unblocked diaphragm*. Tongue-and-groove decks are available in thicknesses up to 1.25 in. and are most often used for floors (with heavier loadings) or for roofs of longer spans where structural members may have spacings greater than 24 in.

TABLE 4.7 Data for Plywood Decks—Face Grain Perpendicular to Supports[1–8]

PANEL SPAN RATING[3]	PLYWOOD THICKNESS (Inch)	ROOF[2]				FLOOR MAXIMUM SPAN[4] (In Inches)
		Maximum Span (In Inches)		Load (In Pounds per Square Foot)		
		Edges Blocked	Edges Unblocked	Total Load	Live Load	
1. 12/0	15/16	12		135	130	0
2. 16/0	5/16, 3/8	16		80	65	0
3. 20/0	5/16, 3/8	20		70	55	0
4. 24/0	3/8	24	16	60	45	0
5. 24/0	15/32, 1/2	24	24	60	45	0
6. 32/16	15/32, 1/2, 19/32, 5/8	32	28	55	35[5]	16[7]
7. 40/20	19/32, 5/8, 23/32, 3/4, 7/8	40	32	40[5]	35[5]	20[7,8]
8. 48/24	23/32, 3/4, 7/8	48	36	40[5]	35[5]	24

[1]These values apply for C-C, C-D, Structural I and II grades only. Spans shall be limited to values shown because of possible effect of concentrated loads.

[2]Uniform load deflection limitations 1/180 of the span under live load plus dead load, 1/240 under live load only. Edges may be blocked with lumber or other approved type of edge support.

[3]Span rating appears on all panels in the construction grades listed in Footnote No. 1.

[4]Plywood edges shall have approved tongue-and-groove joints or shall be supported with blocking unless 1/4-inch minimum thickness underlayment, or 1½ inches of approved cellular or lightweight concrete is placed over the subfloor, or finish floor is 25/32-inch wood strip. Allowable uniform load based on deflection of 1/360 of span is 165 pounds per square foot.

[5]For roof live load of 40 pounds per square foot or total load of 55 pounds per square foot, decrease spans by 13 percent or use panel with next greater span rating.

[6]May be 24 inches if 25/32-inch wood strip flooring is installed at right angles to joists.

[7]May be 24 inches where a minimum of 1½ inches of approved cellular or lightweight concrete is placed over the subfloor and the plywood sheathing is manufactured with exterior glue.

[8]Floor or roof sheathing conforming with this table shall be deemed to meet the design criteria of Section 2516.

Source: Table 25-S-1 from the *Uniform Building Code*, 1988 ed. (Ref. 1), reproduced with permission of the publishers, International Conference of Building Officials.

The all-American sheet size is, of course, 48 × 96 in. If ordered in large enough batches, however, larger sheets can be obtained when the logical structural module simply does not fit in the 2–4–8-ft system.

For decks, ordinary structural plywood is produced in thicknesses from $\frac{5}{16}$ to $\frac{7}{8}$ in. Deck capacities and span limits are rated in industry standards or building codes. The following tables from the *Uniform Building Code* are presented here.

Table 25-J-1, giving capacities for shear in horizontal diaphragm action (our Table 20.1).

Table 25-S-1, giving capacities or span limits for gravity-load spanning decks for roofs or floors with the face grain perpendicular to supports (our Table 4.7).

Table 25-S-2, giving capacities or span limits for roof decks with the face grain parallel to supports (our Table 4.8).

TABLE 4.8 Data for Plywood Decks—Face Grain Parallel to Supports[1,2]

	THICKNESS	NO. OF PLIES	SPAN	TOTAL LOAD	LIVE LOAD
STRUCTURAL I	15/32	4	24	30	20
		5	24	45	35
	1/2	4	24	35	25
		5	24	55	40
Other grades covered in U.B.C. Standard No. 25-9	15/32	5	24	25	20
	1/2	5	24	30	25
	19/32	4	24	35	25
		5	24	50	40
	5/8	4	24	40	30
		5	24	55	45

[1]Uniform load deflection limitations: 1/180 of span under live load plus dead load, 1/240 under live load only. Edges shall be blocked with lumber or other approved type of edge supports.

[2]Roof sheathing conforming with this table shall be deemed to meet the design criteria of Section 2516.

Source: Table 25-S-2 from the *Uniform Building Code*, 1988 ed. (Ref. 1), reproduced with permission of the publishers, International Conference of Building Officials.

Common practice often establishes some minimum usages. The following are some widely used minimum choices for plywood decks in common situations:

1. For roofs of significant slope usually 3 : 12 or greater) with shingle roofing—$\frac{3}{8}$ in.
2. For flat roofs with conventional membrane (three-ply, etc.) roofing and rafters not over 24 in. on center—$\frac{1}{2}$ in.
3. For floors with some permanent structural material between the deck and the finish (concrete fill, $\frac{1}{2}$-in. or thicker particleboard, etc.)—$\frac{5}{8}$ in.
4. For other floors—$\frac{3}{4}$ in.

4.8 Glued Laminated Products

In addition to sheets of plywood, there are a number of other products used for wood structures that are fabricated by gluing together pieces of wood into a solid form. Girders, framed bents, and arch ribs of large size are produced by assembling standard 2-in. nominal lumber (2 × 6, etc.). The resulting thickness of such elements is essentially the width of the standard lumber used, with a small dimensional loss due to finishing. The depth is a multiple of the lumber thickness of 1.5 in.

Availability of glued laminated products should be investigated on a regional basis. Information can be obtained from local suppliers. Fabricators and suppliers also commonly provide some assistance in engineering design when their products are specified for a building. The National Design Specification and most local building codes provide some design requirements and guides for the design of ordinary glued laminated elements fabricated from standard lumber.

4.9 Wood Beam and Deck Systems

Elements of wood are used in a variety of ways for roof and floor framing systems. Probably the most common is the light frame construction in which all major elements (joists, rafters, and wall studs) consist of structural lumber of 2-in. nominal thickness. However, wood roof and floor systems are also used frequently with buildings employing structural walls of masonry or concrete or frameworks of timber or steel. Plywood floor and roof decks are quite often utilized as horizontal diaphragms in conjunction with vertical elements of the lateral bracing systems for resistance of wind and earthquake forces.

Details of the building construction with wood spanning systems are in some ways quite standard, as the products used (structural lumber, nails, fastening devices, etc.) are commonly produced and widely marketed. In terms of the general building construction, however, the variety possible is considerable, reflecting the choices of designers as well as response to a variety of finish materials for exposed building surfaces.

Basic structural design considerations as well as some of the general design factors for development of wood structural systems are developed in the building design examples in Part V.

5

Wood Columns

The wood column that is used most frequently is the *solid-sawn column*. It consists of a single piece of wood, square or rectangular in cross section. Solid columns of round cross section are also considered solid-sawn columns but are used less frequently. A *spaced column* is an assembly of two or more pieces separated at the ends and at intermediate points along their lengths by blocking. Two other types are *built-up columns,* consisting of multiple solid elements fastened together to form a solid mass, and *glued laminated columns*.

5.1 Solid-Sawn Columns

In wood construction the slenderness ratio of a freestanding solid-sawn column is the ratio of the unbraced (laterally unsupported) length to the dimension of its least side, or L/d (Fig. 5.1*a*). When members are braced so that the unsupported length with respect to one face is less than that with respect to the other, L is the distance between the points of support that prevent lateral movement in the direction along which the dimension of the section is measured. This is illustrated in Fig. 5.1*b*. If the section is not square or round, it may be necessary to investigate two L/d

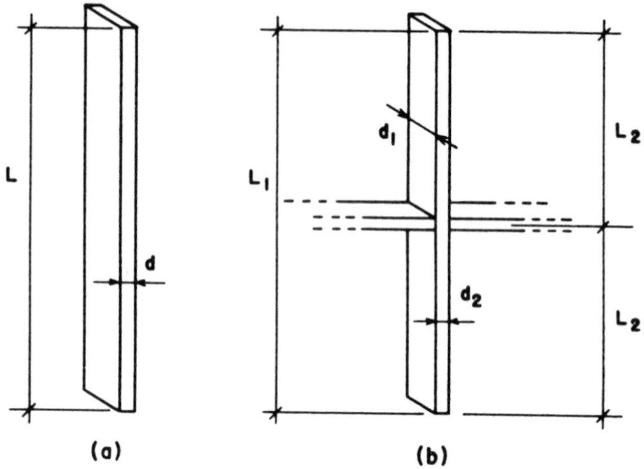

FIGURE 5.1 Determination of slenderness ratio for a wood column.

conditions for such a column to determine which is the limiting one. The slenderness ratio for solid-sawn columns is limited to $L/d = 50$; for spaced columns the limiting ratio is $L/d = 80$.

Figure 5.2 illustrates the typical form of the relationship between axial compression capacity and slenderness for a linear

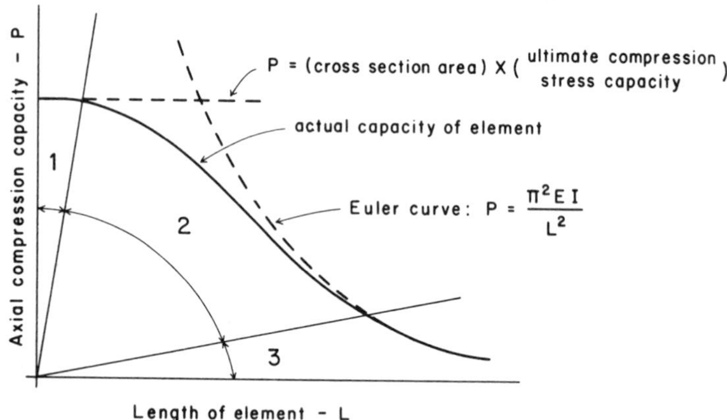

FIGURE 5.2 Relation of member length to axial compression capacity.

compression member (column). The two limiting conditions are those of the very short member and the very long member. The short member (such as a block of wood) fails in crushing, which is limited by the mass of material and the stress limit in compression. The very long member (such as a yardstick) fails in elastic buckling, which is determined by the stiffness of the member; stiffness is determined by a combination of geometric property (shape of the cross section) and material stiffness property (modulus of elasticity). Between these two extremes—which is where most wood compression members fall—the behavior is indeterminate as the transition is made between the two distinctly different modes of behavior.

The National Design Specification currently provides for three separate compression stress calculations, corresponding to the three zones of behavior described in Fig. 5.2. The plot of these three stress formulas, for a specific example wood, is shown in Fig. 5.3. Typical analysis and design procedures for simple solid wood columns are illustrated in the following examples.

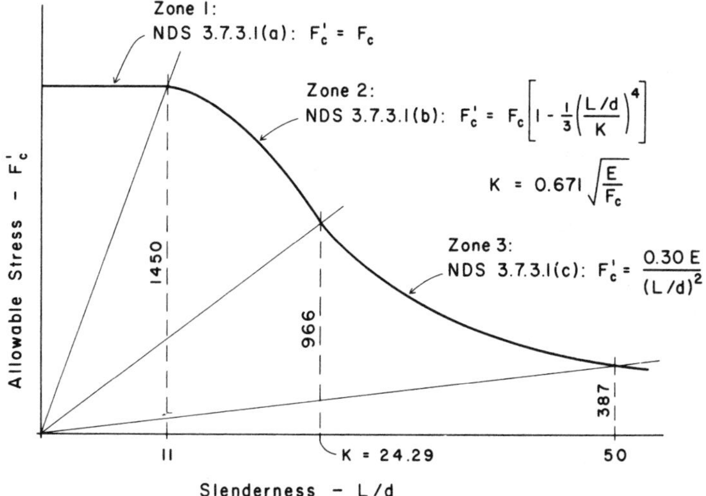

FIGURE 5.3 Allowable axial compression stress as a function of the slenderness ratio *L/d*. NDS requirements for Douglas fir–larch, dense No. 1 grade.

Example 1. A wood compression member consists of a 3 × 6 of Douglas fir–larch, dense No. 1 grade. Find the allowable axial compression force for unbraced lengths of: (1) 2 ft [0.61 m], (2) 4 ft [1.22 m], (3) 8 ft [2.44 m].

Solution: We find from Table 4.1: F_c = 1450 psi [10.0 MPa] and E = 1,900,000 psi [13.1 GPa]. To establish the zone limits we compute the following:

$$11(d) = 11(2.5) = 27.5 \text{ in.}$$

$$50(d) = 50(2.5) = 125 \text{ in.}$$

and

$$K = 0.671\sqrt{\frac{E}{F_c}} = 0.671\sqrt{\frac{1{,}900{,}000}{1450}} = 24.29$$

Thus for (1), L = 24 in., which is in zone 1; $F'_c = F_c$ = 1450 psi; allowable $C = F'_c$ × gross area = 1450 × 13.75 = 19,938 lb [88.7 kN].

For (2), L = 48 in.; $L/d = \frac{48}{2.5} = 19.2$, which is in zone 2.

$$F'_c = F_c\left\{1 - \frac{1}{3}\left(\frac{L/d}{K}\right)^4\right\}$$

$$= 1450\left\{1 - \frac{1}{3}\left(\frac{19.2}{24.29}\right)^4\right\}$$

$$= 1262 \text{ psi}$$

Allowable C = 1262 × (13.75) = 17,353 lb [77.2 kN].

For (3), L = 96 in.; $L/d = \frac{96}{2.5} = 38.4$, which is in zone 3.

$$F'_c = \frac{0.3(E)}{(L/d)^2} = \frac{0.3(1{,}900{,}000)}{(38.4)^2} = 387 \text{ psi}$$

Allowable compression = 387 × (13.75) = 5321 lb [23.7 kN].

Example 2. Wood 2 × 4 elements are to be used as vertical compression members to form a wall (ordinary stud-wall construction). If the wood is Douglas fir–larch, No. 3 grade, and the wall is 8.5 ft high, what is the column load capacity of a single stud?

Solution: In this case it must be assumed that the surfacing materials used for the wall (plywood, drywall, plaster, etc.) will provide adequate bracing for the studs on their weak axis (the 1.5-in. [38-mm] direction). If not, the studs cannot be used, since the specified height of the wall is considerably in excess of the limit for L/d for a solid column (50). We therefore assume the direction of potential buckling to be that of the 3.5-in. [89-mm] dimension. Thus

$$\frac{L}{d} = \frac{8.5 \times 12}{3.5} = 29.1$$

In order to determine which column load formula must be used, we must find the value of K for this wood. From Table 4.1 we find $F_c = 675$ psi [4.65 MPa] and $E = 1{,}500{,}000$ psi [10.3 GPa]. Then

$$K = 0.671 \sqrt{\frac{E}{F_c}} = 0.671 \sqrt{\frac{1{,}500{,}000}{675}} = 31.63$$

We thus establish the condition for the stud as zone 2 (Fig. 5.2), and the allowable compression stress is computed as

$$F'_c = F_c \left\{1 - \frac{1}{3}\left(\frac{L/d}{K}\right)^4\right\}$$

$$= 675 \left\{1 - \frac{1}{3}\left(\frac{29.1}{31.63}\right)^4\right\} = 514 \text{ psi } [3.54 \text{ MPa}]$$

The allowable load for the stud is

$$P = F'_c \times \text{gross area} = 514 \times 5.25 = 2699 \text{ lb } [12.0 \text{ kN}]$$

Example 3. A wood column of Douglas fir–larch, dense No. 1 grade, must carry an axial load of 40 kips [178 kN]. Find the smallest section for unbraced lengths of: (1) 4 ft [1.22 m], (2) 8 ft [2.44 m], (3) 16 ft [4.88 m].

Solution: Since the size of the column is unknown, the values of L/d, F_c, and E cannot be predetermined. Therefore, without design aids (tables, graphs, or computer programs), the process becomes a cut-and-try approach, in which a specific value is assumed for d and the resulting values for L/d, F_c, E, and F'_c are determined. A required area is then determined and the sections with the assumed d compared with the requirement. If an acceptable member cannot be found, another try must be made with a different d. Although somewhat clumsy, the process is usually not all that laborious, since a limited number of available sizes are involved.

We first consider the possibility of a zone 1 stress condition (Fig. 5.2), since this calculation is quite simple. If the maximum $L = 11(d)$, then the minimum $d = (4 \times 12)/11 = 4.36$ in. [111 mm]. This requires a nominal thickness of 6 in., which puts the size range into the "posts and timbers" category in Table 4.1, for which the allowable stress F_c is 1200 psi. The required area is thus

$$A = \frac{\text{load}}{F'_c} = \frac{40{,}000}{1200} = 33.3 \text{ in.}^2 \text{ [21,485 mm}^2\text{]}$$

The smallest section is thus a 6 × 8, with an area of 41.25 in.^{2}, since a 6 × 6 with 30.25 in.^{2} is not sufficient. (See Table 3.6.) If the rectangular-shaped column is acceptable, this becomes the smallest member usable. If a square shape is desired, the smallest size would be an 8 × 8.

If the 6-in. nominal thickness is used for the 8-ft column, we determine that

$$\frac{L}{d} = \frac{8 \times 12}{5.5} = 17.45$$

Since this is greater than 11, the allowable stress is in the next zone, for which

$$F'_c = F_c \left\{1 - \frac{1}{3}\left(\frac{L/d}{K}\right)^4\right\}$$

$$= 1200 \left\{1 - \frac{1}{3}\left(\frac{17.45}{25.26}\right)^4\right\}$$

$$= 1109 \text{ psi } [7.65 \text{ MPa}]$$

in which

$$F_c = 1200 \text{ psi and } E = 1{,}700{,}000 \text{ (from Table 4.1)}$$

and

$$K = 0.671\sqrt{\frac{E}{F_c}} = 0.671\sqrt{\frac{1{,}700{,}000}{1200}} = 25.26$$

The required area is thus

$$A = \frac{\text{load}}{F'_c} = \frac{40{,}000}{1109} = 36.07 \text{ in.}^2 \ [23{,}272 \text{ mm}^2]$$

and the choices remain the same as for the 4-ft column.

If the 6-in. nominal thickness is used for the 16-ft column, we determine that

$$\frac{L}{d} = \frac{16 \times 12}{5.5} = 34.9$$

Since this is greater than the value of K, the stress condition is that of zone 3 (Fig. 5.2), and the allowable stress is

$$F'_c = \frac{0.30E}{(L/d)^2} = \frac{(0.30)(1{,}700{,}000)}{34.9^2} = 419 \text{ psi } [2.89 \text{ MPa}]$$

which requires an area for the column of

$$A = \frac{\text{load}}{F'_c} = \frac{40{,}000}{419} = 95.5 \text{ in.}^2 \ [61{,}617 \text{ mm}^2]$$

This is greater than the area for the largest section with a nominal thickness of 6 in., as listed in Table 3.6. Although larger sections may be available in some areas, it is highly questionable to use a member with these proportions as a column. Therefore, we consider the next larger nominal thickness of 8 in. Then, if

$$\frac{L}{d} = \frac{16 \times 12}{7.5} = 25.6$$

we are still in the zone 3 condition, and the allowable stress is

$$F'_c = \frac{0.30E}{(L/d)^2} = \frac{(0.30)(1{,}700{,}000)}{25.6^2} = 778 \text{ psi} \ [5.36 \text{ MPa}]$$

which requires an area of

$$A = \frac{\text{load}}{F'_c} = \frac{40{,}000}{778} = 51.4 \text{ in.}^2 \ [33{,}163 \text{ mm}^2]$$

The smallest member usable is thus an 8 × 8. It is interesting to note that the required square column remains the same for all the column lengths, even though the stress varies from 1200 psi to 778 psi. This is not uncommon and is simply due to the limited number of sizes available for the square column section.

Note: For the following problems use Douglas fir–larch, No. 1 grade.

Problems 5.1.A,B,C,D. Find the allowable axial compression load for each of the following:

	Nominal Size	Unbraced Length	
	(in.)	(ft)	(m)
A	4 × 4	8	2.44
B	6 × 6	12	3.66
C	8 × 8	18	5.49
D	8 × 8	14	4.27

Problems 5.1.E,F,G,H. Select the smallest square section for each of the following:

	Axial Load		Unbraced Length	
	(kips)	(kN)	(ft)	(m)
E	20	89	8	2.44
F	50	222	12	3.66
G	50	222	20	6.10
H	100	445	16	4.88

5.2 Design Aids for Wood Columns

It should be apparent from these examples that the design of wood columns by these procedures is a laborious task. The working designer, therefore, typically utilizes some design aids in the form of graphs, tables, or computer-aided processes. One should exercise care in using such aids, however, to be sure that any specific values for E or F_c that are used correspond to the true conditions of the design work and that the aids are developed from criteria identical to those in any applicable code for the work.

Figure 5.4 consists of a graph from which the axial compression capacity of solid, square wood columns may be determined. Note that the graph curves are based on a specific species and

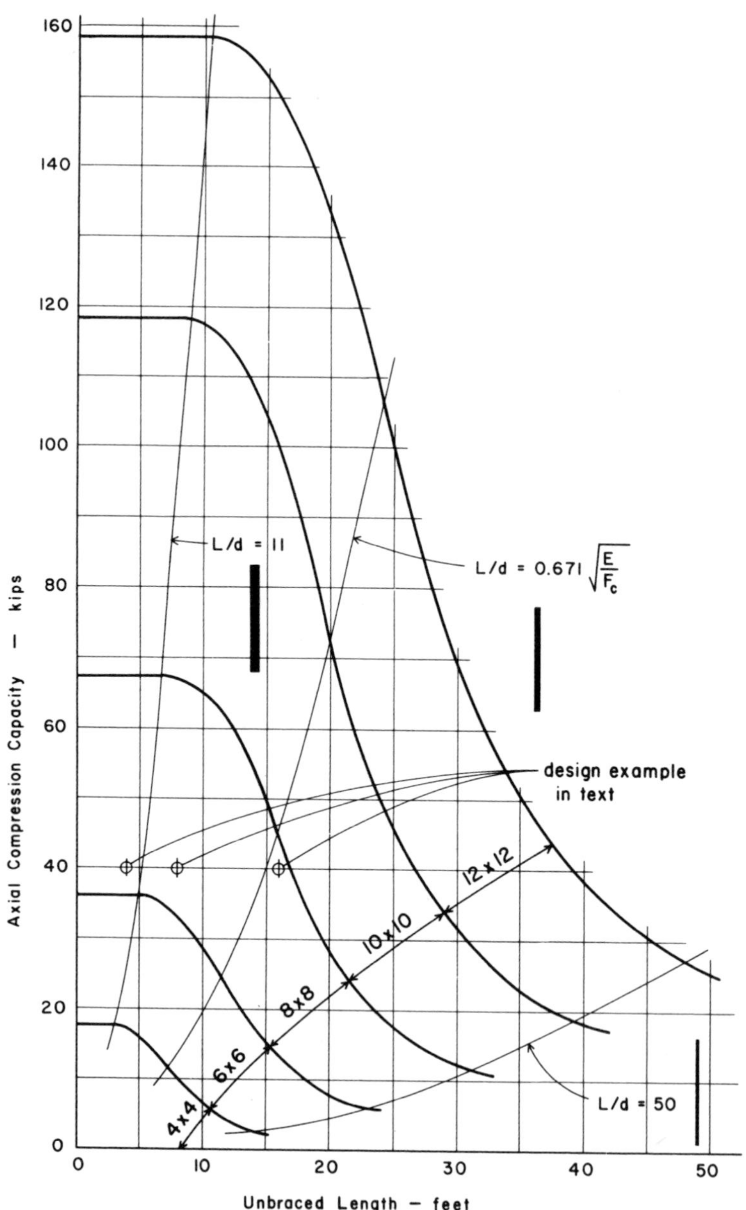

FIGURE 5.4 Axial compression load capacity for wood members of square cross section. Derived from NDS requirements for Douglas fir–larch, dense No. 1 grade.

grade of wood (Douglas fir–larch, dense No. 1 grade). The three circled points on the graph correspond to the design examples in Example 3 of Sec. 5.1.

Table 5.1 gives the axial compression capacity for a range of sizes and unbraced lengths of solid, rectangular wood sections. Note that the design values for elements with nominal thickness of 4 in. and less are different from those with nominal thickness of

TABLE 5.1 Axial Compression Capacity of Solid Wood Elements[a]

Element Size		Unbraced Length (ft)							
Designation	Area of Section (in²)	6	8	10	12	14	16	18	20
2 × 3	3.375	0.8			*L/d* greater than 50				
2 × 4	5.25	1.3							
3 × 4	8.75	6.0	3.4	2.2					
3 × 6	13.75	9.4	5.3	3.4					
4 × 4	12.25	14.7	9.3	5.9	4.1	3.0			
4 × 6	19.25	23.1	14.6	9.3	6.5	4.8			
4 × 8	25.375	30.4	19.2	12.3	8.5	6.3			
6 × 6	30.25	35.4	33.5	29.6	22.5	16.6	12.7	10.0	8.1
6 × 8	41.25	48.3	45.7	40.3	30.6	22.6	17.3	13.6	11.0
6 × 10	52.25	61.2	57.9	51.1	38.8	28.6	21.9	17.2	14.0
6 × 12	63.25	74.1	70.1	61.8	47.0	34.7	26.5	20.9	17.0
8 × 8	56.25	67.5	66.0	63.9	60.0	53.6	43.8	34.6	28.0
8 × 10	71.25	85.5	83.6	80.9	75.9	67.9	55.4	46.9	35.5
8 × 12	86.25	103.5	101.2	98.0	91.9	82.2	67.1	53.0	42.9
8 × 14	101.25	121.5	118.9	115.0	107.9	96.5	78.8	62.3	50.4
10 × 10	90.25	108.3	108.3	106.0	103.6	99.6	93.5	84.5	72.1
10 × 12	109.25	131.1	131.1	128.4	125.4	120.6	113.2	102.4	87.3
10 × 14	128.25	153.9	153.9	150.7	147.2	141.6	132.9	120.2	102.5
10 × 16	147.25	176.7	176.7	173.0	169.0	162.5	152.5	138.0	117.7
12 × 12	132.25	158.7	158.7	158.7	155.5	152.7	148.6	142.5	134.1

[a] Load capacity in kips for solid-sawn sections of dense No. 1 Douglas fir–larch under normal moisture and load duration conditions.

6 in. and over, owing to the different size classifications as given in Table 4.1.

Problems 5.2.A,B,C,D. Select square column sections for the loading and lateral bracing conditions given for Problems 5.1.E,F,G,H, using data from Table 5.1. Note that in the problems in Sec. 5.1, the wood is No. 1 grade, while Table 5.1 uses dense No. 1 grade. It is possible, therefore, that the selections from the table may not agree with those made from the computations.

5.3 Studs

Studs are the vertical elements used for wall framing in the light wood-framed system. They serve utilitarian purposes of providing for attachment of the wall surfacing materials but also serve as vertical columns for support of roof or floor systems for which the wall may serve bearing support functions. The most common stud is the 2 × 4, spaced at 16- or 24-in. centers.

Studs of nominal 2-in.-thick lumber must be braced on their weak axis if used for story-high walls, a simple requirement deriving from the limit for slenderness of the solid-sawn column. Wall finish materials will in most cases serve this bracing function, although some horizontal blocking at midheight can also provide for a reduction of unbraced length on the weak axis, as shown in Fig. 5.1*b*. Where walls are braced by finish on only one side, the blocking is essential.

Studs may also serve other functions, as in the case of exterior walls, where the studs must usually work as vertically spanning elements for lateral loads. For this situation the studs must be designed for the combined actions of axial compression plus bending, as discussed in Sec. 5.4.

In colder climates it is now common to use studs of greater width than the nominal 4 in. to create a larger void space for installation of insulation. This generally results in studs that are quite redundantly strong for the structural tasks of one- or two-story construction. Wall heights may also be increased with the wider studs, where the 2 × 4 limits freestanding unbraced height to about 14 ft.

If loads are high from gravity forces, as with multistory construction or long spans of roofs or floors, it may be necessary to strengthen the stud wall. This can be achieved in a number of ways, such as:

Decreasing the stud spacing to 12-in. centers.

Increasing the stud thickness to 3 in. nominal.

Increasing the stud width to a nominal size greater than 4 in.

Using doubled or tripled studs (or larger-sized members) as posts directly under heavily loaded beams.

It is also sometimes necessary to use thicker studs or to restrict stud spacing when required nailing for plywood shear walls is excessive. This is discussed in Part V.

In general, studs are columns and must comply to the various requirements for design of solid-sawn sections. Any appropriate grade of wood may be used, although special stud grades are commonly used for 2 × 4 members.

5.4 Columns With Bending

In wood structures columns with bending occur most frequently as shown in Fig. 5.5. Studs in exterior walls represent the situation shown in Fig. 5.5*a*, when considered for the case of vertical gravity plus horizontal wind loadings. In various situations, due to framing details, a column carrying only vertical loads may sustain bending if the load is not axial to the column, as shown in Fig. 5.5*b*. Investigation of columns with bending may be quite simple or very complex, depending on a number of qualifying conditions.

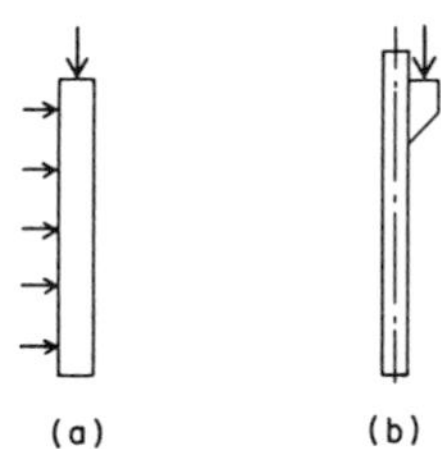

FIGURE 5.5 Columns with combined axial compression plus bending.

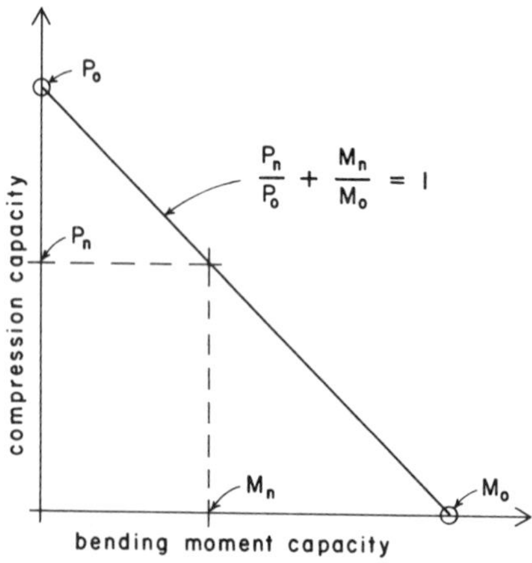

FIGURE 5.6 Column interaction: compression plus bending.

The general consideration for a member subjected to column and beam actions simultaneously is on the basis of *interaction*. The classic form of interaction is represented by the graph in Fig. 5.6. Referring to the notation on the graph:

The maximum axial load capacity of the member with no bending is P_0.

The maximum bending moment capacity of the member without axial compression is M_0.

At some compression load below P_0 (indicated as P_n), the member is assumed to have some tolerance for a bending moment (indicated as M_n) in combination with the axial load.

Combinations of P_n and M_n are assumed to fall on a line connecting P_0 and M_0. The equation of this line has the form

$$\frac{P_n}{P_0} + \frac{M_n}{M_0} = 1$$

A graph similar to that in Fig. 5.6 can be constructed using stresses rather than loads and moments. This is the procedure used with wood and steel members, the investigative equation taking a simple form expressed as

$$\frac{f_a}{F_a} + \frac{f_b}{F_b} \leq 1$$

where f_a = actual stress due to the axial load,
F_a = allowable column-action stress,
f_b = actual stress due to bending,
F_b = allowable beam-action stress in flexure.

The form of this expression permits consideration of the combined actions while allowing for individual treatment of the two separate phenomena of column action and beam action. Current criteria for wood column design uses the simple interaction relationship as a fundamental reference with various adjustments for special cases. The basis of these adjustments is intended primarily to account for two potential complications. The first of these is the *P*-delta effect, which occurs when the deflection due to bending moves the centroid of the column at midheight away from the line of action of the axial compression force. This results in some additional bending moment of a magnitude equal to the product of the compression force (P) times the deflection (delta). This is in general only critical if there is considerable deflection due to a very high span-to-depth ratio of the member or a high magnitude of the lateral loading. In the case of the loading in Fig. 5.5*b*, the entire moment is a *P*-delta effect.

The second adjustment to the simple interaction relationship involves the potential for lateral buckling due to bending. This is usually accounted for by using a value for F_b that is adjusted in the usual manner for a bending member with a critical laterally unsupported length condition.

As presented in the NDS, the general form of the interaction relationship is

$$\frac{f_c}{F'_c} + \frac{f_b}{F'_b - Jf_c} \leq 1$$

in which f_c and f_b are the computed axial compression and bending stresses, respectively.

F'_c is the usual allowable column compression stress as adjusted for the condition of slenderness, as discussed in Sec. 5.1. F'_b is the allowable flexural stress, adjusted if necessary for any stability considerations. J is a factor generally computed by the expression

$$J = \frac{(L_e/d) - 11}{K - 11}$$

in which K is the factor used in determination of F'_c, as discussed in Sec. 5.1.

For the three zones of relative stiffness shown in Fig. 5.3, the use of J is as follows:

Zone 1, $L/d \leq 11$: $J = 0$.
Zone 2, $11 \leq L/d \leq K$: $J =$ the value from the expression.
Zone 3, $L/d \geq K$: $J = 1$.

For slender, unbraced columns it may be necessary to use an adjusted value for F_b. However, based on code qualifications, we note the following two exemptions.

1. For square sections used as beams no lateral support is required and no stress adjustment is made.
2. When the compression edge is continuously supported (as for a typical wall stud) the unsupported length may be considered as zero; thus no stress adjustment is required.

5.5 Miscellaneous Wood Columns

Spaced Columns. Spaced columns consist of multiple wood elements bolted together to form single compression members. They occur most commonly as compression members in large wood trusses. Individually considered, they have a general form as shown in Fig. 5.7. In the direction indicated by the larger

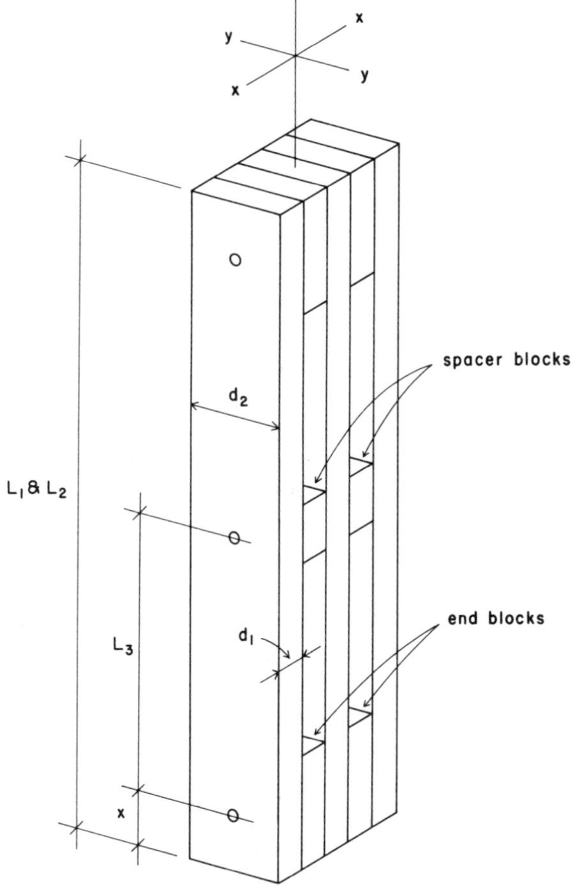

FIGURE 5.7 The spaced column.

dimension of the individual elements (dimension d_2 in Fig. 5.7) they behave simply as multiples of the individual parts, and are limited to compression based on consideration of L/d in the usual manner. In the other direction, however—that relating to the dimension d_1 in the figure—the individual elements behave as fixed-end columns, due to the effect of the bolts and blocking at the column ends. On the axis relating to d_1 they are limited to an overall L/d ratio of 80 (using L_1 in the figure) and an individual L/d of 40 (using L_3 in the figure). The allowable stress based on the d_1 dimension is increased by a factor that depends on the details of the end bolting, notably the ratio of the distance x in the figure to the overall column length.

Built-Up Columns. In various situations columns may consist of multiple elements of solid-sawn sections. Although the description includes glued laminated and spaced columns, the term is generally used for multiple-element columns that do not qualify for those conditions. Glued laminated columns are essentially designed as solid sections. Spaced columns that qualify as such are designed by the criteria discussed in Sec. 5.1.

Built-up columns generally have the elements attached to each other by mechanical devices, such as nails, lag screws, or machine bolts. Unless the particular assembly has been load tested for code approval, it is usually designed on the basis of the single-element capacity. That is, the least load capacity is the sum of the capacities of the individually considered parts.

The most common built-up column is the multiple stud assembly that occurs at corners, wall intersections, and edges of door or window openings in the light wood-framed structure. Since these are braced by wall surfacing in most cases, the aggregate capacity is simply the sum of the individual stud capacities.

When built-up columns occur as freestanding columns, it may be difficult to make a case for rational determination of their capacities, unless single elements have low enough individual slendernesses to qualify for significant capacities. Two types of assembly that have some proven capacity as built-up columns are those shown in Fig. 5.8. In Fig. 5.8a a solid-core column is

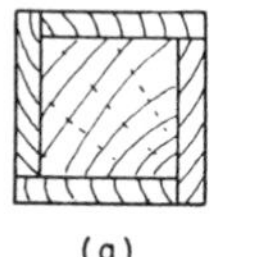

FIGURE 5.8 Sections of built-up columns.

wrapped on all sides by thinner elements. The thinner elements may be assumed to borrow stiffness from the core and an allowable stress for the whole section might be that based on the core's slenderness. In Fig. 5.8*b* a series of thin elements is held together by two cover plates, which tend to restrict the buckling of the core elements on their weak axes. A reasonable assumption for the latter section is to assume the cover plates to brace the core elements so that their slenderness may be considered by using their larger dimension.

Glued Laminated Columns. Columns consisting of multiples of glued laminated 2-in.-nominal-thickness lumber are sometimes used. The advantages of the higher-strength material may be significant—more so if combined compression and flexure must be developed. However, as in other situations, the much higher quality of dimensional stability may be a major factor.

The large glued laminated column section offers the same general advantages of fire resistance that are assigned to heavy timber construction with solid-sawn sections. This does not make it more competitive in comparison to the solid section, but does in comparison to an exposed steel column.

It is also possible to produce columns of great length, tapered form, considerable width, and so on—in other words, all of the potentials offered for other laminated items. In general, laminated columns are used less frequently than laminated beams, and are generally chosen only when some of the limitations of other options for the column are restrictive.

A special application of the laminated column is in built-up sections, such as that shown in Fig. 5.8*b*. In these situations the laminated portion is the functioning structural member and the added solid-sawn elements are limited to decorative functions or use for other construction reasons.

Poles. Poles are round timbers consisting of the peeled logs of coniferous trees. In short lengths they may be of relatively constant diameter, but have a typically tapered profile when long—the natural form of the tree trunk. As columns, poles are designed with the same basic criteria used for rectangular sawn sections. For slenderness considerations the d used for L/d computations is taken as that of a square section's side dimension, the square having an area equal to that of the round section. Thus, calling the pole diameter D,

$$d^2 = \frac{\pi D^2}{4}, \qquad d = 0.886D$$

For the tapered column, a conservative assumption for design is that the critical column diameter is the least diameter at the narrow end. If the column is very short, this is reasonable. However, for a slender column, with buckling occurring in the midheight of the column, this is very conservative and the code provides for some adjustment. Nevertheless, because of a typical lack of initial straightness and presence of numerous flaws, many designers prefer to use the unadjusted small end diameter for design computations.

Poles are used as timber piles and are also used as buried-end posts for fences, signs, and utility transmission lines. Buildings of *pole construction* typically use buried-end posts as building columns. For lateral forces, buried-end posts must be designed for bending, and if sustaining significant vertical loads, as combined action members with bending plus axial compression. For the latter situations it is common to consider the round pole to be equivalent to a square sawn section with the same cross-sectional area and to use the code criteria for solid-sawn sections.

The wood-framed structure utilizing poles has a long history of use and continues to be a practical solution for utilitarian buildings where good poles are readily available. Accepted practices of construction for these buildings are based mostly on experience and do not always yield to highly rational analysis. If it works and many long-standing examples have endured the ravages of time and climate, it is hard to make a case on theory alone.

6

Wood Fastenings

Structures of wood typically consist of large numbers of separate pieces that must be fastened together for the structural action of the whole system. Fastening is sometimes achieved directly, that is, without an intermediate device. Such is the case of the fitted joints used in cabinet and furniture work. However, for building construction, fastening is most often achieved by using some steel device, common ones being nails, screws, bolts, sheet-metal fasteners, and shear developers.

6.1 Bolted Joints in Wood Structures

When steel bolts are used to connect wood members, there are several design considerations. The principal concerns are the following:

1. *Net Stress in Member.* Holes drilled for the placing of bolts reduce the member cross section. For this analysis the hole is assumed to have a diameter $\frac{1}{16}$ in. larger than that of the bolt. The most common situations are those shown in Fig. 6.1. When bolts are staggered, it may be necessary to make two investigations, as shown in the illustration.

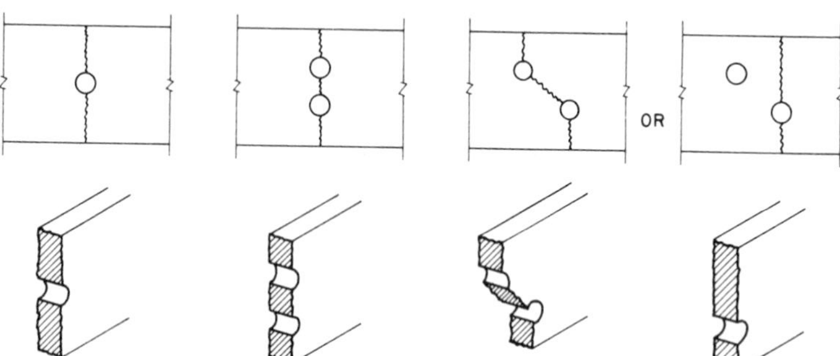

FIGURE 6.1 Effect of bolt holes on reduction of cross section for tension members.

2. *Bearing of the Bolt on the Wood and Bending in the Bolt.* When the members are thick and the bolt thin and long, the bending of the bolt will cause a concentration of stress at the edges of the members. The bearing on the wood is further limited by the angle of the load to the grain, since wood is much stronger in compression in the grain direction.
3. *Number of Members Bolted.* The worst case, as shown in Fig. 6.2, is that of the two-member joint. In this case the lack of symmetry in the joint produces considerable twisting. This situation is referred to as single shear, since the bolt is subjected to shear on a single plane. When more members are joined, this twisting effect is reduced.
4. *Ripping Out the Bolt When Too Close to an Edge.* This problem, together with that of the minimum spacing of the bolts in multiple-bolt joints, is dealt with by using the crite-

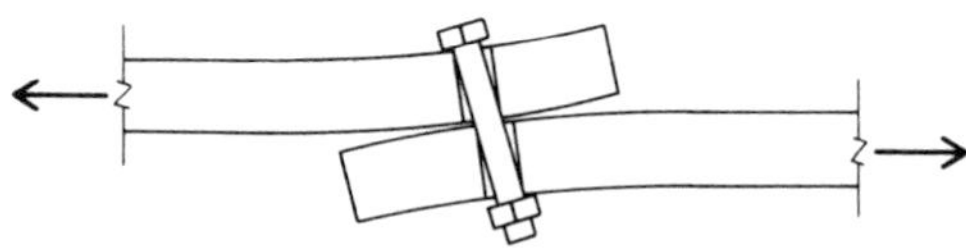

FIGURE 6.2 Behavior of the single lapped joint with the bolt in single shear.

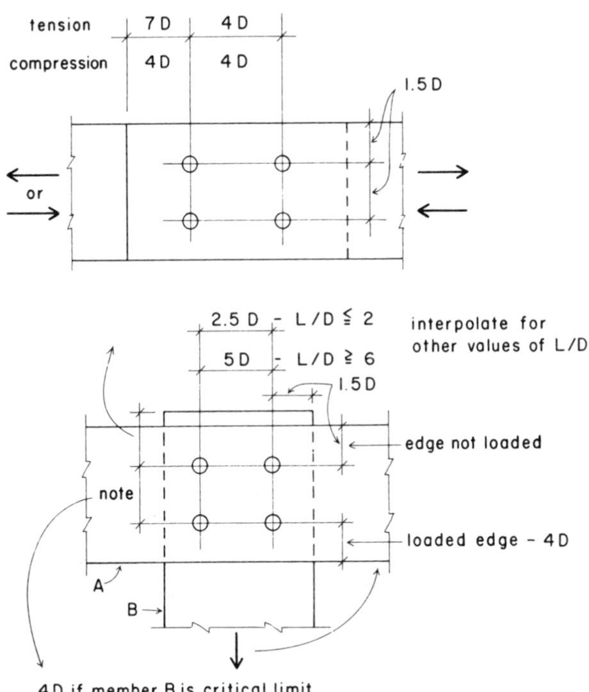

FIGURE 6.3 Edge, end, and spacing distances for bolts in wood structures.

ria given in Fig. 6.3. Note that the limiting dimensions involve the consideration of: the bolt diameter *D;* the bolt length *L;* the type of force—tension or compression; and the angle of load to the grain of the wood.

The bolt design length is established on the basis of the number of members in the joint and the thickness of the wood pieces. There are many possible cases, but the most common are those shown in Fig. 6.4. The critical lengths for these cases are given in Table 6.1. Also given in the table is the factor for determining the

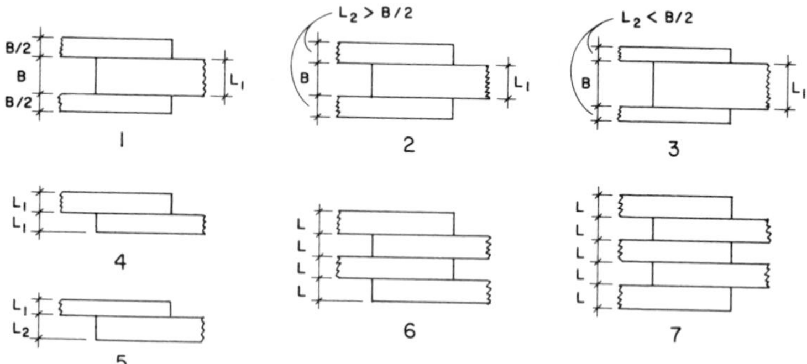

FIGURE 6.4 Various cases of lapped joints with relation to the determination of the critical bolt length.

allowable load on the bolt. Allowable loads ordinarily are tabulated for the three-member joint (case 1, Table 6.1), and the factors represent adjustments for other conditions.

Table 6.2 gives allowable loads for bolts with wood members of dense grades of Douglas fir–larch and Southern pine. The two loads given are that for a load parallel to the grain (*P* load) and that for a load perpendicular to the grain (*Q* load). Fig. 6.5 illustrates these two loading conditions, together with the case of a load at some other angle (Θ). For such cases it is necessary to find the allowable load for the specific angle. Figure 6.5d is an adaptation of the Hankinson graph, which may be used to find values for loads at some angle to the grain.

TABLE 6.1 Design Length for Bolts

Case[a]	Critical Length	Modification Factor
1	L_1	1.0
2	L_1	1.0
3	$2L_2$	1.0
4	L_1	0.5
5	Lesser of L_2 or $2L_1$	0.5
6	L	1.5
7	L	2.0

[a] See Fig. 6.4.

TABLE 6.2 Bolt Design Values for Wood Joints for Douglas Fir–Larch and Southern Pine

Design Length of Bolt (in.)	Diameter of Bolt (in.)	Design Values for One Bolt in Double Shear[a] (lb)			
		Parallel to Grain Load (*P*)		Perpendicular to Grain Load (*Q*)	
		Dense Grades	Ordinary Grades	Dense Grades	Ordinary Grades
1.5	1/2	1100	940	500	430
	5/8	1380	1180	570	490
	3/4	1660	1420	630	540
	7/8	1940	1660	700	600
	1	2220	1890	760	650
2.5	1/2	1480	1260	840	720
	5/8	2140	1820	950	810
	3/4	2710	2310	1060	900
	7/8	3210	2740	1160	990
	1	3680	3150	1270	1080
3.0	1/2	1490	1270	1010	860
	5/8	2290	1960	1140	970
	3/4	3080	2630	1270	1080
	7/8	3770	3220	1390	1190
	1	4390	3750	1520	1300
3.5	1/2	1490	1270	1140	980
	5/8	2320	1980	1330	1130
	3/4	3280	2800	1480	1260
	7/8	4190	3580	1630	1390
	1	5000	4270	1770	1520
5.5	5/8	2330	1990	1650	1410
	3/4	3350	2860	2200	1880
	7/8	4570	3900	2550	2180
	1	5930	5070	2790	2380
	1 1/4	8940	7640	3260	2790
7.5	5/8	2330	1990	1480	1260
	3/4	3350	2860	2130	1820
	7/8	4560	3890	2840	2430
	1	5950	5080	3550	3030
	1 1/4	9310	7950	4450	3800

[a] See Table 6.1 for modification factors for other conditions.

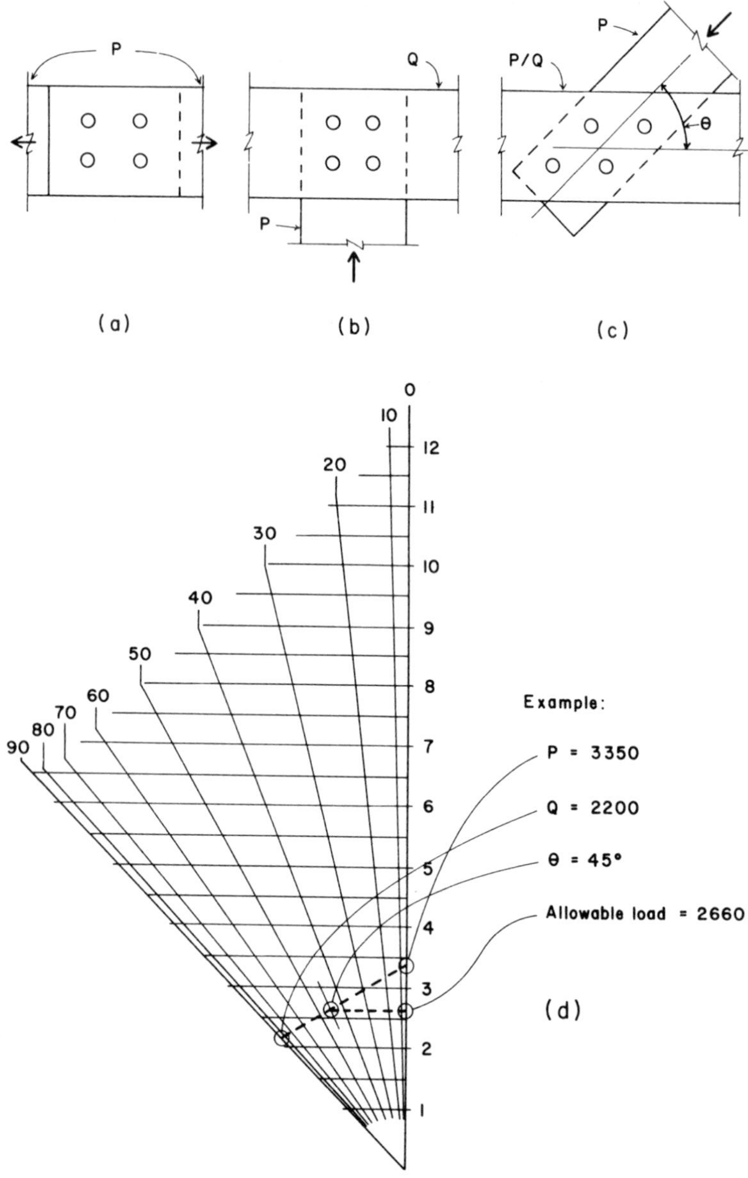

FIGURE 6.5 Relation of load to grain direction in bolted joints. (*a*)–(*c*) show determination of load direction as *P:* parallel to grain, *Q:* perpendicular to grain, or Θ: angle to grain. (*d*) is the Hankinson graph for determination of bolt values for angle to grain loading.

The following examples illustrate the use of the data presented for the design of bolted joints.

Example 1. A three-member bolted joint is made with members of Douglas fir–larch, dense No. 1 grade. The members are loaded in a direction parallel to the grain (Fig. 6.5*a*), with a tension force of 10 kips. The middle member is a 3 × 12 and the outer members are each 2 × 12. If four $\frac{3}{4}$-in. bolts are used for the joint, is the joint capable of carrying the tension load?

Solution: The first step is to identify the critical bolt length (Fig. 6.4) and the load factor (Table 6.1). Since the outer members are greater than one-half the thickness of the middle member, the condition is that of case 2 in Fig. 6.4, and the effective length is the thickness of the middle member: 2.5 in. The load factor from Table 6.1 is 1.0, indicating that the tabulated load may be used with no adjustment.

From Table 6.2 the allowable load per bolt is 2710 lb. (Bolt design length of 2.5 in.; bolt diameter of $\frac{3}{4}$ in.; *P* load; dense grade.) With the four bolts, the total capacity of the bolts is thus 4 × 2710 = 10,840 lb.

For tension stress in the wood, the critical condition is for the middle member, with the net section through the bolts being that shown in Fig. 6.1*b*. With the holes being considered as $\frac{1}{16}$ in. larger than the bolts, the net area for tension stress is thus

$$A = 2.5 \times \left[11.25 - \left(2 \times \frac{13}{16}\right)\right] = 24.06 \text{ in.}^2$$

From Table 4.1 the allowable tension stress is 1200 psi. The maximum tension capacity of the member in tension at the net section is thus

$$T = \text{allowable stress} \times \text{net area} = 1200 \times 24.06 = 28{,}872 \text{ lb}$$

and the joint is adequate for the required load.

Example 2. A bolted two-member joint consists of two 2 × 10 members of Douglas fir–larch, dense No. 2 grade, attached at right angles to each other, as shown in Fig. 6.5*b*. If the joint is made with two $\frac{7}{8}$-in. bolts, what is the maximum compression capacity of the joint?

Solution: This is case 4 in Fig. 6.4, and the effective length is the member thickness of 1.5 in. The modification factor from Table 6.1 is 0.5, and the bolt capacity from Table 6.2 is 700 lb per bolt. (Bolt design length of 1.5 in.; bolt diameter of $\frac{7}{8}$ in.; Q load; dense grade.) The total capacity of the bolts is thus

$$C = 2 \times 700 \times 0.5 = 700 \text{ lb}$$

The net section is not a concern for the compression force, as the capacity of the members would be based on an analysis for the slenderness condition based on the L/d of the members.

Example 3. A three-member bolted joint consists of two outer members, each 2 × 10, and a middle member that is 4 × 12. The outer members are arranged at an angle to the middle member, as shown in Fig. 6.5*c*, such that $\Theta = 60°$. Find the maximum compression force that can be transmitted through the joint by the outer members. Wood is Douglas fir–larch, No. 1 grade. The joint is made with two $\frac{3}{4}$-in. bolts.

Solution: In this case we must investigate both the outer and middle members. For the outer members the effective length is 2 × 1.5 = 3.0 in., and the modification factor is 1.0 (case 3, Fig. 6.4 and Table 6.1). From Table 6.2 the bolt capacity based on the outer members is 2630 lb per bolt. (Bolt design length of 3.0 in.; bolt diameter of $\frac{3}{4}$-in.; P load; ordinary grade.)

For the middle member the effective bolt length is the member thickness of 3.5 in., and the unadjusted load per bolt from Table 6.2 is 2800 lb for the P condition and 1260 lb for the Q condition. If these values are used on the Hankinson graph in Fig. 6.5, the load per bolt for the 60° angle is found to be approximately 1700 lb. Since this value is lower than that found for the outer members, it represents the limit for the joint. The joint capacity based on the bolts is thus 2 × 1700 = 3400 lb.

Note: For all of the following problems, use Douglas fir–larch, dense No. 1 grade.

Problem 6.1.A. A three-member tension joint has 2 × 12 outer members and a 4 × 12 middle member (Fig. 6.5*a*). The joint is made with six $\frac{3}{4}$-in. bolts. Find the capacity of the joint as limited by the bolts and the tension stresses in the members.

Problem 6.1.B. A two-member tension joint consists of 2 × 6 members bolted with two $\frac{7}{8}$-in. bolts (Fig. 6.5*a*). What is the limit for the tension force?

Problem 6.1.C. Two outer members, each 2 × 8, are bolted to a middle member consisting of a 3 × 12. The outer members form an angle of 45° with respect to the middle member (Fig. 6.5*c*; $\Theta = 45°$). What is the maximum compression force that the outer members can transmit to the joint with two $\frac{3}{4}$-in. bolts?

6.2 Nailed Joints

Nails are used in great variety in building construction. For structural fastening, the nail most commonly used is called—appropriately—the *common wire nail.* As shown in Fig. 6.6, the critical concerns for such nails are the following:

1. *Nail Size.* Critical dimensions are the diameter and length. Sizes are specified in pennyweight units, designated as 4d, 6d, and so on, and referred to as four penny, six penny, and so on.
2. *Load Direction.* Pullout loading in the direction of the nail shaft is called *withdrawal;* shear loading perpendicular to the nail shaft is called *lateral load.*
3. *Penetration.* Nailing is typically done through one element and into another, and the load capacity is limited by the

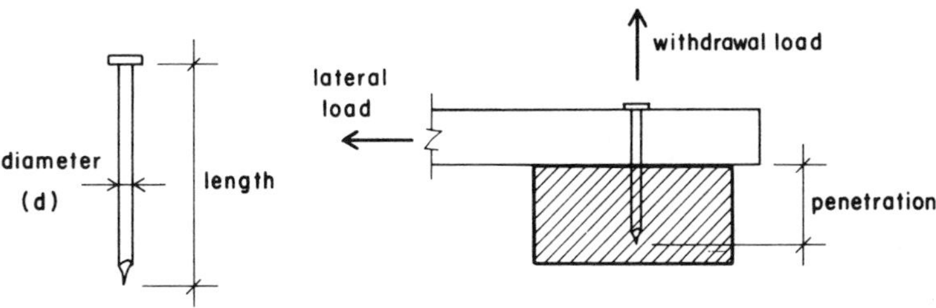

FIGURE 6.6 Typical common wire nail and loading conditions.

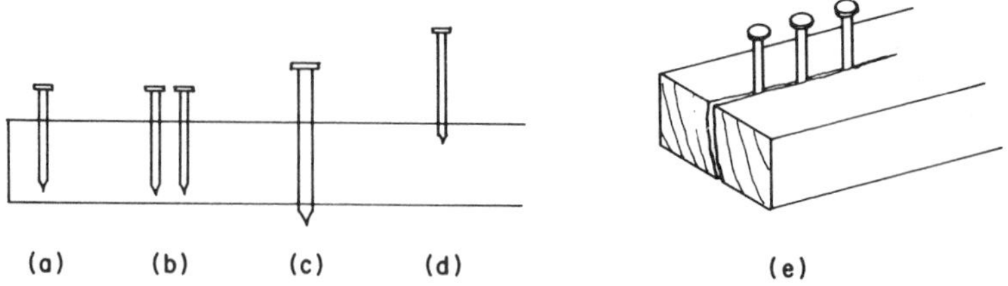

FIGURE 6.7 Poor nailing practices: (*a*) too close to edge, (*b*) nails too closely spaced, (*c*) nail too large for wood piece, (*d*) too little penetration, (*e*) too many nails in a single row parallel to the wood grain.

amount of length of embedment of the nail in the second member. The length of embedment is called the penetration.

4. *Species and Grade of Wood.* The harder, tougher, and heavier the wood, the more the load resistance capability.

Design of good nail joints requires a little engineering and a lot of good carpentry. Some obvious situations to avoid are those shown in Fig. 6.7. A little actual carpentry experience is highly desirable for anyone who designs nailed joints.

Withdrawal load capacities of common wire nails are given in Table 6.3. The capacities are given for both Douglas fir–larch and Southern pine. Note that the table values are given in units of capacity per inch of penetration and must be multiplied by the actual penetration length to obtain the load capacity in pounds. In general, it is best not to use structural joints that rely on withdrawal loading resistance.

Lateral load capacities for common wire nails are given in

TABLE 6.3 Withdrawal Load Capacity of Common Wire Nails (lb/in.)

	Size of Nail				
Pennyweight	6	8	10	12	16
Diameter (in.)	0.113	0.131	0.148	0.148	0.162
Douglas fir–larch	29	34	38	38	42
Southern pine	35	41	46	46	50

TABLE 6.4 Lateral Load Capacity of Common Wire Nails (lb/nail)

	Size of Nail				
Pennyweight	6	8	10	12	16
Diameter (in.)	0.113	0.131	0.148	0.148	0.162
Length (in.)	2.0	2.5	3.0	3.25	3.5
Douglas fir–larch and Southern pine	63	78	94	94	108
Penetration required for 100% of table value[a] (in.)	1.24	1.44	1.63	1.63	1.78
Minimum penetration[b] (in.)	0.42	0.48	0.54	0.54	0.59

[a] Eleven diameters; reduce by straight-line proportion for less penetration.
[b] One-third of that for full value; $\frac{11}{3}$ diameters.

Table 6.4. These values also apply to Douglas fir–larch and Southern pine. Note that a penetration of at least 11 times the nail diameter is required for the development of the full capacity of the nail. A value of one-third of that in the table is permitted with a penetration of one-third of this length, which is the minimum penetration permitted. For actual penetration lengths between these limits, the load capacity may be determined by direct proportion. Orientation of the load to the direction of grain in the wood is not a concern when considering nails in terms of lateral loading.

The following example illustrates the design of a typical nailed joint for a wood truss.

Example. The truss heel joint shown in Fig. 6.8 is made with 2-in. nominal wood elements of Douglas fir–larch, dense No. 1 grade, and gusset plates of $\frac{1}{2}$-in. plywood. Nails are 6d common, with the nail layout shown occurring on both sides of the joint. Find the tension force limit for the bottom chord (load 3 in the illustration).

Solution: The two primary concerns are for the lateral capacity of the nails and the tension, tearing stress in the gussets. For the nails, we observe from Table 6.4 that

Nail length is 2 in. [51 mm].

Minimum penetration for full capacity is 1.24 in. [31.5 mm].

Maximum capacity is 63 lb/nail [0.28 kN].

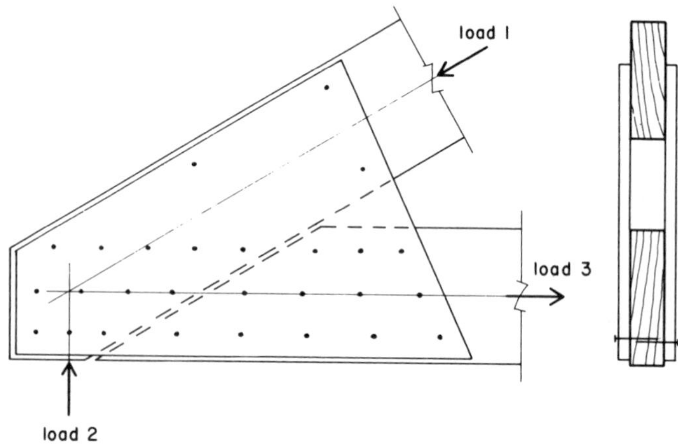

FIGURE 6.8 Truss joint with nails and plywood gusset plates.

From inspection of the joint layout, we see that

$$\begin{aligned}\text{Actual penetration} &= \text{nail length} - \text{plywood thickness}\\ &= 2.0 - 0.5 = 1.5 \text{ in. [38 mm]}\end{aligned}$$

Therefore, we may use the full table value for the nails. With 12 nails on each side of the member, the total capacity is thus

$$F_3 = (24)(63) = 1512 \text{ lb [6.73 kN]}$$

If we consider the cross section of the plywood gussets only in the zone of the bottom chord member, the tension stress in the plywood will be approximately

$$f_t = \frac{1512}{(0.5)(2)(5 \text{ in. of width})} = 302 \text{ psi [2.08 MPa]}$$

which is probably not a critical magnitude for the plywood.

A problem that must be considered in this type of joint is that of the pattern of placement of the nails (commonly called the layout of the nails). In order to accommodate the large number of nails required, they must be quite closely spaced, and since they are close to the ends of the wood pieces, the possibility of splitting the wood is a critical concern. The factors that determine this possibility include the size of the nail (essentially its diameter), the spacing of the nails, the distance of the nails from the end of the piece, and the tendency of the particular wood species to be split. There are no formal guidelines for this problem; it is largely a matter of good carpentry or some experimentation to establish the feasibility of a given layout.

One technique that can be used to reduce the possibility of splitting is to stagger the nails rather than to arrange them in single rows. Another technique is to use a single set of nails for both gusset plates, rather than to nail the plates independently, as shown in Fig. 6.9. The latter procedure consists simply of driving a nail of sufficient length so that its end protrudes from the gusset on the opposite side and then bending the end over—called clinching—so that the nail is anchored on both ends. A single nail may thus be utilized for twice its rated capacity for lateral load. This is similar to the development of a single bolt in double shear in a three-member joint.

It is also possible to glue the gusset plates to the wood pieces and to use the nails essentially to hold the plates in place only until the glue has set and hardened. The adequacy of such joints

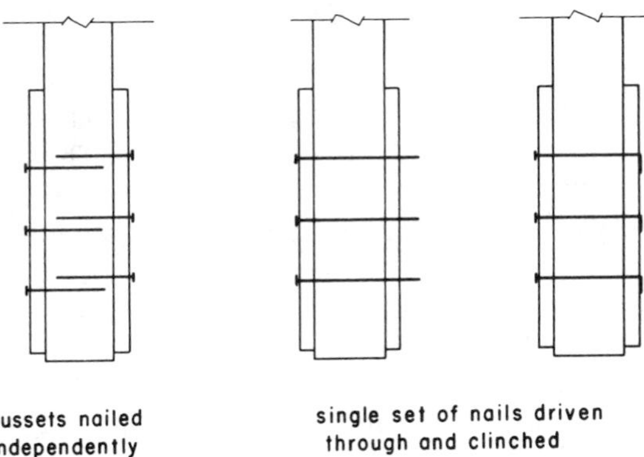

FIGURE 6.9 Nailing techniques for joints with plywood gusset plates.

should be verified by load testing, and the nails should be capable of developing some significant percentage of the design load as a safety backup for the glue.

Problem 6.2.A. Two wood members of 3-in. nominal thickness are attached at right angles to each other by plywood gussets on both sides, to develop a tension force through the joint. The wood is Douglas fir–larch, No. 1 grade; the plywood gussets are $\frac{3}{4}$ in. thick; nails are common wire. Select the size of nail and find the number required if the tension force is 2000 lb [8.90 kN].

6.3 Miscellaneous Fastening Devices

There is a considerable array of hardware items available for assembling of wood structures. The following are some of the commonly employed fastening devices, in addition to steel bolts and common wire nails.

Screws and Lag Bolts. When loosening or popping of nails is a problem, a more positive grabbing of the nail by the surrounding wood can be achieved by having some nonsmooth surface on the nail shaft. Many special nails with formed surfaces are used, but another means for achieving this effect is to use threaded screws in place of nails. Screws can be tightened down to squeeze connected members together in a manner that is not usually possible with nails. For dynamic loading, such as shaking by an earthquake, the tight, positively anchored connection is a distinct advantage.

Screws are produced in great variety. The flat head screw is designed to be driven so that the head sinks entirely into the wood, resulting in a surface with no protrusions. The round head screw is usually used with a washer, or may be used to attach metal objects to wood surfaces. The hex head screw is called a *lag screw* or *lag bolt* and is designed to be tightened by a wrench rather than by a screwdriver. Lag bolts are made in a considerable size range and can be used for some major structural connections.

Screws function essentially the same as nails, being used to resist either withdrawal or lateral shear-type loading. Screws must usually be installed by first drilling a guide hole, called a *pilot hole,* with a diameter slightly smaller than that of the screw shaft. Specifications for structural connections with screws give requirements for the limit of pilot holes and for various other details for proper installation. Capacities for withdrawal and lateral loading are given as a function of the screw size and the type of wood holding the screw.

As with nailed joints, the use of screws involves much judgment that is more craft than science. Choice of the screw type, size, spacing, length, and other details of a good joint may be controlled by some specifications, but it is also a matter of experience.

Although it is generally not recommended that nails be relied on for computed loading in withdrawal, screws are not so limited and are often chosen where the details of the connection require such a loading.

Shear Developers. When wood members are lapped at a joint and bolted, it is often difficult to prevent some joint movement in the form of slipping between the lapped members. If force reversals, such as those that occur with wind or seismic effects, cause back-and-forth stress on the joint, this lack of tightness in the connection may be objectionable. Various types of devices are sometimes inserted between the lapped members so that when the bolts are tightened, there is some form of resistance to slipping besides the simple friction between the lapped members.

Toothed or ridged devices of metal are sometimes used for this development of enhanced shear resistance in the lapped joint. They are simply placed between the lapped wood pieces and the tightening of the bolts causes them to bite into both members. Hardware products of various forms and sizes are available and are commonly used, most notably used for heavy, rough timber construction.

A slightly more sophisticated shear developer for the lapped joint is the split-ring connector consisting of a steel ring that is installed by cutting matching circular grooves in the faces of the lapped pieces of wood. When the ring is inserted in the grooves and the bolt is tightened, the ring is squeezed tightly into the grooves and the resulting connection has a shear resistance considerably greater than that with the bolt alone.

Mechanically Driven Fasteners. Although the hammer, screwdriver, and hand wrench are still in every carpenter's tool kit, many structural fastenings are now routinely achieved with powered devices. In some cases this has produced fasteners that do not fit the old classifications, and codes will eventually probably be more complex to cover the range of fasteners in use. The simple hand staple gun has been extrapolated into a range of devices, including some that can install some significantly strong structural fasteners. Field attachment of plywood roof and floor deck and various wall covering materials is now often accomplished with mechanical driving equipment.

Formed Steel Framing Elements. Formed metal framing devices have been used for many centuries for the assembly of structures of heavy timber. In ancient times elements were formed of bronze or of cast or wrought iron. Later they were formed of forged or bent and welded steel. Some of the devices used today are essentially the same in function and detail to those used long ago. For large timber members, the connections are generally formed of steel plate that is bent and welded to produce the desired form (Fig. 6.10). The ordinary tasks of attaching beams to columns and columns to footings continue to be required and the simple means of achieving the tasks evolve from practical concerns.

For resistance to gravity loads, connections such as those shown in Fig. 6.10 often have no direct structural function. In theory it is possible to simply rest a beam on top of a column as it is done in some rustic construction. However, for resistance to lateral loads from wind or earthquakes, the tying and anchoring functions of these connecting devices are often quite essential. They also serve a practical function in simply holding the structure in position during the construction process.

A development of more recent times is the extension of the use of metal devices for the assembly of light wood-framed construction. Devices of thin sheet metal, such as those shown in Fig. 6.11, are now quite commonly used for stud and joist construction employing predominantly wood members of 2-in. nominal thickness. As with the devices used for heavy timber construc-

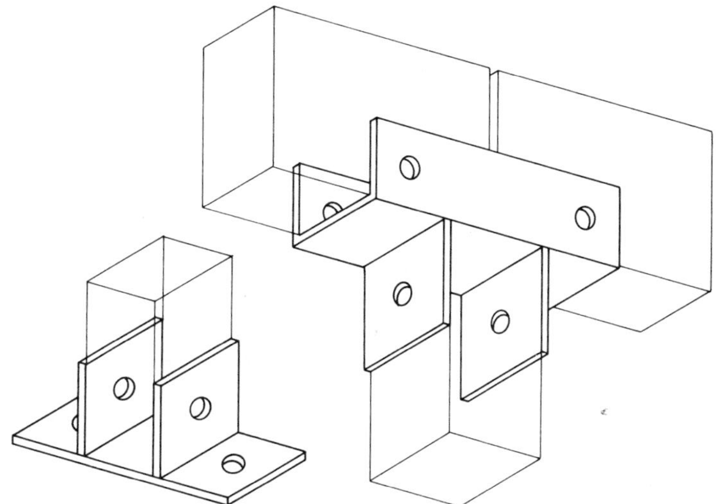

FIGURE 6.10 Formed steel connecting devices: bent and welded steel plate.

tion, these lighter connectors often serve useful functions of tying and anchoring the structure. Load transfers between basic elements of a building's lateral bracing system are often achieved with these elements. (See discussion in Chapter 20.)

Commonly used connection devices of both the light sheet steel type and the heavier steel plate type are readily available from building materials suppliers. Many of these devices are ap-

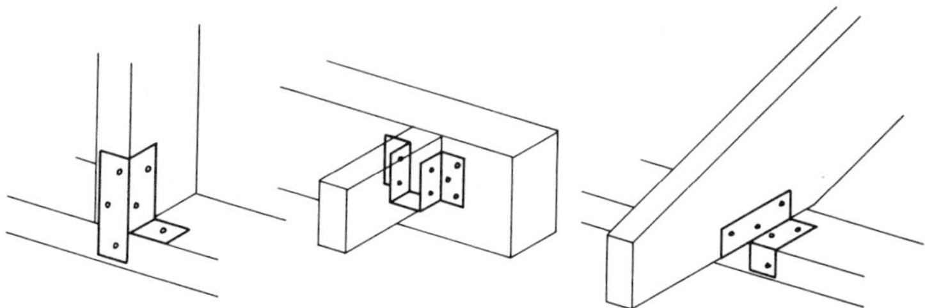

FIGURE 6.11 Formed steel connecting devices: light gage sheet steel.

proved by local building codes and have quantified load ratings which are available from the manufacturers. If these ratings are approved by the building code with jurisdiction for a particular building, the devices can be used for computed structural load transfers. Information should be obtained from local suppliers for these items.

For special situations it may be necessary to design a custom-formed framing device. Such devices can often be quickly and easily made up by local metal fabricating shops. However, the catalogs of manufacturers of these devices are filled with a considerable variety of products for all kinds of situations, and it is wise to determine first that the required device is not available as a standard hardware item before proceeding to design it as a special, custom-made item—which is sure to be more expensive in most cases.

Concrete and Masonry Anchors. Wood members supported by concrete or masonry structures must usually be anchored directly or through some intermediate device. The most common attachment is with steel bolts cast into the concrete or masonry; however, there is also a wide variety of cast-in, drilled-in, or shot-in devices that may be used for various situations. The latter types of anchoring elements are manufactured products, and any data for load capacities or required installation details must be obtained from the manufacturer or supplier.

Two common situations are those shown in Fig. 6.12. The sill member for a wood stud wall is typically attached to a supporting concrete base through steel anchor bolts cast into the concrete. These bolts serve essentially simply to hold the wall securely in position during the construction process. However, they may also serve to anchor the wall against lateral or uplift forces, as is discussed in Chapter 20. For lateral force, the load limit is typically based on the bolt-to-wood limit, as discussed in Sec. 6.1. For uplift the limit may be based on tension stress in the bolt or the pullout limit for the bolt in the encasing concrete or masonry.

Figure 6.12*b* shows a common situation in which a wood-framed roof or floor is attached to a masonry wall through a member bolted flat to the wall surface, called a ledger. For verti-

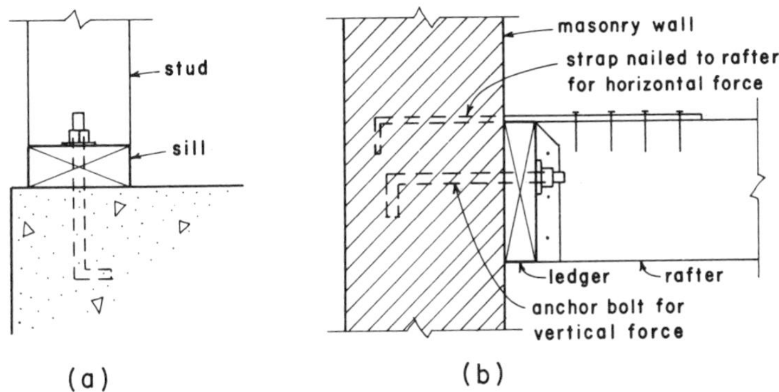

FIGURE 6.12 Anchoring devices for concrete and masonry.

cal load transfer, the shear effect on the bolt is essentially as described in Sec. 6.1. For lateral force the problem is one of pullout or tension stress in the bolt, although a critical concern is for the cross-grain bending effect in the ledger. In zones of high seismic risk, it is usually required to have a separate horizontal anchor to avoid the cross-grain bending in the ledger, although the anchor bolts can still be used for gravity loads.

Plywood Gussets. Cut pieces of plywood are sometimes used as connecting devices, although the increasing variety of formed products for all conceivable purposes has reduced such usage. Trusses consisting of single wood members of a constant thickness are sometimes assembled with gussets of plywood with joints, as shown in Fig. 6.8. Although such connections may have considerable load resistance, it is best to be conservative in using them for computed structural forces, especially with regard to tension stress in the plywood.

STEEL CONSTRUCTION

Steel is used in a wide variety of forms for many tasks in building construction. Wood, concrete, and masonry structures require many steel objects. This part of the book, however, deals with steel as a structural material for the production of components and systems of steel. This usage includes some very common ones and an endless range of special possibilities. Our concentration here will be on ordinary uses.

7

Steel Structural Products

For the assemblage of structures, building components of steel consist mostly of standard forms of industrially produced products. The most commonly and widely used of these products are in forms that have been developed and produced for a long time. Production methods and assemblage and erection techniques change continuously, but the basic forms of ordinary steel structures are pretty much as they have been for many years.

7.1 Materials for Steel Products

Steel is an immensely variable material and is actually produced in hundreds of grades. For structural use, however, it commonly conforms to a limited number of highly controlled grades for specific product applications.

The most widely used source for information for design of steel structures is the *Manual of Steel Construction,* published by the American Institute of Steel Construction. It is commonly referred to simply as the AISC Manual, and we will refer to it as such here. The current edition is the eighth edition, copyrighted in 1980, and it contains the 1978 Specification for the Design, Fabrication, and Erection of Structural Steel for Buildings. This specification, in its various editions, has been generally adopted for reference by code-enforcing agencies in the United States. When referring to this document, we will call it simply the AISC Specification.

Although the AISC is the principal service organization in the area of steel construction, there are many other industry and professional organizations that provide materials for the designer. The American Society for Testing and Materials (ASTM) establishes widely used standard specifications for types of steels, for welding and connector materials, and for various production and fabrication processes. Standard grades of steel and other materials are quite commonly referred to by short versions of their ASTM designation or identification codes. Some other organizations of note are the Steel Deck Institute (SDI), the Steel Joist Institute (SJI), and the American Iron and Steel Institute (AISI).

Steel meeting the requirements of ASTM Specification A36 is the grade commonly used for rolled steel elements utilized in building construction. This steel is required to have a minimum yield point of 36 ksi and a minimum ultimate tensile strength of 58–80 ksi. It may be used for bolted, riveted, or welded fabrication. Other steels are available when special properties are desired, such as higher strength or increased corrosion resistance.

The structural property of primary concern is the yield point, or yield stress, which is designated F_y. Most allowable design stresses are based on this value. The other limiting stress is the ultimate tensile stress, designated F_u, on which a few design stresses are based. For some grades the ultimate stress is given as a range rather than as a single value, in which case it is advisable to use the lower value for design unless a higher value can be verified by a specific supplier for a particular rolled product.

7.2 Rolled Structural Shapes

The process of rolling is used to produce a variety of linear elements, including rods, bars, and plates. If the rolled element has a more complex cross-sectional form it is referred to as a rolled shape.

The rolled structural steel shapes that are most commonly used in building construction are the wide flanges (W shapes), American Standard I-beams (S shapes), channels (C shapes), angles (L shapes), and plates. Tables 3.1–3.5 give properties for some of these products. Complete tables for all available rolled shapes are given in the AISC Manual.

A study of the tables will reveal that wide flange shapes have greater flange widths and thinner webs than standard I-beams and are further characterized by parallel flange surfaces, as contrasted with the tapered inside flange surfaces of the S shapes. In W, M, and S shapes, most of the material is concentrated in the flanges, which makes them especially efficient in resisting bending. W shapes are specified by a nominal dimension of depth (out-to-out of the flanges) which usually varies slightly from the actual depth. S and C shapes have their nominal and actual depths the same.

7.3 Miscellaneous Steel Products

Industrially produced structural products are developed in response to market potentials and the variety of the production processes. In addition to the rolling process, in which a heat-softened mass of steel is progressively squeezed through rollers to work it into a defined shape, some common production processes include:

1. *Wire.* This is formed by pulling (called drawing) the steel through a small opening.
2. *Extrusion.* This is similar to drawing, although the sections produced are other than simple round shapes. This process is not much used for steel products of the sizes used for buildings.
3. *Casting.* This is done by pouring the molten steel into a form (mold). This is limited to objects of a three-dimensional form, and is also not common for building construction elements.

4. *Forging.* This consists of pounding the softened steel into a mold until it takes the shape of the mold. This is preferred to casting because of the effects of the working on the properties of the finished material.

The raw stock steel elements produced by the basic forming processes may be reworked by various means, such as:

1. *Cutting.* Shearing, sawing, punching, or flame cutting can be used to trim and shape specific forms.
2. *Machining.* This may consist of drilling, planing, grinding, routing, or turning on a lathe.
3. *Bending.* Sheets, plates, or linear elements may be bent if made from steel with a ductile character.
4. *Stamping.* This is similar to forging; in this case sheet steel is punched into a mold that forms it into some three-dimensional shape, such as a hemisphere.
5. *Rerolling.* This consists of reworking a linear element into a curved form (arched) or of forming a sheet or flat strip into a formed cross section.

Finally, raw stock or reformed elements can be assembled by various means into objects of multiple parts, such as a manufactured truss or a prefabricated wall panel. Basic means of assemblage include:

1. *Fitting.* Threaded parts may be screwed together or various interlocking techniques may be used, such as the tongue-and-groove joint or the bayonet twist lock.
2. *Friction.* Clamping, wedging, or squeezing with high-tensile bolts may be used to resist the sliding of parts in surface contact.
3. *Pinning.* Overlapping flat elements may have matching holes through which a pin-type device (bolt, rivet, or actual pin) is placed to prevent slipping of the parts at the contact face.

4. *Nailing, Screwing.* Thin elements—mostly with some preformed holes—may be attached by nails or screws.
5. *Welding.* Gas or electric arc welding may be used to produce a bonded connection, achieved partly by melting the contacting elements together at the contact point.
6. *Adhesive Bonding.* This usually consists of some form of chemical bonding that results in some fusion of the materials of the connected parts.

We are dealing here with industrial processes which at any given time relate to the state of development of the technology, the availability of facilities, the existence of the necessary craft, and competition with other materials and products.

7.4 Allowable Stresses for Structural Steel

The designation "structural steel" is given in general to rolled products used for major structural components. The design of structures of this type is done with data and specifications supplied by the steel companies through the AISC.

For structural steel the AISC Specification expresses the allowable unit stresses in terms of some percent of the yield stress F_y or the ultimate stress F_u. Selected allowable unit stresses used in design are listed in Table 7.1. Specific values are given for ASTM A36 steel, with values of 36 ksi for F_y and 58 ksi for F_u. This is not a complete list, but it generally includes the stresses used in the examples in this book. Reference is made to the more complete descriptions in the AISC Specification which is included in the AISC Manual (Ref. 2). There are in many cases a number of qualifying conditions, some of which are discussed in other portions of this book. Table 7.1 gives the location of some of these discussions.

7.5 Nomenclature

The standard symbols—called the general nomenclature—that are used in steel design work are those established in the AISC

TABLE 7.1 Allowable Unit Stresses for Structural Steel: ASTM A36[a]

Type of Stress and Conditions	See Discussion in This Book in:	Stress Designation	AISC Specification	Allowable Stress ksi	Allowable Stress MPa
Tension					
1. On the gross (unreduced) area	Sec. 10-3	F_t	$0.60F_y$	22	150
2. On the effective net area, except at pinholes	Sec. 10-3		$0.50F_u$	29	125
3. Threaded rods on net area at thread			$0.33F_u$	19	80
Compression	Chapter 9	F_a	See discussion		
Shear					
1. Except at reduced sections	Sec. 8-6	F_v	$0.40F_y$	14.5	100
2. At reduced sections	Sec. 10-3		$0.30F_u$	17.4	120
Bending					
1. Tension and compression on extreme fibers of compact members braced laterally, symmetrical about and loaded in the plane of their minor axis	Sec. 8-2	F_b	$0.66F_y$	24	165
2. Tension and compression on extreme fibers of other rolled shapes braced laterally			$0.60F_y$	22	150
3. Tension and compression on extreme fibers of solid round and square bars, on solid rectangular sections bent on their weak axis, on qualified doubly symmetrical I and H shapes bent about their minor axis			$0.75F_y$	27	188
Bearing					
1. On contact area of milled surfaces		F_p	$0.90F_y$	32.4	225
2. On projected area of bolts and rivets in shear connections	Chapter 10		$1.50F_u$	87	600

[a] $F_y = 36$ ksi; assume that $F_u = 58$ ksi; some table values rounded off as permitted in the AISC Manual (Ref. 2). For SI units $F_y = 250$ MPa, $F_u = 400$ MPa.

Specification. Although these are in part special to the area of steel, the attempt is being made increasingly to standardize the nomenclature among all fields. The following is a list of the symbols used in this section, which is an abridged version of a more extensive list in the AISC Manual.

A	Cross-sectional area (sq in.) Gross area of an axially loaded compression member (sq in.)
A_e	Effective net area of an axially loaded tension member (sq in.)
A_n	Net area of an axially loaded tension member (sq in.)
C_c	Column slenderness ratio separating elastic and inelastic buckling
E	Modulus of elasticity of steel (29,000 ksi)
F_a	Axial compressive stress permitted in a prismatic member in the absence of a bending moment (ksi)
F_b	Bending stress permitted in a prismatic member in the absence of axial force (ksi)
F'_e	Euler stress for a prismatic member divided by the factor of safety (ksi)
F_p	Allowable bearing stress (ksi)
F_t	Allowable axial tensile stress (ksi)
F_u	Specified minimum tensile strength of the type of steel or fastener being used (ksi)
F_v	Allowable shear stress (ksi)
F_y	Specified minimum yield stress of the type of steel being used (ksi); as used in this book, *yield stress* denotes either the specified minimum yield point (for those steels that have a yield point) or specified minimum yield strength (for those steels that have no yield point)
I	Moment of inertia of a section (in.^{4})
I_x	Moment of inertia of a section about the X–X axis (in.^{4})
I_y	Moment of inertia of a section about the Y–Y axis (in.^{4})
J	Torsional constant of a cross section (in.^{4})
K	Effective length factor for a prismatic member
L	Span length (ft) Length of connection angles (in.)
L_c	Maximum unbraced length of the compression flange at which the allowable bending stress may be taken at $0.66F_y$

or as determined by AISC Specification Formula (1.5-5*a*) or Formula (1.5-5*b*), when applicable (ft)
Unsupported length of a column section (ft)

L_u Maximum unbraced length of the compression flange at which the allowable bending stress may be taken at $0.6F_y$ (ft)

M Moment (kip-ft)

M_D Moment produced by a dead load

M_L Moment produced by a live load

M_p Plastic moment (kip-ft)

M_R Beam resisting moment (kip-ft)

N Length of base plate (in.)
Length of bearing of applied load (in.)

P Applied load (kips)
Force transmitted by a fastener (kips)

S Elastic section modulus (in.^{3})

S_x Elastic section modulus about the X–X (major) axis (in.^{3})

S_y Elastic section modulus about the Y–Y (minor) axis (in.^{3})

V Maximum permissible web shear (kips)
Statical shear on a beam (kips)

Z Plastic section modulus (in.^{3})

Z_x Plastic section modulus with respect to the major (X–X) axis (in.^{3})

Z_y Plastic section modulus with respect to the minor (Y–Y) axis (in.^{3})

b_f Flange width of a rolled beam or plate girder (in.)

d Depth of a column, beam, or girder (in.)
Nominal diameter of a fastener (in.)

e_o Distance from the outside face of a web to the shear center of a channel section (in.)

f_a Computed axial stress (ksi)

f_b Computed bending stress (ksi)

f'_c Specified compression strength of concrete at 28 days (ksi)

f_p Actual bearing pressure on a support (ksi)

f_t	Computed tensile stress (ksi)
f_v	Computed shear stress (ksi)
g	Transverse spacing locating fastener gage lines (in.)
k	Distance from the outer face of a flange to the web toe of a fillet of rolled shape or equivalent distance on a welded section (in.)
l	For beams, the distance between cross sections braced against twist or lateral displacement of the compression flange (in.) For columns, the actual unbraced length of a member (in.) Length of a weld (in.)
m	Cantilever dimension of a base plate (in.)
n	Number of fasteners in one vertical row Cantilever dimension of a base plate (in.)
r	Governing radius of gyration (in.)
r_x	Radius of gyration with respect to the X–X axis (in.)
r_y	Radius of gyration with respect to the Y–Y axis (in.)
s	Longitudinal center-to-center spacing (pitch) of any two consecutive holes (in.)
t	Girder, beam, or column web thickness (in.) Thickness of a connected part (in.) Wall thickness of a tubular member (in.) Angle thickness (in.)
t_f	Flange thickness (in.)
t_w	Web thickness (in.)
x	Subscript relating a symbol to strong axis bending
y	Subscript relating a symbol to weak axis bending
D	Beam deflection (in.)

8

Steel Beams and Framing Elements

There are many steel elements that can be used for the basic function of spanning, including rolled shapes, cold-formed shapes, and fabricated beams and trusses. This chapter deals with some of the fundamental considerations of use of steel elements for beams, with an emphasis on the rolled shapes. For simplicity, it is assumed that all the rolled shapes used for the work in this chapter are of ASTM A36 steel with F_y = 36 ksi [250 MPa].

8.1 Factors in Beam Design

The complete design of a beam includes considerations of bending strength, shear resistance, deflection, lateral support, web crippling, and support details. Design for bending is usually the primary concern, but all other factors must be considered.

The current AISC Specification includes special factors that influence design for bending. The availability of higher-strength steels and attendant higher allowable stresses has led to the classification of structural shapes for beams as *compact* or *noncom-*

pact sections. Consequently, the shape employed, the laterally unsupported length of the span, and the grade of steel must be known in order to establish the allowable stress for bending. Unless otherwise stated, the examples and problems presented in this book are based on the assumption that A36 steel is used. The design procedures and specification formulas, however, are applicable to any grade of steel by selecting the appropriate values for the yield stress F_y or the ultimate strength F_u.

Compact and Noncompact Sections. In order to qualify for use of the maximum allowable bending stress of 0.66 F_y, a beam consisting of a rolled shape must satisfy several qualifications. The principal ones are that:

1. The beam section must be symmetrical about its minor (Y–Y) axis, and the plane of the loading must coincide with the plane of this axis, otherwise a torsional twist will be developed along with the bending.
2. The web and flanges of the section must have width–thickness ratios that qualify the section as *compact.*
3. The compression flange of the beam must be adequately braced against lateral buckling.

The criteria for establishing whether or not a section is compact include as a variable the F_y of the steel. It is therefore not possible to identify sections for this condition strictly on the basis of their geometric properties. The yield stress limits for the qualification of sections as compact for bending and those for combined actions of compression and bending are given, together with other properties, in the tables in the *Manual of Steel Construction,* published by the American Institute of Steel Construction. For beams of A36 steel, all S and M shapes and all wide flange shapes, except the W 6 × 15, qualify as compact when used for bending alone. When sections do not qualify as compact, the allowable bending stress must be reduced with the use of formulas given in the AISC Specification.

Lateral Support of Beams. A beam may fail by sideways (lateral) buckling of the compression flange when lateral deflection is not prevented. The tendency to buckle increases as the compressive bending stress in the flange increases and also as the unbraced length of the span increases. The full value of the allowable bending stress, $F_b = 0.66F_y$, can be used only when the compression flange is adequately braced. As for the compact section, the value of this required bracing length includes the variable of the F_y of the beam steel. (See Sec. 8.5.)

8.2 Design for Bending

The design of a beam for bending consists of applying the flexure formula, $S = M/F_b$. Before starting computations, it is necessary to decide on the grade of steel to be used and to determine the laterally unsupported length of the compression flange. The design procedure involves the steps listed below:

1. Determine the maximum bending moment.
2. Compute the required section modulus.
3. Refer to the tables of properties of steel sections, and select a beam with a section modulus equal to or greater than that which is required. In general, the lightest-weight section is the most economical. Table 8.1 will be found very useful in this operation because the beams are listed in the order of decreasing section moduli. This arrangement and the listing of corresponding values of L_c and L_u are presented in the AISC Manual in a much more extensive table called Allowable Stress Design Selection Table for Shapes Used as Beams.

A complete design would require investigation for shear and deflection. In the following example the procedure outlined above is applied to a simple beam with a more complicated loading.

Example. A girder has a span of 18 ft with floor beams framing into it from both sides at 6-ft intervals. The reaction of each floor beam is 4000 lb, so the girder receives two concentrated loads of 8000 lb each at the third points of the span. The framing of the beams provides lateral support for the girder at 6-ft intervals. In

TABLE 8.1 Section Modulus and Moment of Resistance for Selected Rolled Structural Shapes

S_x	Shape	F_y = 36 ksi		
		L_c	L_u	M_R
In.³		Ft.	Ft.	Kip-ft.
1110	**W 36x300**	**17.6**	**35.3**	**2220**
1030	**W 36x280**	**17.5**	**33.1**	**2060**
953	**W 36x260**	**17.5**	**30.5**	**1910**
895	**W 36x245**	**17.4**	**28.6**	**1790**
837	**W 36x230**	**17.4**	**26.8**	**1670**
829	W 33x241	16.7	30.1	1660
757	**W 33x221**	**16.7**	**27.6**	**1510**
719	**W 36x210**	**12.9**	**20.9**	**1440**
684	**W 33x201**	**16.6**	**24.9**	**1370**
664	**W 36x194**	**12.8**	**19.4**	**1330**
663	W 30x211	15.9	29.7	1330
623	**W 36x182**	**12.7**	**18.2**	**1250**
598	W 30x191	15.9	26.9	1200
580	**W 36x170**	**12.7**	**17.0**	**1160**
542	**W 36x160**	**12.7**	**15.7**	**1080**
539	W 30x173	15.8	24.2	1080
504	**W 36x150**	**12.6**	**14.6**	**1010**
502	W 27x178	14.9	27.9	1000
487	W 33x152	12.2	16.9	974
455	W 27x161	14.8	25.4	910
448	**W 33x141**	**12.2**	**15.4**	**896**
439	**W 36x135**	**12.3**	**13.0**	**878**
414	W 24x162	13.7	29.3	828
411	W 27x146	14.7	23.0	822
406	**W 33x130**	**12.1**	**13.8**	**812**
380	W 30x132	11.1	16.1	760
371	W 24x146	13.6	26.3	742
359	**W 33x118**	**12.0**	**12.6**	**718**
355	W 30x124	11.1	15.0	710
329	**W 30x116**	**11.1**	**13.8**	**658**
329	W 24x131	13.6	23.4	658
329	W 21x147	13.2	30.3	658
299	**W 30x108**	**11.1**	**12.3**	**598**
299	W 27x114	10.6	15.9	598
295	W 21x132	13.1	27.2	590
291	W 24x117	13.5	20.8	582
273	W 21x122	13.1	25.4	546
269	**W 30x 99**	**10.9**	**11.4**	**538**
267	W 27x102	10.6	14.2	534
258	W 24x104	13.5	18.4	516
249	W 21x111	13.0	23.3	498
243	**W 27x 94**	**10.5**	**12.8**	**486**
231	W 18x119	11.9	29.1	462
227	W 21x101	13.0	21.3	454
222	**W 24x 94**	**9.6**	**15.1**	**444**
213	**W 27x 84**	**10.5**	**11.0**	**426**
204	W 18x106	11.8	26.0	408
196	**W 24x 84**	**9.5**	**13.3**	**392**
192	W 21x 93	8.9	16.8	384
190	W 14x120	15.5	44.1	380
188	W 18x 97	11.8	24.1	376
176	**W 24x 76**	**9.5**	**11.8**	**352**
175	W 16x100	11.0	28.1	350
173	W 14x109	15.4	40.6	346
171	W 21x 83	8.8	15.1	342
166	W 18x 86	11.7	21.5	332
157	W14x 99	15.4	37.0	314
155	W 16x 89	10.9	25.0	310
154	**W 24x 68**	**9.5**	**10.2**	**308**
151	W 21x 73	8.8	13.4	302
146	W 18x 76	11.6	19.1	292
143	W 14x 90	15.3	34.0	286
140	**W 21x 68**	**8.7**	**12.4**	**280**
134	W 16x 77	10.9	21.9	268
131	**W 24x 62**	**7.4**	**8.1**	**262**
127	**W 21x 62**	**8.7**	**11.2**	**254**
127	W 18x 71	8.1	15.5	254
123	W 14x 82	10.7	28.1	246
118	W 12x 87	12.8	36.2	236
117	W 18x 65	8.0	14.4	234
117	W 16x 67	10.8	19.3	234
114	**W 24x 55**	**7.0**	**7.5**	**228**
112	W 14x 74	10.6	25.9	224
111	W 21x 57	6.9	9.4	222
108	W 18x 60	8.0	13.3	216
107	W 12x 79	12.8	33.3	214
103	W 14x 68	10.6	23.9	206
98.3	**W 18x 55**	**7.9**	**12.1**	**197**
97.4	W 12x 72	12.7	30.5	195

Source: Reproduced from the *Manual of Steel Construction*, 8th ed. (Ref. 2), with permission of the publishers, American Institute of Steel Construction.

TABLE 8.1 (*Continued*)

S_x	Shape	F_y = 36 ksi		
		L_c	L_u	M_R
In.^{3}		Ft.	Ft.	Kip-ft.
94.5	**W 21x50**	**6.9**	**7.8**	**189**
92.2	W 16x57	7.5	14.3	184
92.2	W 14x61	10.6	21.5	184
88.9	**W 18x50**	**7.9**	**11.0**	**178**
87.9	W 12x65	12.7	27.7	176
81.6	**W 21x44**	**6.6**	**7.0**	**163**
81.0	W 16x50	7.5	12.7	162
78.8	W 18x46	6.4	9.4	158
78.0	W 12x58	10.6	24.4	156
77.8	W 14x53	8.5	17.7	156
72.7	W 16x45	7.4	11.4	145
70.6	W 12x53	10.6	22.0	141
70.3	W 14x48	8.5	16.0	141
68.4	**W 18x40**	**6.3**	**8.2**	**137**
66.7	W 10x60	10.6	31.1	133
64.7	**W 16x40**	**7.4**	**10.2**	**129**
64.7	W 12x50	8.5	19.6	129
62.7	W 14x43	8.4	14.4	125
60.0	W 10x54	10.6	28.2	120
58.1	W 12x45	8.5	17.7	116
57.6	**W 18x35**	**6.3**	**6.7**	**115**
56.5	W 16x36	7.4	8.8	113
54.6	W 14x38	7.1	11.5	109
54.6	W 10x49	10.6	26.0	109
51.9	W 12x40	8.4	16.0	104
49.1	W 10x45	8.5	22.8	98
48.6	**W 14x34**	**7.1**	**10.2**	**97**
47.2	**W 16x31**	**5.8**	**7.1**	**94**
45.6	W 12x35	6.9	12.6	91
42.1	W 10x39	8.4	19.8	84
42.0	**W 14x30**	**7.1**	**8.7**	**84**
38.6	**W 12x30**	**6.9**	**10.8**	**77**
38.4	**W 16x26**	**5.6**	**6.0**	**77**
35.3	**W 14x26**	**5.3**	**7.0**	**71**
35.0	W 10x33	8.4	16.5	70
33.4	**W 12x26**	**6.9**	**9.4**	**67**
32.4	W 10x30	6.1	13.1	65
31.2	W 8x35	8.5	22.6	62

S_x	Shape	F_y = 36 ksi		
		L_c	L_u	M_R
In.^{3}		Ft.	Ft.	Kip-ft.
29.0	**W 14x22**	**5.3**	**5.6**	**58**
27.9	W 10x26	6.1	11.4	56
27.5	W 8x31	8.4	20.1	55
25.4	**W 12x22**	**4.3**	**6.4**	**51**
24.3	W 8x28	6.9	17.5	49
23.2	**W 10x22**	**6.1**	**9.4**	**46**
21.3	**W 12x19**	**4.2**	**5.3**	**43**
21.1	**M 14x18**	**3.6**	**4.0**	**42**
20.9	W 8x24	6.9	15.2	42
18.8	W 10x19	4.2	7.2	38
18.2	W 8x21	5.6	11.8	36
17.1	**W 12x16**	**4.1**	**4.3**	**34**
16.7	W 6x25	6.4	20.0	33
16.2	W 10x17	4.2	6.1	32
15.2	W 8x18	5.5	9.9	30
14.9	**W 12x14**	**3.5**	**4.2**	**30**
13.8	W 10x15	4.2	5.0	28
13.4	W 6x20	6.4	16.4	27
13.0	M 6x20	6.3	17.4	26
12.0	**M 12x11.8**	**2.7**	**3.0**	**24**
11.8	W 8x15	4.2	7.2	24
10.9	W 10x12	3.9	4.3	22
10.2	W 6x16	4.3	12.0	20
10.2	W 5x19	5.3	19.5	20
9.91	W 8x13	4.2	5.9	20
9.72	W 6x15	6.3	12.0	19
9.63	M 5x18.9	5.3	19.3	19
8.51	W 5x16	5.3	16.7	17
7.81	**W 8x10**	**4.2**	**4.7**	**16**
7.76	**M 10x 9**	**2.6**	**2.7**	**16**
7.31	W 6x12	4.2	8.6	15
5.56	**W 6x 9**	**4.2**	**6.7**	**11**
5.46	W 4x13	4.3	15.6	11
5.24	M 4x13	4.2	16.9	10
4.62	**M 8x 6.5**	**2.4**	**2.5**	**9**
2.40	**M 6x 4.4**	**1.9**	**2.4**	**5**

addition, the girder has a uniformly distributed load of 400 lb/ft, including its own weight, extending over the full span. Design the girder for bending, assuming that A36 steel is used.

Solution: (1) Because of the symmetry of loading, we know that the maximum bending moment will occur at the center of the span and that each reaction will equal half the total load. Then

$$R = \tfrac{1}{2}[8000 + 8000 + (400 \times 18)] = 11{,}600 \text{ lb}$$

$$M_{(x=9)} = (11{,}600 \times 9) - [(8000 \times 3) + (400 \times 9 \times 4.5)]$$

$$= 64{,}200 \text{ ft-lb}$$

(2) Assuming a compact shape will be used, and therefore F_b = 24 ksi (Table 7.1), the required section modulus is

$$S = \frac{M}{F_b} = \frac{64{,}200 \times 12}{24{,}000} = 32.1 \text{ in.}^3$$

(3) From Table 3.1 we determine some possible choices to be

$$\text{W } 10 \times 30, \quad S = 32.4 \text{ in.}^3$$

$$\text{W } 12 \times 26, \quad S = 33.4 \text{ in.}^3$$

$$\text{W } 14 \times 26, \quad S = 35.3 \text{ in.}^3$$

Selection on the basis of required section modulus is made easier by the use of Table 8.1, in which the shapes most used for beams are listed in order of their S_x values. Moment values given in the table represent the total resisting moments of the shapes for A36 steel, based on a maximum bending stress of 24 ksi [165 MPa] for compact sections and 22 ksi [150 MPa] for noncompact shapes.

To assist in the identification of situations in which buckling is a concern, the table gives the limiting values L_c and L_u—discussed in Sec. 8.5—for each shape. The maximum bending stress of 0.66 F_y is permitted for unsupported lengths only up to L_c. For lengths between L_c and L_u, the allowable stress is reduced to 0.60

F_y. For laterally unsupported lengths in excess of L_u, design is more practically achieved by use of the charts from the AISC Manual, of which Fig. 8.1 is an example.

In the preceding example the three possible choices could have been more quickly determined from Table 8.1 than from scanning Table 3.1.

In the following problems assume that A36 steel is used and that the beams are supported laterally throughout their length, permitting the use of the full allowable bending stress of 24 ksi.

Problem 8.2.A. Design for flexure a simple beam 14 ft [4.3 m] in length and having a total uniformly distributed load of 19.8 kips [88 kN].

Problem 8.2.B. Design for flexure a beam having a span of 16 ft [4.9 m] with a concentrated load of 12.4 kips [55 kN] at the center of the span.

Problem 8.2.C. A beam 15 ft [4.6 m] in length has three concentrated loads of 4 kips, 5 kips, and 6 kips at 4 ft, 10 ft, and 12 ft [17.8 kN, 22.2 kN, and 26.7 kN at 1.2 m, 3 m, and 3.6 m], respectively, from the left-hand support. Design the beam for flexure.

Problem 8.2.D. A beam 30 ft [9 m] long has concentrated loads of 9 kips [40 kN] each at the third points and also a total uniformly distributed load of 30 kips [133 kN]. Design the beam for flexure.

Problem 8.2.E. Design for flexure a beam 12 ft [3.6 m] in length, having a uniformly distributed load of 2 kips/ft [29 kN/m] and a concentrated load of 8.4 kips [37.4 kN] a distance of 5 ft [1.5 m] from one support.

Problem 8.2.F. A beam of 19 ft [5.8 m] in length has concentrated loads of 6 kips [26.7 kN] and 9 kips [40 kN] at 5 ft [1.5 m] and 13 ft [4 m], respectively, from the left-hand support. In addition, there is a uniformly distributed load of 1.2 kip-ft [17.5 kN/m] beginning 5 ft [1.5 m] from the left support and continuing to the right support. Design the beam for flexure.

Problem 8.2.G. A steel beam 16 ft [4.9 m] long has two uniformly distributed loads, one of 200 lb/ft [2.92 kN/m] extending 10 ft [3 m] from the left support and the other of 100 lb/ft [1.46 kN/m] extending over the remainder of the beam. In addition, there is a concentrated load of 8 kips [35.6 kN] at 10 ft [3 m] from the left support. Design the beam for flexure.

Problem 8.2.H. Design for flexure a simple beam 12 ft [3.7 m] in length, having two concentrated loads of 12 kips [53.4 kN] each, one 4 ft [1.2 m] from the left end and the other 4 ft [1.2 m] from the right end.

Problem 8.2.I. A cantilever beam 8 ft [2.4 m] long has a uniformly distributed load of 1600 lb/ft [23.3 kN/m]. Design the beam for flexure.

Problem 8.2.J. A cantilever beam 6 ft [1.8 m] long has a concentrated load of 12.3 kips [54.7 kN] at its unsupported end. Design the beam for flexure.

8.3 Safe Load Tables

The simple span beam loaded entirely with uniformly distributed load occurs so frequently in steel structural systems that it is useful to have a rapid design method for quick selection of shapes for a given load and span condition. Use of Table 8.2 allows for a simple design procedure where design conditions permit its use. When beams of A36 steel are loaded in the plane of their minor axis ($Y–Y$) and have lateral bracing spaced not farther than L_c, they may be selected from Table 8.2 after determination of only the total load and the beam span.

For a check, the values of L_c are given for each shape in the table. If the actual distance between points of lateral support for the compression (top) flange exceeds L_c, the table values must be reduced, as discussed in Sec. 8.1.

Deflections may be determined by using the factors given in the table or the graphs in Fig. 8.2. Both of these are based on a maximum bending stress of 24 ksi [165 MPa], and deflections for other stresses may be proportioned from these values if necessary.

The loads in the tables will not result in excessive shear stress on the beam webs if the full beam depth is available for stress development. Where end framing details result in some reduction of the web area, investigation of the stress on the reduced section may be required.

The following examples illustrate the use of Table 8.2 for some common design situations.

Example 1. A simple span beam of A36 steel is required to carry a total uniformly distributed load of 40 kips [178 kN] on a span of 30 ft [9.14 m]. Find (1) the lightest shape permitted and (2) the shallowest (least-deep) shape permitted.

TABLE 8.2 Load-Span Values for Beams[a]

		Span (ft)										
		8	10	12	14	16	18	20	22	24	26	28
		Deflection Factor[b]										
Shape	L_c[c] (ft)	1.59	2.48	3.58	4.87	6.36	8.05	9.93	12.0	14.3	16.8	19.5
8 × 6.5	2.4	9.24	7.39	6.16	5.28	4.62	4.11					
10 × 9	2.6	15.5	12.4	10.3	8.87	7.76	6.90	6.21	5.64			
8 × 10	4.2	15.6	12.5	10.4	8.92	7.81	6.94					
8 × 13	4.2	19.9	15.9	13.2	11.3	9.91	8.81					
10 × 12	3.9	21.8	17.4	14.5	12.5	10.9	9.69	8.72	7.93			
8 × 15	4.2	23.6	18.9	15.7	13.5	11.8	10.5					
12 × 11.8	2.7	24.0	19.2	16.0	13.7	12.0	10.7	9.60	8.73	8.00	7.38	6.86
10 × 15	4.2	27.6	22.1	18.4	15.8	13.8	12.3	11.0	10.0			
12 × 14	3.5	29.8	23.8	19.9	17.0	14.9	13.2	11.9	10.8	9.93	9.17	8.51
8 × 18	5.5	30.4	24.3	20.3	17.4	15.2	13.5					
10 × 17	4.2	32.4	25.9	21.6	18.5	16.2	14.4	13.0	11.8			
12 × 16	4.1	34.2	27.4	22.8	19.5	17.1	15.2	13.7	12.4	11.4	10.5	9.77
8 × 21	5.6	36.4	29.1	24.3	20.8	18.2	16.2					
10 × 19	4.2	37.6	30.1	25.1	21.5	18.8	16.7	15.0	13.7			
8 × 24	6.9	41.8	33.4	27.9	23.9	20.9	18.6					
14 × 18	3.6	42.2	33.8	28.1	24.1	21.1	18.7	16.9	15.3	14.1	13.0	12.0
12 × 19	4.2	42.6	34.1	28.4	24.3	21.3	18.9	17.0	15.5	14.2	13.1	12.2
10 × 22	6.1	46.4	37.1	30.9	26.5	23.2	20.6	18.5	16.9			
8 × 28	6.9	48.6	38.9	32.4	27.8	24.3	21.6					

		Span (ft)										
		12	14	16	18	20	22	24	26	28	30	32
		Deflection Factor[b]										
Shape	L_c[c] (ft)	3.58	4.87	6.36	8.05	9.93	12.0	14.3	16.8	19.5	22.3	25.4
12 × 22	4.3	33.9	29.0	25.4	22.6	20.3	18.5	16.9	15.6	14.5		
10 × 26	6.1	37.2	31.9	27.9	24.8	22.3	20.3					
14 × 22	5.3	38.7	33.1	29.0	25.8	23.2	21.1	19.3	17.8	16.6	15.5	14.5
10 × 30	6.1	43.2	37.0	32.4	28.8	25.9	23.6					
12 × 26	6.9	44.5	38.2	33.4	29.7	26.7	24.3	22.3	20.5	19.1		
10 × 33	8.4	46.7	40.0	35.0	31.0	28.0	25.4					
14 × 26	5.3	47.1	40.3	35.3	31.4	28.2	25.7	23.5	21.7	20.2	18.8	17.6
16 × 26	5.6	51.2	43.9	38.4	34.1	30.7	27.9	25.6	23.6	21.9	20.5	19.2
12 × 30	6.9	51.5	44.1	38.6	34.3	30.9	28.1	25.7	23.8	22.0		
14 × 30	7.1	56.0	48.0	42.0	37.3	33.6	30.5	28.0	25.8	24.0	22.4	21.0
10 × 39	8.4	56.1	48.1	42.1	37.4	33.7	30.6					
12 × 35	6.9	60.8	52.1	45.6	40.5	36.5	33.2	30.4	28.1	26.0		
16 × 31	5.8	62.9	53.9	47.2	41.9	37.8	34.3	31.5	29.0	27.0	25.2	23.6
14 × 34	7.1	64.8	55.5	48.6	43.2	38.9	35.3	32.4	29.9	27.8	25.9	24.3
10 × 45	8.5	65.5	56.1	49.1	43.6	39.3	35.7					

TABLE 8.2 (*Continued*)

		Span (ft)										
		16	18	20	22	24	26	28	30	32	34	36
		Deflection Factor[b]										
Shape	L_c[c] (ft)	6.36	8.05	9.93	12.0	14.3	16.8	19.5	22.3	25.4	28.7	32.2
W 12 × 40	8.4	51.9	46.1	41.5	37.7	34.6	31.9	29.6				
W 14 × 38	7.1	54.6	48.5	43.7	39.7	36.4	33.6	31.2	29.1	27.3		
W 16 × 36	7.4	56.5	50.2	45.2	41.1	37.7	34.8	32.3	30.1	28.2	26.6	25.1
W 18 × 35	6.3	57.8	51.4	46.2	42.0	38.5	35.6	33.0	30.8	28.9	27.2	25.7
W 12 × 45	8.5	58.1	51.6	46.5	42.2	38.7	35.7	33.2				
W 14 × 43	8.4	62.7	55.7	50.1	45.6	41.8	38.6	35.8	33.4	31.3		
W 12 × 50	8.5	64.7	57.5	51.7	47.0	43.1	39.8	37.0				
W 16 × 40	7.4	64.7	57.5	51.7	47.0	43.1	39.8	37.0	34.5	32.3	30.4	28.7
W 18 × 40	6.3	68.4	60.8	54.7	49.7	45.6	42.1	39.1	36.5	34.2	32.2	30.4
W 14 × 48	8.5	70.3	62.5	56.2	51.1	46.9	43.3	40.2	37.5	35.1		
W 12 × 53	10.6	70.6	62.7	56.5	51.3	47.1	43.4	40.3				
W 16 × 45	7.4	72.7	64.6	58.2	52.9	48.5	44.7	41.5	38.8	36.3	34.2	32.3
W 14 × 53	8.5	77.8	69.1	62.2	56.6	51.9	47.9	44.4	41.5	38.9		
W 18 × 46	6.4	78.8	70.0	63.0	57.3	52.5	48.5	45.0	42.0	39.4	37.1	35.0
W 16 × 50	7.5	81.0	72.0	64.8	58.9	54.0	49.8	46.3	43.2	40.5	38.1	36.0

		Span (ft)										
		16	18	20	22	24	27	30	33	36	39	42
		Deflection Factor[b]										
Shape	L_c[c] (ft)	6.36	8.05	9.93	12.0	14.3	18.1	22.3	27.0	32.2	37.8	43.8
W 21 × 44	6.6	81.6	72.5	65.3	59.3	54.4	48.3	43.5	39.6	36.3	33.5	31.1
W 18 × 50	7.9	88.9	79.0	71.1	64.6	59.3	52.7	47.4	43.1	39.5	36.5	
W 14 × 61	10.6	92.2	81.9	73.8	67.0	61.5	54.6	49.2	44.7			
W 16 × 57	7.5	92.2	81.9	73.8	67.0	61.5	54.6	49.2	44.7	41.0		
W 21 × 50	6.9	94.5	84.0	75.6	68.7	63.0	56.0	50.4	45.8	42.0	38.8	36.0
W 18 × 55	7.9	98.3	87.4	78.6	71.5	65.5	58.2	52.4	47.7	43.7	40.3	
W 18 × 60	8.0	108	96.0	86.4	78.5	72.0	64.0	57.6	52.4	48.0	44.3	
W 21 × 57	6.9	111	98.7	88.6	80.7	74.0	65.8	59.2	53.8	49.3	45.5	42.3
W 24 × 55	7.0	114	101	91.2	82.9	76.0	67.5	60.8	55.3	50.7	46.8	43.4
W 16 × 67	10.8	117	104	93.6	85.1	78.0	69.3	62.4	56.7	52.0		
W 18 × 65	8.0	117	104	93.6	85.1	78.0	69.3	62.4	56.7	52.0	48.0	
W 18 × 71	8.1	127	113	102	92.4	84.7	72.2	67.7	61.5	56.4	52.1	
W 21 × 62	8.7	127	113	102	92.4	84.7	72.2	67.7	61.5	56.4	52.1	48.4
W 24 × 62	7.4	131	116	105	95.3	87.3	77.6	69.9	63.5	58.2	53.7	49.9
W 16 × 77	10.9	134	119	107	97.4	89.3	79.4	71.5	65.0	59.5		
W 21 × 68	8.7	140	124	112	102	93.3	83.0	74.7	67.9	62.2	57.4	53.3
W 18 × 76	11.6	146	130	117	106	97.3	86.5	77.9	70.8	64.9	59.9	
W 21 × 73	8.8	151	134	121	110	101	89.5	80.5	73.2	67.1	61.9	57.5
W 24 × 68	9.5	154	137	123	112	103	91.2	82.1	74.7	68.4	63.2	58.7
W 18 × 86	11.7	166	147	133	121	111	98.4	88.5	80.5	73.8	68.1	
W 21 × 83	8.8	171	152	137	124	114	101	91.2	82.9	76.0	70.1	65.1

ABLE 8.2 *(Continued)*

Shape	$L_c{}^c$ (ft)	Span (ft)										
		24	27	30	33	36	39	42	45	48	52	56
		Deflection Factor[b]										
		14.3	18.1	22.3	27.0	32.2	37.8	43.8	50.3	57.2	67.1	77.9
24 × 76	9.5	117	104	93.9	85.3	78.2	72.2	67.0	62.6	58.7		
21 × 93	8.9	128	114	102	93.1	85.3	78.8	73.1	68.3			
24 × 84	9.5	131	116	104	95.0	87.1	80.4	74.7	69.7	65.3		
27 × 84	10.5	142	126	114	103	94.7	87.4	81.1	75.7	71.0	65.5	60.8
24 × 94	9.6	148	131	118	108	98.7	91.1	84.6	78.9	74.0		
21 × 101	13.0	151	134	121	110	101	93.1	86.5	80.7			
27 × 94	10.5	162	144	130	118	108	99.7	92.6	86.4	81.0	74.8	69.4
24 × 104	13.5	172	153	138	125	115	106	98.3	91.7	86.0		
27 × 102	10.6	178	158	142	129	119	109	102	94.9	89.0	82.1	76.3
30 × 99	10.9	179	159	143	130	120	110	102	95.6	89.7	82.8	76.9
24 × 117	13.5	194	172	155	141	129	119	111	103	97.0		
27 × 114	10.6	199	177	159	145	133	123	114	106	99.7	92.0	85.4
30 × 108	11.1	199	177	159	145	133	123	114	106	99.7	92.0	85.4
30 × 116	11.1	219	195	175	159	146	135	125	117	110	101	94.0
30 × 124	11.1	237	210	189	172	158	146	135	126	118	109	101

Shape	$L_c{}^c$ (ft)	Span (ft)										
		30	33	36	39	42	45	48	52	56	60	65
		Deflection Factor[b]										
		22.3	27.0	32.2	37.8	43.8	50.3	57.2	67.1	77.9	89.4	105
33 × 118	12.0	191	174	159	147	137	128	120	110	103	95.7	88.4
30 × 132	11.1	203	184	169	156	145	135	127	117	109	101	
33 × 130	12.1	216	197	180	166	155	144	135	125	116	108	99.9
27 × 146	14.7	219	199	183	169	156	146	137	126	117		
36 × 135	12.3	234	213	195	180	167	156	146	135	125	117	108
33 × 141	12.2	239	217	199	184	171	159	149	138	128	119	110
33 × 152	12.2	260	236	216	200	185	173	162	150	139	130	120
36 × 150	12.6	269	244	224	207	192	179	168	155	144	134	124
30 × 173	15.8	287	261	239	221	205	192	180	166	154	144	
36 × 160	12.7	289	263	241	222	206	193	181	167	155	144	133
36 × 170	12.7	309	281	258	238	221	206	193	178	166	155	143
30 × 191	15.9	319	290	268	245	228	213	199	184	171	159	
36 × 182	12.7	332	302	277	256	237	221	208	192	178	166	153
36 × 194	12.8	354	322	295	272	253	236	221	204	190	177	163
33 × 201	16.6	365	332	304	281	260	243	228	210	195	182	168
36 × 210	12.9	383	349	319	295	274	256	240	221	205	192	177
33 × 221	16.7	404	367	336	310	288	269	252	233	216	202	186
33 × 241	16.7	442	402	368	340	316	295	276	255	237	221	204
36 × 230	17.4	446	406	372	343	319	298	279	257	239	223	206
36 × 245	17.4	477	434	398	367	341	318	298	275	256	239	220
36 × 260	17.5	508	462	423	391	363	339	318	293	272	254	234
36 × 280	17.5	549	499	458	422	392	366	343	317	294	275	253
36 × 300	17.6	592	538	493	455	423	395	370	341	317	296	273

Note: Total allowable uniformly distributed load is in kips for simple span beams of A36 steel with yield stress of 36 ksi 50 MPa].

Maximum deflection in inches at the center of the span may be obtained by dividing this factor by the depth of the beam in ches. This is based on a maximum bending stress of 24 ksi [165 MPa].

Maximum permitted distance between points of lateral support. If distance exceeds this, use the charts in the AISC anual. (See Fig. 8.1 for example.)

Solution: From Table 8.2 we find the following:

Shape	Allowable Load (kips)
W 21 × 44	43.5
W 18 × 46	42.0
W 16 × 50	43.2
W 14 × 53	41.5

Thus the lightest shape is the W 21 × 44, and the shallowest is the W 14 × 53.

Example 2. A simple span beam of A36 steel is required to carry a total uniformly distributed load of 25 kips [111 kN] on a span of 24 ft [7.32 m] while sustaining a maximum deflection of not more than $\frac{1}{360}$ of the span. Find the lightest shape permitted.

Solution: From Table 8.2 we find the lightest shape that will carry this load to be the W 16 × 26. For this beam the deflection will be

$$D = \frac{25}{25.6} \times \frac{14.3}{16} = \frac{\text{(actual load)}}{\text{(table load)}} \times \frac{\text{(deflection factor)}}{\text{(beam depth)}}$$

$$= 0.873 \text{ in.}$$

which exceeds the allowable of $\dfrac{24 \times 12}{360} = 0.80$ in.

The next heaviest shape from Table 8.2 is a W 16 × 31, for which the deflection will be

$$D = \frac{25}{31.5} \times \frac{14.3}{16} = 0.709 \text{ in.}$$

which is less than the limit, so this shape is the lightest choice.

Problems 8.3.A,B,C,D,E,F. For each of the following conditions find (a) the lightest permitted shape and (b) the shallowest permitted shape of A36 steel:

	Span (ft)	Total Uniformly Distributed Load (kips)	Deflection Limited to $\frac{1}{360}$ of the Span
A	16	10	No
B	20	30	No
C	36	40	No
D	18	16	Yes
E	32	20	Yes
F	42	50	Yes

8.4 Equivalent Tabular Loads

The safe loads shown in Table 8.2 are uniformly distributed loads on simple span beams. By use of coefficients it is possible to convert other types of loading to equivalent uniform loads and thereby greatly extend the usefulness of Table 8.2.

For a simple beam with a uniformly distributed load, the maximum moment is $WL/8$. For a simple beam with two equal loads at the third points of the span, the maximum moment is $PL/3$. These values are shown in case 2 and case 3, Fig. 2.27, respectively. Equating these moment values, we can write

$$\frac{WL}{8} = \frac{PL}{3} \quad \text{and} \quad W = 2.67 \times P$$

which demonstrates that a total uniformly distributed load 2.67 times the value of one of the concentrated loads will produce the same maximum moment on the beam as that caused by the concentrated loads. Thus in order to use Table 8.2 for design of a beam with this loading, we merely multiply the concentrated load by the ETL (equivalent tabular load) factor and look for the new load value in the table.

Coefficients for other loadings are also given in Fig. 2.27. It is important to remember that an ETL loading does not include the weight of the beam, for which an estimated amount should be added. Also the shear values and reactions for the beam must be determined from the actual loading and not from the ETL.

8.5 Design of Laterally Unsupported Beams

As discussed in Sec. 7.4, the allowable bending stress for beams of A36 steel is 24 ksi [165 MPa] (0.66 F_y) when the compression flanges of compact shapes are braced laterally at intervals not greater than L_c. For beams supported laterally at intervals greater than L_c but not greater than L_u, the allowable bending stress is reduced to 22 ksi [152 MPa] (0.60 F_y, rounded off). Both L_c and L_u values are given in Table 8.1, so when the lateral support interval falls between these values, the various table values based on a stress of 24 ksi may simply be reduced by the proportion 22/24 = 0.917.

When the distance between points of lateral support exceeds L_u, the AISC Specification provides an equation for the determination of the allowable stress that includes the value of the unsupported length. This is quite cumbersome to use for design, so a series of charts is provided in the AISC Manual that contain plots of the relationship of the moment capacity to the laterally unsupported length for commonly used beam shapes. Figure 8.1 is a reproduction of one of these charts. In order to use the chart, you must determine the maximum bending moment in the beam and the distance between points of lateral support. These two values are used to find a point on the chart; any beam whose graph lies above or to the right of this point is adequate, the nearest solid-line graph representing the shape of least weight. The following example illustrates the use of the chart in Fig. 8.1.

Example. A simple beam carries a total uniform load of 22 kips [98 kN], including its own weight, over a span of 24 ft [7.32 m]. Other elements of the framing provide lateral support at intervals of 8 ft [2.44 m]. Select a beam of A36 steel for these conditions.

Solution: For this loading the maximum moment is

$$M = \frac{WL}{8} = \frac{22 \times 24}{8} = 66 \text{ kip-ft} \left[\frac{98 \times 7.32}{8} = 89.7 \text{ kN-m}\right]$$

Using this moment and the unsupported length, we locate the point on the chart in Fig. 8.1. The nearest line above and to the right of this point is that for a W 10 × 33, which is an acceptable

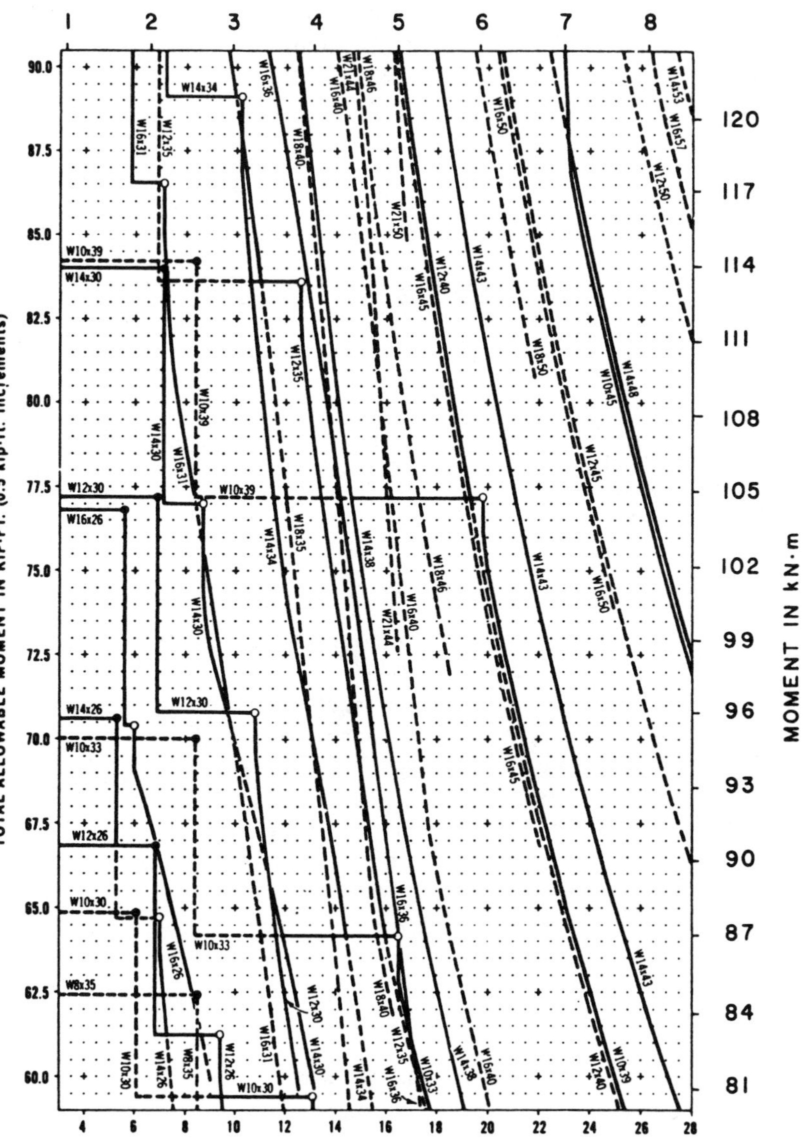

FIGURE 8.1 Allowable bending moment for beams with various unbraced lengths. F_y = 36 ksi [250 MPa]. Adapted from the *Manual of Steel Construction*, 8th ed. (Ref. 2), with permission of the publishers, American Institute of Steel Construction.

beam. However, the line for this beam is dashed, indicating that there is a lighter possible choice. Proceeding on the chart, upward and to the right, the first solid line we encounter is that for a W 12 × 30, which is also acceptable and represents the lightest possible shape.

The chart in Fig. 8.1 does not incorporate consideration of shear or deflection, which should also be investigated in real design situations.

Problem 8.5.A. A W 14 × 34 is used as a simple beam to carry a total uniformly distributed load, including the beam weight, of 40 kips [178 kN] on a span of 16 ft [4.88 m]. Lateral support is provided only at the ends and at the midspan. Is the shape an adequate choice for these conditions?

Problem 8.5.B. A simple beam is required for a span of 24 ft [7.32 m]. The load, including the beam weight, is uniformly distributed and consists of a total of 28 kips [125 kN]. If lateral support is provided only at the ends and at the midspan, and the deflection under full load is limited to 0.6 in. [15.2 mm], find an adequate rolled shape for the beam.

8.6 Shear in Wide-Flange Beams

Shear stress in a steel beam is seldom a factor in determining its size. If, however, the beam has a relatively short span with a large load or has any large load placed near one support, the bending moment may be low and the shear relatively high. For W, S, and C shape beams, the shear stress is assumed to be developed by the beam web only, with the web area being determined as the product of the beam depth times the web thickness. For this situation the AISC Specification allows a maximum stress of 0.40 F_y rounded off to 14.5 ksi [100 MPa] for A36 steel. Shear stress is thus investigated with the following formula:

$$f_v = \frac{V}{A_w} = \frac{V}{dt_w}$$

in which f_v = actual unit shear stress,

V = maximum vertical shear,

A_w = gross area of the beam web,

d = overall depth of the beam,

t_w = thickness of the beam web.

Example. A simple beam of A36 steel is 6 ft [1.83 m] long and has a concentrated load of 36 kips [160 kN] applied 1 ft [0.3 m] from the left end. It is found that a W 8 × 24 is large enough to sustain the resulting bending moment. Investigate the beam for shear.

Solution: We find the two reactions for this loading to be 30 kips [133 kN] and 6 kips [27 kN]. The maximum vertical shear is thus equal to the value of the larger reaction—the usual case for a simple span beam.

From Table 3.1 we find that $d = 7.93$ in. and $t_w = 0.245$ in. for the W 8 × 24. Then

$$A_w = d \times t_w = 7.93 \times 0.245 = 1.94 \text{ in.}^2$$

and

$$f_v = \frac{V}{A_w} = \frac{30}{1.94} = 15.5 \text{ ksi}$$

Since this exceeds the allowable value of 14.5 ksi, the W 8 × 24 is not acceptable.

It may be noted that S shapes in general have rather thicker webs than W shapes of corresponding depth. We may thus consider the use of an S shape with the necessary value for S_x, for which the AISC Manual yields an S 8 × 23 with a depth of 8 in. and web thickness 0.441 in. Then

$$f_v = \frac{V}{A_w} = \frac{30}{8 \times 0.441} = 8.50 \text{ ksi}$$

which is less than the allowable stress.

Problems 8.6.A,B,C,D. Compute the maximum permissible web shears for the following beams of A36 steel:

A W 24 × 84
B W 12 × 40
C W 10 × 19
D C 10 × 20

8.7 Deflection

In addition to resisting bending and shear, beams must not deflect excessively. Floor and ceiling cracks may result if the beams are not stiff enough, and beams should be investigated to see that the deflection does not exceed $\frac{1}{360}$ of the span, the generally accepted limit with plastered ceilings. The current AISC Specification requirement is that steel beams and girders supporting plastered ceilings be of such dimensions that the maximum *live load* deflection will not exceed $\frac{1}{360}$ of the span. It frequently happens that a beam may be of adequate dimensions to resist bending and shear but, on investigation, may be found to deflect more than the maximum permitted by building codes.

For typical beams and loads the actual deflection may be computed from the formulas given in Fig. 2.27, but in using these formulas note carefully that l, in the term l^3, is in inches, not feet. For a simple beam with a uniformly distributed load, the deflection is found by the formula

$$D = \frac{5}{384} \times \frac{Wl^3}{EI} \qquad \text{(case 2, Fig. 2.27)}$$

in which D = maximum deflection in inches,
W = total uniformly distributed load in pounds or kips,
l = length of the span *in inches,*
E = modulus of elasticity of the beam in psi or ksi (for structural steel $E = 29{,}000{,}000$ psi),
I = moment of inertia of the cross section of the beam in inches to the fourth power.

For a beam on which the loading is not typical, we may find W, the uniformly distributed load that would produce the same bending moment, and then apply the foregoing formula to find the approximate deflection. When the maximum deflection occurs at the center of the span, it is sometimes convenient to compute the deflections due to individual loads on the beam; their sum will be the total deflection.

For a uniformly loaded simple span beam (case 2, Fig. 2.27) with a fixed value for the maximum bending stress, a formula may be derived that expresses the maximum deflection in terms of the span and the beam depth. For a stress value of 24 ksi, the formula has the form

$$D = \frac{0.02483 \times L^2}{d}$$

in which D = maximum deflection in inches,
L = span length in feet,
d = beam depth in inches.

(*Note:* For a derivation of this formula, see Ref. 15.)

Figure 8.2 presents graphs of this equation for various values of beam depth from 6 in. to 36 in. Deflections may be obtained directly from these graphs for the particular case of a beam of A36 steel with maximum bending stress of 24 ksi. Since both the bending stress and the deflection are directly proportional to the load, deflections for stress values of other magnitudes may be obtained by simple proportion, as indicated in the following example.

Example. A W 12 × 22 is used for a simple span beam to carry a total uniformly distributed load of 10 kips, including the beam weight. If the span is 16 ft, find the maximum deflection.

Solution: We first determine the actual maximum bending stress as follows:

$$M = \frac{W \times L}{8} = \frac{10 \times 16}{8} = 20 \text{ kip-ft}$$

From Table 3.1 $S_x = 25.4 \text{ in.}^3$ for the shape. Then

$$f_b = \frac{M}{S} = \frac{20 \times 12}{25.4} = 9.449 \text{ ksi}$$

We next use Fig. 8.2 to obtain a deflection value of approximately 0.53 in. for the 12-in.-deep beam. The true deflection is

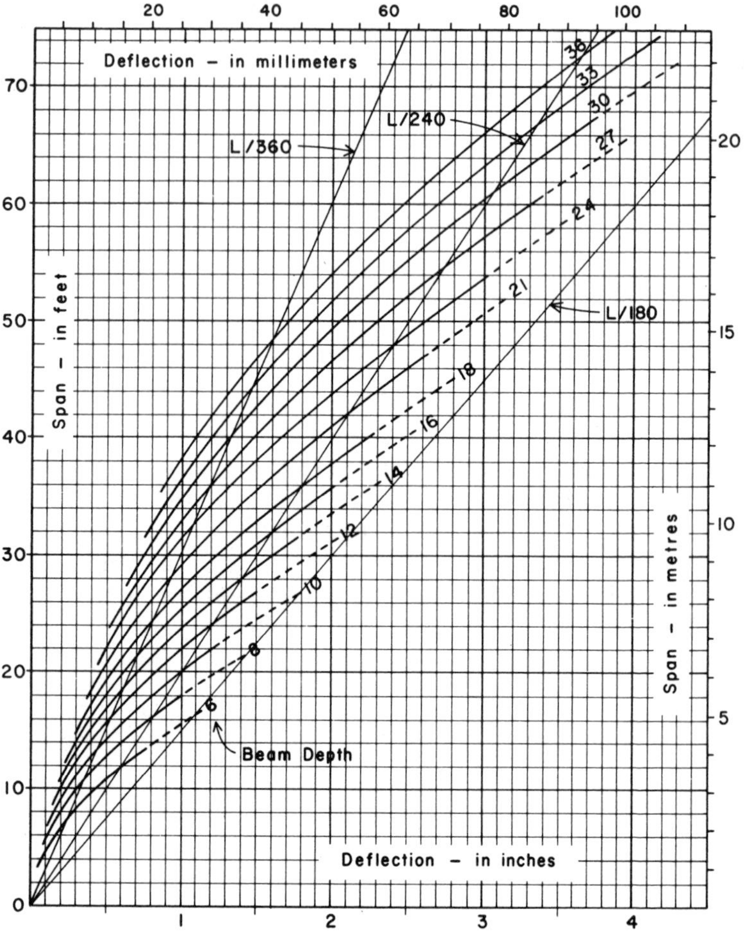

FIGURE 8.2 Deflection of steel beams with bending stress of 24 ksi [165 MPa].

then determined as

$$D = \frac{9.449}{24} \times 0.53 = 0.209 \text{ in.}$$

Not much accuracy can be expected with this process, but it is usually sufficient, since many conditions in real design situations make deflection calculations subject to speculation.

When SI units are used, the formula used for the graphs in Fig. 8.2 takes the form

$$D = \frac{0.1719 \times L^2}{d}$$

in which L = the span length in meters,
d = the beam depth in millimeters,

and the deflection is determined in millimeters.

Another method for finding deflections for the simple span, uniformly loaded beam is with the use of the factors from Table 8.2. For the beam in this example it may be noted that Table 8.2 yields a load of 25.4 kips and a deflection factor of 6.36 for the 16-ft span. The deflection of the beam is found by dividing the deflection factor by the beam depth in inches. (We will use the nominal beam depth of 12 in., although a more precise value is obtainable from Table 3.1.) This will, however, yield the deflection only for the full table load of 25.4 kips. To obtain the deflection for the 10-kip load, we simply use a correction factor consisting of the ratio of the loads, since the deflection is directly proportional to the loading. Thus

$$D = \frac{10}{25.4} \times \frac{6.36}{12} = 0.209 \text{ in.}$$

Finally, of course, we may apply the equation for deflection for this load case. Obtaining a value of $I = 156 \text{ in.}^4$ for the W 12 × 22 from Table 3.1, we find

$$D = \frac{5}{384}\frac{WL^3}{EI} = \frac{5}{384}\frac{10(16 \times 12)^3}{29{,}000(156)} = 0.204 \text{ in.}$$

If the computation is correct, the formula will yield the most accurate value, but the level of accuracy that is significant for real design work can be obtained by any of the methods illustrated.

Problems 8.7.A,B,C. Find the maximum deflection for the following uniformly loaded beams of A36 steel. Find deflection values in inches, using Fig. 8.2, and verify the values by use of the equation for case 2, Fig. 2.27.

A W 10 × 33, span = 18 ft, total load = 30 kips [5.5 m, 133 kN]
B W 16 × 36, span = 20 ft, total load = 50 kips [6 m, 222 kN]
C W 18 × 46, span = 24 ft, total load = 55 kips [7.3 m, 245 kN]

8.8. Fireproofing for Steel Beams

In fire-resistive construction, some insulating material must be placed around the structural steel. Materials commonly used for this purpose are concrete, masonry, or lath and plaster. There are also fibrous and cementitious coatings that can be sprayed directly on the surfaces of steel members or on the undersides of steel roofs and floor decks. When a poured concrete deck is used with a steel framing system, the steel beams are sometimes fireproofed by encasement in poured concrete. This results in considerable dead weight for the beam, which should be given consideration in the design of the beam.

Fireproofing is presently more often accomplished by use of sprayed materials or encasement in plaster or drywall materials, resulting in weights considerably less than those obtained with concrete encasement.

8.9 Crippling of Beam Webs

An excessive end reaction on a beam or an excessive concentrated load at some point along the interior of the span may cause crippling, or localized yielding, of the beam web. The AISC Spec-

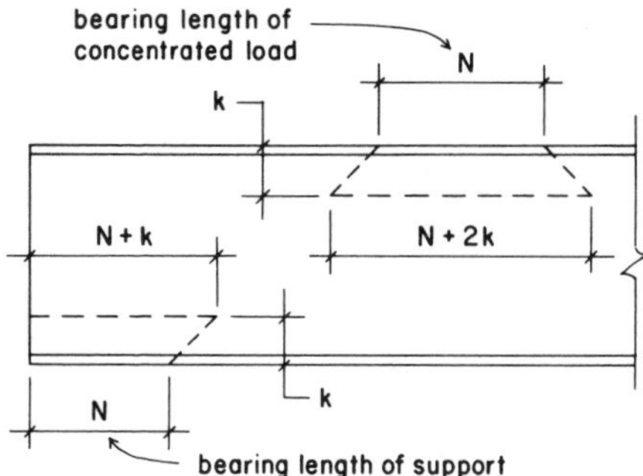

FIGURE 8.3 Determination of effective length for computation of web crippling.

ification requires that end reactions or concentrated loads for beams without stiffeners or other web reinforcement shall not exceed the following (Fig. 8.3):

$$\text{Maximum end reaction} = 0.75F_y t(N + k)$$
$$\text{Maximum interior load} = 0.75F_y t(N + 2k)$$

where t = thickness of the beam *web*, in inches,
N = length of the bearing or length of the concentrated load (not less than k for end reactions), in inches,
k = distance from the outer face of the flange to the web toe of the fillet, in inches,
$0.75Fy$ = 27 ksi for A36 steel [186 MPa].

When these values are exceeded, the webs of the beams should be reinforced with stiffeners, the length of bearing increased, or a beam with a thicker web selected.

Example 1. A W 21 × 57 beam of A36 steel has an end reaction that is developed in bearing over a length of N = 10 in. [254 mm]. Check the beam for web crippling if the reaction is 44 kips [196 kN].

Solution: In Table 3.1 we find that k = 1.38 in. [35 mm] and the web thickness is 0.405 in. [10 mm]. To check for web crippling, we find the maximum end reaction permitted and compare it with the actual value for the reaction. Thus

$$R = F_p \times t \times (N + k) = 27 \times 0.405 \times (10 + 1.38)$$
$$= 124 \text{ kips (the allowable reaction)}$$
$$\left[R = \frac{186 \times 10 \times (254 + 35)}{10^3} = 538 \text{ kN}\right]$$

Because this is greater than the actual reaction, the beam is not critical with regard to web crippling.

Example 2. A W 10 × 26 of A36 steel supports a column load of 70 kips [311 kN] at the center of the span. The bearing length of the column on the beam is 10 in. [254 mm]. Investigate the beam for web crippling under this concentrated load.

Solution: In Table 3.1 we find that $k = 0.88$ in. [22 mm] and the web thickness is 0.260 in. [6.60 mm]. The allowable load that can be supported on the given bearing length is

$$P = F_P \times t \times (N + 2k) = 27 \times 0.260 \times [10 + (2 \times 0.88)]$$
$$= 82.6 \text{ kips [366 kN]}$$

which exceeds the required load, so the beam is safe from web crippling.

Problem 8.9A. Compute the maximum allowable reaction with respect to web crippling for a W 14 × 30 of A36 steel with an 8-in. [203-mm] bearing plate length.

Problem 8.9B. A column load of 81 kips [360 kN] with a bearing plate length of 11 in. [279 mm] is placed on top of the beam in the preceding problem. Are web stiffeners required to prevent the web crippling?

8.10 Beam Bearing Plates

Beams that are supported on walls or piers of masonry or concrete usually rest on steel bearing plates. The purpose of the plate is to provide an ample bearing area. The plate also helps to seat the beam at its proper elevation. Bearing plates provide a level

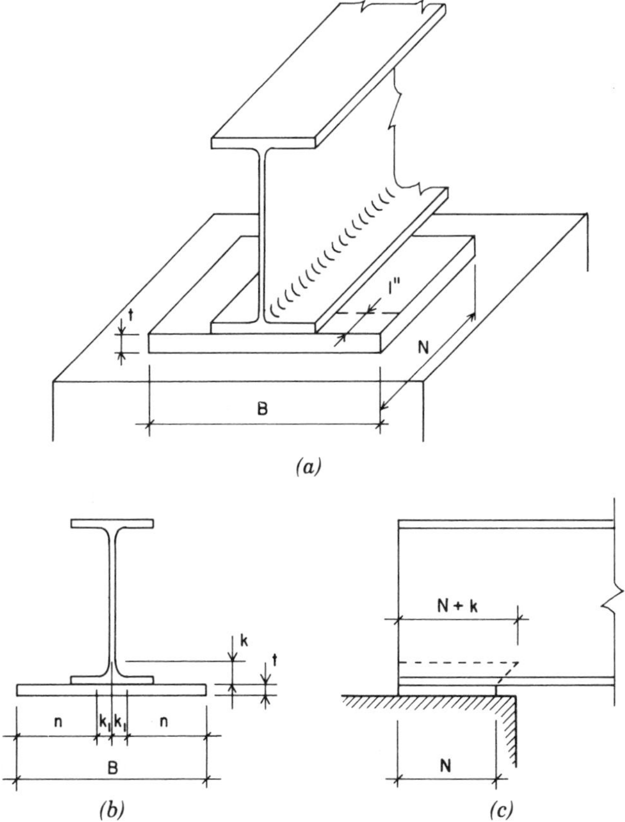

FIGURE 8.4 Reference dimensions for beam end bearing plates.

surface for a support and, when properly placed, afford a uniform distribution of the beam reaction over the area of contact with the supporting material.

By reference to Fig. 8.4, the area of the bearing plate is $B \times N$. It is found by dividing the beam reaction by F_p, the allowable bearing value of the supporting material. Then

$$A = \frac{R}{F_p}$$

where $A = B \times N$, the area of the plate in square inches,
R = reaction of the beam in pounds of kips,
F_p = allowable bearing pressure on the supporting material in psi or ksi (see Table 8.3).

The thickness of the wall generally determines N, the dimension of the plate parallel to the length of the beam. If the load from the beam is unusually large, the dimension B may become excessive. For such a condition, one or more shallow-depth I-beams, placed parallel to the wall length, may be used instead of a plate. The dimensions B and N are usually in even inches, and a great variety of thicknesses is available.

The thickness of the plate is determined by considering the projection n (Fig. 8.4*b*) as an inverted cantilever; the uniform bearing pressure on the bottom of the plate tends to curl it upward about the beam flange. The required thickness may be computed readily by the following formula, which does not involve direct

TABLE 8.3 Allowable Bearing Pressure on Masonry and Concrete

Type of Material and Conditions	Allowable Unit Stress in Bearing F_p (psi)	(kPa)
Solid brick, unreinforced, type S mortar		
f'_m = 1500 psi	170	1200
f'_m = 4500 psi	338	2300
Hollow unit masonry, unreinforced, type S mortar, f'_m = 1500 psi (on net area of masonry)	225	1500
Concrete[a]		
(1) Bearing on full area of support		
f'_c = 2000 psi	500	3500
f'_c = 3000 psi	750	5000
(2) Bearing on $\frac{1}{3}$ or less of support area		
f'_c = 2000 psi	750	5000
f'_c = 3000 psi	1125	7500

[a] Stresses for areas between these limits may be determined by direct proportion.

computation of bending moment and section modulus:

$$t = \sqrt{\frac{3f_p n^2}{F_b}}$$

where t = thickness of the plate in inches,

f_p = actual bearing pressure of the plate on the masonry, in psi or ksi,

F_b = allowable bending stress in the plate (the AISC Specification gives the value of F_b as 0.75 F_y; for A36 steel F_y = 36 ksi; therefore $F_b = 0.75 \times 36 =$ 27 ksi),

$n = \frac{B}{2} - k_1$, in inches (see Fig. 8.4*b*),

k_1 = the distance from the center of the web to the toe of the fillet; values of k_1 for various beam sizes may be found in the tables in the AISC Manual.

The foregoing formula is derived by considering a strip of plate 1 in. wide (Fig. 8.4*a*) and t in. thick, with a projecting length of n inches, as a cantilever. Because the upward pressure on the steel strip is f_p, the bending moment at distance n from the edge of the plate is

$$M = f_p n \times \frac{n}{2} = \frac{f_p n^2}{2}$$

For this strip with rectangular cross section,

$$\frac{I}{c} = \frac{bd^2}{6} \qquad \text{(Sec. 3.5)}$$

and because b = 1 in. and $d = t$ in.,

$$\frac{I}{c} = \frac{1 \times t^2}{6} = \frac{t^2}{6}$$

Then from the beam formula,

$$\frac{M}{F_b} = \frac{I}{c} \qquad \text{(Sec. 2.10)}$$

Substituting the values of M and I/c determined above,

$$\frac{f_p n^2}{2} \times \frac{1}{F_b} = \frac{t^2}{6}$$

and

$$t^2 = \frac{6 f_p n^2}{2 F_b} \quad \text{or} \quad t = \sqrt{\frac{3 f_p n^2}{F_b}}$$

When the dimensions of the bearing plate are determined, the beam should be investigated for web crippling on the length $(N + k)$ shown in Figure 8.4c. This is explained in Sec. 8.9.

Example 1. A W 21 × 57 of A36 steel transfers an end reaction of 44 kips [196 kN] to a wall built of solid brick by means of a bearing plate of A36 steel. Assume type S mortar and a brick with $f'_m = 1500$ psi. The N dimension of the plate (see Fig. 8.4) is 10 in. [254 mm]. Design the bearing plate.

Solution: In the AISC Manual we find that k_1 for the beam is 0.875 in. [22 mm]. From Table 8.3 the allowable bearing pressure F_p for this wall is 170 psi [1200 kPa]. The required area of the plate is then

$$A = \frac{R}{F_p} = \frac{44{,}000}{170} = 259 \text{ in.}^2\ [163 \times 10^3 \text{ mm}^2]$$

Then, because $N = 10$ in. [254 mm],

$$B = \frac{259}{10} = 25.9 \text{ in. [643 mm]}$$

which is rounded off to 26 in. [650 mm].

With the true dimensions of the plate, we now compute the true bearing pressure:

$$f_p = \frac{R}{A} = \frac{44{,}000}{10 \times 26} = 169 \text{ psi } [1.19 \text{ MPa}]$$

To find the thickness, we first determine the value of n.

$$n = \frac{B}{2} - k_1 = \frac{26}{2} - 0.875 = 12.125 \text{ in. } [303 \text{ mm}]$$

Then

$$t = \sqrt{\frac{3f_p n^2}{F_b}} = \sqrt{\frac{3 \times 169 \times (12.125)^2}{27{,}000}} = 1.66 \text{ in. } [42 \text{ mm}]$$

The complete design for this problem would include a check of the web crippling in the beam. This has been done already as Example 1 in Sec. 8.9.

Problem 8.10.A. A W 14 × 30 with a reaction of 20 kips [89 kN] rests on a brick wall with brick of f'_m = 1500 psi and type S mortar. The beam has a bearing length of 8 in. [203 mm] parallel to the length of the beam. If the bearing plate is A36 steel, determine its dimensions. (k_1 = 0.625 in.)

Problem 8.10.B. A wall of brick with f'_m of 1500 psi and type S mortar supports a W 18 × 55 of A36 steel. The beam reaction is 25 kips [111 kN], and the bearing length N is 9 in. [229 mm]. Design the beam bearing plate. (k_1 = 0.8125 in.)

8.11 Decks

Figure 8.5 shows four possibilities for a floor deck, all of which may be used with a steel framing system. When a wood deck is used, it is usually nailed to a series of wood joists or trusses, which are supported by the steel beams. However, in some cases wood nailers may be bolted to the tops of steel beams, and the deck can then be directly attached to the beams.

When a concrete slab is poured at the site, it may be combined with poured concrete fireproofing. When the beams are fire-

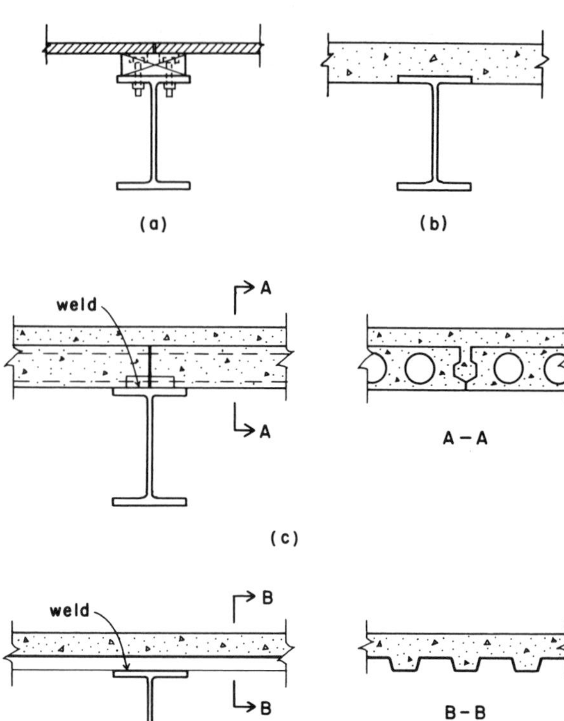

FIGURE 8.5 Typical floor decks: (*a*) plywood, (*b*) cast-in-place concrete, (*c*) precast concrete, (*d*) concrete fill on formed sheet steel units.

proofed by other means, it is common to use the detail shown in Fig. 8.5*b*. Concrete may also be used in the form of precast deck units that are welded to the steel beams using steel devices cast into the units. A concrete fill is normally used on top of precast units in order to provide a smooth top surface.

A floor deck system used widely with steel framing is that of corrugated or fluted steel units with concrete fill, as shown in Fig. 8.5*d*. For low-slope roofs the decking problem is essentially similar to that for floors, and many of the same types of deck may be used. However, there are some different issues involved in roof construction, and therefore some products are used only for roofs (see Fig. 8.6).

Steel deck units are available from a large number of manufacturers. Specific information regarding the type of deck available, possible range of sizes, rated load-carrying capacities, and so on,

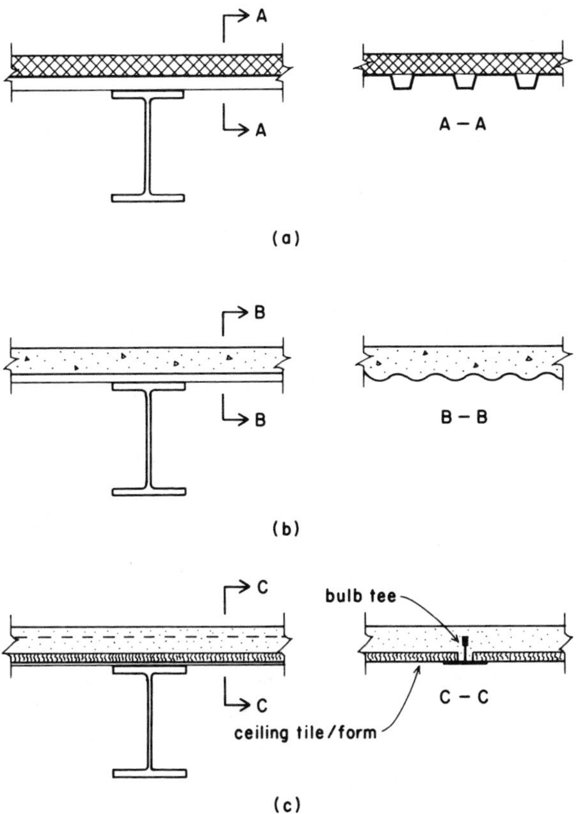

FIGURE 8.6 Typical roof decks: (*a*) formed sheet steel units with rigid insulating units, (*b*) formed sheet steel forming units with lightweight concrete fill, (*c*) gypsum concrete fill on combined form and ceiling finish units.

TABLE 8.4 Load Capacity of Formed Steel Roof Deck

Deck Type[a]	Span Condition	Weight[b] (psf)	Total (Dead & Live) Safe Load[c] for Spans Indicated in ft-in.												
			4-0	4-6	5-0	5-6	6-0	6-6	7-0	7-6	8-0	8-6	9-0	9-6	10-0
NR22	Simple	1.6	73	58	47										
NR20		2.0	91	72	58	48	40								
NR18		2.7	121	95	77	64	54	46							
NR22	Two	1.6	80	63	51	42									
NR20		2.0	96	76	61	51	43								
NR18		2.7	124	98	79	66	55	47	41						
NR22	Three or More	1.6	100	79	64	53	44								
NR20		2.0	120	95	77	63	53	45							
NR18		2.7	155	123	99	82	69	59	51	44					
IR22	Simple	1.6	86	68	55	45									
IR20		2.0	106	84	68	56	47	40							
IR18		2.7	142	112	91	75	63	54	46	40					
IR22	Two	1.6	93	74	60	49	41								
IR20		2.0	112	88	71	59	50	42							
IR18		2.7	145	115	93	77	64	55	47	41					
IR22	Three or More	1.6	117	92	75	62	52	44							
IR20		2.0	140	110	89	74	62	53	46	40					
IR18		2.7	181	143	116	96	81	69	59	52	45	40			
WR22	Simple	1.6			(89)	(70)	(56)	(46)							
WR20		2.0			(112)	(87)	(69)	(57)	(47)	(40)					
WR18		2.7			(154)	(119)	(94)	(76)	(63)	(53)	(45)				
WR22	Two	1.6			98	81	68	58	50	43					
WR20		2.0			125	103	87	74	64	55	49	43			
WR18		2.7			165	137	115	98	84	73	65	57	51	46	41
WR22	Three or More	1.6			122	101	85	72	62	54	(46)	(40)			
WR20		2.0			156	129	108	92	80	(67)	(57)	(49)	(43)		
WR18		2.7			207	171	144	122	105	(91)	(76)	(65)	(57)	(50)	(44)

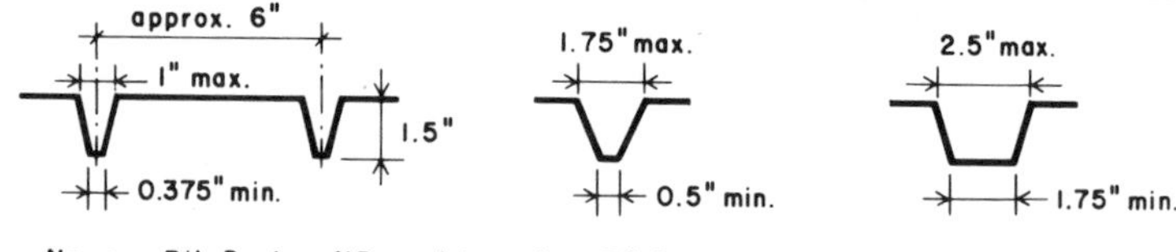

Source: Adapted from the *Steel Deck Institute Design Manual for Composite Decks, Form Decks, and Roof Decks,* 1981–1982 issue (Ref. 9), with permission of the publishers, the Steel Deck Institute. May not be reproduced without express permission of the Steel Deck Institute.

[a] Letters refer to rib type (see key). Numbers indicate gage (thickness) of steel.

[b] Approximate weight with paint finish; other finishes also available.

[c] Total safe allowable load in lb/sq ft. Loads in parentheses are governed by live load deflection not in excess of $\frac{1}{240}$ of the span, assuming a dead load of 10 psf.

should be obtained directly from those who supply these products to the region in which a proposed building is to be built. The information in Table 8.4 is provided by the national organization that provides standards to the deck manufacturers and indicates one widely used type of deck.

Problem 8.11.A,B,C. Using data from Table 8.4, select the lightest steel deck for the following:

A Simple span of 7 ft, total load of 45 psf

B Two-span condition, span of 8.5 ft, total load of 45 psf

C Three-span condition, span of 6 ft, total load of 50 psf

8.12 Prefabricated Joists

The truss is a highly efficient spanning device, especially when spans are somewhat longer and loads are relatively light. Roof trusses—particularly of the gabled, triangular profile—were one of the first systems developed and popularized when steel framing (preceded by iron) was first exploited in the last century. Today, many trusses for buildings are predesigned and shop-fabricated as standard manufactured products. The truss form in widest use is the parallel-chorded type shown in Fig. 8.7*a*, known as an open-web joist. For slightly longer spans, the form shown in Fig. 8.7*b*, may also be used for a joist or for a truss-girder to carry smaller, shorter-spanning open-web joists. Other popular forms for light steel trusses are the classic triangular-profiled Fink truss in Fig. 8.7*c* and the three-chorded truss called a delta truss (for the triangular cross section—resembling the Greek letter delta Δ) shown in Fig. 8.7*d*.

Shop-fabricated parallel-chorded trusses are produced in a wide range of sizes with various fabrication details by a number of manufacturers. Most producers comply with the regulations developed by the major industry organization in this area: the Steel Joist Institute. Publications of the institute (called the SJI) are a chief source of design information (see Ref. 8), although the products of individual manufacturers vary some, so that more information may be provided by suppliers or the manufacturers themselves.

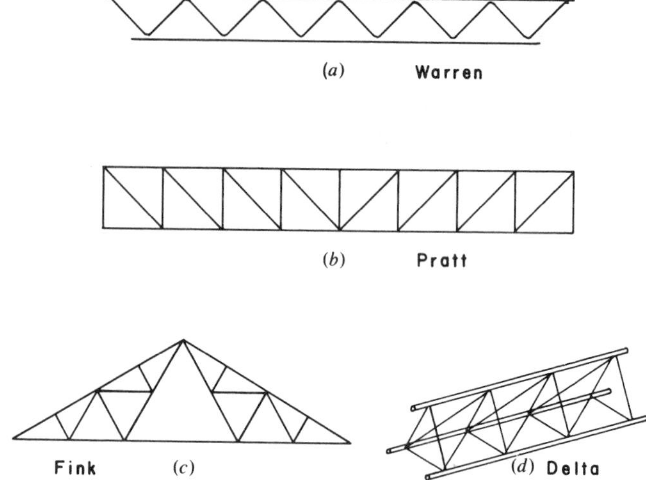

FIGURE 8.7 Common forms of light steel trusses.

The smallest and lightest members produced, called open-web joists, are used for the direct support of roof and floor decks, sustaining essentially only uniformly distributed loads on their top chords. A popular form for these is that shown in Fig. 8.8, with chords of cold-formed sheet steel and webs of steel rods. Chords may also be double angles, with the rods sandwiched between the angles at joints.

Table 8.5 is adapted from the standard tables of the Steel Joists Institute. This table lists the range of joist sizes available in the basic K series. (*Note:* A few of the heavier sizes have been omitted to shorten the table). Joists are identified by a three-unit designation. The first number indicates the overall depth of the joist,

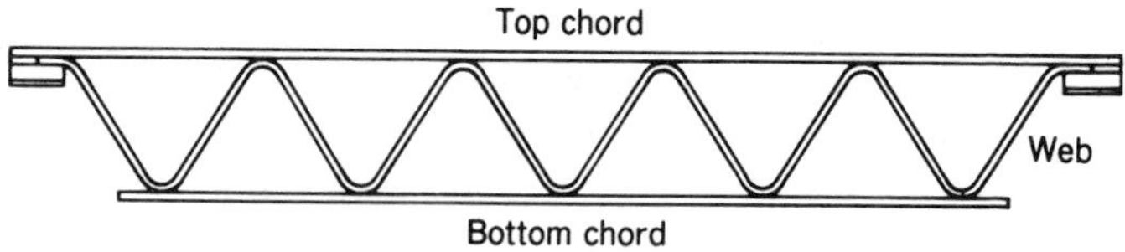

FIGURE 8.8 Form of light open-web steel joist.

the letter tells the series, and the second number gives the class of size of the members used; the higher the number, the heavier the joist.

Table 8.5 can be used to select the proper joist for a determined load and span condition. There are usually two entries in the table for each span; the first number represents the total load capacity of the joist, and the number in parentheses identifies the load that will produce a deflection of $\frac{1}{360}$ of the span. The following examples illustrate the use of the table data for some typical design situations.

Example 1. Open-web steel joists are to be used to support a roof with a unit live load of 20 psf and a unit dead load of 15 psf (not including the weight of the joists) on a span of 40 ft. Joists are spaced at 6 ft center to center. Select the lightest joist if deflection under live load is limited to $\frac{1}{360}$ of the span.

Solution: We first determine the load per foot on the joist:

Live load:	$6 \times 20 =$	120 lb/ft
Dead load:	$6 \times 15 =$	90 lb/ft
Total load:	$=$	210 lb/ft

We then scan the entries in Table 8.5 for the joists that will just carry these loads, noting that the joist weight must be deducted from the entry for total capacity. The possible choices for this example are summarized in Table 8.6. Although the joist weights are all very close, the 24K6 is the lightest choice.

Example 2. Open-web steel joists are to be used for a floor with a unit live load of 75 psf and a unit dead load of 40 psf (not including the joists) on a span of 30 ft. Joists are 2 ft center to center, and deflection is limited to $\frac{1}{360}$ of the span under live load only and to $\frac{1}{240}$ of the span under the total load. Determine the lightest joist possible and the joist with the least depth possible.

Solution: As in Example 1, we first find the loads on the joist:

Live load:	$2 \times 75 =$	150 lb/ft
Dead load:	$2 \times 40 =$	80 lb/ft
Total load:	$=$	230 lb/ft

TABLE 8.5 Allowable Loads for Open-Web Steel Joists[a]

Joist Designation	12K1	12K3	12K5	14K1	14K3	14K4	14K6	16K2	16K3	16K4	16K6	18K3	18K4	18K5	18K7	20K3	20K4	20K5	20K7
Weight (lb/ft)	5.0	5.7	7.1	5.2	6.0	6.7	7.7	5.5	6.3	7.0	8.1	6.6	7.2	7.7	9.0	6.7	7.6	8.2	9.3
Span (ft)																			
20	241 (142)	302 (177)	409 (230)	284 (197)	356 (246)	428 (287)	525 (347)	368 (297)	410 (330)	493 (386)	550 (426)	463 (423)	550 (490)	550 (490)	550 (490)	517 (517)	550 (550)	550 (550)	550 (550)
22	199 (106)	249 (132)	337 (172)	234 (147)	293 (184)	353 (215)	432 (259)	303 (222)	337 (247)	406 (289)	498 (351)	382 (316)	460 (370)	518 (414)	550 (438)	426 (393)	514 (461)	550 (490)	550 (490)
24	166 (81)	208 (101)	282 (132)	196 (113)	245 (141)	295 (165)	362 (199)	254 (170)	283 (189)	340 (221)	418 (269)	320 (242)	385 (284)	434 (318)	526 (382)	357 (302)	430 (353)	485 (396)	550 (448)
26				166 (88)	209 (110)	251 (129)	308 (156)	216 (133)	240 (148)	289 (173)	355 (211)	272 (190)	328 (222)	369 (249)	448 (299)	304 (236)	366 (277)	412 (310)	500 (373)
28				143 (70)	180 (88)	216 (103)	265 (124)	186 (106)	207 (118)	249 (138)	306 (168)	234 (151)	282 (177)	318 (199)	385 (239)	261 (189)	315 (221)	355 (248)	430 (298)
30								161 (86)	180 (96)	216 (112)	266 (137)	203 (123)	245 (144)	276 (161)	335 (194)	227 (153)	274 (179)	308 (201)	374 (242)
32								142 (71)	158 (79)	190 (92)	233 (112)	178 (101)	215 (118)	242 (132)	294 (159)	199 (126)	240 (147)	271 (165)	328 (199)
36												141 (70)	169 (82)	191 (92)	232 (111)	157 (88)	189 (103)	213 (115)	259 (139)
40																127 (64)	153 (75)	172 (84)	209 (101)

Joist Designation	22K4	22K5	22K6	22K9	24K4	24K5	24K6	24K9	26K5	26K6	26K9	28K6	28K7	28K8	28K10	30K7	30K8	30K9	30K12
Weight (lb/ft)	8.0	8.8	9.2	11.3	8.4	9.3	9.7	12.0	9.8	10.6	12.2	11.4	11.8	12.7	14.3	12.3	13.2	13.4	17.6
Span (ft)																			
28	348 (270)	392 (302)	427 (328)	550 (413)	381 (323)	429 (362)	467 (393)	550 (456)	466 (427)	508 (464)	550 (501)	548 (541)	550 (543)	550 (543)	550 (543)				
30	302 (219)	341 (245)	371 (266)	497 (349)	331 (262)	373 (293)	406 (319)	544 (419)	405 (346)	441 (377)	550 (459)	477 (439)	531 (486)	550 (500)	550 (500)	550 (543)	550 (543)	550 (543)	550 (543)
32	265 (180)	299 (201)	326 (219)	436 (287)	290 (215)	327 (241)	357 (262)	478 (344)	356 (285)	387 (309)	519 (407)	418 (361)	466 (400)	515 (438)	549 (463)	501 (461)	549 (500)	549 (500)	549 (500)
36	209 (126)	236 (141)	257 (153)	344 (201)	229 (150)	258 (169)	281 (183)	377 (241)	280 (199)	305 (216)	409 (284)	330 (252)	367 (280)	406 (306)	487 (366)	395 (323)	436 (353)	475 (383)	487 (392)
40	169 (91)	190 (102)	207 (111)	278 (146)	185 (109)	208 (122)	227 (133)	304 (175)	227 (145)	247 (157)	331 (207)	266 (183)	297 (203)	328 (222)	424 (284)	319 (234)	353 (256)	384 (278)	438 (315)
44	139 (68)	157 (76)	171 (83)	229 (109)	153 (82)	172 (92)	187 (100)	251 (131)	187 (108)	204 (118)	273 (155)	220 (137)	245 (152)	271 (167)	350 (212)	263 (176)	291 (192)	317 (208)	398 (258)
48					128 (63)	144 (70)	157 (77)	211 (101)	157 (83)	171 (90)	229 (119)	184 (105)	206 (117)	227 (128)	294 (163)	221 (135)	244 (148)	266 (160)	365 (216)
52									133 (65)	145 (71)	195 (93)	157 (83)	175 (92)	193 (100)	250 (128)	188 (106)	208 (116)	226 (126)	336 (184)
56												135 (66)	151 (73)	166 (80)	215 (102)	162 (84)	179 (92)	195 (100)	301 (153)
60																141 (69)	156 (75)	169 (81)	262 (124)

Source: Data adapted from more extensive tables in the *Standard Specifications, Load Tables, and Weight Tables for Steel Joists and Joist Girders*, 1986 ed. (Ref. 8), with permission of the publishers, Steel Joist Institute. The Steel Joist Institute publishes both specifications and load tables; each of these contains standards that are to be used in conjunction with one another.

[a] Loads in pounds per foot of joist span; first entry represents the total joist capacity; entry in parentheses is the load that produces a maximum deflection of $\frac{1}{360}$ of the span. See Fig. 8.9 for definition of span.

TABLE 8.6 Possible Choices for the Roof Joist

	Load per Foot for the Indicated Joists			
Load Condition	22K9	24K6	26K5	28K6
Total capacity (from Table 8.5)	278	227	227	266
Joist weight (from Table 8.5)	11.3	9.7	9.8	11.4
Net usable capacity	266.7	217.3	217.2	254.6
Load for deflection of $L/360$ (from Table 8.5)	146	133	145	183

To satisfy the deflection criteria, we must find a table entry in parentheses of 150 lb/ft (for live load only) or $\frac{240}{360} \times 230 = 153$ lb/ft (for total load). Scanning Table 8.5 we observe

The lightest joist is the 20K4.
The shallowest joist is the 18K5.

In real situations there may be compelling reasons for selection of a deeper joist, even though its load capacity may be redundant. One such case occurs when the top chord of the joist supports a roof or floor deck and the bottom chord is used directly for attachment of ceiling construction.

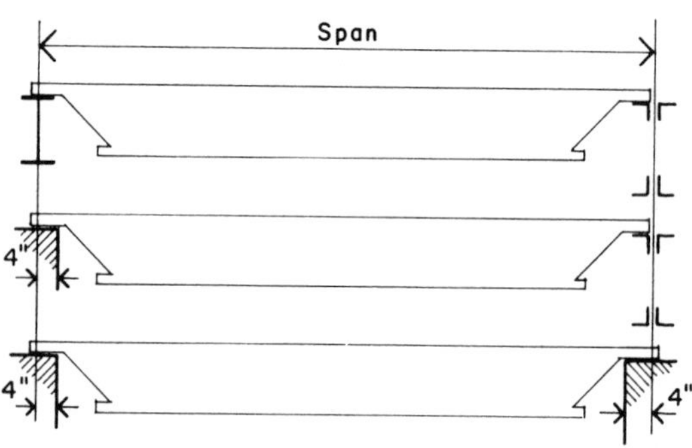

FIGURE 8.9 Definition of span for open web joists—as given in Ref. 8.

For heavier loads, and longer spans, trusses are produced in series described as long span and deep long span, the latter achieving depths of 7 ft and spans approaching 150 ft. In some situations the particular loading and span may clearly indicate the choice of the series, as well as the specific size of member. In many cases, however, the separate series overlap in capabilities, making the choice dependent on the product costs.

Open-web, long-span, and deep long-span trusses are all essentially designed for the uniformly loaded condition. This load may be due only to a roof or floor deck on the top, or may include a ceiling attached to the bottom. For roofs, an often used potential is that for sloping the top chord to facilitate drainage, while maintaining the flat bottom for a ceiling. Relatively small concentrated loads may be tolerated, the more so if they are applied close to the truss joints.

For development of a complete truss system, a special type of truss available is that described as a joist girder. This truss is specifically designed to carry the spaced, concentrated loads consisting of the end supports of directly loaded joists.

Load tables for all the standard products, as well as specifications for lateral bracing, support details, and so on, are available from the SJI (see Ref. 8). Planar trusses, like very thin beams, need considerable lateral support. Attached roof or floor decks and even ceilings may provide the necessary support if the attachment and the stiffness of the bracing construction are adequate. For trusses where no ceiling exists, bridging or other forms of lateral bracing must be used.

Development of bracing must be carefully studied in many cases. This is a three-dimensional problem, involving not only the stability of the trusses, but the general development of the building construction. Elements used for bracing can often do service as supports for ducts, lighting, building equipment, catwalks, and so on.

Problem 8.12.A. Open-web steel joists are to be used for a roof with a live load of 25 psf and a dead load of 20 psf (not including joists) on a span of 48 ft. Joists are 4 ft center to center, and the deflection under live load is limited to $\frac{1}{360}$ of the span. Select the lightest joist possible.

Problem 8.12.B. Open-web steel joists are to be used for a roof with a live load of 30 psf and a dead load of 18 psf (not including joists) on a span of 44 ft. Joists are 5 ft center to center, and deflection under live load is limited to $\frac{1}{360}$ of the span. Select the lightest joist.

Problem 8.12.C. Open-web steel joists are to be used for a floor with a live load of 50 psf and a dead load of 45 psf (not including joists) on a span of 36 ft. Joists are 2 ft center to center, and deflection is limited to $\frac{1}{360}$ of the span under live load and to $\frac{1}{240}$ of the span under total load. Select (a) the lightest possible joist and (b) the shallowest possible joist.

Problem 8.12.D. Repeat Problem 8.12C except that the live load is 100 psf, the dead load 35 psf, and the span 26 ft.

8.13 Steel Framing Systems

This chapter has presented various elements of steel that are commonly used as components of framing systems for building roofs and floors. These elements and systems may be used in conjunction with a general steel-framed structure or as the spanning systems for buildings with structural wall systems of masonry or concrete. The variety of potential combinations for different situations, locations, types of buildings, and general differences in design requirements is quite extensive. Some of the general considerations for system design are discussed in Chapter 19, and a few specific situations of design are illustrated in the example building designs in Chapters 20–22.

Choice of alternatives for individual elements and for general systems is a major part of the general effort for structural design. The structure must perform its structural tasks in a safe and reasonably efficient manner, be economical and relatively easy to build, satisfy any legal code requirements, and be competitive with alternative forms of construction. The judgments necessary to assure all of that must be based on good information, careful analysis of the design problems, and no little experience.

9

Steel Columns

Steel compression members range from small, single-piece columns and truss members to huge, built-up sections for high-rise buildings and large tower structures. The basic column function is one of simple compressive force resistance ($f = P/A$), but is complicated by the effects of buckling and the possible presence of bending actions. This chapter presents various issues relating to design of building columns of steel, with an emphasis on the common, single-piece section.

9.1 Column Shapes

For modest load conditions, the most frequently used shapes are the round pipe, the rectangular tube, and the H-shaped wide flange. Considerations of accommodation of beam framing connections favor the use of approximately square 10-, 12-, and 14-in. wide-flange sections, which are available in a considerable range of weights, up to the heaviest rolled section: the W 14 × 730.

For various reasons, it is sometimes necessary to make a column section by assembling two or more individual steel elements. Figure 9.1 shows some commonly used assemblages

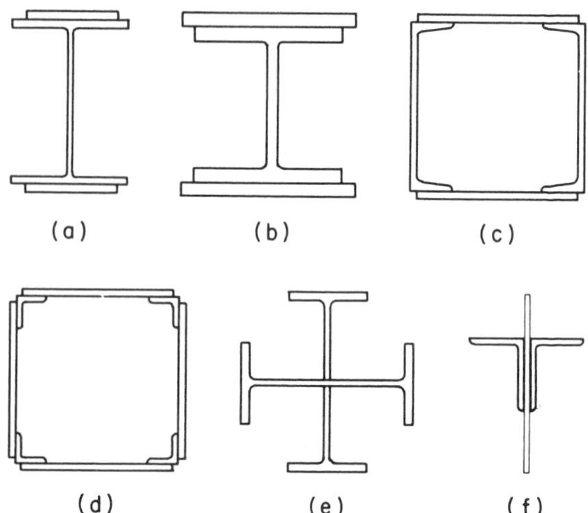

FIGURE 9.1 Built-up steel column sections.

which are used for special purposes. These are often required where a particular size or shape is not available from the inventory of stock rolled sections, where a particular structural task is required, or where details of the framing require a particular form. These are somewhat less used now, as the range of size and shape of stock sections has steadily increased. They are now mostly used for exceptionally large columns or for special framing problems. The customized fabrication of built-up sections is usually costly, so a single piece is typically favored if one is available.

One widely used built-up section is that of the double angle, shown in Fig. 9.1*f*. This occurs most often as a member of a truss or as a bracing member in a frame, the general stability of the paired members being much better than that of the single angle. This section is not much used as a building column.

9.2 Slenderness and End Conditions

The general effect of column slenderness and its effect on axial compression capacity are discussed in Chapter 5 and illustrated in

Fig. 5.1. For steel columns, the value of the allowable stress in compression is determined from formulas in the AISC Specification that include variables of the yield stress, modulus of elasticity, relative column stiffness, and special conditions of restraint of the column. As a measure of resistance to buckling, the basic property of a column is its slenderness, computed as L/r, in which L is the unbraced length and r the radius of gyration of the column section with respect to the direction of potential buckling. Effects of end restraint are considered by use of a modifying factor (K) resulting in some reduced or magnified value for L (see Fig. 9.2). The slenderness is thus commonly expressed as KL/r.

Figure 9.3 is a graph of the allowable axial compressive stress for a column of A36 steel (F_y = 36 ksi [165 MPa]) as determined from the AISC formulas. Values for full number increments of

	(a)	(b)	(c)	(d)	(e)	(f)
Buckled shape of column is shown by dashed line	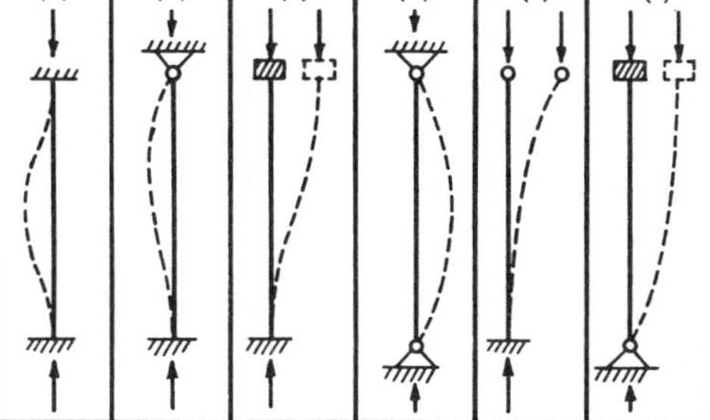					
Theoretical K value	0.5	0.7	1.0	1.0	2.0	2.0
Recommended design value when ideal conditions are approximated	0.65	0.80	1.2	1.0	2.10	2.0
End condition code		Rotation fixed and translation fixed Rotation free and translation fixed Rotation fixed and translation free Rotation free and translation free				

FIGURE 9.2 Determination of effective unbraced column length. Reprinted from the *Manual of Steel Construction*, 8th ed. (Ref. 2), with permission of the publishers, American Institute of Steel Construction.

TABLE 9.1 Allowable Unit Stresses for Columns of A36 Steel (in ksi)

Main and Secondary Members Kl/r not over 120						Main Members Kl/r 121 to 200				Secondary Members[a] l/r 121 to 200			
$\frac{Kl}{r}$	F_a (ksi)	$\frac{Kl}{r}$	F_a (ksi)	$\frac{Kl}{r}$	F_a (ksi)	$\frac{Kl}{r}$	F_a (ksi)	$\frac{Kl}{r}$	F_a (ksi)	$\frac{l}{r}$	F_{as} (ksi)	$\frac{l}{r}$	F_{as} (ksi)
1	21.56	41	19.11	81	15.24	121	10.14	161	5.76	121	10.19	161	7.25
2	21.52	42	19.03	82	15.13	122	9.99	162	5.69	122	10.09	162	7.20
3	21.48	43	18.95	83	15.02	123	9.85	163	5.62	123	10.00	163	7.16
4	21.44	44	18.86	84	14.90	124	9.70	164	5.55	124	9.90	164	7.12
5	21.39	45	18.78	85	14.79	125	9.55	165	5.49	125	9.80	165	7.08
6	21.35	46	18.70	86	14.67	126	9.41	166	5.42	126	9.70	166	7.04
7	21.30	47	18.61	87	14.56	127	9.26	167	5.35	127	9.59	167	7.00
8	21.25	48	18.53	88	14.44	128	9.11	168	5.29	128	9.49	168	6.96
9	21.21	49	18.44	89	14.32	129	8.97	169	5.23	129	9.40	169	6.93
10	21.16	50	18.35	90	14.20	130	8.84	170	5.17	130	9.30	170	6.89
11	21.10	51	18.26	91	14.09	131	8.70	171	5.11	131	9.21	171	6.85
12	21.05	52	18.17	92	13.97	132	8.57	172	5.05	132	9.12	172	6.82
13	21.00	53	18.08	93	13.84	133	8.44	173	4.99	133	9.03	173	6.79
14	20.95	54	17.99	94	13.72	134	8.32	174	4.93	134	8.94	174	6.76
15	20.89	55	17.90	95	13.60	135	8.19	175	4.88	135	8.86	175	6.73
16	20.83	56	17.81	96	13.48	136	8.07	176	4.82	136	8.78	176	6.70
17	20.78	57	17.71	97	13.35	137	7.96	177	4.77	137	8.70	177	6.67
18	20.72	58	17.62	98	13.23	138	7.84	178	4.71	138	8.62	178	6.64
19	20.66	59	17.53	99	13.10	139	7.73	179	4.66	139	8.54	179	6.61
20	20.60	60	17.43	100	12.98	140	7.62	180	4.61	140	8.47	180	6.58
21	20.54	61	17.33	101	12.85	141	7.51	181	4.56	141	8.39	181	6.56
22	20.48	62	17.24	102	12.72	142	7.41	182	4.51	142	8.32	182	6.53
23	20.41	63	17.14	103	12.59	143	7.30	183	4.46	143	8.25	183	6.51
24	20.35	64	17.04	104	12.47	144	7.20	184	4.41	144	8.18	184	6.49
25	20.28	65	16.94	105	12.33	145	7.10	185	4.36	145	8.12	185	6.46
26	20.22	66	16.84	106	12.20	146	7.01	186	4.32	146	8.05	186	6.44
27	20.15	67	16.74	107	12.07	147	6.91	187	4.27	147	7.99	187	6.42
28	20.08	68	16.64	108	11.94	148	6.82	188	4.23	148	7.93	188	6.40
29	20.01	69	16.53	109	11.81	149	6.73	189	4.18	149	7.87	189	6.38
30	19.94	70	16.43	110	11.67	150	6.64	190	4.14	150	7.81	190	6.36
31	19.87	71	16.33	111	11.54	151	6.55	191	4.09	151	7.75	191	6.35
32	19.80	72	16.22	112	11.40	152	6.46	192	4.05	152	7.69	192	6.33
33	19.73	73	16.12	113	11.26	153	6.38	193	4.01	153	7.64	193	6.31
34	19.65	74	16.01	114	11.13	154	6.30	194	3.97	154	7.59	194	6.30
35	19.58	75	15.90	115	10.99	155	6.22	195	3.93	155	7.53	195	6.28
36	19.50	76	15.79	116	10.85	156	6.14	196	3.89	156	7.48	196	6.27
37	19.42	77	15.69	117	10.71	157	6.06	197	3.85	157	7.43	197	6.26
38	19.35	78	15.58	118	10.57	158	5.98	198	3.81	158	7.39	198	6.24
39	19.27	79	15.47	119	10.43	159	5.91	199	3.77	159	7.34	199	6.23
40	19.19	80	15.36	120	10.28	160	5.83	200	3.73	160	7.29	200	6.22

Source: Reproduced from the *Manual of Steel Construction*, 8th ed. (Ref. 2), with permission of the publishers, American Institute of Steel Construction.

[a] K taken as 1.0 for secondary members.

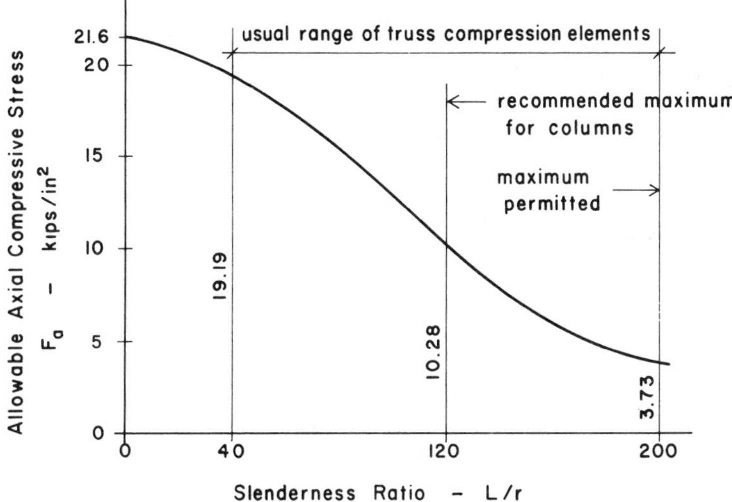

FIGURE 9.3 Allowable compression stress as a function of the slenderness ratio *L/r*. AISC requirements for A36 steel.

KL/r are also given in Table 9.1. Note that the table has two parts for values of *KL/r* over 120. Building columns would be considered as "main members" in this classification.

It is generally recommended that building columns not have an *L/r* in excess of 120. This is a matter of some concern for efficient use of the material, but is also a desire not to have excessively skinny columns, as the column profiles in Fig. 9.2 indicate.

9.3 Allowable Column Loads

The allowable axial compression load for a column is computed by multiplying the allowable stress (F_a) by the cross-sectional area of the column. The following examples demonstrate the process. For single-piece columns, the most direct method is through the use of column load tables, as discussed in the next section. For built-up sections, it is necessary to compute the properties of the section and to use the process demonstrated in Example 3.

Example 1. A W 12 × 58 is used as a column with an unbraced length of 16 ft [4.88 m]. Compute the allowable load.

Solution: Referring to Table 3.1, we find that $A = 17.0$ in.^{2} [10,968 mm^2], $r_x = 5.28$ in. [134 mm], and $r_y = 2.51$ in. [63.8 mm]. If the column is unbraced on both axes, it is limited by the lower r value for the weak axis. With no qualifying end conditions, we assume an end condition, as shown in Fig. 9.2*d*, for which $K = 1$. Thus the relative slenderness is computed as

$$\frac{Kl}{r} = \frac{1(16)(12)}{2.51} = 76.5$$

In design work it is usually considered acceptable to round the value for the slenderness ratio off to the next highest figure in front of the decimal point. Therefore we consider the Kl/r ratio to be 77, and find from Table 9.1 that the allowable stress (F_a) is 15.69 ksi [108.2 MPa]. The allowable load for the column is thus

$$P = AF_a = 17.0(15.69) = 266.7 \text{ kips [1187 kN]}$$

Example 2. Compute the allowable load for the column in Example 1 if the top is pinned and the bottom is fixed.

Solution: Referring to Fig. 9.2, the recommended K factor for this condition is 0.8. The modified stiffness is thus

$$\frac{KL}{r} = \frac{0.8[16(12)]}{2.51} = 61.2, \text{ say } 61$$

and the allowable stress from Table 9.1 is 17.33 ksi [119.5 MPa]. The allowable load is thus

$$P = AF_a = 17.0(17.33) = 294.6 \text{ kips [1311 kN]}$$

Conditions frequently exist to cause different modification of the basic column action on the two axes of a wide-flange column. In this event it may be necessary to investigate the conditions relating to the separate axes in order to determine the limiting condition. The following example illustrates such a case.

Example 3. Figure 9.4*a* shows an elevation of the steel framing at the location of an exterior wall. The column is laterally restrained but rotationally free at the top and bottom. (End conditions as in case *d* in Fig. 9.2.) With respect to the *x*-axis of the section, the column is laterally unbraced for its full height. However, the existence of horizontal framing in the wall plane provides lateral bracing at a point between the top and bottom with respect to the *y*-axis of the section. If the column is a W 12 × 58 of A36 steel, L_1 is 30 ft, and L_2 is 18 ft (see Fig. 9.4*b*), what is the allowable compression load?

Solution: With respect to the *x*-axis, the column functions as a pin-ended member for its full height (Fig. 9.2*d*). However, with

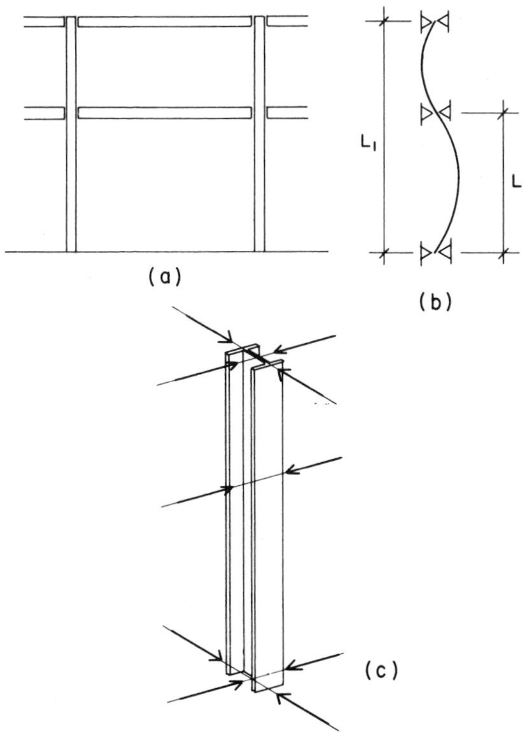

FIGURE 9.4 Biaxial bracing of columns.

respect to the y-axis, the form of buckling is as shown in Fig. 9.4b, and the laterally unsupported height is 18 ft [5.49 m]. For both conditions the K factor is 1, and the investigation is as follows (see Example 1 for data for the section):

$$x\text{-axis:} \quad \frac{KL}{r} = \frac{30(12)}{5.28} = 68.2$$

$$y\text{-axis:} \quad \frac{KL}{r} = \frac{18(12)}{2.51} = 86.1, \text{ say } 86$$

Despite the bracing, the column is still critical on its y-axis. From Table 9.1 we obtain F_a = 14.67 ksi [101 MPa] and the allowable load is thus

$$P = AF_a = 17.0(14.67) = 249.4 \text{ kips [1108 kN]}$$

Note: For the following problems use A36 steel.

Problem 9.3.A. Determine the allowable axial compression load for a W 10 × 60 column with an unbraced height of 15 ft [4.57 m]. Assume $K = 1$.

Problem 9.3.B. Determine the allowable axial compression load for a W 12 × 106 column with a height of 22 ft [6.71 m], if both ends are fixed against rotation and horizontal movement.

Problem 9.3.C. Determine the allowable load for the column in Problem 9.3.A if the conditions are as shown in Fig. 9.4, with L_1 = 15 ft and L_2 = 8 ft.

9.4 Design With Tabular Data

As demonstrated in the examples in Sec. 9.3 the determination of the allowable compression load for a column is quite simple once the necessary data are known. When designing columns, however, the process is complicated by the fact that the allowable stress is unknown if the section is undefined. This results in a pick-and-try method of design unless some aids are used. Since there are a fixed number of stock sections for ordinary use as single-piece columns, it is possible to determine their allowable loads for various unbraced lengths and record the data in tabular form. Extensive tables exist in the AISC Manual (Ref. 2) for single-piece columns as well as for commonly used combinations of double angles.

Selected data from the AISC tables are displayed in Tables 9.2 and 9.4–9.6. Table 9.2 lists allowable axial compression loads for the range of W and M shapes commonly used for columns, due to their approximately square shape. Verification of any of the table entries may be made by the process illustrated in Sec. 9.3. The significance of the bending factors given at the extreme right of the table is considered in Sec. 9.5. The following example illustrates the use of the table for a simple design problem.

Example 1. A column is pinned at top and bottom and has a laterally unsupported height of 20 ft [6.10 m]. Select a wide-flange shape for an axial load of 200 kips [890 kN].

Solution: Scanning down the table column for 20 ft, we find the possible choices listed in Table 9.3. While all of these shapes are adequate for the load, other considerations in the development of the structure and building construction would probably point to the preference for one size over others. The L/r values are not taken from the table, but were determined from the r_y properties of the shapes. The table cuts off at the maximum value of $L/r = 200$, but most designers would prefer to stay below a value of 120 for major building columns.

Table 9.4 gives allowable axial loads for standard weight steel pipe columns with a yield point of 36 ksi [250 MPa]. The outside diameters at the head of the table are *nominal* dimensions that designate the pipe sizes. True outside diameters are slightly larger and can be found from the Table 3.5. The AISC Manual contains additional tables that list allowable loads for the two heavier weight groups of steel pipe: extra strong and double-extra strong.

Example 2. Same data as for Example 1; select a pipe of A36 steel.

Solution: From Table 9.4 the minimum size is the 10-in.-nominal-diameter pipe. It should be noted that Table 3.4 gives the actual outside diameter as 10.75 in. Table 3.4 also gives properties for the two heavier classes of pipe, although Table 9.4 gives

TABLE 9.2 Allowable Column Loads for Selected W and M Shapes[a]

Shape	Effective length (KL) in feet										Bending factor	
	8	9	10	11	12	14	16	18	20	22	B_x	B_y
M 4 × 13	48	42	35	29	24	18					0.727	2.228
W 4 × 13	52	46	39	33	28	20	16				0.701	2.016
W 5 × 16	74	69	64	58	52	40	31	24	20		0.550	1.560
M 5 × 18.9	85	78	71	64	56	42	32	25			0.576	1.768
W 5 × 19	88	82	76	70	63	48	37	29	24		0.543	1.526
W 6 × 9	33	28	23	19	16	12					0.482	2.414
W 6 × 12	44	38	31	26	22	16					0.486	2.367
W 6 × 16	62	54	46	38	32	23	18				0.465	2.155
W 6 × 15	75	71	67	62	58	48	38	30	24	20	0.456	1.424
M 6 × 20	98	92	87	81	74	61	47	37	30	25	0.453	1.510
W 6 × 20	100	95	90	85	79	67	54	42	34	28	0.438	1.331
W 6 × 25	126	120	114	107	100	85	69	54	44	36	0.440	1.308
W 8 × 24	124	118	113	107	101	88	74	59	48	39	0.339	1.258
W 8 × 28	144	138	132	125	118	103	87	69	56	46	0.340	1.244
W 8 × 31	170	165	160	154	149	137	124	110	95	80	0.332	0.985
W 8 × 35	191	186	180	174	168	155	141	125	109	91	0.330	0.972
W 8 × 40	218	212	205	199	192	127	160	143	124	104	0.330	0.959
W 8 × 48	263	256	249	241	233	215	196	176	154	131	0.326	0.940
W 8 × 58	320	312	303	293	283	263	240	216	190	162	0.329	0.934
W 8 × 67	370	360	350	339	328	304	279	251	221	190	0.326	0.921
W 10 × 33	179	173	167	161	155	142	127	112	95	78	0.277	1.055
W 10 × 39	213	206	200	193	186	170	154	136	116	97	0.273	1.018
W 10 × 45	247	240	232	224	216	199	180	160	138	115	0.271	1.000
W 10 × 49	279	273	268	262	256	242	228	213	197	180	0.264	0.770
W 10 × 54	306	300	294	288	281	267	251	235	217	199	0.263	0.767
W 10 × 60	341	335	328	321	313	297	280	262	243	222	0.264	0.765
W 10 × 68	388	381	373	365	357	339	320	299	278	255	0.264	0.758
W 10 × 77	439	431	422	413	404	384	362	339	315	289	0.263	0.751
W 10 × 88	504	495	485	475	464	442	417	392	364	335	0.263	0.744
W 10 × 100	573	562	551	540	428	503	476	446	416	383	0.263	0.735
W 10 × 112	642	631	619	606	593	565	535	503	469	433	0.261	0.726
W 12 × 40	217	210	203	196	188	172	154	135	114	94	0.227	1.073
W 12 × 45	243	235	228	220	211	193	173	152	129	106	0.227	1.065
W 12 × 50	271	263	254	246	236	216	195	171	146	121	0.227	1.058
W 12 × 53	301	295	288	282	275	260	244	227	209	189	0.221	0.813
W 12 × 58	329	322	315	308	301	285	268	249	230	209	0.218	0.794
W 12 × 65	378	373	367	361	354	341	326	311	294	277	0.217	0.656
W 12 × 72	418	412	406	399	392	377	361	344	326	308	0.217	0.651
W 12 × 79	460	453	446	439	431	415	398	379	360	339	0.217	0.648
W 12 × 87	508	501	493	485	477	459	440	420	398	376	0.217	0.645
W 12 × 96	560	552	544	535	526	506	486	464	440	416	0.215	0.635
W 12 × 106	620	611	602	593	583	561	539	514	489	462	0.215	0.633

TABLE 9.2 (*Continued*)

Shape	Effective Length (KL) in Feet										Bending Factor	
	8	10	12	14	16	18	20	22	24	26	B_x	B_y
W 12 × 120	702	692	660	636	611	584	555	525	493	460	0.217	0.630
W 12 × 136	795	772	747	721	693	662	630	597	561	524	0.215	0.621
W 12 × 152	891	866	839	810	778	745	710	673	633	592	0.214	0.614
W 12 × 170	998	970	940	908	873	837	798	757	714	668	0.213	0.608
W 12 × 190	1115	1084	1051	1016	978	937	894	849	802	752	0.212	0.600
W 12 × 210	1236	1202	1166	1127	1086	1042	995	946	894	840	0.212	0.594
W 12 × 230	1355	1319	1280	1238	1193	1145	1095	1041	985	927	0.211	0.589
W 12 × 252	1484	1445	1403	1358	1309	1258	1203	1146	1085	1022	0.210	0.583
W 12 × 279	1642	1600	1554	1505	1452	1396	1337	1275	1209	1141	0.208	0.573
W 12 × 305	1799	1753	1704	1651	1594	1534	1471	1404	1333	1260	0.206	0.564
W 12 × 336	1986	1937	1884	1827	1766	1701	1632	1560	1484	1404	0.205	0.558
W 14 × 43	230	215	199	181	161	140	117	96	81	69	0.201	1.115
W 14 × 48	258	242	224	204	182	159	133	110	93	79	0.201	1.102
W 14 × 53	286	268	248	226	202	177	149	123	104	88	0.201	1.091
W 14 × 61	345	330	314	297	278	258	237	214	190	165	0.194	0.833
W 14 × 68	385	369	351	332	311	289	266	241	214	186	0.194	0.826
W 14 × 74	421	403	384	363	341	317	292	265	236	206	0.195	0.820
W 14 × 82	465	446	425	402	377	351	323	293	261	227	0.196	0.823
W 14 × 90	536	524	511	497	482	466	449	432	413	394	0.185	0.531
W 14 × 99	589	575	561	546	529	512	494	475	454	433	0.185	0.527
W 14 × 109	647	633	618	601	583	564	544	523	501	478	0.185	0.523
W 14 × 120	714	699	682	663	644	623	601	578	554	528	0.186	0.523
W 14 × 132	786	768	750	730	708	686	662	637	610	583	0.186	0.521
W 14 × 145	869	851	832	812	790	767	743	718	691	663	0.184	0.489
W 14 × 159	950	931	911	889	865	840	814	786	758	727	0.184	0.485
W 14 × 176	1054	1034	1011	987	961	933	904	874	842	809	0.184	0.484
W 14 × 193	1157	1134	1110	1083	1055	1025	994	961	927	891	0.183	0.477
W 14 × 211	1263	1239	1212	1183	1153	1121	1087	1051	1014	975	0.183	0.477
W 14 × 233	1396	1370	1340	1309	1276	1241	1204	1165	1124	1081	0.183	0.472
W 14 × 257	1542	1513	1481	1447	1410	1372	1331	1289	1244	1198	0.182	0.470
W 14 × 283	1700	1668	1634	1597	1557	1515	1471	1425	1377	1326	0.181	0.465
W 14 × 311	1867	1832	1794	1754	1711	1666	1618	1568	1515	1460	0.181	0.459
W 14 × 342		2022	1985	1941	1894	1845	1793	1738	1681	1621	0.181	0.457
W 14 × 370		2181	2144	2097	2047	1995	1939	1881	1820	1756	0.180	0.452
W 14 × 398		2356	2304	2255	2202	2146	2087	2025	1961	1893	0.178	0.447
W 14 × 426		2515	2464	2411	2356	2296	2234	2169	2100	2029	0.177	0.442
W 14 × 455		2694	2644	2589	2430	2467	2401	2332	2260	2184	0.177	0.441
W 14 × 500		2952	2905	2845	2781	2714	2642	2568	2490	2409	0.175	0.434
W 14 × 550		3272	3206	3142	3073	3000	2923	2842	2758	2670	0.174	0.429
W 14 × 605		3591	3529	3459	3384	3306	3223	3136	3045	2951	0.171	0.421
W 14 × 665		3974	3892	3817	3737	3652	3563	3469	3372	3270	0.170	0.415
W 14 × 730		4355	4277	4196	4100	4019	3923	3823	3718	3609	0.168	0.408

Source: Adapted from data in the *Manual of Steel Construction*, 8th ed. (Ref. 2), with permission of the publishers, American Institute of Steel Construction.

[a] Loads in kips for shapes of steel with yield stress of 36 ksi [250 MPa], based on buckling with respect to the *y*-axis.

TABLE 9.3 Alternate Choices for the Column: Example 1

Shape	Table Load (kips)	Actual L/r_y
W 8 × 67	221	113
W 10 × 54	217	94
W 12 × 53	209	97
W 14 × 61	237	98

loads only for standard pipe. When size is critical, it is possible to use the heavier pipe, for which loads are given in the AISC Manual.

Steel columns are fabricated from structural tubing in both square and rectangular shapes. Square tubing is available in sizes of 2–16 in. and rectangular sizes range from 3 × 2 to 20 × 12 in. Sections are produced with various wall thicknesses, thus allowing a considerable range of structural capacities. Although round pipe is specified by a nominal outside dimension, tubing is specified by its actual outside dimensions.

The AISC Manual contains safe load tables for square and rectangular tubing based on F_y = 46 ksi [317 MPa]. Table 9.5 is a reproduction of one of these tables—for 3- and 4-in.-square tubing—and is presented to illustrate the form of the tables.

Example 3. Using Table 9.5, select a structural tubing column to carry a load of 41 kips [182 kN] if the unbraced height is 12 ft [3.66 m].

Solution: Entering Table 9.5 with an effective length of 12 ft, we find that a load of 43 kips can be supported by a square section with 4-in. sides and a wall thickness of $\frac{3}{16}$ in. (designated TS 4 × 4 × $\frac{3}{16}$).

Both pipe and tubing may be available in various steel strengths. We have used the properties in these examples because they appear in the AISC Manual. The choice between round pipe or rectangular tubing for a column is usually made for reasons other than simple structural efficiency. Freestanding columns are often round, but when built into wall construction, the rectangular shapes are often preferred.

Two angle sections, separated by the thickness of a connection

TABLE 9.4 Allowable Column Loads for Standard Steel Pipe[a]

Nominal Dia.		12	10	8	6	5	4	3½	3
Wall Thickness		0.375	0.365	0.322	0.280	0.258	0.237	0.226	0.216
Weight per Foot		49.56	40.48	28.55	18.97	14.62	10.79	9.11	7.58
F_y		36 ksi							
Effective length in feet KL with respect to radius of gyration	0	315	257	181	121	93	68	58	48
	6	303	246	171	110	83	59	48	38
	7	301	243	168	108	81	57	46	36
	8	299	241	166	106	78	54	44	34
	9	296	238	163	103	76	52	41	31
	10	293	235	161	101	73	49	38	28
	11	291	232	158	98	71	46	35	25
	12	288	229	155	95	68	43	32	22
	13	285	226	152	92	65	40	29	19
	14	282	223	149	89	61	36	25	16
	15	278	220	145	86	58	33	22	14
	16	275	216	142	82	55	29	19	12
	17	272	213	138	79	51	26	17	11
	18	268	209	135	75	47	23	15	10
	19	265	205	131	71	43	21	14	9
	20	261	201	127	67	39	19	12	
	22	254	193	119	59	32	15	10	
	24	246	185	111	51	27	13		
	25	242	180	106	47	25	12		
	26	238	176	102	43	23			
	28	229	167	93	37	20			
	30	220	158	83	32	17			
	31	216	152	78	30	16			
	32	211	148	73	29				
	34	201	137	65	25				
	36	192	127	58	23				
	37	186	120	55	21				
	38	181	115	52					
	40	171	104	47					
Properties									
Area A (in.^{2})		14.6	11.9	8.40	5.58	4.30	3.17	2.68	2.23
I (in.^{4})		279	161	72.5	28.1	15.2	7.23	4.79	3.02
r (in.)		4.38	3.67	2.94	2.25	1.88	1.51	1.34	1.16
B } Bending factor		0.333	0.398	0.500	0.657	0.789	0.987	1.12	1.29
* a		41.7	23.9	10.8	4.21	2.26	1.08	0.717	0.447

* Tabulated values of a must be multiplied by 10^6.
Note: Heavy line indicates Kl/r of 200.

Source: Reproduced from the *Manual of Steel Construction,* 8th ed. (Ref. 2), with permission of the publishers, American Institute of Steel Construction.

[a] Loads in kips for axially loaded columns with yield stress of 36 ksi [250 MPa].

TABLE 9.5 Allowable Column Loads for Steel Structural Tubing[a]

Nominal Size	4 x 4					3 x 3		
Thickness	1/2	3/8	5/16	1/4	3/16	5/16	1/4	3/16
Wt./ft.	21.63	17.27	14.83	12.21	9.42	10.58	8.81	6.87
F_y	46 ksi							
Effective length in feet *KL* with respect to radius of gyration								
0	176	140	120	99	76	86	71	56
2	168	134	115	95	73	80	67	53
3	162	130	112	92	71	77	64	50
4	156	126	108	89	69	73	61	48
5	150	121	104	86	67	68	57	45
6	143	115	100	83	64	63	53	42
7	135	110	95	79	61	57	49	39
8	126	103	90	75	58	51	44	35
9	117	97	84	70	55	44	38	31
10	108	89	78	65	51	37	33	27
11	98	82	72	60	47	31	27	22
12	87	74	65	55	43	26	23	19
13	75	65	58	49	39	22	19	16
14	65	57	51	43	35	19	17	14
15	57	49	44	38	30	16	15	12
16	50	43	39	33	27	14	13	11
17	44	38	34	29	24	13	11	9
18	39	34	31	26	21		10	8
19	35	31	28	24	19			
20	32	28	25	21	17			
21	29	25	23	19	16			
22	26	23	21	18	14			
23	24	21	19	16	13			
24		19	17	15	12			
25				14	11			
Properties								
A (in.^{2})	6.36	5.08	4.36	3.59	2.77	3.11	2.59	2.02
I (in.^{4})	12.3	10.7	9.58	8.22	6.59	3.58	3.16	2.60
r (in.)	1.39	1.45	1.48	1.51	1.54	1.07	1.10	1.13
B Bending factor	1.04	0.949	0.910	0.874	0.840	1.30	1.23	1.17
*a	1.83	1.59	1.43	1.22	0.983	0.533	0.470	0.387

* Tabulated values of a must be multiplied by 10^6.
Note: Heavy line indicates Kl/r of 200.

Source: Reproduced from the *Manual of Steel Construction*, 8th ed. (Ref. 2), with permission of the publishers. American Institute of Steel Construction.

[a] Loads in kips for axially loaded columns with yield stress of 46 ksi [317 MPa].

plate at each end and fastened together at intervals by fillers and welds or bolts, are commonly used as compression members in roof trusses. (See Fig. 9.1*f*.) These members, whether or not in a vertical position, are called struts; their size is determined in accordance with the requirements and formulas for columns in Sec. 9.3). To ensure that the angles act as a unit, the intermittent connections are made at intervals such that the slenderness ratio l/r of either angle between fasteners does not exceed the governing slenderness ratio of the built-up member. The least radius of gyration r is used in computing the slenderness ratio of each angle.

The AISC Manual contains safe load tables for struts of two angles with $\frac{3}{8}$ in. separation back to back. Three series are given: equal-leg angles, unequal-leg angles with short legs back to back, and unequal-leg angles with long legs back to back. Table 9.6 has been abstracted from the latter series and lists allowable loads with respect to the X–X and Y–Y axes. The smaller (least) radius of gyration gives the smaller allowable load, and unless the member is braced with respect to the weaker axis, this is the tabular load to be used. The usual practice is to assume K equal to 1.0. The following example shows how the loads in the table are computed.

Example 4. Two $5 \times 3\frac{1}{2} \times \frac{1}{2}$-in. angle sections spaced with their long legs $\frac{3}{8}$ in. back to back are used as a compression member. If the member is A36 steel and has an effective length of 10 ft, compute the allowable axial load.

Solution: From Table 9.6 we find that the area of the two-angle member is 8.0 in.^{2} and that the radii of gyration are $r_x = 1.58$ in. and $r_y =$ in 1.49 in. Using the smaller r, the slenderness ratio is

$$\frac{Kl}{r} = 1 \times \frac{10 \times 12}{1.49} = 80.5, \text{ say } 81$$

Referring to Table 9.1, we find that $F_a = 15.24$ ksi, making the allowable load

$$P = A \times F_a = 8.0 \times 15.24 = 121.9 \text{ kips}$$

TABLE 9.6 Allowable Axial Compression for Double-Angle Struts[a]

Size (in.)		8 × 6	
Thickness (in.)		3/4	1/2
Weight (lb/ft)		67.6	46.0
Area (in²)		19.9	13.5
r_x (in.)		2.53	2.56
r_y (in.)		2.48	2.44
Effective Length (*KL*) with Respect to Indicated Axis			
X-X Axis	0	430	266
	10	370	231
	12	353	222
	14	334	211
	16	315	200
	20	271	175
	24	222	148
	28	168	117
	32	129	90
	36	102	71
Y-Y Axis	0	430	266
	10	368	229
	12	351	219
	14	332	207
	16	311	195
	20	266	169
	24	216	139
	28	162	106
	32	124	81
	36	98	64

Size (in.)		6 × 4		
Thickness (in.)		5/8	1/2	3/8
Weight (lb/ft)		40.0	32.4	24.6
Area (in²)		11.7	9.50	7.22
r_x (in.)		1.90	1.91	1.93
r_y (in.)		1.67	1.64	1.62
X-X Axis	0	253	205	142
	8	214	174	122
	10	200	163	115
	12	185	151	107
	14	168	137	99
	16	150	123	89
	20	110	90	69
	24	76	62	48
	28	56	46	36
Y-Y Axis	0	253	205	142
	6	222	179	125
	8	207	167	117
	10	190	153	108
	12	171	137	97
	14	151	120	86
	16	129	102	74
	20	85	66	49
	24	59	46	34

Size (in.)		5 × 3 1/2	
Thickness (in.)		1/2	3/8
Weight (lb/ft)		27.2	20.8
Area (in²)		8.00	6.09
r_x (in.)		1.58	1.60
r_y (in.)		1.49	1.46
X-X Axis	0	173	129
	4	159	119
	6	150	113
	8	139	105
	10	126	96
	12	113	86
	14	97	75
	16	81	63
	20	52	40
Y-Y Axis	0	173	129
	4	158	118
	6	148	110
	8	136	101
	10	122	91
	12	107	79
	14	90	67
	16	72	53
	20	46	34

Size (in.)		5 × 3		
Thickness (in.)		1/2	3/8	5/16
Weight (lb/ft)		25.6	19.6	16.4
Area (in²)		7.50	5.72	4.80
r_x (in.)		1.59	1.61	1.61
r_y (in.)		1.25	1.23	1.22
X-X Axis	0	162	121	94
	4	149	112	88
	6	141	106	83
	8	130	98	77
	10	119	90	71
	12	106	81	64
	14	92	70	57
	16	76	59	49
	20	49	38	32
Y-Y Axis	0	162	121	94
	4	145	108	85
	6	132	99	78
	8	118	88	69
	10	101	75	60
	12	82	61	49
	14	62	46	38
	16	47	35	29
	20	30	22	19

Source: Abstracted from data in the *Manual of Steel Construction,* 8th ed. (Ref. 2), with permission of the publishers, American Institute of Steel Construction.

[a] Loads in kips for angles with long legs back to back with $\frac{3}{8}$-in. separation and steel with F_y = 36 ksi [250 MPa].

This value is, of course, readily verified by entering Table 9.6 under "*Y–Y* Axis," with an effective length of 10 ft and then finding the column of loads for the 5 × $3\frac{1}{2}$ × $\frac{1}{2}$-in. angle, and proceeding down to the entry for an effective length of 10 ft.

The following example illustrates the use of the data from Table 9.6 for a simple design selection.

Example 5. A double-angle member is to be used for a 12 ft-long compression member carrying 60 kips. The member is unbraced

TABLE 9.6 *(Continued)*

	4 × 3		
	1/2	3/8	5/16
	22.2	17.0	14.4
	6.50	4.97	4.18
	1.25	1.26	1.27
	1.33	1.31	1.30
0	140	107	90
2	134	103	86
4	126	96	81
6	115	88	74
8	102	78	66
10	88	67	57
12	71	55	47
14	54	42	36
16	41	32	27
18	33	25	22
20	26	20	17
0	140	107	90
2	135	103	86
4	127	97	81
6	117	89	74
8	105	80	67
10	92	70	58
12	77	58	48
14	61	45	37
16	47	35	29
18	37	27	23
20	30	22	18

	3 1/2 × 2 1/2		
	3/8	5/16	1/4
	14.4	12.2	9.8
	4.22	3.55	2.88
	1.10	1.11	1.12
	1.11	1.10	1.09
0	91	77	60
2	86	73	57
4	80	67	53
6	71	60	48
8	61	52	41
10	50	42	34
12	37	31	26
14	27	23	19
16	21	18	15
18	16	14	12
0	91	77	60
2	87	73	57
4	80	67	53
6	72	60	47
8	62	52	41
10	50	42	33
12	37	31	25
14	28	23	18
16	21	17	14
18	17	14	11

	3 × 2		
	3/8	5/16	1/4
	11.8	10.0	8.2
	3.47	2.93	2.38
	0.940	0.948	0.957
	0.917	0.903	0.891
0	75	63	51
2	70	59	48
3	67	57	46
4	63	54	44
6	55	46	38
8	44	38	31
10	32	27	23
12	22	19	16
14	16	14	12
0	75	63	51
2	70	59	48
3	67	56	46
4	63	53	43
6	54	45	36
8	43	36	28
10	30	25	20
12	21	17	14
14	15	13	10

	2 1/2 × 2		
	3/8	5/16	1/4
	10.6	9.0	7.2
	3.09	2.62	2.13
	0.768	0.776	0.784
	0.961	0.948	0.935
0	67	57	46
2	61	52	42
3	58	49	40
4	53	45	37
5	48	41	34
6	42	36	30
8	30	26	21
10	19	16	14
12	13	11	9
0	67	57	46
2	63	53	43
3	60	51	41
4	57	48	39
6	49	41	33
8	40	34	27
10	30	24	19
12	21	17	13
14	15	12	10

and has a K factor of 1.0 on both axes. Select an appropriate angle size for the member.

Solution: In this case it is necessary to read both portions of the table to determine the lowest value for the member. Possible choices from those listed are given in Table 9.7. The actual choice would most likely be influenced by other concerns, most notably those relating to the development of the framing connections. It should be noted that the data in Table 9.6 is a quite tiny sample of the material in the reference.

Note: For the following problems, use data from the appropriate tables and assume $K = 1$.

Problem 9.4.A. Using Table 9.2, select the lightest column section to support an axial load of 148 kips [658 kN] if the unbraced height is 12 ft [3.66 m].

TABLE 9.7 Alternate Choices for the Strut: Example 5

Angle Size	Load for x-axis (kips)	Load for y-axis (kips)	Weight (lb)
$4 \times 3 \times \frac{1}{2}$	71	77	22.2
$5 \times 3 \times \frac{3}{8}$	81	61	19.6

Problem 9.4.B. Same data as in Problem 9.4.A, except that the load is 258 kips [1148 kN] and the height is 16 ft [4.88 m].

Problem 9.4.C. Same data as in Problem 9.4.A, except that the load is 355 kips [1579 kN] and the height is 20 ft [6.10 m].

Problems 9.4.D,E,F,G. Select the minimum nominal-size round pipe from Table 9.4 for an axial load of 50 kips [222 kN] and the following unbraced heights:

D 8 ft [2.44 m]
E 12 ft [3.66 m]
F 18 ft [5.49 m]
G 25 ft [7.62 m]

Problems 9.4.H,I,J. Select the lightest-weight steel structural tubing section from Table 9.5 for an axial load of 40 kips [178 kN] and the following unbraced heights:

H 8 ft [2.44 m]
I 12 ft [3.66 m]
J 16 ft [4.88 m]

Problems 9.4.K,L,M. Select the lightest-weight double-angle members from Table 9.6 for an axial load of 100 kips [445 kN] and the following unbraced lengths:

K 6 ft [1.83 m]
L 10 ft [3.05 m]
M 16 ft [4.88 m]

9.5 Columns With Bending

Many steel columns must sustain bending in addition to the usual axial compression. Figure 9.5 shows three of the most common

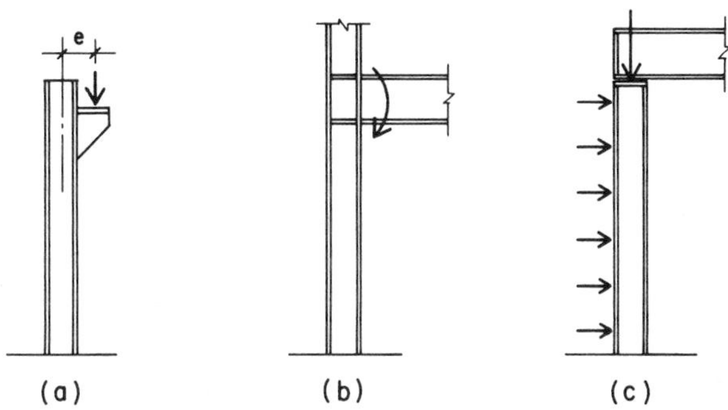

FIGURE 9.5 Columns with bending.

situations that result in this combined effect. When framing members are supported at the column face or on a bracket, the compression load may actually occur with some eccentricity, as shown in Fig. 9.5*a*. When moment-resistive connections are used, and the column becomes a member of a rigid-frame bent, moments will be induced in the ends of the columns, as shown in Fig. 9.5*b*. Columns in exterior walls frequently function as part of the general wall framing; if vertical spanning for wind load is involved, the columns may receive a direct beam loading, as shown in Fig. 9.5*c*.

The fundamental relationship for a column with bending is one of interaction, the simplest form of which is expressed by the straight-line interaction formula:

$$\frac{f_a}{F_a} + \frac{f_b}{F_b} = 1$$

In reality, the problem has the potential for great complexity. The usual problems of dealing with buckling effects of the column must be combined with that of the laterally unsupported beam and some possible synergetic behaviors, such as the *P*-delta effect. While the basic form of the interaction formula is used for investigation, numerous adjustments must often be made for various situations. The AISC Specifications are quite extensive and complex and not very self-explanatory with regard to this problem. For a full treatment of the topic the reader is referred to one of the major textbooks on steel design, such as *Steel Buildings: Analysis and Design*, by S. W. Crawley and R. M. Dillon (Ref. 12).

For use in preliminary design work, or to obtain a first trial section to be used in a more extensive design investigation, a procedure may be used that involves the determination of an equivalent total design load that incorporates the bending effects. This is done by using the bending factors, B_x and B_y, which are listed in Table 9.2. The equivalent design load is determined as

$$P' = P + B_x M_x + B_y M_y$$

where P' = equivalent axial load,
P = actual axial load,
B_x = bending factor for the section's x-axis,
M_x = bending moment about the x-axis,
B_y = bending factor for the section's y-axis,
M_y = bending moment about the y-axis.

The following example illustrates the use of the method.

Example 1. It is desired to use a 10-in. W shape for a column in a situation such as that shown in Fig. 9.6. The axial compression load from above is 120 kips and the beam load is 24 kips, with the beam attached at the column face. The column is 16 ft high and has a K factor of 1.0. Select a trial section for the column.

Solution: Since bending occurs only about the x-axis, we use only the B_x factor for this case. Scanning the column of B_x factors in Table 9.2, we observe that the factor for 10 W sections varies from 0.261 to 0.277. As we have not yet determined the section to be used, it is necessary to make an assumption for the factor and to verify the assumption after the selection is made. Let us assume a B_x of 0.27, with which we find

$$P' = P + B_x M_x = (120 + 24) + 0.27(24)(5)$$
$$= 144 + 32.4 = 176.4 \text{ kips}$$

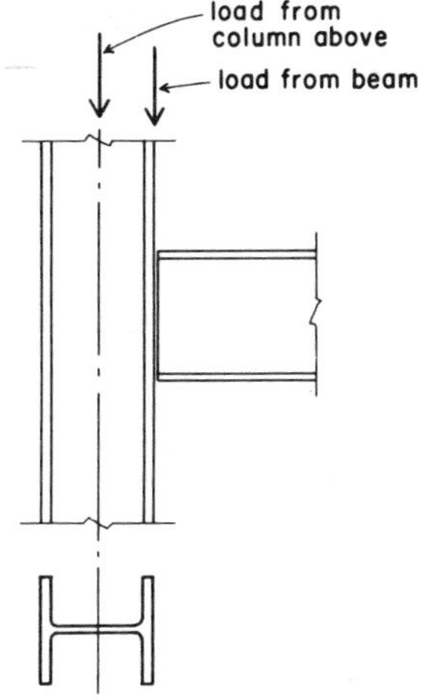

FIGURE 9.6

From Table 9.2, for a KL of 16 ft, we obtain a W 10 × 45. For this shape the B_x factor is 0.271, which is very close to our assumption.

For most situations use of the bending factors in the manner just demonstrated will result in conservative selections. If the designer intends to use the section thus obtained in a more thorough investigation, it is probably wise to reduce the size slightly before proceeding with the work.

Example 2. It is desired to use a 12-in. W section for a column that sustains an axial compression plus bending about both axes. Select a column for the following data: axial load of 60 kips, bending about the x-axis of 40 kip-ft, bending about the y-axis of 32 kip-ft, column unbraced height of 12 ft.

Solution: In Table 9.2 we observe that in the midrange of sizes for 12-in. sections approximate values for bending factors are $B_x = 0.215$ and $B_y = 0.63$. Thus

$$\begin{aligned} P' &= P + B_xM_x + B_yM_y \\ &= 60 + (0.215)(40 \times 12) + (0.63)(32 \times 12) \\ &= 60 + 103 + 242 = 405 \text{ kips} \end{aligned}$$

From Table 9.2, for the 12-ft height, the lightest section is a W 12 × 79, with an allowable load of 431 kips and bending factors of $B_x = 0.217$ and $B_y = 0.648$. As these bending factors slightly exceed those assumed, a new design value for P' should be found to verify the section. Thus

$$P' = 60 + (0.217)(40 \times 12) + (0.648)(32 \times 12) = 413 \text{ kips}$$

and the section is still a valid choice.

Problem 9.5.A. It is desired to use a 12-in. W section for a column to support a beam, as shown in Fig. 9.6. Select a trial size for the column for the following data: column axial load = 200 kips [890 kN], beam reaction = 30 kips [133 kN], unbraced height of column = 14 ft.

Problem 9.5.B. A 14-in. W section is to be used for a column that sustains bending about both axes. Select a trial section for the column for the following data: axial load = 160 kips [712 kN], bending about the x-axis = 65 kip-ft [88 kN-m], bending about the y-axis = 45 kip-ft [61 kN-m], unbraced column height = 16 ft [4.88 m].

A major occurrence of the condition of a column with bending is that of the case of a column in a rigid-frame bent. For steel structures this most often occurs when a steel frame is developed as a moment-resisting space frame in three dimensions with the rigid-frame action being used for lateral bracing. If the frame is made rigid (by using moment-resistive connections between columns and beams), both vertical gravity and horizontal wind or earthquake loads will result in bending and shear in the columns. Multiple-bayed, multistoried frames are highly indeterminate, requiring investigative procedures beyond the scope of the work developed here. Some aspects of frame behavior are developed in the examples of simple frames in Chapter 2. Discussion of rigid-frame actions and an example of approximate analysis of frame behavior are given in Chapter 22.

9.6. Column Framing

Connection details for columns must be developed with considerations of the column shape and size, the shape and orientation of other framing, and the particular structural functions of the joints. Some common connections are shown in Fig. 9.7. When beams

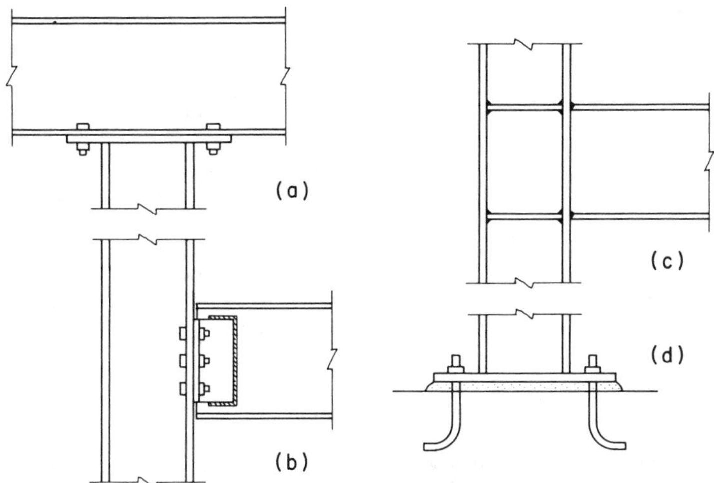

FIGURE 9.7 Typical column connections: lightly loaded frames.

sit directly on the top of a column, the usual solution is to fasten a bearing plate on top of the column, with provision for the attachment of the beam to the plate. The plate serves no specific structural purpose in this case, functioning essentially only as an attachment device, assuming that the load transfer is one of simple vertical bearing (see Fig. 9.7*a*).

In many situations beams must frame into the side of a column. If simple transfer of vertical load is all that is required, a common solution is the connection shown in Fig. 9.7*b*, in which a pair of steel angles are used to connect the beam web to the column flange or the column web. This type of connection is discussed in Sec. 10-7. If moment must be transferred between the columns and beams, as in the development of rigid-frame bents, the most common solution involves the use of welding to achieve a direct connection between the members, as shown in Fig. 9.7*c*. The details of rigid connections vary considerably, mostly on the basis of the size of the members and the magnitude of forces.

Another common connection is that of the bottom of a column to its supports. Figure 9.7*d* shows a typical solution for transfer of bearing to a concrete footing, consisting of a steel plate welded to the column bottom and bolted to preset anchor bolts. In this case the plate has a definite structural function, serving to transform the highly concentrated, punching effect of the column force into a low-valued bearing stress on the much softer concrete. Thus while the general form of the connection is similar, the role of the plates in Fig. 9.7*a* and *d* are considerably different.

Part 4 of the AISC Manual provides extensive data relating to the development of connections for structural steel members. Some of the material is presented in the discussions in this book in Chapters 10 and 11. Issues of specific concern are also discussed in some of the design examples in Part V.

9.7 Column Base Plates

Column base plates vary from relatively modest ones for small, lightly loaded columns to huge thick ones for the heavy 14 W shapes in high-strength steel grades. The following procedure is based on the specifications and design illustrations in the AISC Manual (Ref. 2) and the recommendations in the ACI Code (Ref. 3).

The plan area required for the base plate is determined as

$$A_1 = \frac{P}{F_p}$$

where A_1 = plan area of the bearing plate,
P = compression load from the column,
F_p = allowable bearing stress on the concrete.

Allowable bearing is based on the concrete design strength f'_c. It is limited to a value of $0.3f'_c$ when the plate covers the entire area of the support member, which is seldom the case. If the support member has a larger plan area A_2, the allowable bearing stress may be increased by a factor of $\sqrt{A_2/A_1}$, but not greater than two. For modest-sized columns supported on footings, the maximum factor is most likely to be used. For large columns or those supported on pedestals or piers, the adjustment may be less.

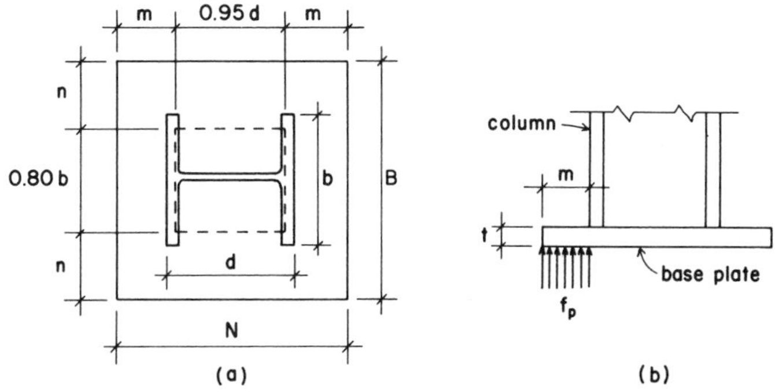

FIGURE 9.8 Reference dimensions for column base plates.

For a W-shaped column, the basis for determination of the thickness of the plate due to bending is shown in Fig. 9.8. Once the required value for A_1 is found, the dimensions B and N are established so that the projections m and n are approximately equal. Choice of dimensions must also relate to the locations of anchor bolts and to any details for development of the attachment of the plate to the column. The required plate thickness is determined with the formula

$$t = \sqrt{\frac{3f_p m^2}{F_b}} \text{ or } t = \sqrt{\frac{3f_p n^2}{F_b}}$$

where t = thickness of the bearing plate, inches,
f_p = actual bearing pressure: P/A_1,
F_b = allowable bending stress in the plate: $0.75F_y$.

The following example illustrates the process for a column with a relatively light load.

Example 1. Design a base plate of A36 steel for a W 12 × 58 column with a load of 250 kips. The column bears on a concrete footing with f'_c = 3 ksi.

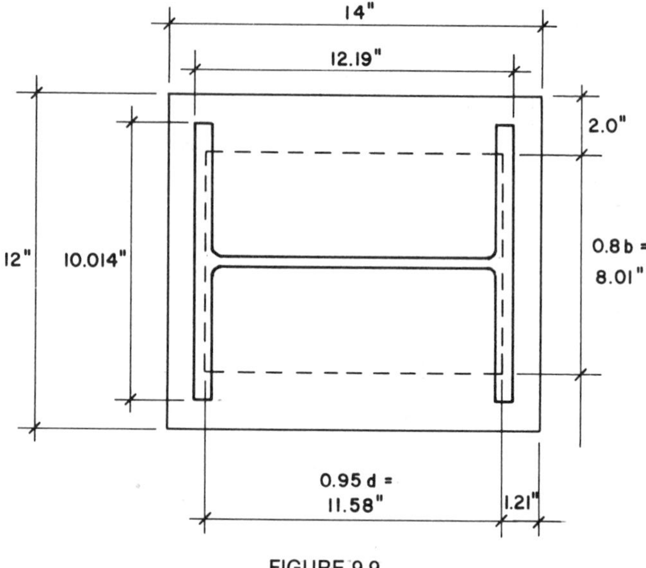

FIGURE 9.9

Solution: We assume the footing area to be considerably larger than the plate area; thus $F_p = 0.6f'_c = 1.8$ ksi. Then

$$A_1 = \frac{P}{F_p} = \frac{250}{1.8} = 138.9 \text{ in.}^2$$

If the plate is square,

$$B = N = \sqrt{138.9} = 11.8 \text{ in.}$$

Since this is almost the same size as the column, we will assume the plan size layout shown in Fig. 9.9, which allows the welding of the plate to the column and the placing of the anchor bolts. It may be observed that the dimension labeled n in Fig. 9.8 is critical in this case, and the plate thickness is thus found as

$$t = \sqrt{\frac{3f_p n^2}{F_b}}$$

for which

$$f_p = \frac{P}{A_1} = \frac{250}{12(14)} = 1.49 \text{ ksi}$$

and

$$t = \sqrt{\frac{3(1.49)(2)^2}{0.75(36)}} = \sqrt{0.662} = 0.814 \text{ in.}$$

Plates are usually specified in thickness increments of $\frac{1}{8}$ in., so the minimum thickness would be $\frac{7}{8}$ in. (0.875 in.).

Problem 9.7.A. Design a column base plate for a W 8 × 31 column that is supported on concrete for which the allowable bearing capacity is 750 psi [5000 kPa]. The load on the column is 178 kips [792 kN].

Problem 9.7.B. Design a column base plate for the W 8 × 31 in Problem 9.7.A if the bearing pressure allowed on the concrete is 1125 psi [7800 kPa].

10

Bolted Connections

Elements of structural steel are often connected by mating flat parts with common holes and inserting a pin-type device to hold them together. In times past the pin device was a rivet; today it is usually a bolt. A great number of types and sizes of bolt are available, as are many connections in which they are used. The material in this chapter deals with a few of the common bolting methods used in building structures.

10.1 Structural Bolts

The diagrams in Fig. 10.1 show a simple connection between two steel bars that functions to transfer a tension force from one bar to another. Although this is a tension-transfer connection, it is also referred to as a shear connection because of the manner in which the connecting device (the bolt) works in the connection (see Fig. 10.1*b*). If the bolt tension (due to tightening of the nut) is relatively low, the bolt serves primarily as a pin in the matched holes, bearing against the sides of the holes, as shown in Fig. 10.1*d*. In addition to these functions, the bars develop tension stress that will be a maximum at the section through the bolt holes.

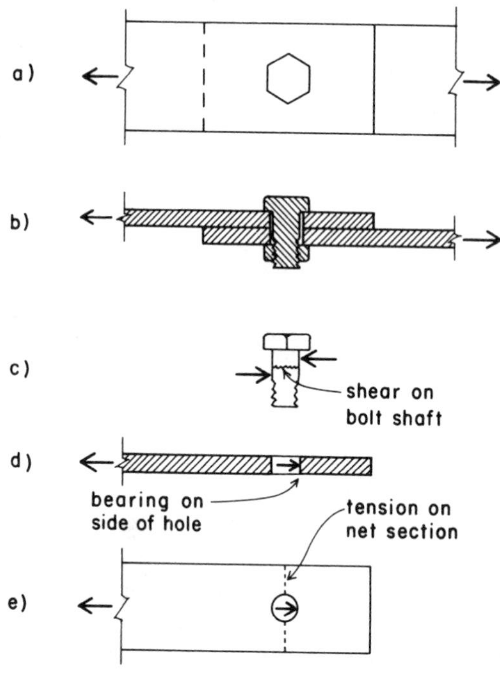

FIGURE 10.1 Actions of bolted joints.

In the connection shown in Fig. 10.1, the failure of the bolt involves a slicing (shear) failure that is developed as a shear stress on the bolt cross section. The resistance of the bolt can be expressed as an allowable shear stress F_v times the area of the bolt cross section, or

$$R = F_v \times A$$

With the size of the bolt and the grade of steel known, it is a simple matter to establish this limit. In some types of connections, it may be necessary to slice the same bolt more than once to separate the connected parts. This is the case in the connection shown in Fig. 10.2, in which it may be observed that the bolt must be sliced twice to make the joint fail. When the bolt develops

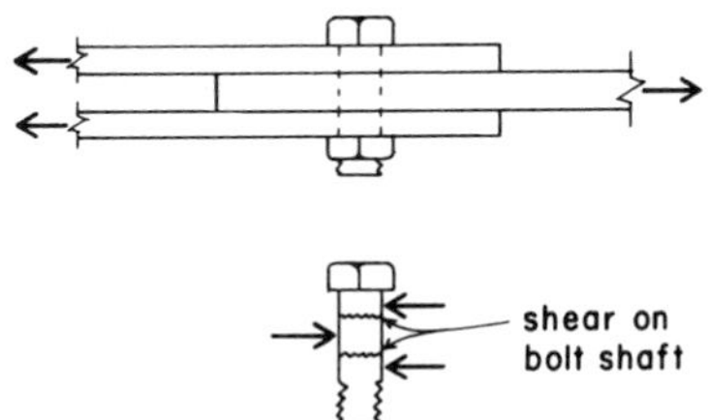

FIGURE 10.2 Bolted joint with double shear.

shear on only one section (Fig. 10.1), it is said to be in *single shear*; when it develops shear on two sections (Fig. 10.2), it is said to be in *double shear*.

When the bolt diameter is large or the bolt is made of strong steel, the connected parts must be sufficiently thick if they are to develop the full capacity of the bolts. The maximum bearing stress permitted for this situation by the AISC Specification is $F_p = 1.5F_u$, where F_u is the ultimate tensile strength of the steel in the part in which the hole occurs.

Bolts used for the connection of structural steel members come in two types. Bolts designated A307 and called *unfinished* have the lowest load capacity of the structural bolts. The nuts for these bolts are tightened just enough to secure a snug fit of the attached parts; because of this, plus the oversizing of the holes, there is some movement in the development of full resistance. These bolts are generally not used for major connections, especially when joint movement or loosening under vibration or repeated loading may be a problem.

Bolts designated A325 or A490 are called *high-strength bolts*. The nuts of these bolts are tightened to produce a considerable tension force, which results in a high degree of friction resistance between the attached parts. High-strength bolts are further designated as F, N, or X. The F designation denotes bolts for which the limiting resistance is that of friction. The N designation denotes bolts that function ultimately in bearing and shear but for which the threads are not excluded from the bolt shear planes. The X designation denotes bolts that function like the N bolts but for which the threads are excluded from the shear planes.

TABLE 10.1 Capacity of Structural Bolts (in kips)

ASTM Designation	Connection Type*	Loading Condition**	Nominal Diameter (in.)							
			5/8	3/4	7/8	1	1-1/8	1-1/4	1-3/8	1-1/2
			Area, Based on Nominal Diameter (in^2)							
			0.3068	0.4418	0.6013	0.7854	0.9940	1.227	1.485	1.767
A307		S	3.1	4.4	6.0	7.9	9.9	12.3	14.8	17.7
		D	6.1	8.8	12.0	15.7	19.9	24.5	29.7	35.3
		T	6.1	8.8	12.0	15.7	19.9	24.5	29.7	35.3
A325	F	S	5.4	7.7	10.5	13.7	17.4	21.5	26.0	30.9
		D	10.7	15.5	21.0	27.5	34.8	42.9	52.0	61.8
	N	S	6.4	9.3	12.6	16.5	20.9	25.8	31.2	37.1
		D	12.9	18.6	25.3	33.0	41.7	51.5	62.4	74.2
	X	S	9.2	13.3	18.0	23.6	29.8	36.8	44.5	53.0
		D	18.4	26.5	36.1	47.1	59.6	73.6	89.1	106.0
	All	T	13.5	19.4	26.5	34.6	43.7	54.0	65.3	77.7
A490	F	S	6.7	9.7	13.2	17.3	21.9	27.0	32.7	38.9
		D	13.5	19.4	26.5	34.6	43.7	54.0	65.3	77.7
	N	S	8.6	12.4	16.8	22.0	27.8	34.4	41.6	49.5
		D	17.2	24.7	33.7	44.0	55.7	68.7	83.2	99.0
	X	S	12.3	17.7	24.1	31.4	39.8	49.1	59.4	70.7
		D	24.5	35.3	48.1	62.8	79.5	98.2	119.0	141.0
	All	T	16.6	23.9	32.5	42.4	53.7	66.3	80.2	95.4

Source: Reproduced from data in the *Manual of Steel Construction*, 8th ed. (Ref. 2), with permission of the publishers, American Institute of Steel Construction.

* F = friction; N = bearing, threads not excluded; X = bearing, threads excluded.

** S = single shear; D = double shear; T = tension.

When bolts are loaded in tension, their capacities are based on the development of the ultimate resistance in tension stress at the reduced section through the threads. When loaded in shear, bolt capacities are based on the development of shear stress in the bolt shaft. The shear capacity of a single bolt is further designated as *S* for single shear (Fig. 10.1) or *D* for double shear (Fig. 10.2). The capacities of structural bolts in both tension and shear are given in Table 10.1. The size range given in the table—$\frac{5}{8}$–$1\frac{1}{2}$ in.—is that listed in the AISC Manual. However, the most commonly used sizes for structural steel framing are $\frac{3}{4}$ and $\frac{7}{8}$ in.

Bolts are ordinarily installed with a washer under both head and nut. Some manufactured high-strength bolts have specially formed heads or nuts that in effect have self-forming washers, eliminating the need for a separate, loose washer. When a washer is used, it is sometimes the limiting dimensional factor in detailing for bolt placement in tight locations, such as close to the fillet (inside radius) of angles or other rolled shapes.

For a given diameter of bolt, there is a minimum thickness required for the bolted parts in order to develop the full shear capacity of the bolt. This thickness is based on the bearing stress between the bolt and the side of the hole, which is limited to a maximum of $F_p = 1.5F_u$. The stress limit may be established by either the bolt steel or the steel of the bolted parts.

Steel rods are sometimes threaded for use as anchor bolts or tie rods. When they are loaded in tension, their capacities are usually limited by the stress on the reduced section at the threads. Tie rods are sometimes made with *upset ends,* which consist of larger diameter portions at the ends. When these enlarged ends are threaded, the net section at the thread is the same as the gross section in the remainder of the rods; the result is no loss of capacity for the rod.

10.2 Layout of Bolted Connections

Design of bolted connections generally involves a number of considerations in the dimensioned layout of the bolt-hole patterns for the attached structural members. Although we cannot develop all the points necessary for the production of structural steel con-

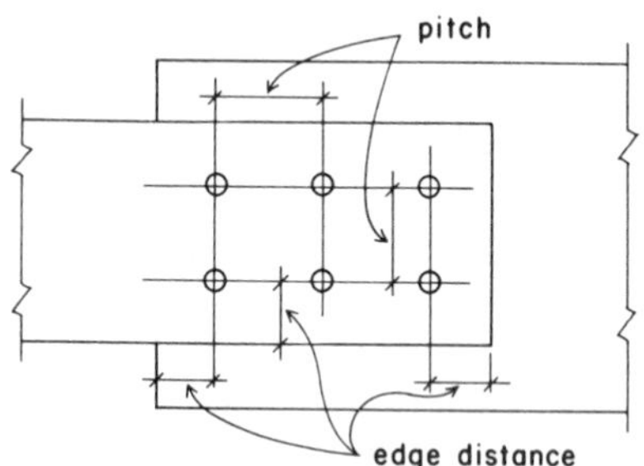

FIGURE 10.3 Pitch and edge distances for bolts.

struction and fabrication details, the material in this section presents basic factors that often must be included in the structural calculations.

Figure 10.3 shows the layout of a bolt pattern with bolts placed in two parallel rows. Two basic dimensions for this layout are limited by the size (nominal diameter) of the bolt. The first is the center-to-center spacing of the bolts, usually called the *pitch*. The AISC Specification limits this dimension to an absolute minimum of $2\frac{2}{3}$ times the bolt diameter. The preferred minimum, however, which is used in this book, is 3 times the diameter.

The second critical layout dimension is the *edge distance*, which is the distance from the center line of the bolt to the nearest edge. There is also a specified limit for this as a function of bolt size. This dimension may also be limited by edge tearing, which is discussed in Sec. 10.4.

Table 10.2 gives the recommended limits for pitch and edge distance for the bolt sizes used in ordinary steel construction.

In some cases bolts are staggered in parallel rows (Fig. 10.4). In this case the diagonal distance, labeled m in the illustration, must also be considered. For staggered bolts the spacing in the direction of the rows is usually referred to as the pitch; the spacing of the rows is called the gage. The reason for staggering the bolts is that sometimes the rows must be spaced closer (gage spacing) than the minimum spacing required for the bolts selected. Table 10.3 gives the pitch required for a given gage spac-

TABLE 10.2 Pitch and Edge Distances for Bolts

Rivet or Bolt Diameter d (in.)	Minimum Edge Distance for Punched, Reamed, or Drilled Holes (in.)		Pitch, Center to Center (in.)	
	At Sheared Edges	At Rolled Edges of Plates, Shapes, or Bars, or Gas-Cut Edges[a]	Minimum $2\frac{2}{3}d$	Recommended $3d$
$\frac{5}{8}$	1.125	0.875	1.67	1.875
$\frac{3}{4}$	1.25	1	2	2.25
$\frac{7}{8}$	1.5[b]	1.125	2.33	2.625
1	1.75b	1.25	2.67	3

Source: Reproduced from data in the *Manual of Steel Construction*, 8th ed. (Ref. 2), with permission of the publishers, American Institute of Steel Construction.

[a] May be reduced $\frac{1}{8}$ in. when the hole is at a point where stress does not exceed 25% of the maximum allowed in the connected element.

[b] May be $1\frac{1}{4}$ in. at the ends of beam connection angles.

ing to keep the diagonal spacing m within the recommended diameter limit.

Location of bolt lines is often related to the size and type of structural members being attached. This is especially true of bolts placed in the legs of angles or in the flanges of W, M, S, C, and structural tee shapes. Figure 10.5 shows the placement of bolts in

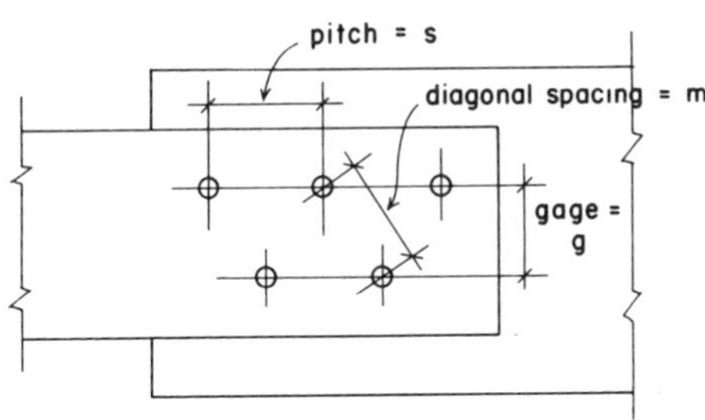

FIGURE 10.4 Standard reference dimensions for layout of bolted joints.

TABLE 10.3 Minimum Pitch to Maintain Three Diameters Center to Center of Holes

Diameter of Rivet	m	Distance g (in.) 1	$1\frac{1}{4}$	$1\frac{1}{2}$	$1\frac{3}{4}$	2	$2\frac{1}{4}$	$2\frac{1}{2}$	$2\frac{3}{4}$	3
$\frac{5}{8}$	$1\frac{7}{8}$	$1\frac{5}{8}$	$1\frac{3}{8}$	$1\frac{1}{8}$	$\frac{5}{8}$	0				
$\frac{3}{4}$	$2\frac{1}{4}$	2	$1\frac{7}{8}$	$1\frac{5}{8}$	$1\frac{3}{8}$	1	0			
$\frac{7}{8}$	$2\frac{5}{8}$	$2\frac{1}{2}$	$2\frac{3}{8}$	$2\frac{1}{8}$	2	$1\frac{3}{4}$	$1\frac{3}{8}$	$\frac{3}{4}$	0	
1	3	$2\frac{7}{8}$	$2\frac{3}{4}$	$2\frac{5}{8}$	$2\frac{1}{2}$	$2\frac{1}{4}$	2	$1\frac{3}{8}$	$1\frac{1}{8}$	0

Source: Reproduced from data in the *Manual of Steel Construction*, 8th ed. (Ref. 2), with permission of the publishers, American Institute of Steel Construction. (See Fig. 10.4.)

the legs of angles. When a single row is placed in a leg, its recommended location is at the distance labeled g from the back of the angle. When two rows are used, the first row is placed at the distance g_1, and the second row is spaced a distance g_2 from the first. Table 10.4 gives the recommended values for these distances.

When placed at the recommended locations in rolled shapes, bolts will end up a certain distance from the edge of the part. Based on the recommended edge distance for rolled edges given in Table 10.2, it is thus possible to determine the maximum size of bolt that can be accommodated. For angles, the maximum fastener may be limited by the edge distance, especially when two rows are used: however, other factors may in some cases be more

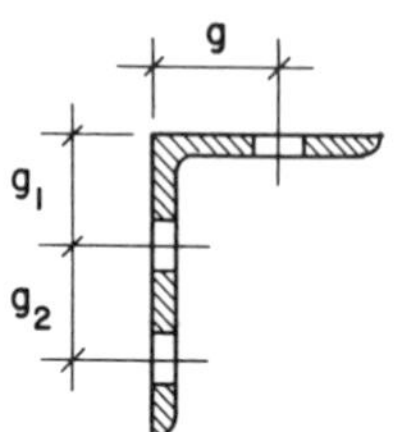

FIGURE 10.5 Gage dimensions for angles.

TABLE 10.4 Usual Gage Dimensions for Angles (in.)

Gage Dimension	Width of Angle Leg								
	8	7	6	5	4	$3\frac{1}{2}$	3	$2\frac{1}{2}$	2
g	$4\frac{1}{2}$	4	$3\frac{1}{2}$	3	$2\frac{1}{2}$	2	$1\frac{3}{4}$	$1\frac{3}{8}$	$1\frac{1}{8}$
g_1	3	$2\frac{1}{2}$	$2\frac{1}{4}$	2					
g_2	3	3	$2\frac{1}{2}$	$1\frac{3}{4}$					

Source: Reproduced from data in the *Manual of Steel Construction,* 8th ed. (Ref. 2), with permission of the publishers, American Institute of Steel Construction.

critical. The distance from the center of the bolts to the inside fillet of the angle may limit the use of a large washer where one is required. Another consideration may be the stress on the net section of the angle, especially if the member load is taken entirely by the attached leg.

10.3 Tension Connections

When tension members have reduced cross sections, two stress investigations must be considered. This is the case for members with holes for bolts or for bolts or rods with cut threads. For the member with a hole (Fig. 10.1*d*), the allowable tension stress at the reduced cross section through the hole is $0.50F_u$, where F_u is the ultimate tensile strength of the steel. The total resistance at this reduced section (also called the net section) must be compared with the resistance at other, unreduced sections at which the allowable stress is $0.60\ F_y$.

For threaded steel rods the maximum allowable tension stress at the threads is $0.33\ F_u$. For steel bolts the allowable stress is specified as a value based on the type of bolt. The load capacity of various types and sizes of bolt is given in Table 10.1.

When tension elements consist of W, M, S, and tee shapes, the tension connection is usually not made in a manner that results in the attachment of all the parts of the section (e.g., both flanges plus the web for a W). In such cases the AISC Specification requires the determination of a reduced effective net area A_e that consists of

$$A_e = C_t A_n$$

where A_n = actual net area of the member,
C_t = reduction coefficient.

Unless a larger coefficient can be justified by tests, the following values are specified:

1. For W, M, or S shapes with flange widths not less than two-thirds the depth and structural tees cut from such shapes, when the connection is to the flanges and has at least three fasteners per line in the direction of stress, $C_t = 0.90$.
2. For W, M, or S shapes not meeting the above conditions and for tees cut from such shapes, provided the connection has not fewer than three fasteners per line in the direction of stress, $C_t = 0.85$.
3. For all members with connections that have only two fasteners per line in the direction of stress, $C_t = 0.75$.

Angles used as tension members are often connected by only one leg. In a conservative design, the effective net area is only that of the connected leg, less the reduction caused by bolt holes. Rivet and bolt holes are punched larger in diameter than the nominal diameter of the fastener. The punching damages a small amount of the steel around the perimeter of the hole; consequently the diameter of the hole to be deducted in determining the net section is $\frac{1}{8}$ in. greater than the nominal diameter of the fastener.

When only one hole is involved, as in Fig. 10.1, or in a similar connection with a single row of fasteners along the line of stress, the net area of the cross section of one of the plates is found by multiplying the plate thickness by its net width (width of member minus diameter of hole).

When holes are staggered in two rows along the line of stress (Fig. 10.6), the net section is determined somewhat differently.

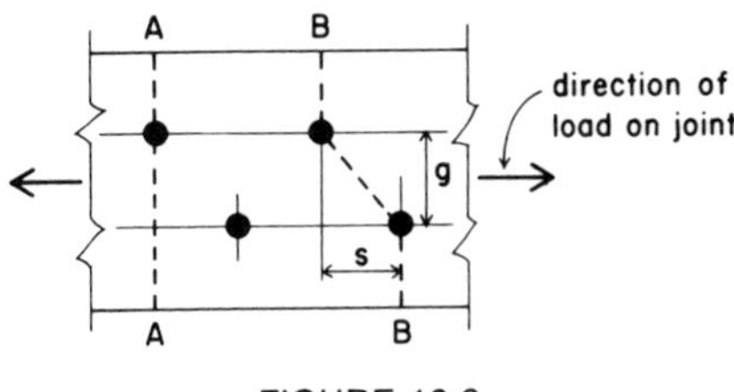

FIGURE 10.6

The AISC Specification reads:

> In the case of a chain of holes extending across a part in any diagonal or zigzag line, the net width of the part shall be obtained by deducting from the gross width the sum of the diameters of all the holes in the chain and adding, for each gage space in the chain, the quantity $s^2/4g$, where
>
> s = longitudinal spacing (pitch) in inches or any two successive holes
>
> and
>
> g = transverse spacing (gage) in inches for the same two holes
>
> The critical net section of the part is obtained from that chain which gives the least net width.

The AISC Specification also provides that in no case shall the net section through a hole be considered as more than 85% of the corresponding gross section.

10.4 Tearing in Bolted Connections

One possible form of failure in a bolted connection is that of tearing out the edge of one of the attached members. The diagrams in Fig. 10.7 show this potentiality in a connection between two plates. The failure in this case involves a combination of

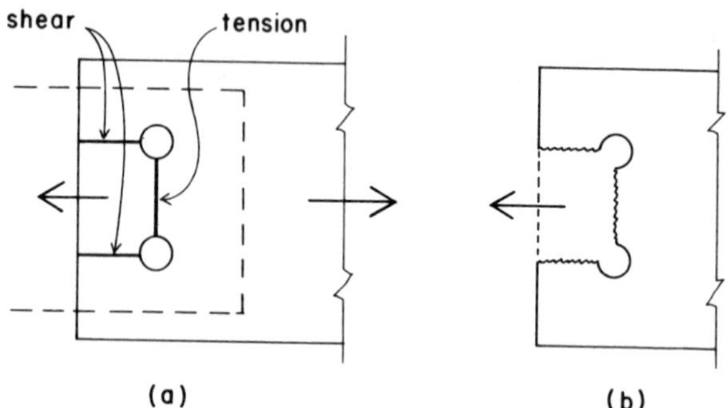

FIGURE 10.7 Tearing in a bolted tension joint.

shear and tension to produce the torn-out form shown. The total tearing force is computed as the sum required to cause both forms of failure. The allowable stress on the net tension area is specified as $0.50F_u$, where F_u is the maximum tensile strength of the steel. The allowable stress on the shear areas is specified as $0.30\ F_u$. With the edge distance, hole spacing, and diameter of the holes known, the net widths for tension and shear are determined and multiplied by the thickness of the part in which the tearing occurs. These areas are then multiplied by the appropriate stresses to find the total tearing force that can be resisted. If this force is

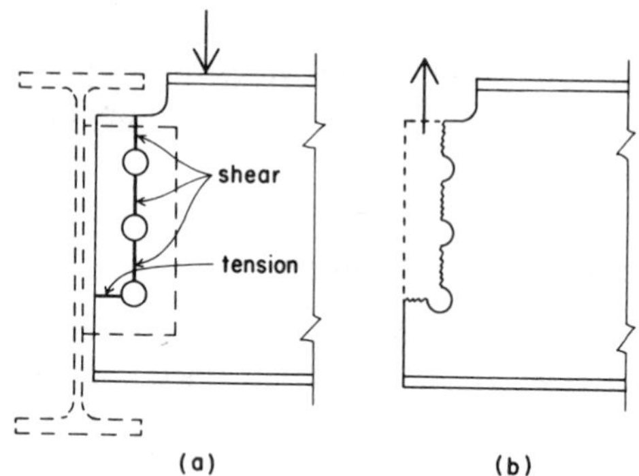

FIGURE 10.8 Tearing in a bolted beam connection.

greater than the connection design load, the tearing problem is not critical.

Another case of potential tearing is shown in Figure 10.8. This is the common situation for the end framing of a beam in which support is provided by another beam, whose top is aligned with that of the supported beam. The end portion of the top flange of the supported beam must be cut back to allow the beam web to extend to the side of the supporting beam. With the use of a bolted connection, the tearing condition shown is developed.

10.5 Design of a Bolted Tension Connection

The issues raised in several of the preceding sections are illustrated in the following design example. Before proceeding with the problem data, we should consider some of the general requirements for this joint.

If friction-type bolts are used, the surfaces of the connected parts must be cleaned and made reasonably true. If high-strength bolts are used, the determination to exclude threads from the shear failure planes must be established.

The AISC Specification has a number of general requirements for connections:

1. Need for a minimum of two bolts per connection.
2. Need for a minimum connection capacity of 6 kips.
3. Need for the connection to develop at least 50% of the full potential capacity of the member (for trusses only).

Although a part of the design problem may be the selection of the type of fastener or the required strength of steel for the attached parts, we provide this as given data in the example problem.

Example. The connection shown in Fig. 10.9 consists of a pair of narrow plates that transfer a load of 100 kips [445 kN] in tension to a single 10-in. [254-mm] wide plate. The plates are A36 steel with F_u = 58 ksi [400 MPa] and are attached with $\frac{3}{4}$-in. A325F bolts placed in two rows. Determine the number of bolts required, the width and thickness of the narrow plates, the thickness of the wide plate, and the layout of the bolts.

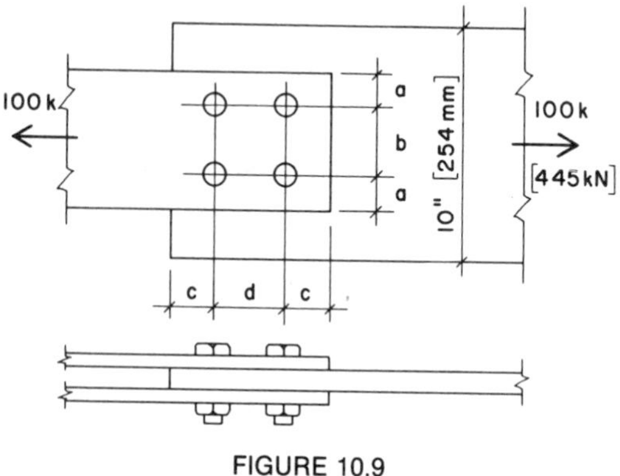

FIGURE 10.9

Solution: From Table 10.1 we find the double-shear (D) capacity for one bolt is 15.5 kips [69 kN]. The required number of bolts is thus

$$n = \frac{\text{connection load}}{\text{bolt capacity}} = \frac{100}{15.5} = 6.45$$

and the minimum number for a symmetrical connection is eight.

With eight bolts used, the load on one bolt is

$$P = \frac{100}{8} = 12.5 \text{ kips [55.6 kN]}$$

According to Table 10.2, the $\frac{3}{4}$-in. bolts require a minimum edge distance of 1.25 in. (at a sheared edge) and a recommended pitch of 2.25 in. The minimum width for the narrow plates is therefore (see Fig. 10.9)

$$w = b + 2(a)$$

$$w = 2.25 + 2(1.25) = 4.75 \text{ in. [121 mm]}$$

With no other constraining conditions given, we arbitrarily select a width of 6 in. [152.4 mm] for the narrow plates. Checking first for the requirement of a maximum tension stress of 0.60 F_y

on the gross area, we find

$$F_t = 0.60F_y = 0.60(36) = 21.6 \text{ ksi } [149 \text{ MPa}]$$

(*Note:* AISC permits rounding off to 22 ksi.)

$$A_{req} = \frac{100}{22} = 4.55 \text{ in.}^2 \ [2928 \text{ mm}^2]$$

and the required thickness with the width selected is

$$t = \frac{4.55}{2(6)} = 0.38 \text{ in. } [9.6 \text{ mm}]$$

We therefore select a minimum thickness of $\frac{7}{16}$ in. (0.4375 in.) [11 mm]. The next step is to check the stress condition on the net section through the holes, for which the allowable stress is $0.50F_u$. For the computations, we assume a hole diameter $\frac{1}{8}$ in. [3.18 mm] larger than the bolt. Thus

$$\text{Hole size} = 0.875 \text{ in. } [22.23 \text{ mm}]$$

$$\text{Net width} = 2\{6 - (2 \times 0.875)\} = 8.5 \text{ in. } [216 \text{ mm}]$$

and the stress on the net section of the two plates is

$$f_t = \frac{100}{0.4375 \times 8.5} = 26.89 \text{ ksi } [187 \text{ MPa}]$$

This computed stress is compared with the specified allowable stress of

$$F_t = 0.50F_u = 0.50 \times 58 = 29 \text{ ksi } [200 \text{ MPa}]$$

Bearing stress is computed by dividing the load on a single bolt by the product of the bolt diameter and the plate thickness. Thus

$$f_p = \frac{12.5}{2 \times 0.75 \times 0.4375} = 19.05 \text{ ksi } [146 \text{ MPa}]$$

This is compared with the allowable stress of

$$F_p = 1.5F_u = 1.5 \times 58 = 87 \text{ ksi [600 MPa]}$$

For the middle plate the procedure is essentially the same except that, in this case, the plate width is given. As before, on the basis of stress on the unreduced section, we determine that the total area required is 4.55 in^2 [2928 mm^2]. Thus the thickness required is

$$t = \frac{4.55}{10} = 0.455 \text{ in. [11.6 mm]}$$

We therefore select a minimum thickness of $\frac{1}{2}$ in. (0.50 in.) [13 mm]. We then proceed as before to check the stress on the net width. The net width through the two holes is

$$w = 10 - (2 \times 0.875) = 8.25 \text{ in. [210 mm]}$$

and the tension stress on this net cross section is

$$f_t = \frac{100}{8.25 \times 0.5} = 24.24 \text{ ksi [177 MPa]}$$

which is less than the allowable stress of 29 ksi [200 MPa] determined previously.

The computed bearing stress on the wide plate is

$$f_p = \frac{12.5}{0.75 \times 0.50} = 33.3 \text{ ksi [243 MPa]}$$

which is considerably less than the allowable determined before, $F_p = 87$ ksi [600 MPa].

In addition to the layout restrictions given in Sec. 10.2, the AISC Specification requires that the minimum spacing in the direction of the load be

$$\frac{2P}{F_u t} + \frac{D}{2} \qquad (D \text{ is bolt diameter})$$

and that the minimum edge distance in the direction of the load be

$$\frac{2P}{F_u t} \qquad \text{(dimension } c \text{ in Fig. 10.9)}$$

where P = force transmitted by one fastener to the critical connected part,
F_u = specified minimum (ultimate) tensile strength of the connected part,
t = thickness of the critical connected part.

For our case

$$\frac{2P}{F_u t} = \frac{2 \times 12.5}{58 \times 0.5} = 0.862 \text{ in.}$$

which is considerably less than the specified edge distance listed in Table 10.2 for a $\frac{3}{4}$-in. bolt at a sheared edge: 1.25 in.

For the spacing

$$\frac{2P}{F_u t} + \frac{D}{2} = 0.862 + 0.375 = 1.237 \text{ in.}$$

which is also not critical.

A final problem that must be considered is the potential of tearing out the two bolts at the ends of the plates. Because the combined thickness of the two outer plates is greater than that of the middle plate, the critical case in this connection is that of the middle plate. Figure 10.10 shows the condition for the tearing, which involves tension on the section labeled "1" and shear on the two sections labeled "2."

For the tension section

$$w_{(net)} = 3 - 0.875 = 2.125 \text{ in. [54.0 mm]}$$

$$F_t = 0.50F_u = 29 \text{ ksi [200 MPa]}$$

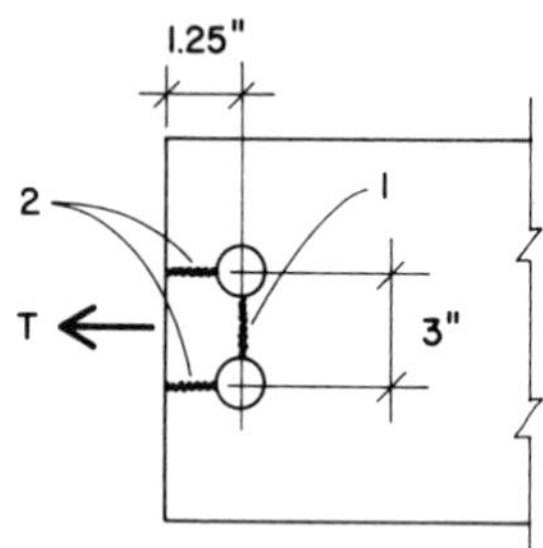

FIGURE 10.10

For the shear sections

$$w_{(net)} = 2\left(1.25 - \frac{0.875}{2}\right) = 1.625 \text{ in. [41.3 mm]}$$

$$F_v = 0.30F_u = 17.4 \text{ ksi [120 MPa]}$$

The total resistance to tearing is

$$T = (2.125 \times 0.5 \times 29) + (1.625 \times 0.5 \times 17.4)$$
$$= 44.95 \text{ kips [205 kN]}$$

Because this is greater than the combined load of 25 kips [111.2 kN] on the two bolts, the problem is not critical.

Connections that transfer compression between the joined parts are essentially the same with regard to the bolt stresses and bearing on the parts. Stress on the net section is less likely to be critical because the compression members will usually be designed for column action, with a considerably reduced value for the allowable compression stress.

Problem 10.5.A. A bolted connection of the general form shown in Fig. 10.9 is to be used to transmit a tension force of 200 kips [890 kN] by using $\frac{7}{8}$-in. A490N bolts and plates of A36 steel. The outer plates are to be 8 in. [200 mm] wide, and the center plate is to be 12 in. [300 mm] wide. Find the required thicknesses of the plates and the number of bolts needed if the bolts are placed in two rows. Sketch the bolt layout with the necessary dimensions.

Problem 10.5.B. Design a connection for the data in Problem 10.5.A except that the bolts are 1-in. A325N, the outside plates are 9 in. wide, and the bolts are placed in three rows.

10.6 Framing Connections

The joining of structural steel members in a structural system generates a wide variety of situations, depending on the form of the connected parts, the type of connecting device used, and the nature and magnitude of the forces that must be transferred between the members. Figure 10.11 shows a number of common connections that are used to join steel columns and beams consisting of rolled shapes.

In the joint shown in Fig. 10.10*a*, a steel beam is connected to a supporting column by the simple means of resting it on top of a steel plate that is welded to the top of the column. The bolts in this case carry no computed loads if the force transfer is limited to that of the vertical end reaction of the beam. The only computed stress condition that is likely to be of concern in this situation is that of crippling the beam web (Sec. 8.9). This is a situation in which the use of unfinished bolts is indicated.

The remaining details in Fig. 10.11 illustrate situations in which the beam end reactions are transferred to the supports by attachment to the beam web. This is, in general, an appropriate form of force transfer because the vertical shear at the end of the beam is resisted primarily by the beam web. The most common form of connection is that which uses a pair of angles (Fig. 10.11*b*). The two most frequent examples of this type of connection are the joining of a steel beam to the side of a column (Fig. 10.11*b*) or to the side of another beam (Fig. 10.11*d*). A beam may also be joined to the web of a W-shaped column in this manner if the column depth provides enough space for the angles.

An alternative to this type of connection is shown in Fig. 10.11*c*, where a single plate is welded to the side of a column, and the beam web is bolted to one side of the plate. This is generally acceptable only when the magnitude of the load on the beam is low because the one-sided connection experiences some torsion.

When the two intersecting beams must have their tops at the

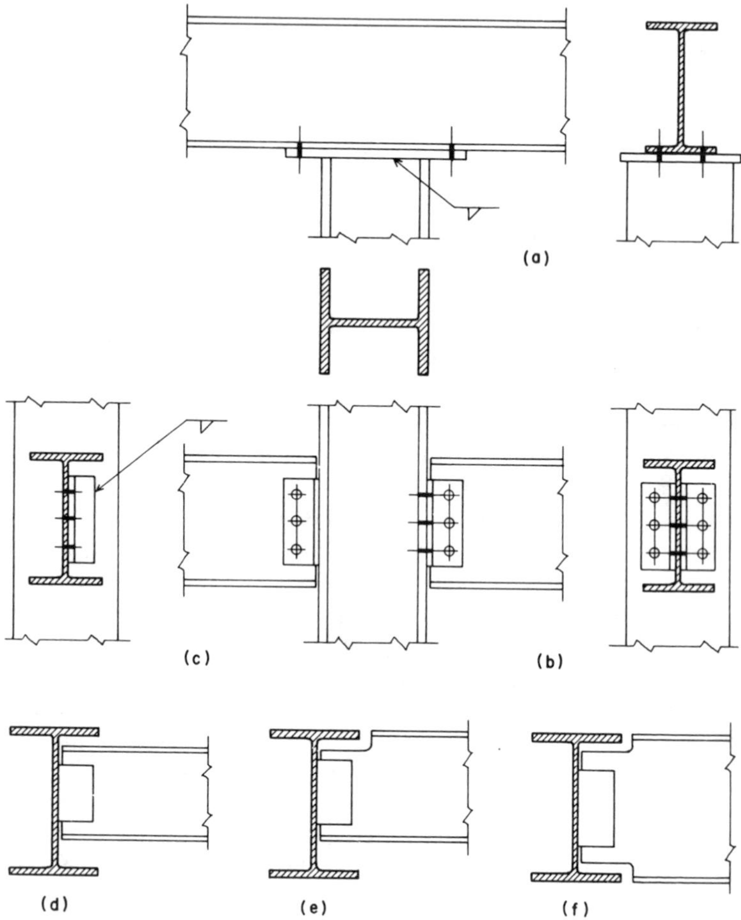

FIGURE 10.11 Typical bolted framing connections for steel structures.

same level, the supported beam must have its top flange cut back, as shown at Fig. 10.11*e*. This is to be avoided, if possible, because it represents an additional cost in the fabrication and also reduces the shear capacity of the beam. Even worse is the situation in which the two beams have the same depth and which requires cutting both flanges of the supported beam (see Fig. 10.11*f*).

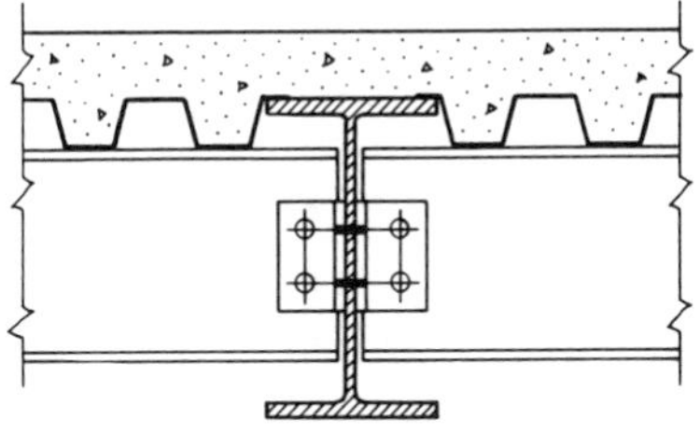

FIGURE 10.12

When these conditions produce critical shear in the beam web, it will be necessary to reinforce the beam end.

Alignment of the tops of beams is usually done to simplify the installation of decks on top of the framing. When steel deck is used it may be possible to adopt some form of the detail shown in Fig. 10.12, which permits the beam tops to be offset by the depth of the deck ribs. Unless the flange of the supporting beam is quite thick, it will probably provide sufficient space to permit the connection shown, which does not require cutting the flange of the supported beam.

Figure 10.13 shows additional framing details that may be used in special situations. The technique described in Fig. 10.13*a* is sometimes used when the supported beam is shallow. The vertical load in this case is transferred through the seat angle, which may be bolted or welded to the carrying beam. The connection to the web of the supported beam merely provides additional resistance to roll-over, or torsional rotation, on the part of the beam. Another reason for favoring this detail is the possibility that the seat angle may be welded in the shop and the web connection made with small unfinished bolts in the field, which greatly simplifies the field work.

Figure 10.13*b* shows the use of a similar connection for joining a beam and column. For heavy beam loads the seat angle may be braced with a stiffening plate. Another variation of this detail involves the use of two plates rather than the angle which may be used if more than four bolts are required for attachment to the column.

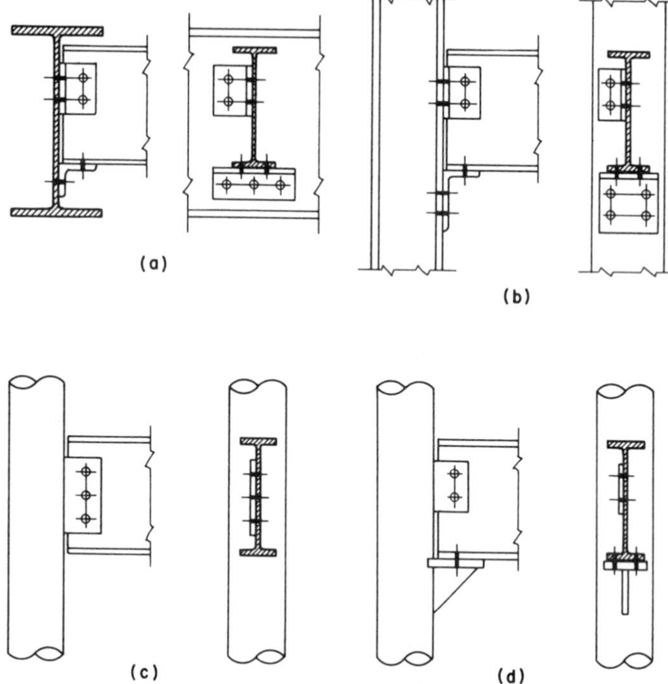

FIGURE 10.13 Bolted connections for special situations.

Figure 10.13*c* and *d* shows connections commonly used when pipe or tube columns carry the beams. Because the one-sided connection in Fig. 10.13*c* produces some torsion in the beam, the seat connection is favored when the beam load is high.

Framing connections quite commonly involve the use of welding and bolting in a single connection, as illustrated in the figures. In general, welding is favored for fabrication in the shop and bolting, for erection in the field. If this practice is recognized, the connections must be developed with a view to the overall fabrication and erection process and some decision made regarding what is to be done where. With the best of designs, however, the contractor who is awarded the job may have some of his own ideas about these procedures and may suggest alterations in the details.

Development of connection details is particularly critical for structures in which a great number of connections occur. The truss is one such structure.

10.7 Framed Beam Connections

The connection shown in Fig. 10.11*b* is the type used most frequently in the development of structures that consist of I-shaped beams and H-shaped columns. This device is referred to as a *framed beam connection*, for which there are several design considerations:

1. *Type of Fastening*. This may be accomplished with rivets or with any of the several types of structural bolt. The angles may also be welded in place, as described in Chapter 11. The most common practice is to weld the angles to the beam web in the fabricating shop and to bolt them to the supports in the field.
2. *Number of Fasteners*. This refers to the number of bolts used on the beam web; there are twice this number in the outstanding legs of the angles. The capacities are matched, however, because the web bolts are in double shear, the others in single shear.
3. *Size of the Angles*. This depends on the size of the fasteners, the magnitude of the loads, and the size of the support, if it is a column with a particular limiting dimension. Two sizes used frequently are 4 × 3 in. and 5 × $3\frac{1}{2}$ in. Thickness of the angle legs is usually based on the size and type of the fastener.
4. *Length of the Angles*. This is primarily a function of the size of the fasteners. As shown in Fig. 10.14, typical dimensions are an end distance of 1.25 in. and a pitch of 3 in. In special situations, however, smaller dimensions may be used with bolts of 1 in. or smaller diameter.

The AISC Manual (Ref. 2) provides considerable information to assist in the design of this type of connection in both the bolted

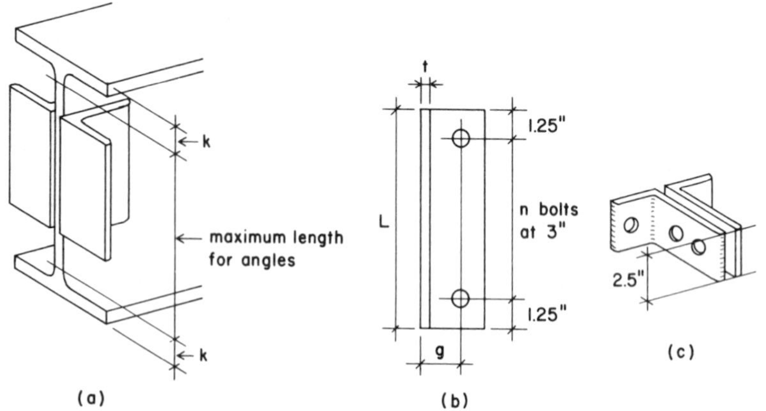

FIGURE 10.14 Framed beam connections using steel angles.

and welded versions. A sample for bolted connections that use A325F bolts and angles of A36 steel is given in Table 10.5. The angle lengths in the table are based on the standard dimensions, as shown in Fig. 10.14. For a given beam shape the maximum size of connection (designated by the number of bolts) is limited by the dimension of the flat portion of the beam web. By referring to Table 3.1 we can determine this dimension for any beam designation.

Although there is no specified limit for the minimum size of a framed connection to be used with a beam, the general rule is to choose one with an angle length of at least one-half the beam depth. This is intended in the most part to ensure some rotational stability for the beam end.

The one-bolt connection with an angle length of only 2.5 in. (Fig. 10.14*c*) is the shortest. This special connection has double-gage spacing of bolts in the beam web to ensure its stability.

The following example illustrates the general design procedure for a framed beam connection. In practice, this process can be shortened because experience permits judgments that will eventually make some of the steps unnecessary. Other design aids in the AISC Manual (Ref. 2) will shorten the work for some computations.

TABLE 10.5 Framed Beam Connections with A325F Bolts and A36 Angles

No. of Bolts n	Angle Length L (in.) (Fig. 10.14)	Total Shear Capacity of Bolts (kips): Bolt Diameter d (in.) $\frac{3}{4}$; Usual Angle Thickness t (in.) $\frac{1}{4}$	$\frac{7}{8}$; $\frac{5}{16}$	1; $\frac{1}{2}$	Use With the Following Rolled Shapes
10	$29\frac{1}{2}$	155	210	275	W 36
9	$26\frac{1}{2}$	139	189	247	W 36, 33
8	$23\frac{1}{2}$	124	168	220	W 35, 33, 30
7	$20\frac{1}{2}$	108	147	192	W 36, 33, 30, 27, 24, S 24
6	$17\frac{1}{2}$	92.8	126	165	W 36, 33, 30, 27, 24, 21, S 24
5	$14\frac{1}{2}$	77.3	105	137	W 30, 27, 24, 21, 18, S 24, 20, 18, C 18
4	$11\frac{1}{2}$	61.9	84.2	110	W 24, 21, 18, 16, S 24, 20, 18, 15, C 18, 15
3	$8\frac{1}{2}$	46.4	61.9[a]	82.5	W 18, 16, 14, 12, 10, S 18, 15, 12, 10, C 18, 15, 12, 10
2	$5\frac{1}{2}$	30.9	39.4[a]	55.0	W 12, 10, 8, S 12, 10, 8, C 12, 10, 9, 8
1	$2\frac{1}{2}$	15.4	21.0	27.5	W 6, 5, M 6, 5, C 7, 6, 5

Source: Adapted from data in the *Manual of Steel Construction*, 8th ed. (Ref. 2), with permission of the publishers, the American Institute of Steel Construction.

[a] Limited by shear on the angles.

Example. A beam consists of a W 27 × 84 of A36 steel with F_u of 58 ksi [400 MPa] that is needed to develop an end reaction of 80 kips [356 kN]. Design a standard framed beam connection with A325F bolts and angles of A36 steel.

Solution: A scan of Table 10.5 reveals that the range of possible connections for a W 27 is $n = 5$ to $n = 7$. For the required load possible choices are

$n = 6$, $\frac{3}{4}$-in. bolts, angle $t = \frac{1}{4}$ in., load = 92.8 kips [413 kN]

$n = 5$, $\frac{7}{8}$-in. bolts, angle $t = \frac{5}{16}$ in., load = 105 kips [467 kN]

Bolt size is ordinarily established for a series of framing rather than for each element. Having no other criterion, we make an arbitrary choice of the connection with $\frac{7}{8}$-in. bolts.

The bolt capacity in double shear is the primary consideration in the development of data in Table 10.5. We must make a separate investigation of the bearing on the beam web because it is not incorporated in the table data. It is actually seldom a problem except in heavily loaded beams, but the following procedure should be used:

From Table 3.1 the thickness of the beam web is 0.460 in. [11.7 mm]. The total bearing capacity of the five bolts is

$$V = n \times (\text{bolt diameter}) \times (\text{web } t) \times 1.5F_u$$
$$= 5 \times 0.875 \times 0.460 \times 87 = 175.1 \text{ kips [780 kN]}$$

which is considerably in excess of the required load of 80 kips [356 kN].

Another concern in the typical situation is that for the shear stress through the net section of the web, reduced by the chain of bolt holes. If the connection is made as shown in Fig. 10.11*b* or *d*, this section is determined as the full web width (beam depth), less the sum of the hole diameters, times the web thickness, and the allowable stress is specified as $0.40F_y$. From Table 3.1 for the W 27, $d = 26.92$ in. [684 mm]. The net shear width through the bolt holes is thus

$$w = 26.92 - (5 \times 0.9375) = 22.23 \text{ in. [565 mm]}$$

and the computed stress due to the load is

$$f_v = \frac{80}{22.23 \times 0.460} = 7.82 \text{ ksi [54 MPa]}$$

which is less than the allowable of $0.40 \times 36 = 14.4$ ksi [100 MPa].

If the top flange of the beam is cut back to form the type of connection shown in Fig. 10.11*e*, a critical condition that must be investigated is that of tearing out the end portion of the beam web, as discussed in Sec. 10.4. This is also called *block shear*,

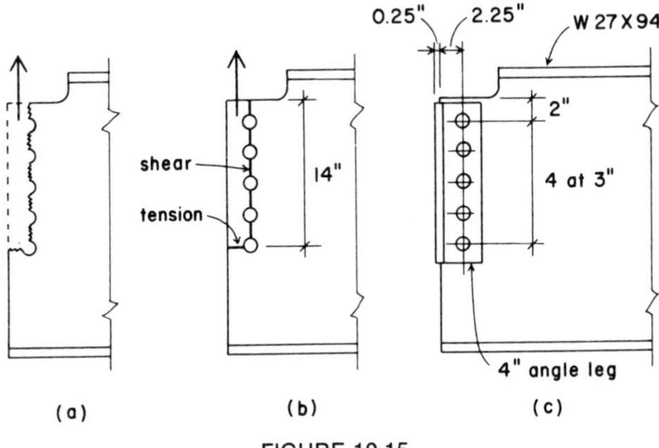

FIGURE 10.15

which refers to the form of the failed portion (Fig. 10.15). If the angles are placed with the edge distances shown in Fig. 10.15, this failure block will have the dimensions of 14 × 2.25 in. [356 × 57 mm]. The tearing force V is resisted by a combination of tension stress and shearing stress. The allowable stresses for this situation are $0.30F_u$ for shear and $0.50F_u$ for tension. Next find the net widths of the sections, multiply them by the web thickness to obtain the areas, and multiply by the allowable stresses to obtain the total resisting forces.

For the tension resistance.

$$w = 2.25 - \frac{0.9375}{2} = 1.78 \text{ in. [45.1 mm]}$$

For the shear resistance

$$w = 14 - (4\tfrac{1}{2} \times 0.9375) = 9.78 \text{ in. [249 mm]}$$

For the total resisting force

$$\begin{aligned} V &= (\text{tension } w \times t_w) \times (0.50F_u) + (\text{shear } w \times t_w) \times (0.30F_u) \\ &= (1.78 \times 0.46 \times 29) + (9.78 \times 0.46 \times 17.4) \\ &= 23.7 + 78.3 = 102 \text{ kips} \end{aligned}$$

Because this potential total resistance exceeds the load required, the tearing is not critical.

Problem 10.7.A. A W 30 × 108 of A36 steel with F_u = 58 ksi [400 MPa] is required to develop an end reaction of 120 kips [534 kN]. Determine the possible choices for a framed beam connection with A325F bolts and A36 angles, Investigate the resistance with the smallest bolts to bearing and shear if the beam flange is not cut back, and the resistance to tearing if the connection is installed as shown in Fig. 10.15.

Problem 10.7.B. Proceed as in the preceding problem, except that the beam is a W 24 × 69 and the reaction is 100 kips [445 kN].

Problem 10.7.C. Proceed as in Problem 10.7.A, except that the beam is a W 16 × 45 and the reaction is 50 kips [222 kN].

10.8. Bolted Truss Connections

A major factor in the design of trusses is the development of the truss joints. Since a single truss typically has several joints, the joints must be relatively easy to produce and economical, especially if there are many trusses of a single type in the building structural system. Considerations involved in the design of connections for the joints include the truss configuration, member shapes and sizes, and the fastening method—usually welding or high-strength bolts.

In most cases the preferred method of fastening for connections made in the fabricating shop is welding. In most cases trusses will be shop-fabricated in the largest units possible, which means the whole truss for modest span trusses or the maximum-sized unit that can be transported for large trusses. Bolting is mostly used for connections made at the building site. For the small truss, the only bolting is usually done for the connections to supports and to supported elements or bracing. For the large truss, bolting may also be done at splice points between shop-fabricated units. All of this is subject to many considerations relating to the nature of the rest of the building structure, the particular location of the site, and the practices of local fabricators and erectors.

Two common forms for light steel trusses are shown in Fig. 10.16. In Fig. 10.16*a* the truss members consist of pairs of angles

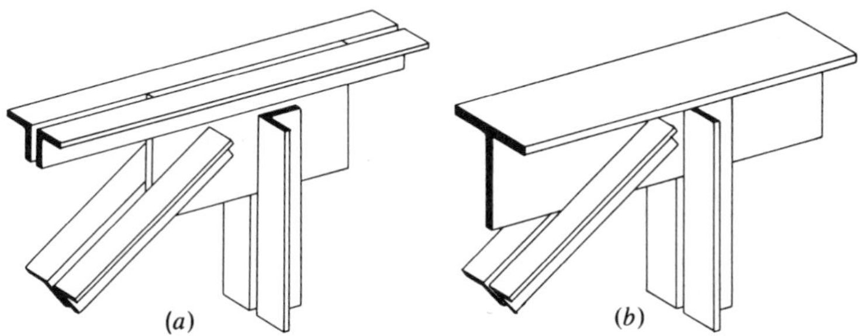

FIGURE 10.16 Typical framing details for light steel trusses.

and the joints are achieved by using steel gusset plates to which the members are attached. For top and bottom chords the angles are often made continuous through the joint, reducing the number of connectors required and the number of separate cut pieces of the angles. For flat-profiled, parallel-chorded trusses of modest size, the chords are sometimes made from tees, with interior members fastened to the tee web (Fig. 10.16*b*).

Figure 10.17 shows a layout for several joints of a light roof truss, employing the system shown in Fig. 10.16*a*. This is a form commonly used in the past for roofs with high slopes, with many short-span trusses fabricated in a single piece in the shop, usually with riveted joints. Trusses of this form are now mostly welded or use high-strength bolts as shown in Fig. 10.17.

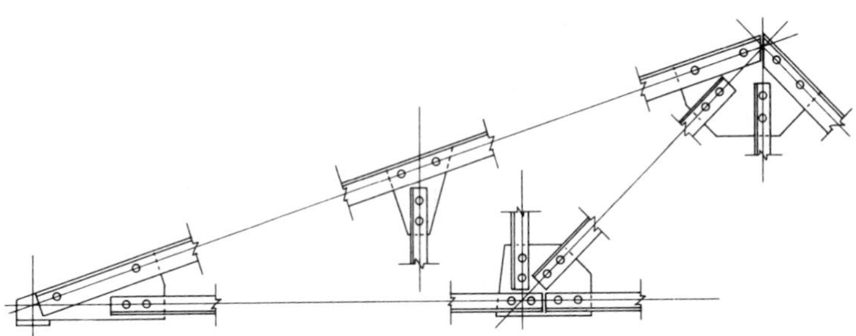

FIGURE 10.17 Typical light steel truss with bolted joints.

Development of the joint designs for the truss shown in Fig. 10.17 would involve many considerations, including:

1. *Member Size and Load Magnitude.* This determines primarily the size and type of connector (bolt) required, based on individual connector capacity.
2. *Angle Leg Size.* This relates to the maximum diameter of bolt that can be used, based on minimum edge distances.
3. *Thickness and Profile Size of Gusset Plates.* The preference is to have the lightest weight added to the structure (primarily for the cost per pound of the steel), which is achieved by reducing the plates to a minimum thickness and general minimum size.
4. *Layout of Members at Joints.* The general attempt is to have the action lines of the forces (vested in the rows of bolts) all meet at a single point, thus avoiding twisting in the joint.

Many of the points mentioned are determined by data. Minimum edge distances for bolts (Table 10.2) can be matched to usual gage dimensions for angles (Table 10.4). Forces in members can be related to bolt capacities in Table 10.1, the general intent being to keep the number of bolts to a minimum in order to make the required size of the gusset plate smaller.

Other issues involve some judgment or skill in the manipulation of the joint details. For really tight or complex joints, it is often necessary to study the form of the joint with carefully drawn large-scale layouts. Actual dimensions and form of the member ends and the gusset plates may be derived from these drawings.

The truss shown in Fig. 10.17 has some features that are quite common for small trusses. All member ends are connected by only two bolts, the minimum required by the specifications. This simply indicates that the minimum-sized bolt chosen has sufficient capacity to develop the forces in all members with only two bolts. At the top chord joint between the support and the peak, the top chord member is shown as being continuous (uncut) at the joint. This is quite common where the lengths of members available are greater than the joint-to-joint distances in the truss, a cost savings in member fabrication as well as connection.

If there are only one or a few of the trusses as shown in Fig. 10.17 to be used in a building, the fabrication may indeed be as shown in the illustration. However, if there are many such trusses, or the truss is actually a manufactured, standardized product, it is much more likely to be fabricated with joints as described previously, employing welding for shop work and bolting only for field connections. Possible forms of joints for such trusses are discussed in Chapter 11.

11

Welded Connections

11.1 Introduction

Welding is in some instances an alternative means of making connections in a structural joint, the other principal option being structural bolts. A common situation is that of a connecting device (bearing plate, framing angles, etc.) that is welded to one member in the shop and fastened by bolting to a connecting member in the field. However, there are also many instances of joints that are fully welded, whether done in the shop or at the site of the building construction. For some situations the use of welding may be the only reasonable means of making an attachment for a joint. As in many other situations, the design of welded joints requires considerable awareness of the problems encountered by the welder and the fabricator of the welded parts. This chapter presents some of the problems and potential uses of welding in building structures.

One advantage of welding is that if offers the possibility for direct connection of members, often eliminating the need for intermediate devices, such as gusset plates or framing angles. Another advantage is the lack of need for holes (required for bolts), which permits development of the capacity of the unreduced cross section of tension members. Welding also offers the possibility of developing exceptionally rigid joints, an advantage in moment-resistive connections or generally nondeforming connections.

11.2 Electric Arc Welding

Although there are many welding processes, electric arc welding is the one generally used in steel building construction. In this type of welding, an electric arc is formed between an electrode and the two pieces of metal that are to be joined. The intense heat melts a small portion of the members to be joined, as well as the end of the electrode or metallic wire. The term *penetration* is used to indicate the depth from the original surface of the base metal to the point at which fusion ceases. The globules of melted metal from the electrode flow into the molten seat and, when cool, are united with the members that are to be welded together. *Partial penetration* is the failure of the weld metal and base metal to fuse at the root of a weld. It may result from a number of items, and such incomplete fusion produces welds that are inferior to those of full penetration (called *complete penetration* welds).

11.3 Welded Joints

When two members are to be joined, the ends may or may not be grooved in preparation for welding. In general, there are three classifications of joints: *butt joints, tee joints,* and *lap joints.* The selection of the type of weld to use depends on the magnitude of the load requirement, the manner in which it is applied, and the cost of preparation and welding. Several joints are shown in Fig. 11.1. The type of joint and preparation permit a number of variations. In addition, welding may be done from one or both sides. The scope of this book prevents a detailed discussion of the many joints and their uses and limitations.

A weld commonly used for structural steel in building construction is the *fillet weld.* It is approximately triangular in cross section and is formed between the two intersecting surfaces of the joined members (see Fig. 11.2*a* and *b*). The *size* of a fillet weld is

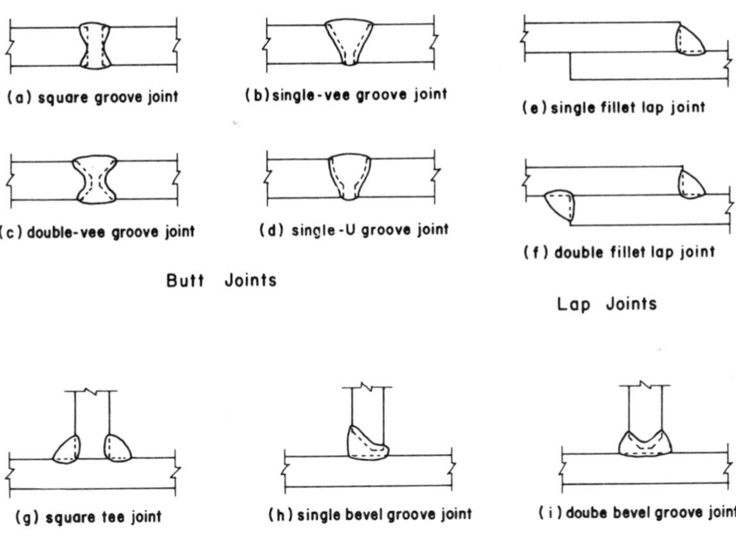

FIGURE 11.1 Typical welded joints.

the leg length of the largest inscribed isosceles right triangle, AB or BC (see Fig. 11.2*a*). The *root* of the weld is the point at the bottom of the weld, point B in Fig. 11.2*a*. The *throat* of a fillet weld is the distance from the root to the hypotenuse of the largest isosceles right triangle that can be inscribed within the weld cross section, distance BC in Fig. 11.2*a*. The exposed surface of a weld is not the plane surface indicated in Fig. 11.2*a* but is usually

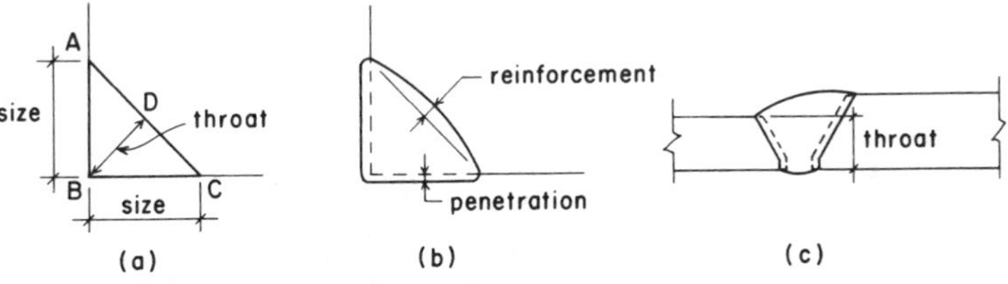

FIGURE 11.2 Properties of welded joints.

somewhat convex, as shown in Fig. 11.2*b*. Therefore the actual throat may be greater than that shown in Fig. 11.2*a*. This additional material is called *reinforcement*. It is not included in determining the strength of a weld.

A single-vee groove weld between two members of unequal thickness is shown in Fig. 11.2*c*. The *size* of a butt weld is the thickness of the thinner part joined, with no allowance made for the weld reinforcement.

11.4 Stresses in Fillet Welds

If the dimension (size) of AB in Fig. 11.2*a* is one unit in length, $(AD)^2 + (BD)^2 = 1^2$. Because AD and BD are equal, $2(BD)^2 = 1^2$, and $BD = \sqrt{0.5}$, or 0.707. Therefore the throat of a fillet weld is equal to the *size* of the weld multiplied by 0.707. As an example, consider a $\frac{1}{2}$-in. fillet weld. This would be a weld with dimensions AB or BC equal to $\frac{1}{2}$ in. In accordance with the above, the throat would be 0.5×0.707, or 0.3535 in. Then, if the allowable unit shearing stress on the throat is 21 ksi, the allowable working strength of a $\frac{1}{2}$-in. fillet weld is $0.3535 \times 21 = 7.42$ kips *per lin in. of weld*. If the allowable unit stress is 18 ksi, the allowable working strength is $0.3535 \times 18 = 6.36$ kips *per lin in. of weld*.

The permissible unit stresses used in the preceding paragraph are for welds made with E 70 XX- and E 60 XX-type electrodes on A36 steel. Particular attention is called to the fact that *the stress in a fillet weld is considered as shear on the throat, regardless of the direction of the applied load*. Neither plug nor slot welds shall be assigned any values in resistances other than shear. The allowable working strengths of fillet welds of various sizes are given in Table 11.1 with values rounded to $\frac{1}{10}$ kip.

The stresses allowed for the metal of the connected parts (known as the *base metal*) apply to complete penetration groove welds that are stressed in tension or compression parallel to the axis of the weld or are stressed in tension perpendicular to the effective throat. They apply also to complete or partial penetration groove welds stressed in compression normal to the effective throat and in shear on the effective throat. Consequently, allowable stresses for butt welds are the same as for the base metal.

TABLE 11.1 Allowable Working Strength of Fillet Welds

	Allowable Load (kips/in.)		Allowable Load (kN/mm)		
Size of Weld (in.)	E 60 XX Electrodes $F_{vw} = 18$ (ksi)	E 70 XX Electrodes $F_{vw} = 21$ (ksi)	E 60 XX Electrodes $F_{vw} = 124$ (MPa)	E 70 XX Electrodes $F_{vw} = 145$ (MPa)	Size of Weld (mm)
$\frac{3}{16}$	2.4	2.8	0.42	0.49	4.76
$\frac{1}{4}$	3.2	3.7	0.56	0.65	6.35
$\frac{5}{16}$	4.0	4.6	0.70	0.81	7.94
$\frac{3}{8}$	4.8	5.6	0.84	0.98	9.52
$\frac{1}{2}$	6.4	7.4	1.12	1.30	12.7
$\frac{5}{8}$	8.0	9.3	1.40	1.63	15.9
$\frac{3}{4}$	9.5	11.1	1.66	1.94	19.1

The relation between the weld size and the maximum thickness of material in joints connected only by fillet welds is shown in Table 11.2. The maximum size of a fillet weld applied to the square edge of a plate or section that is $\frac{1}{4}$ in. or more in the thickness should be $\frac{1}{16}$ in. less than the nominal thickness of the edge. Along edges of material less than $\frac{1}{4}$ in. thick, the maximum size may be equal to the thickness of the material.

The effective area of butt and fillet welds is considered to be the effective length of the weld multiplied by the effective throat thickness. The minimum effective length of a fillet weld should not be less than four times the weld size. For starting and stop-

TABLE 11.2 Relation Between Material Thickness and Minimum Size of Fillet Welds

Material Thickness of the Thicker Part Joined		Minimum Size of Fillet Weld	
(in.)	(mm)	(in.)	(mm)
To $\frac{1}{4}$ inclusive	To 6.35 inclusive	$\frac{1}{8}$	3.18
Over $\frac{1}{4}$ to $\frac{1}{2}$	Over 6.35 to 12.7	$\frac{3}{16}$	4.76
Over $\frac{1}{2}$ to $\frac{3}{4}$	Over 12.7 to 19.1	$\frac{1}{4}$	6.35
Over $\frac{3}{4}$	Over 19.1	$\frac{5}{16}$	7.94

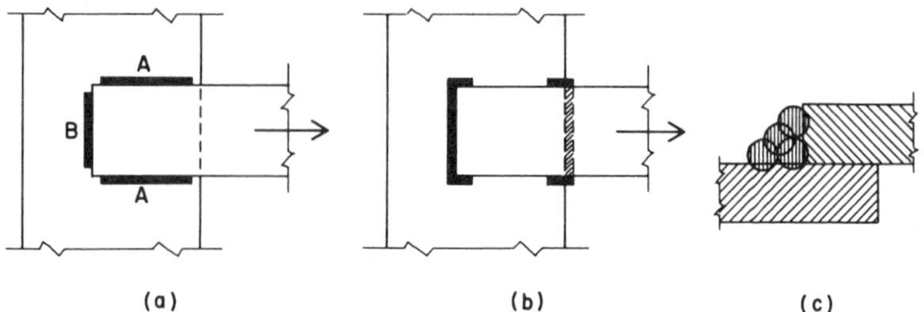

FIGURE 11.3 Welding of lapped plates.

ping the arc, approximately $\frac{1}{4}$ in. should be added to the design length of fillet welds.

Figure 11.3*a* represents two plates connected by fillet welds. The welds marked *A* are longitudinal; *B* indicates a transverse weld. If a load is applied in the direction shown by the arrow, the stress distribution in the longitudinal weld is not uniform, and the stress in the transverse weld is approximately 30% higher per unit of length.

Added strength is given to a transverse fillet weld that terminates at the end of a member, as shown in Fig. 11.3*b*, if the weld is returned around the corner for a distance not less than twice the weld size. These end returns, sometimes called *boxing*, afford considerable resistance to the tendency of tearing action on the weld.

The $\frac{1}{4}$-in. fillet weld is considered to be the minimum practical size, and a $\frac{5}{16}$-in. weld is probably the most economical size that can be obtained by one pass of the electrode. A small continuous weld is generally more economical than a larger discontinuous weld if both are made in one pass. Some specifications limit the single-pass fillet weld to $\frac{5}{16}$ in. Large fillet welds require two or more passes (multipass welds) of the electrode, as shown in Fig. 11.3*c*.

11.5 Design of Welded Joints

The most economical weld to use for a given condition depends on several factors. It should be borne in mind that members to be connected by welding must be firmly clamped or held rigidly in position during the welding process. When welding a beam to a column, you must provide a support to keep the beam in position for the welding work. The seat angle is not considered as adding strength to the connection. The designer must have in mind the actual conditions during erection and must provide for economy and ease in working the welds. Seat angles or similar members used to facilitate erection are *shop-welded* before the material is sent to the site. The welding done during erection is called *field welding*. In preparing welding details, the designer indicates shop or field welds on the drawings. Conventional welding symbols are used to identify the type, size, and position of the various welds. Only engineers or architects experienced in the design of welded connections should design or supervise welded construction. It is apparent that a wide variety of connections is possible; experience is the best aid in determining the most economical and practical connection.

The following examples illustrate the basic principles on which welded connections are designed.

Example 1. A bar of A36 steel, $3 \times \frac{7}{16}$ in. [76.2 × 11 mm] in cross section, is to be welded with E 70 XX electrodes to the back of a channel so that the full tensile strength of the bar may be developed. What is the size of the weld? (See Fig. 11.4.)

Solution: The area of the bar is 3 × 0.4375 = 1.313 in.^{2} [76.2 × 11 = 838.2 mm^2]. Because the allowable unit tensile stress of the steel is 22 ksi (Table 7.1), the tensile strength of the bar is $F_t \times A$ = 22 × 1.313 = 28.9 kips [152 × 838.2/10^3 = 127 kN]. The weld must be of ample dimensions to resist a force of this magnitude.

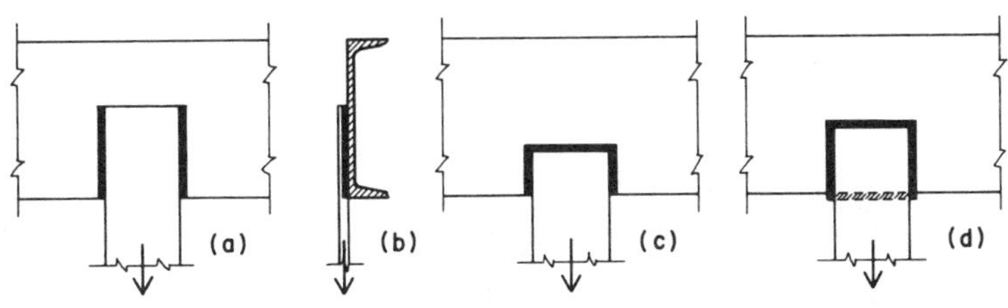

FIGURE 11.4

A $\frac{3}{8}$-in. [9.52-mm] fillet weld will be used. Table 11.1 gives the allowable working strength as 5.6 kips/in. [0.98 kN/mm]. Hence the required length of weld to develop the strength of the bar is 28.9 ÷ 5.6 = 5.16 in. [127 ÷ 0.98 = 130 mm]. The position of the weld with respect to the bar has several options, three of which are shown in Fig. 11.4*a*, *c*, and *d*.

Example 2. A $3\frac{1}{2} \times 3\frac{1}{2} \times \frac{5}{16}$-in. [89 × 89 × 7.94-mm] angle of A36 steel subjected to a tensile load is to be connected to a plate by fillet welds, using E 70 XX electrodes. What should the dimensions of the welds be to develop the full tensile strength of the angle?

Solution: We shall use a $\frac{1}{4}$-in. fillet weld which has an allowable working strength of 3.7 kips/in. [0.65 kN/mm] (Table 11.1). From Table 3.3 the cross-sectional area of the angle is 2.09 in.^{2} [1348 mm^2]. By using the allowable tension stress of 22 ksi [152 MPa] for A36 steel (Table 7.1), the tensile strength of the angle is 22 × 2.09 = 46 kips [152 × 1348/10^3 = 205 kN]. Therefore the required total length of weld to develop the full strength of the angle is 46 ÷ 3.7 = 12.4 in. [205 ÷ 0.65 = 315 mm].

An angle is an unsymmetrical cross section, and the welds marked L_1 and L_2 in Fig. 11.5 are made unequal in length so that their individual resistance will be proportioned in accordance to the distributed area of the angle. From Table 3.3 we find that the centroid of the angle section is 0.99 in. [25 mm] from the back of the angle; hence the two welds are 0.99 in. [25 mm] and 2.51 in. [64 mm] from the centroidal axis, as shown in Fig. 11.5. The

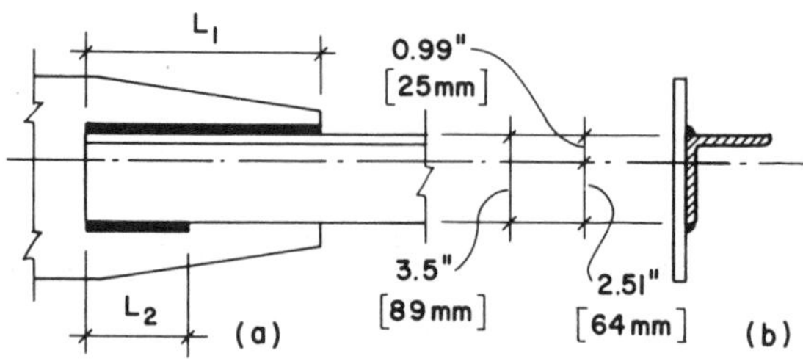

FIGURE 11.5

lengths of welds L_1 and L_2 are made inversely proportional to their distances from the axis, but the sum of their lengths is 12.4 in. [315 mm]. Therefore

$$L_1 = \frac{2.51}{3.5} \times 12.4 = 8.9 \text{ in. } [227 \text{ mm}]$$

and

$$L_2 = \frac{0.99}{3.5} \times 12.4 = 3.5 \text{ in. } [88 \text{ mm}]$$

These are the design lengths required, and as noted earlier, each weld would actually be made $\frac{1}{4}$ in. [6.4 mm] longer than its computed length.

When angle shapes are used as tension members and connected by fastening only one leg, it is questionable to assume a stress distribution of equal magnitude on the entire cross section. Some designers therefore prefer to ignore the stress in the unconnected leg and to limit the capacity of the member in tension to the force obtained by multiplying the allowable stress by the area of the connected leg only. If this is done, it is logical to use welds of equal length on each side of the leg, as in Example 1.

Problem 11.5.A. A $4 \times 4 \times \frac{1}{2}$-in. angle of A36 steel is to be welded to a plate with E 70 XX electrodes to develop the full tensile strength of the angle. Using $\frac{3}{8}$-in. fillet welds, compute the design lengths L_1 and L_2, as shown in Fig. 11.5, assuming the development of tension on the entire cross section of the angle.

Problem 11.5.B. Redesign the welded connection in Problem 11.5.A assuming that the tension force is developed only by the connected leg of the angle.

11.6 Miscellaneous Welded Connections

Plug and Slot Welds. One method of connecting two overlapping plates uses a weld in a hole made in one of the two plates (see Fig. 11.6). Plug and slot welds are those in which the entire area of the hole or slot receives weld metal. The maximum and minimum diameters of plug and slot welds and the maximum length of

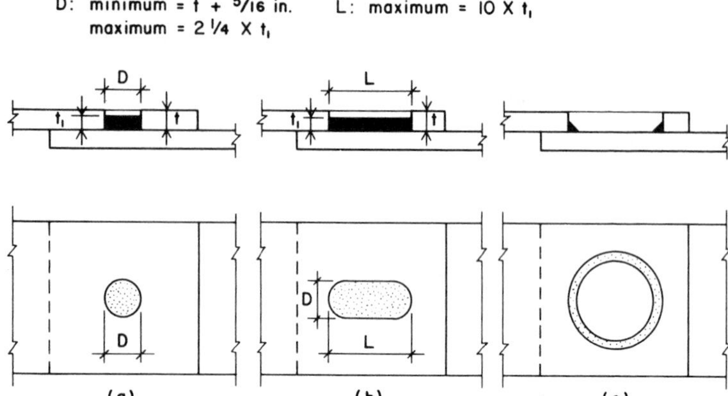

FIGURE 11.6 Welds in holes: (*a*) plug weld, (*b*) slot weld, (*c*) fillet weld in a large hole.

slot welds are shown in Fig. 11.6. If the plate containing the hole is not more than $\frac{5}{8}$ in. thick, the hole should be filled with weld metal. If the plate is more than $\frac{5}{8}$ in. thick, the weld metal should be at least one-half the thickness of the material but not less than $\frac{5}{8}$ in.

The stress in a plug or slot weld is considered to be shear on the area of the weld at the plane of contact of the two plates being connected. The allowable unit shearing stress, when E 70 XX electrodes are used, is 21 ksi [145 MPa].

A somewhat similar weld consists of a continuous fillet weld at the circumference of a hole, as shown in Fig. 11.6*c*. This is not a plug or slot weld and is subject to the usual requirements for fillet welds.

Framing Connections. Part 4 of the AISC Manual contains a series of tables that pertain to the design of welded connections. The tables cover free-end as well as moment-resisting connections. In addition, suggested framing details are shown for various situations.

A few common connections are shown in Fig. 11.7. As an aid to erection, certain parts are welded together in the shop before

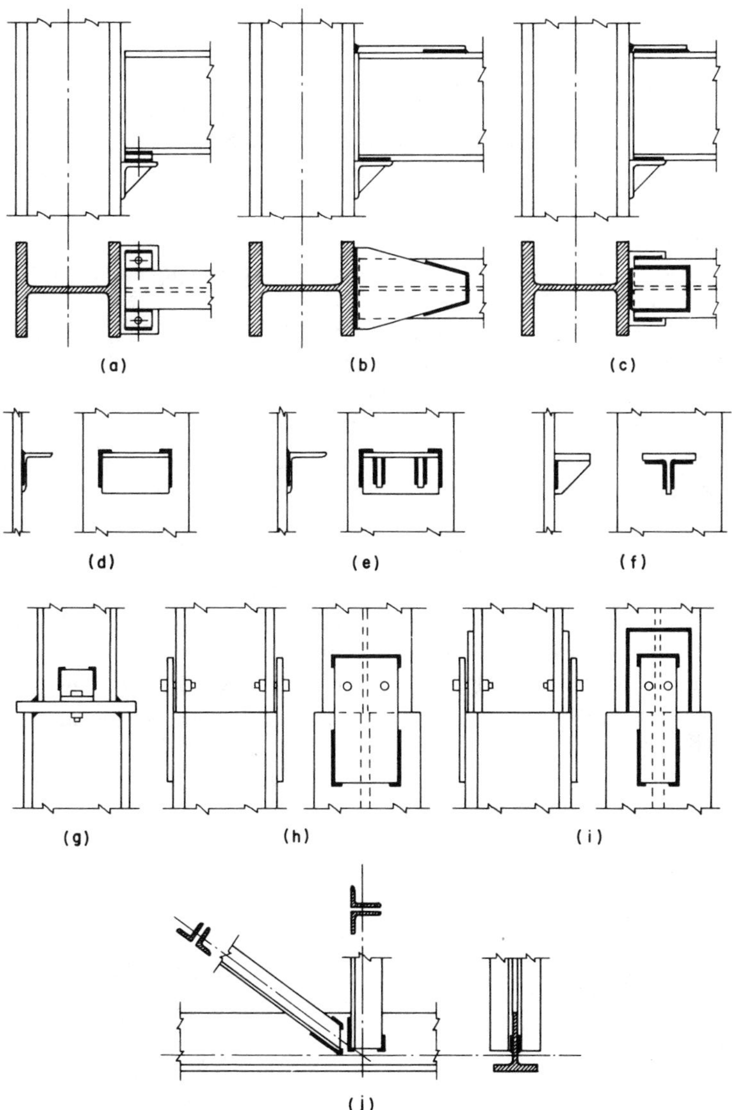

FIGURE 11.7 Welded framing connections.

being sent to the site. Connection angles may be shop-welded to beams and the angles field-welded or field-bolted to girders or columns. The beam connection in Fig. 11.7*a* shows a beam supported on a seat that has been shop-welded to the column. A small connection plate is shop-welded to the lower flange of the beam, and the plate is bolted to the beam seat. After the beams have been erected and the frame plumbed, the beams are field-welded to the seat angles. This type of connection provides no degree of continuity in the form of moment transfer between the beam and column.

The connections shown in Fig. 11.7*b* and *c* are designed to develop some moment transfer between the beam and its supporting column. Auxiliary plates are used to make the connection at the upper flanges.

Beam seats shop-welded to columns are shown in Fig. 11.7*d–f*. A short length of angle welded to the column with no stiffeners is shown in Fig. 11.7*d*. Stiffeners consisting of triangular plates are welded to the legs of the angles shown in Fig. 11.7*e* and add materially to the strength of the seat. Another method of forming a seat, using a short piece of structural tee, is shown in Fig. 11.7*f*.

Various types of column splice are shown in Fig. 11.7*g–i*. The auxiliary plates and angles are shop-welded to the columns and provide for bolted connections in the field before the permanent welds are made.

Figure 11.7*j* shows a type of welded construction used in light trusses in which the lower chord consists of a structural tee.

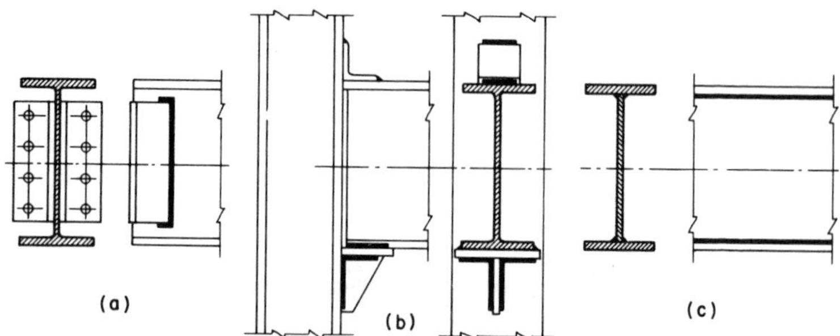

FIGURE 11.8 Additional welded framing connections.

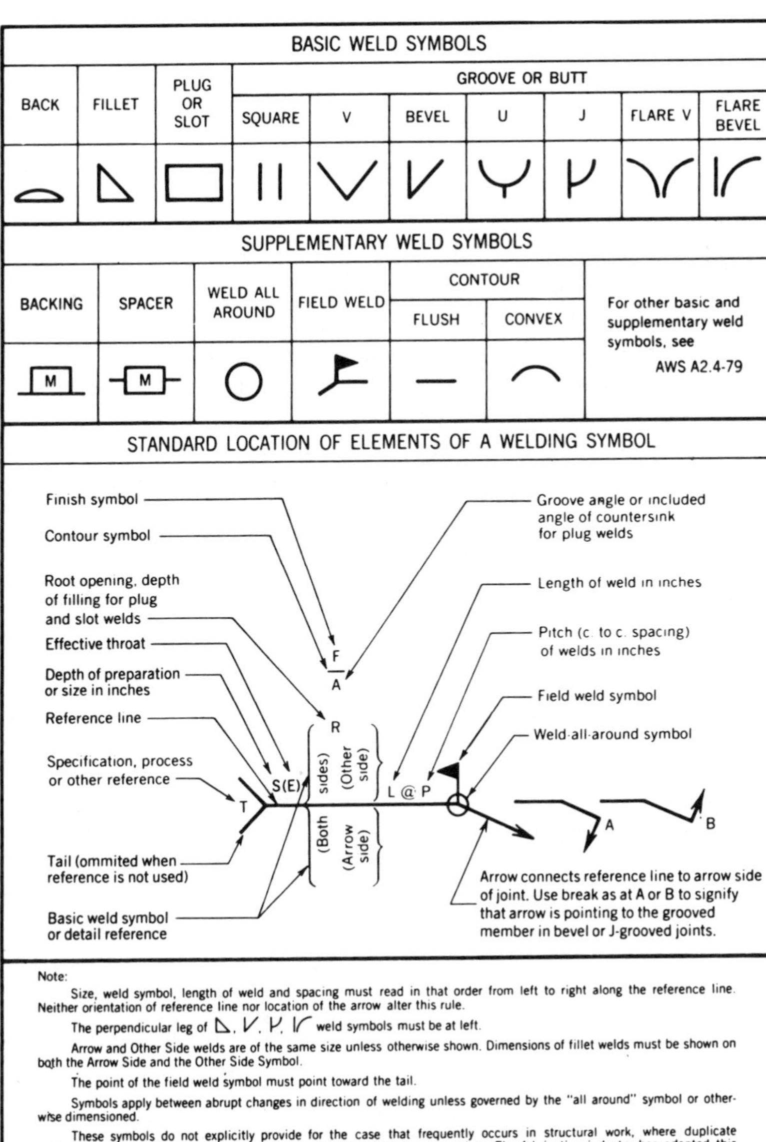

Note:

Size, weld symbol, length of weld and spacing must read in that order from left to right along the reference line. Neither orientation of reference line nor location of the arrow alter this rule.

The perpendicular leg of ◺, V, ⊬, |⌐ weld symbols must be at left.

Arrow and Other Side welds are of the same size unless otherwise shown. Dimensions of fillet welds must be shown on both the Arrow Side and the Other Side Symbol.

The point of the field weld symbol must point toward the tail.

Symbols apply between abrupt changes in direction of welding unless governed by the "all around" symbol or otherwise dimensioned.

These symbols do not explicitly provide for the case that frequently occurs in structural work, where duplicate material (such as stiffeners) occurs on the far side of a web or gusset plate. The fabricating industry has adopted this convention: that when the billing of the detail material discloses the existence of a member on the far side as well as on the near side, the welding shown for the near side shall be duplicated on the far side.

FIGURE 11.9 Standard weld symbols used on construction drawings. Reproduced from the *Manual of Steel Construction,* 8th ed. (Ref. 2), with permission of the publishers, American Institute of Steel Construction.

Truss web members consisting of pairs of angles are welded to the stem of the tee chord. Welding may also be used for the connections for trusses using gusset plates, as shown in Fig. 10.17. In general, shop fabrication is most often achieved with welding; thus the various seats and bracing clips shown in Figs. 10.13 and 10.14 would most likely be attached to members in the shop by welding, even though field connections are achieved with bolts.

Some additional connection details are given in Fig. 11.8. The detail in Fig. 11.8*a* is an arrangement for framing a beam to a girder, in which welds are substituted for bolts or rivets. In this figure welds replace the fasteners that secure the connection angles to the web of the supported beam.

A welded connection for a stiffened seated beam connection to a column is shown in Fig. 11.8*b*. Figure 11.8*c* shows the simplicity of welding in connecting the upper and lower flanges of a plate girder to the web plate.

11.7 Symbols for Welds

Standard symbols are used in detail drawings of welded connections of structural elements. In addition to the type of weld, other information to be conveyed includes size, exact location, and finishes. Figure 11.9, reproduced from the AISC Manual, gives the standard symbols for welded joints. It will be noted that the

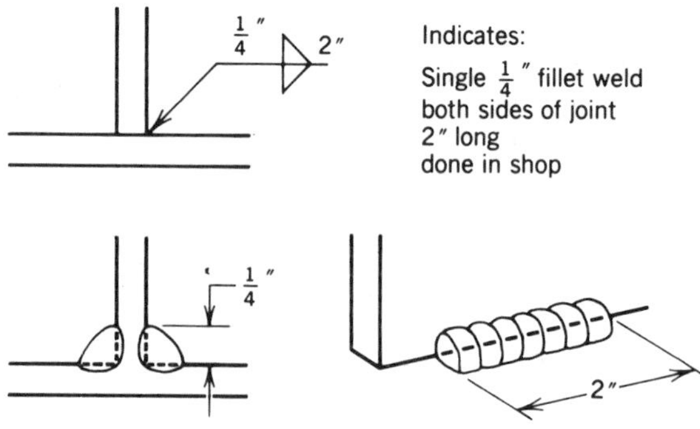

FIGURE 11.10 Use of standard weld symbols.

symbol for a fillet weld is a triangle; this is drawn below the horizontal line if the weld is on the near side, above if it is on the far side; two triangles, one above and one below, are drawn for welds on both sides of the joint. The size of the weld is placed to the left of the vertical line of the triangle and the length to the right side of the hypotenuse. An example of the use of standard symbols for a simple fillet weld is shown in Fig. 11.10.

12

Plastic Behavior and Strength Design

12.1 Plastic Versus Elastic Behavior

Up to this point the discussions of the design of members in bending have been based on bending stresses well within the yield-point stress. In general, allowable stresses are based on the *theory of elastic behavior*. However, it has been found by tests that steel members can carry loads much higher than anticipated, even when the yield-point stress is reached at sections of maximum bending moment. This is particularly evident in continuous beams and in structures with rigid connections. An inherent property of structural steel is its ability to resist large deformations without failure. These large deformations occur chiefly in the *plastic range,* with no increase in the magnitude of the bending stress. Because of this phenomenon, the *plastic design theory,* sometimes called the *ultimate-strength design theory* (or more recently *strength design*), has been developed.

Figure 12.1 represents the typical form of a load-test response for a specimen of ductile steel. The graph shows that up to a stress f_y, the yield point, the deformations are directly propor-

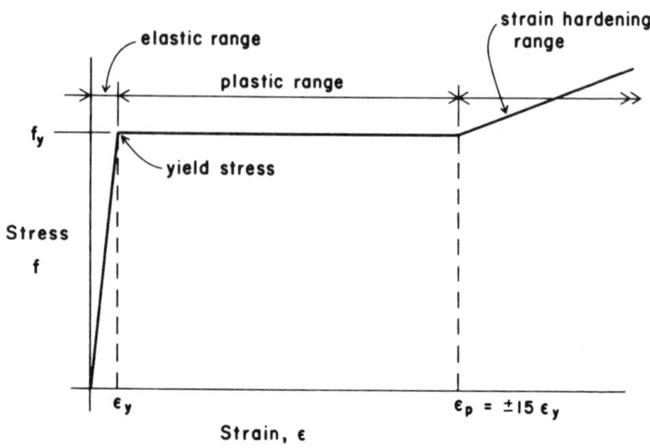

FIGURE 12.1 Idealized form of the stress–strain response for ductile steel.

tional to the applied stresses and that beyond the yield point there is a deformation without an increase in stress. For A36 steel this additional deformation, called the *plastic range,* is approximately 15 times that produced elastically. This magnitude of the plastic range is the basis for qualification of a metal as *ductile*. Note that beyond this range *strain hardening* (loss of ductility) begins, when further deformation can occur only with an increase in stress.

For plastic behavior to be significant, the extent of the plastic range of deformation must be several times the elastic deformation. As the yield point is increased in magnitude, this ratio of deformations decreases, which is to say that higher-strength steels tend to be less ductile. At present, the theory of plastic design is generally limited to steels with a yield point of not more than 65 ksi [450 MPa].

12.2 Plastic Moment and the Plastic Hinge

Sec. 3.1 explains the design of members in bending in accordance with the theory of elasticity. When the extreme fiber stress does not exceed the elastic limit, the bending stresses in the cross

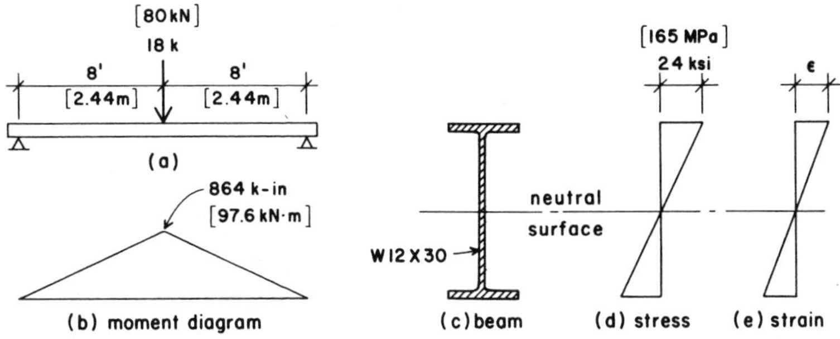

FIGURE 12.2 Elastic behavior of the beam.

section of a beam are directly proportional to their distances from the neutral surface. In addition, the strains (deformations) in these fibers are also proportional to their distances from the neutral surface. Both stresses and strains are zero at the neutral surface, and both increase to maximum magnitudes at the fibers farthest from the neutral surface.

The following example illustrates the analysis of a steel beam for bending, according to the theory of elastic behavior.

Example. A simple steel beam has a span of 16 ft [4.88 m] with a concentrated load of 18 kips [80 kN] at the center of the span. The section used is a W 12 × 30, the beam is adequately braced throughout its length, and the beam weight is ignored in the computations. Let us compute the maximum extreme fiber stress (see Fig. 12.2*d*).

Solution: To do this we use the flexure formula

$$f = \frac{M}{S} \qquad \text{(Sec. 2.1)}$$

Then

$$M = \frac{PL}{4} = \frac{18 \times 16}{4} = 72 \text{ kip-ft [98 kN-m]}$$

which is the maximum bending moment. In Table 3.1 we find S = 38.6 in.^{3} [632 × 10^3 mm^3]. Thus

$$f = \frac{M}{S} = \frac{72 \times 12}{38.6} = 22.4 \text{ ksi [154 MPa]}$$

which is the stress on the fiber farthest from the neutral surface (see Fig. 12.2*d*).

Note that this stress occurs only at the beam section at the center of the span, where the bending moment has its maximum value. Figure 12.2*e* shows the deformations that accompany the stresses shown in Fig. 12.2*d*. Note that both stresses and deformations are directly proportional to their distances from the neutral surface in elastic analysis.

When a steel beam is loaded to produce an extreme fiber stress in excess of the yield point, the property of the material's ductility affects the distribution of the stresses in the beam cross section. Elastic analysis does not suffice to explain this phenomenon because the beam will experience some plastic deformation.

Assume that the bending moment on a beam is of such magnitude that the extreme fiber stress is f_y, the yield stress. Then if M_y is the elastic bending moment at the yield stress, $M = M_y$, and the distribution of the stresses in the cross section is as shown in Fig. 12.3*a*, the maximum bending stress f_y is at the extreme fiber.

Next consider that the loading and the resulting bending moment have been increased; M is now greater than M_y. The stress on the extreme fiber is still f_y, but *the material has yielded* and a

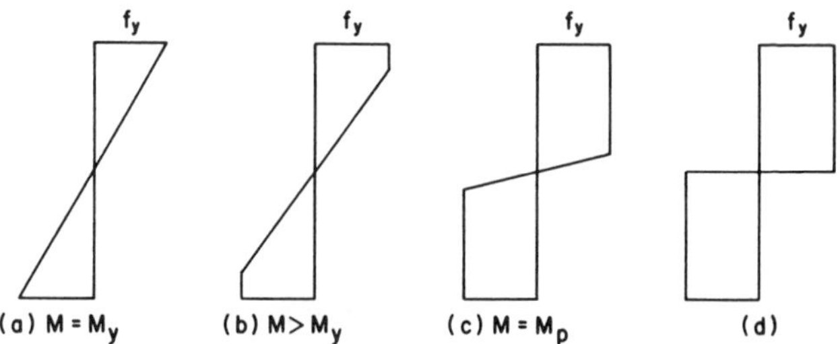

FIGURE 12.3 Progression of stress response: elastic to plastic.

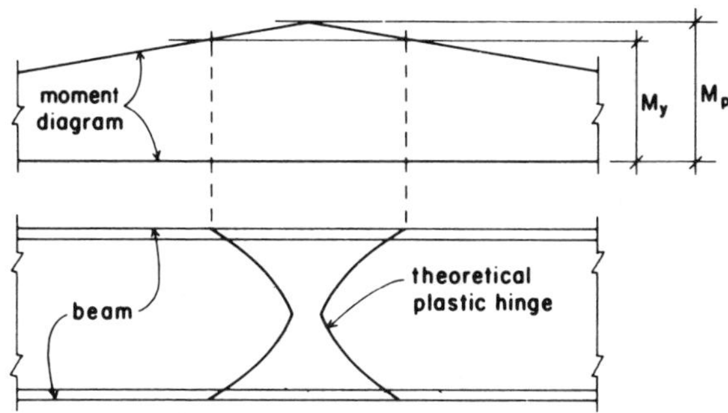

FIGURE 12.4 Development of the plastic hinge.

greater area of the cross section is also stressed to f_y. The stress distribution is shown in Fig. 12.3*b*.

Now imagine that the load is further increased. The stress on the extreme fiber is still f_y and, theoretically, *all fibers in the cross section are stressed to* f_y. This idealized plastic stress distribution is shown in Fig. 12.3*d*. The bending moment that produces this condition is M_p, the plastic bending moment. In reality about 10% of the central portion of the cross section continues to resist in an elastic manner, as indicated in Fig. 12.3*c*. This small resistance is quite negligible, and we assume that the stresses on all fibers of the cross section are f_y, as shown in Fig. 12.3*d*. The section is now said to be fully plastic, and any further increase in load will result in large deformations; the beam acts as if it were hinged at this section. We call this a plastic hinge, at which free rotation is permitted only after M_p has been attained (see Fig. 12.4). At sections of a beam in which this condition prevails, the bending resistance of the cross section has been exhausted.

12.3 Plastic Section Modulus

In elastic design the moment that produces the maximum allowable resisting moment may be found by the flexure formula

$$M = f \times S$$

where M = maximum allowable bending moment in inch-pounds,
f = maximum allowable bending stress in pounds per square inch,
S = section modulus in inches to the third power.

If the extreme fiber is stressed to the yield stress,

$$M_y = f_y \times S$$

where M_y = elastic bending moment at the yield stress,
f_y = yield stress in pounds per square inch,
S = section modulus in inches to the third power.

Now let us find a similar relation between the plastic moment and its plastic resisting moment. Refer to Fig. 12.5, which shows the cross section of a W or S section in which the bending stress f_y, the yield stress, is constant over the cross section. In the figure,

A_u = upper area of the cross section above the neutral axis, in square inches,
y_u = distance of the centroid of A_u from the neutral axis,
A_l = lower area of the cross section below the neutral axis, in square inches,
y_l = distance of the centroid of A_l from the neutral axis.

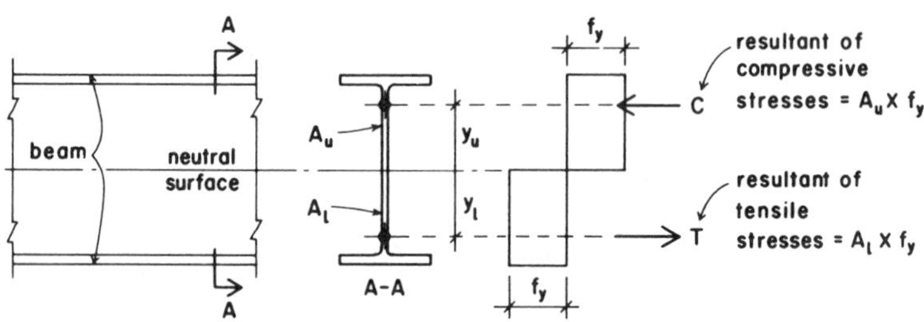

FIGURE 12.5 Development of the plastic resisting moment.

For equilibrium the algebraic sum of the horizontal forces must be zero. Then

$$\sum H = 0$$

or

$$[A_u \times (+f_y)] + [A_l \times (-f_y)] = 0$$

and

$$A_u = A_l$$

This shows that the neutral axis divides the cross section into equal areas, which is apparent in symmetrical sections, but it applies to unsymmetrical sections as well. Also the bending moment equals the sum of the moments of the stresses in the section. Thus for M_p, the plastic moment,

$$M_p = (A_u \times f_y \times y_u) + (A_l \times f_y \times y_l)$$

or

$$M_p = f_y[(A_u \times y_u) + (A_l \times y_l)]$$

and

$$M_p = f_y \times Z$$

The quantity $(A_u y_u + A_l y_l)$ is called the *plastic section modulus* of the cross section and is designated by the letter Z; because it is an area multiplied by a distance; it is in units to the third power. If the area is in units of square inches and the distance is in linear inches, Z, the section modulus, is in units of inches to the third power.

The plastic section modulus is always larger than the elastic section modulus.

It is important to note that in plastic design the neutral axis for unsymmetrical cross sections does not pass through the centroid of the section. In plastic design the neural axis divides the cross section into *equal* areas.

12.4 Computation of the Plastic Section Modulus

The notation used in Sec. 12.3 is appropriate for both symmetrical and unsymmetrical sections. Consider now a symmetrical section such as a W or S shape, as shown in Fig. 12.5]. $A_u = A_l$, $y_u = y_l$, and $A_u + A_l = A$, the total area of the cross section. Then

$$M_p = (A_u \times f_y \times y_u) + (A_l \times f_y \times y_l)$$

and

$$M_p = f_y \times A \times y \quad \text{or} \quad M_p = f_y \times Z$$

where f_y = yield stress,
A = total area of the cross section,
y = distance from the neutral axis to the centroid of the portion of the area on either side of the neutral axis,
Z = plastic modulus of the section (in in.^3 or mm^3).

Now, because $Z = A \times y$, we can readily compute the value of the plastic section modulus of a given cross section.

Consider a W 16 × 45. In Table 3.1 we find that its total depth is 16.13 in. [410 mm] and its cross-sectional area is 13.3 in.^2 [8581 mm^2]. From the AISC Manual we find that a WT 8 × 22.5 (which is one-half a W 16 × 45) has its centroid located 1.88 in. [48 mm] from the outside of the flange. Therefore the distance from the centroid of either half of the W shape to the neutral axis is one-half the beam depth less the distance obtained for the tee. Thus the distance y is (see Fig. 12.6)

$$y = \left(\frac{16.13}{2}\right) - 1.88 = 6.185 \text{ in. } [157 \text{ mm}]$$

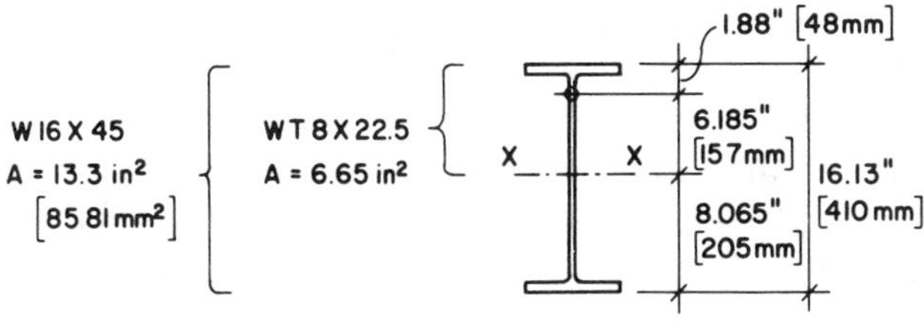

FIGURE 12.6

Then the plastic modulus of the W 16 × 45 is

$$Z = A \times y = 13.3 \times 6.185 = 82.26 \text{ in.}^3 \ [1347 \times 10^3 \text{ mm}^3]$$

Use of the full value of the plastic hinge moment requires the shape to have limited values for the width–thickness ratio of the flanges and the depth–thickness ratio of the web. These requirements are given in Section 2-7 of the AISC Specification.

Problem 12.4.A. Find the value for the plastic section modulus for a W 14 × 68. Distance y, as shown in Fig. 12.5, is 5.73 in. [146 mm].

12.5 Factored Loads

Consider a beam of A36 steel laterally supported throughout its length. Its span is 24 ft [7.315 m], and its carries a concentrated load of 42 kips [186.8 kN] at the center. Let us determine the size of the beam in accordance with the theory of elastic behavior.

The maximum bending moment for this beam is

$$M = \frac{PL}{4} = \frac{42 \times 24}{4} = 252 \text{ kip-ft } [342 \text{ kN-m}]$$

and the required section modulus is

$$S = \frac{M}{f_b} = \frac{252 \times 12}{24} = 126 \text{ in.}^3 \ [2070 \times 10^3 \text{ mm}^3]$$

Table 3.1 shows that a W 21 × 62 has a section modulus of 127 in.^{3} and is acceptable.

Now let us compute the magnitude of the concentrated load at the center of the span that would produce a bending moment equal to the plastic resisting moment. In the AISC Manual, the plastic modulus for the W 21 × 62 is 144 in.^{3}. Thus the plastic moment is

$$M_p = F_y \times Z_x = 36 \times 144 = 5184 \text{ kip-in., or } 432 \text{ kip-ft} \quad [585 \text{ kN-m}]$$

Then the load that corresponds to this moment is

$$M_p = 432 = \frac{PL}{4} = \frac{P \times 24}{4}$$

and

$$P = \frac{4 \times 432}{24} = 72 \text{ kips } [320 \text{ kN}]$$

This load would produce a plastic hinge at the center of the span and a slight increase in load would result in failure.

The term *load factor* is given to the ratio of the ultimate load to the design load. In this example it is 72/42 = 1.714.

In elastic design the allowable bending stress F_b is decreased to a fraction of F_y, the yield stress. For compact sections, $F_b = 0.66$ F_y. Therefore the implied factor of safety against yielding is 1/0.66, or 1.5. This factor is higher, of course, if the beam's limiting capacity is taken as the plastic moment.

In plastic design the concept of allowable stress is not used, and computations are based strictly on the limit of the yield stress. Safety is produced by the use of the load factor, by which the beam is literally designed to fail but at a load larger than that it actually must sustain. For simple and continuous beams, the load factor is specified as 1.7; for rigid frames it is 1.85.

12.6 Design of a Simple Beam

The design of simple beams by the elastic or plastic theory will usually result in the same size beam, as illustrated in the following examples.

Example 1. A simple beam of A36 steel has a span of 20 ft [6.1 m] and supports a uniformly distributed load of 4.8 kips/ft [70 kN/m], including its own weight. Design this beam in accordance with the elastic theory, assuming that it is laterally supported throughout its length.

Solution: The maximum bending moment is $wL^2/8$. Then

$$M = \frac{4.8 \times (20)^2}{8} = 240 \text{ kip-ft } [326 \text{ kN-m}]$$

By use of the allowable bending stress of 24 ksi [165 MPa] for a compact section, the required section modulus is

$$S = \frac{M}{F_b} = \frac{240 \times 12}{24} = 120 \text{ in.}^3 \ [1973 \times 10^3 \text{ mm}^3]$$

In Table 3.1 we find a W 21 × 68 with $S_x = 140$ in.^{3}. (*Note:* Use of the AISC Manual will yield a W 21 × 62 with $S_x = 127$ in.^{3}—see Table 8.1.)

Example 2. Design the same beam in accordance with the plastic theory.

Solution: We adjust the load with the load factor, which is given in Sec. 12.5 as 1.7. Thus

$$w_p = 4.8 \times 1.7 = 8.16 \text{ kips/ft } [119 \text{ kN/m}]$$

and the maximum bending moment is

$$M_p = \frac{w_p L^2}{8} = \frac{8.16 \times (20)^2}{8} = 408 \text{ kip-ft } [554 \text{ kN-m}]$$

and

$$M_p = F_y \times Z, Z = \frac{M_p}{F_y} = \frac{408 \times 12}{36} = 136 \text{ in.}^3$$

$$[2232 \times 10^3 \text{ mm}^3]$$

which is the minimum plastic section modulus. From the AISC Manual select a W 21 × 62 for which $Z_x = 144$ in.^{3}. Note that the elastic and plastic theories have yielded the same result in this example, which is common for simple beams.

Problem 12.6.A. A steel beam of A36 steel is used to support a load of 12 kips [54 kN], including its own weight, on a span of 32 ft [9.75 m]. Select the lightest-weight W section by (a) use of elastic theory and (b) use of the factored load and strength theory.

IV

CONCRETE AND MASONRY CONSTRUCTION

13

General Considerations for Concrete Structures

13.1 Introduction

Concrete is made by mixing a paste of cement and water with sand and crushed stone, gravel, or other inert material. The sand and other inert materials are called the *aggregate*. After this plastic mixture is placed in forms, a group of chemical reactions called hydration takes place and the mass hardens. Concrete, although strong in compression, is relatively weak in resisting tensile and shearing stresses which develop in structural members. To overcome this lack of resistance, steel bars are placed in the concrete at the proper positions; the result is *reinforced concrete*. In beams and slabs the principal function of the concrete is to resist compressive stresses, whereas the steel bars resist tensile stresses.

13.2 Design Methods

The design of reinforced concrete structural members may be accomplished by two different methods. The first, called *working stress design,* is the principal method used in this part of the book; the second method is known as *strength design*. At present the principal reference for the analysis and design of reinforced concrete members is *Building Code Requirements for Reinforced Concrete* (ACI 318-83) (Ref. 3), published by the American Concrete Institute, commonly referred to as the ACI Code. Most of the material in the present edition of the ACI Code (1983) is based on strength design methods. A limited treatment of working stress design is given in Appendix B of the code, under the title "Alternative Design Method." The last edition of the code to fully develop the working stress design method was the 1963 edition (ACI 318-63). The discussion of reinforced concrete design in Chapters 13–16 is keyed primarily to the 1963 code but also draws on the alternative method of the 1983 code. A brief introduction to strength design methods is presented in Sections 13.8 and 13.9.

The design of reinforced concrete members must, of course, be carried out in compliance with the building code that has jurisdiction in your locality. The ACI Code is extensive, and those who desire more complete information should examine it in detail. In this elementary book, space permits discussions of only the basic structural members, and many of the items referred to in the code must necessarily be omitted.

13.3 Strength of Concrete

The designer of a reinforced concrete structure bases his or her computations on the use of concrete having a specified compressive strength (2500, 3000, 3500, etc., psi) at the end of a 28-day curing period. The symbol for this specified compressive strength is f'_c. Concretes of different strengths are produced by varying the proportions of cement, fine aggregate (sand), coarse aggregate, and water in the mix. The general theory in establishing the proportions of fine and coarse aggregates is that the voids in the coarse aggregate should be filled with the cement paste and fine aggregate. The proportioning of concrete mixes and the attendant procedures for strength verification are not discussed in this book. Readers interested in studying this aspect of concrete manufacture should consult *Recommended Practice for Selecting Proportions for Normal and Heavy Weight Concrete* (ACI 211.1-74) and Section 4-2 of the 1983 ACI Code.

Very little concrete is proportioned and mixed at the building site today. Central or ready-mixed concrete is used whenever it is available. The use of a concrete mixed under ideal controlled conditions at a central plant affords many advantages. It is delivered to the building site in a revolving mixer, the proportions of cement, aggregate, and water are maintained accurately, any desired strength may be ordered, and the concrete thus provided is uniform in quality.

Table 13.1 gives allowable stresses in flexure (bending), shear, and bearing for concretes of four different specified compressive strengths. Note that the individual values are functions of f'_c.

An important factor affecting the strength of concrete is the *water–cement ratio*. This is expressed as the number of pounds of water per pound of cement used in the mix (or the number of gallons of water for each 94-lb bag of cement). The relationship is an inverse one, that is, lower values of the ratio produce higher strengths. However, in order to produce freshly mixed concrete that possesses *workability*—the property that controls the ease with which it can be placed in the forms and around the reinforcing bars— more water must be used than the amount required for hydration of the cement. The use of too much water may cause segregation of the mix components, thereby producing nonuniform concrete. To control this situation, building codes specify the maximum permissible water–cement ratios for concretes of specified design strengths.

13.4 Cement

The cement used most extensively in building construction is *portland cement*. Of the five types of standard portland cement generally available in the United States and for which the American Society for Testing and Materials has established specifications, two types account for most of the cement used in buildings. These are ASTM Type I, a general-purpose cement for use in concrete designed to reach its required strength in about 28 days,

TABLE 13.1 Allowable Stresses in Concrete

Description (All Values Given for Normal-Weight Concrete, 145 lb/ft³)		Allowable Stresses (psi): Based on Specified Concrete Strength	For Strength of Concrete f'_c Shown Below: 2000	3000	4000	5000	Values in MPa Based on Specified Concrete Strength
Modulus of elasticity	E_c	$57{,}000\sqrt{f'_c}$	2.55×10^6	3.12×10^6	3.60×10^6	4.03×10^6	$4730\sqrt{f'_c}$
Modular ratio: $n = \frac{E_s}{E_c}$	n	$\frac{29{,}000}{57\sqrt{f'_c}}$	11.3	9.2	8.0	7.2	$\frac{200{,}000}{4730\sqrt{f'_c}}$
Flexure							
Extreme fiber stress in compression	f_c	$0.45 f'_c$	900	1350	1800	2250	$0.45 f'_c$
Extreme fiber stress in tension, plain concrete walls and footings	f_c	$1.6\sqrt{f'_c}$	72	88	102	113	$0.13\sqrt{f'_c}$

Shear (carried by concrete)							
Beams, walls, one-way slabs	v_c	$1.1 \sqrt{f'_c}$	49	60	70	78	$0.09 \sqrt{f'_c}$
Joists	v_c	$1.2 \sqrt{f'_c}$	54	66	77	86	$0.10 \sqrt{f'_c}$
Two-wall slabs and footings (peripheral shear)	v_c	$2 \sqrt{f'_c}$	89	110	126	141	$0.17 \sqrt{f'_c}$
Bearing							
On full area	f_c	$0.3 f'_c$	600	900	1200	1500	$0.3 f'_c$
When supporting surface A_2 is wider on all sides than loaded area A_1	f_c	$0.3 f'_c \times \sqrt{A_2/A_1}$ (but not more than $0.6 f'_c$)					

Source: Most tabulated stresses are those specified in Appendix B, "Alternate Design Methods," *Building Code Requirements for Reinforced Concrete* (ACI 318-83) (Ref. 3), abstracted with permission of the publishers, American Concrete Institute. Extreme fiber stresses in tension in plain concrete walls and footings were abstracted from *Building Code Requirements for Reinforced Concrete* (ACI 318-63) with permission of the publishers, American Concrete Institute.

and ASTM Type III, a high-early-strength cement for use in concrete that attains its design strength in a period of a week or less. All portland cements set and harden by reacting with water, and this hydration process is accompanied by generation of heat. In massive concrete structures such as dams, the resulting temperature rise of the materials becomes a critical factor in both design and construction, but the problem is not usually significant in building construction. ASTM Type IV, a low-heat cement, is designed for use where the heat rise during hydration is a critical factor. It is, of course, essential that the cement actually used in construction corresponds to that employed in designing the mix, to produce the specified compressive strength of the concrete.

Air-entrained concrete is produced by using special cement and by introducing an additive during mixing of the concrete. In addition to improving workability (mobility of the wet mix), entrainment permits lower water–cement ratios and significantly improves the durability of the concrete. Air-entraining agents produce billions of microscopic air cells throughout the concrete mass. These minute voids prevent accumulation of water in cracks and other large voids which, on freezing, would permit the water to expand and result in spalling away of the exposed surface of the concrete.

13.5 Reinforcement

The steel used in reinforced concrete consists of round bars, mostly of the deformed type, with lugs or projections on their surfaces. The surface deformations help to develop a greater bond between the concrete and steel. The most common grades of reinforcing steel are Grade 60 and Grade 40, having yield strengths of 60,000 psi [414 MPa] and 40,000 psi [276 MPa] respectively. Table 13.2 gives the properties for standard deformed reinforcing bars.

Ample concrete protection, called *cover*, must be provided for the steel reinforcing. Cover is measured as the distance from the outside face of the concrete to the edge of a reinforcing bar. For reinforcement near surfaces not exposed to the ground or to weather, cover should be not less than $\frac{3}{4}$ in. [19 mm] for slabs,

ABLE 13.2 Properties of Standard Deformed Reinforcing Bars

	Nominal Dimensions					
	U.S. Units			SI Units		
Bar esignation No.	Diameter (in.)	Cross-Sectional Area (in.^{2})	Perimeter (in.)	Diameter (mm)	Cross-Sectional Area (mm^2)	Perimeter (mm)
3	0.375	0.11	1.178	9.52	71	29.9
4	0.500	0.20	1.571	12.70	129	39.9
5	0.625	0.31	1.963	15.88	200	49.9
6	0.750	0.44	2.356	19.05	284	59.8
7	0.875	0.60	2.749	22.22	387	69.8
8	1.000	0.79	3.142	25.40	510	79.8
9	1.128	1.00	3.544	28.65	645	90.0
10	1.270	1.27	3.990	32.26	819	101.3
11	1.410	1.56	4.430	35.81	1006	112.5
14	1.693	2.25	5.32	43.00	1452	135.1
18	2.257	4.00	7.09	57.33	2581	180.1

walls, and joists, and 1.5 in. [38 mm] for beams, girders, and columns. Where formed surfaces are exposed to earth or weather, the cover should be 1.5 in. [3.8 mm] for No. 5 bars and smaller and 2 in. [51 mm] for No. 6–No. 18 bars. For foundation construction poured directly against ground without forms, cover should be 3 in. [76 mm].

Where multiple bars are used in members (which is the common situation), there are both upper and lower limits for the spacing of the bars. Lower limits are intended to permit adequate development of the concrete-to-steel stress transfers and to facilitate the flow of the wet concrete during pouring. For columns, the minimum clear distance between bars is specified as 1.5 times the bar diameter or a minimum of 1.5 in. For other situations, the minimum is one bar diameter or a minimum of 1 in.

For walls and slabs, maximum center-to-center bar spacing is specified as three times the wall or slab thickness or a maximum of 18 in. This applies to reinforcement required for computed stresses. For reinforcement that is required for control of cracking due to shrinkage or temperature change, the maximum spacing is five times the wall or slab thickness or a maximum of 18 in.

For adequate placement of the concrete, the largest size of the coarse aggregate should be not greater than three-quarters of the clear distance between bars.

The essential purpose of steel reinforcing is to prevent the cracking of the concrete due to tension stresses. In the design of concrete structures, investigation is made for the anticipated structural actions that will produce tensile stress: primarily the actions of bending, shear, and torsion. However, tension can also be induced by shrinkage of the concrete during its drying out after the initial pour. Temperature variations may also induce tension in various situations. To provide for these latter actions, the ACI Code requires a minimum amount of reinforcing in members such as walls and slabs even when structural actions do not indicate any need. These requirements are discussed in the sections that deal with the design of these members.

In the design of most reinforced concrete members the amount of steel reinforcing required is determined from computations and represents the amount determined to be necessary to resist the required tensile force in the member. In various situations, however, there is a minimum amount of reinforcing that is desirable, which may on occasion exceed that determined by the computations. The ACI Code makes provisions for such minimum reinforcing in columns, beams, slabs, and walls. The minimum reinforcing may be specified as a minimum percentage of the member cross-sectional area, as a minimum number of bars, or as a minimum bar size. These requirements are discussed in the sections that deal with the design of the various types of members.

13.6 Modulus of Elasticity

The modulus of elasticity E_c of hardened concrete is a measure of its resistance to deformation. The magnitude of E_c depends on w, the weight of the concrete, and on f'_c, its strength. Its value may be determined from the expression $E_c = w^{1.5}33\sqrt{f'_c}$ for values of w between 90 and 155 lb/ft^3. For normal weight concrete (145 lb/ft^3), E_c may be considered as equal to $57{,}000\sqrt{f'_c}$. [$E_c = w^{1.5}\,0.043\sqrt{f'_c}$

for values of w between 1440 and 2480 kg/m^3. For normal-weight concrete (2320 kg/m^3), E_c may be considered as equal to 4730 $\sqrt{f'_c}$.]

In the design of reinforced concrete members, we employ the term n. This is the ratio of the modulus of elasticity of steel to that of concrete, or $n = E_s/E_c$. E_s is taken as 29,000 ksi [200,000 MPa].

Consider a concrete for which f'_c is 4000 psi and w is 145 lb/ft^3. Then $E_c = 57{,}000\sqrt{f'_c} = 57{,}000\sqrt{4000} = 3{,}600{,}000$ psi and $n = E_s/E_c = 29{,}000/3{,}600 = 8.055$. The values for n for four different strengths of concrete are given in Table 13.1 As is the usual practice, the values for n are rounded off to those given in the table.

13.7 Beams: Working Stress Method

Figure 13.1*a* represents a simple beam loaded with two concentrated loads. The reactions at the left and right ends are, respectively, R_1 and R_2. In accordance with the laws of equilibrium the algebraic sum of the vertical forces equals zero, the algebraic sum of the horizontal forces equals zero, and the algebraic sum of the

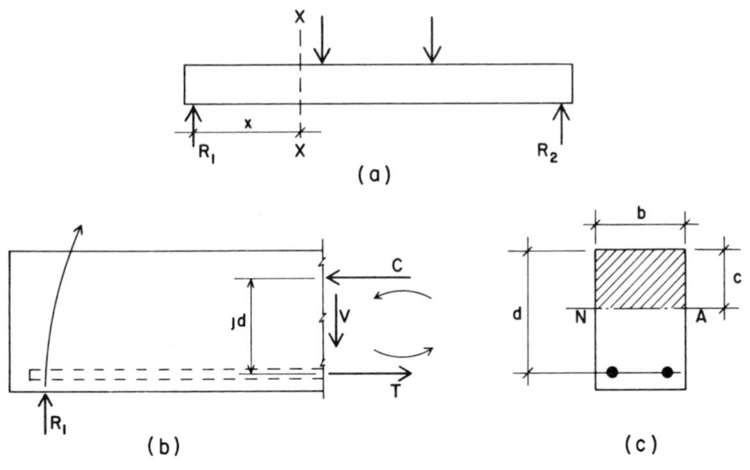

FIGURE 13.1

moments of the forces equals zero. These three laws are expressed in this manner: $\sum V = 0$, $\sum H = 0$, and $\sum M = 0$. For the present we are concerned only with the law relating to moments.

The resisting moment in a reinforced concrete beam is created by the development of internal stresses which may be represented as a resultant tension T and a resultant compression C, acting on a section X–X taken through the beam (Fig. 13.1*b*). These two forces constitute a resisting mechanical couple that tends to rotate the beam at section X–X in a counterclockwise direction, thereby opposing the tendency of the bending moment $R_1 \times x$ to cause clockwise rotation. Under equilibrium conditions the resisting moment is always equal to the bending moment. Consequently, if we wish to design a beam for a given loading condition, we arrange its concrete dimensions and the steel reinforcement so that it is capable of developing a resisting moment equal to the maximum bending moment caused by the loads.

Because the reinforcing steel is assumed to carry all the tension, T is located at the centroid of the reinforcement. The force C, however, is the resultant of compressive stresses in the concrete distributed over some portion of the beam depth (c in Fig. 13.1*c*), which will be established later. The *effective depth* of the beam is the distance from the top (compression) face to the centroid of the steel and is denoted by the symbol d. The additional depth below the reinforcement is not considered in calculations; it provides fire and moisture protection for the steel and assists in developing bond between the concrete and steel. The distance between C and T is called the *arm* of the resisting couple; it is represented by the symbol jd. The shear force V indicated in Fig. 13.1*b* is considered in Sec. 13.13.

We know from structural mechanics that when a simple beam is subjected to flexure, as indicated in Fig. 13.1, the upper portion of the member is in compression and the lower portion in tension. Furthermore, there is a horizontal plane separating the compressive and tensile stresses known as the *neutral surface;* at this plane the value of the bending stress is zero. The line in which the neutral surface intersects the beam cross section is called the *neutral axis* (*NA* in Fig. 13.1*c*) and its distance from the top of the beam is denoted by c.

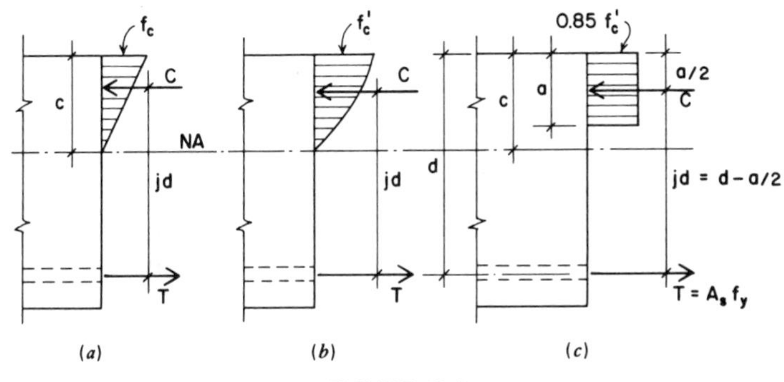

FIGURE 13.2

Deferring consideration of the way in which the value of c is determined, let us direct out attention to Fig. 13.2, which shows three different assumed distributions of stress on the compression side of the neutral axis. The diagram in Fig. 13.2*a* illustrates straight-line distribution in accordance with elastic theory. The stress varies directly as the distance from the neutral axis, at which it is zero, and increases to a maximum value at the compression face of the beam. This value is called the *extreme fiber stress* and is denoted by the symbol f_c. Figure 13.2*b* illustrates a parabolic distribution when the value of the extreme fiber stress has reached f'_c, the *specified compressive strength* (ultimate strength) of the concrete. This corresponds to inelastic behavior of concrete with an assumed parabolic stress–strain diagram. Figure 13.2*c* shows a rectangular compressive stress distribution which is assumed to be equilvalent in its static effect to the parabolic pattern. The rectangular "stress block" is based on the assumption that a concrete stress of $0.85\ f'_c$ is uniformly distributed over a part of the compression zone with dimensions equal to the beam width b and the distance a which locates a line parallel to and above the neutral axis. This is the stress distribution most often used in design under the ultimate strength theory.

In working stress design a maximum allowable (working) value for the extreme fiber stress is established (Table 13.1) and the formulas are predicated on elastic behavior of the reinforced concrete member under service load. The straight-line distribution of compressive stress is valid at working stress levels because the stresses developed vary approximately with the distance from the neutral axis, in accordance with elastic theory. Shrinkage and cracking of the concrete, however, together with the phenomenon of creep under sustained loading, complicate the stress distribution. Over time, stresses computed in reinforced concrete members on the basis of elastic theory are not realistic. Generally speaking, the acceptable safety of the working stress design method is maintained by the differentials provided between the allowable compressive stress f_c and the specified compressive strength of the concrete f'_c and between the allowable tensile stress f_s and the yield strength of the steel reinforcement f_y. These, in effect, are measures of safety.

The 1963 Code of the American Concrete Institute contained separate sections that covered working stress design and ultimate strength design, but the 1983 Code, promulgated as *Building Code Requirements for Reinforced Concrete* (ACI 318-83) (Ref. 3), is built primarily around ultimate strength design. Among reasons contributing to the displacement of WSD in favor of USD is the greater uniformity in safety factors for beams and columns with different loading conditions that is yielded by the latter method. The 1983 ACI Code uses the designation *strength design method* for USD.

The 1983 Code also permits an *alternate design method* that is similar to many of the working stress design procedures of the 1963 Code. For ordinary beams and girders flexural computations are identical with those for WSD under the 1963 Code. There are significant differences, however, in other areas such as design for shear, anchorage length of reinforcement, and design of columns.

The ACI Code is extensive and covers many aspects of ultimate strength design. Only a limited number of its provisions, however, can be discussed in a book of this nature. The treatment herein focuses on design procedures for the common structural elements that occur frequently in building construction, and many items referred to in the Code must necessarily be omitted.

Flexural Formulas: Working Stress Method. The following is a presentation of the formulas and procedures used in the work-

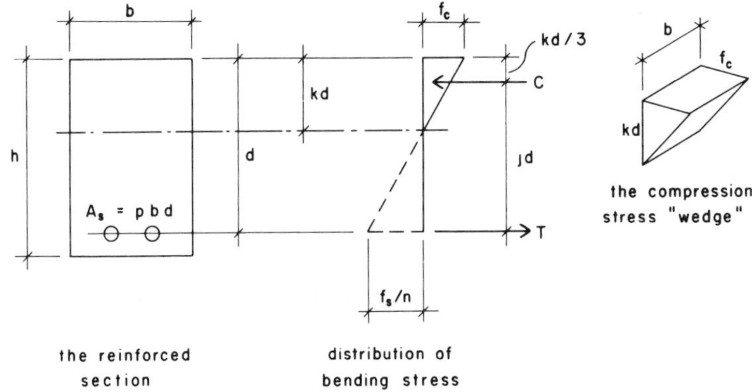

FIGURE 13.3 Moment of resistance of a rectangular concrete section with tension reinforcement.

ing stress method. The discussion is limited to a rectangular beam section with tension reinforcing only.

Referring to Fig. 13.3, the following are defined:

b = width of the concrete compression zone,

d = effective depth of the section for stress analysis; from the centroid of the steel to the edge of the compression zone,

A_s = cross-sectional area of the reinforcing,

p = perentage of reinforcing, defined as

$$p = \frac{A_s}{bd}$$

n = elastic ratio = $\dfrac{E \text{ of the steel reinforcing}}{E \text{ of the concrete}}$,

kd = height of the compression stress zone; used to locate the neutral axis of the stressed section; expressed as a percentage (k) of d,

jd = internal moment arm, between the net tension force and the net compression force; expressed as a percentage (j) of d,

f_c = maximum compressive stress in the concrete,
f_s = tensile stress in the reinforcing.

The compression force C may be expressed as the volume of the compression stress "wedge," as shown in the figure.

$$C = \tfrac{1}{2}(kd)(b)(f_c) = \tfrac{1}{2}kf_cbd$$

Using the compression force, we may express the moment resistance of the section as

$$M = Cjd = (\tfrac{1}{2}kf_cbd)(jd) = \tfrac{1}{2}kjf_cbd^2 \quad \text{(Formula 13.7.1)}$$

This may be used to derive an expression for the concrete stress:

$$f_c = \frac{2M}{kjbd^2} \quad \text{(Formula 13.7.2)}$$

The resisting moment may also be expressed in terms of the steel and the steel stress as

$$M = Tjd = (A_s)(f_s)(jd)$$

This may be used for determination of the steel stress or for finding the required area of steel:

$$f_s = \frac{M}{A_sjd} \quad \text{(Formula 13.7.3)}$$

$$A_s = \frac{M}{f_sjd} \quad \text{(Formula 13.7.4)}$$

A useful reference is the so-called balanced section, which occurs when use of the exact amount of reinforcing results in the simultaneous limiting stresses in the concrete and steel. The properties which establish this relationship may be expressed as follows:

$$\text{Balanced } k = \frac{1}{1 + f_s/nf_c} \quad \text{(Formula 13.7.5)}$$

$$j = 1 - \frac{k}{3} \qquad \text{(Formula 13.7.6)}$$

$$p = \frac{f_c k}{2f_s} \qquad \text{(Formula 13.7.7)}$$

$$M = Rbd^2 \qquad \text{(Formula 13.7.8)}$$

in which

$$R = \tfrac{1}{2}\, kjf_c \qquad \text{(Formula 13.7.9)}$$

derived from Formula (1). If the limiting compression stress in the concrete ($f_c = 0.45\, f'_c$) and the limiting stress in the steel are entered in Formula (5), the balanced section value for k may be found. Then the corresponding values for j, p, and R may be found. The balanced p may be used to determine the maximum amount of tensile reinforcing that may be used in a section without the addition of compressive reinforcing. If less tensile reinforcing is used, the moment will be limited by the steel stress, the maximum stress in the concrete will be below the limit of $0.45\, f'_c$, the value of k will be slightly lower than the balanced value, and the value of j will be slightly higher than the balanced value. These relationships are useful in design for the determination of approximate requirements for cross sections.

Table 13.3 gives the balanced section properties for various combinations of concrete strength and limiting steel stress. The values of n, k, j, and p are all without units. However, R must be expressed in particular units; the unit used in the table is kip-inches (kip-in.).

When the area of steel used is less than the balanced p, the true value of k may be determined by the following formula:

$$k = \sqrt{2np - (np)^2} - np \qquad \text{(Formula 13.7.10)}$$

Figure 13.4 may be used to find approximate k values for various combinations of p and n. Beams with reinforcement less than that required for the balanced moment are called *underbalanced sections* or *under-reinforced sections*. If a beam must carry bending moment in excess of the balanced moment for the section, it is

TABLE 13.3 Balanced Section Properties for Rectangular Concrete Sections with Tension Reinforcing Only

f_s (ksi)	f_s (MPa)	f'_c (ksi)	f'_c (MPa)	n	k	j	p	R (kip-in.)	R (kN-m)
16	110	2.0	13.79	11.3	0.389	0.870	0.0109	0.152	1045
		2.5	17.24	10.1	0.415	0.862	0.0146	0.201	1382
		3.0	20.68	9.2	0.437	0.854	0.0184	0.252	1733
		4.0	27.58	8.0	0.474	0.842	0.0266	0.359	2468
20	138	2.0	13.79	11.3	0.337	0.888	0.0076	0.135	928
		2.5	17.24	10.1	0.362	0.879	0.0102	0.179	1231
		3.0	20.68	9.2	0.383	0.872	0.0129	0.226	1554
		4.0	27.58	8.0	0.419	0.860	0.0188	0.324	2228
24	165	2.0	13.79	11.3	0.298	0.901	0.0056	0.121	832
		2.5	17.24	10.1	0.321	0.893	0.0075	0.161	1107
		3.0	20.68	9.2	0.341	0.886	0.0096	0.204	1403
		4.0	27.58	8.0	0.375	0.875	0.0141	0.295	2028

necessary to provide some compressive reinforcement, as discussed in Sec. 13.11. The balanced section is not necessarily a design ideal, but is useful in establishing the limits for the section.

In the design of concrete beams, there are two situations that commonly occur. The first occurs when the beam is entirely undetermined; that is, the concrete dimensions and the reinforcing are unknown. The second occurs when the concrete dimensions are given, and the required reinforcing for a specific bending moment must be determined. The following examples illustrate the use of the formulas just developed for each of these problems.

Example 1. A rectangular concrete beam of concrete with f'_c of 3000 psi [20.7 MPa] and steel reinforcing with f_s = 20 ksi [138 MPa] must sustain a bending moment of 200 kip-ft [271 kN-m]. Select the beam dimensions and the reinforcing for a section with tension reinforcing only.

Solution: (1) With tension reinforcing only, the minimum size beam will be a balanced section, since a smaller beam would have to be stressed beyond the capacity of the concrete to develop the

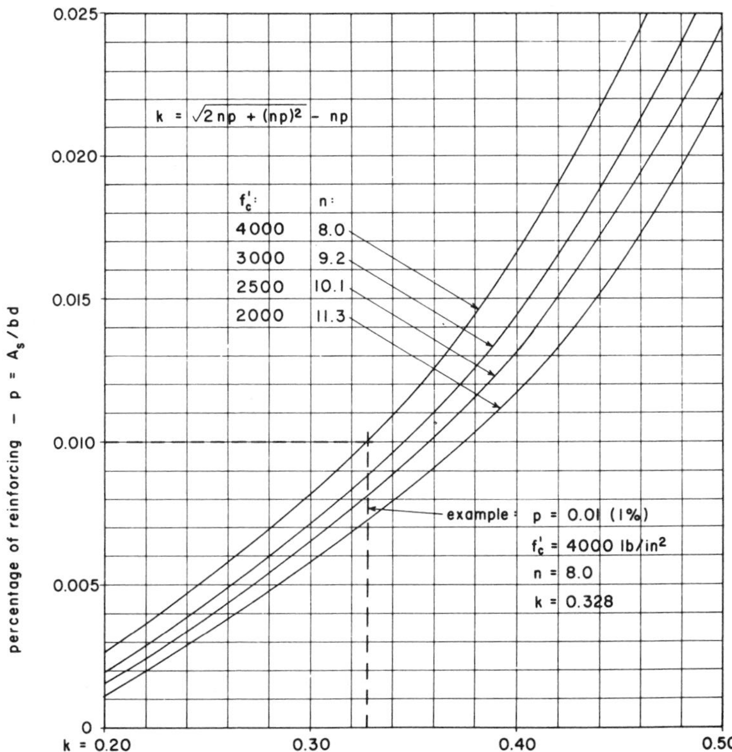

FIGURE 13.4 *k* factors for rectangular concrete sections with tension reinforcing—as a function of *p* and *n*.

required moment. Using Formula (8),

$$M = Rbd^2 = 200 \text{ kip-ft } [271 \text{ kN-m}]$$

Then from Table 13.3, for f'_c of 3000 psi and f_s of 20 ksi,

$$R = 0.226 \text{ (in units of kip-in.) } [1554 \text{ in units of kN-m}]$$

Therefore

$$M = 200 \times 12 = 0.226(bd^2), \text{ and } bd^2 = 10{,}619$$

(2) Various combinations of b and d may be found; for example,

$$b = 10 \text{ in.}, \quad d = \sqrt{\frac{10{,}619}{10}}$$

$$= 32.6 \text{ in. } [b = 0.254 \text{ m}, d = 0.829 \text{ m}]$$

$$b = 15 \text{ in.}, \quad d = \sqrt{\frac{10{,}619}{15}}$$

$$= 26.6 \text{ in. } [b = 0.381 \text{ m}, d = 0.677 \text{ m}]$$

Although they are not given in this example, there are often some considerations other than flexural behavior alone that influence the choice of specific dimensions for a beam. These situations are discussed in Chapter 14. If the beam is of the ordinary form shown in Fig. 13.3, the specified dimension is usually that given as h. Assuming the use of a No. 3 U-stirrup, a cover of 1.5 in. [38 mm], and an average-size reinforcing bar of 1-in. [25-mm] diameter (No. 8 bar), the design dimension d will be less than h by 2.375 in. [60 mm]. Lacking other considerations, we will assume a b of 15 in. [380 mm] and an h of 29 in. [740 mm], with the resulting d of 29 − 2.375 = 26.625 in. [680 mm].

(3) We next use the specific value for d with Formula (4) to find the required area of steel A_s. Since our selection is very close to the balanced section, we may use the value of j from Table 13.3. Thus

$$A_s = \frac{M}{f_s jd} = \frac{200 \times 12}{20 \times 0.872 \times 26.625} = 5.17 \text{ in.}^2 \ [3312 \text{ mm}^2]$$

Or using the formula for the definition of p and the balanced p value from Table 13.3,

$$A_s = pbd = 0.0129(15 \times 26.625) = 5.15 \text{ in.}^2 \ [3333 \text{ mm}^2]$$

(4) We next select a set of reinforcing bars to obtain this area. As with the beam dimensions, there are other concerns, as discussed in Chapter 14. For the purpose of our example, if we

select bars all of a single size (see Table 13.2), the number required will be:

$$\text{For No. 6 bars, } \frac{5.17}{0.44} = 11.75, \text{ or } 12 \quad \left[\frac{3312}{284} = 11.66\right]$$

$$\text{For No. 7 bars, } \frac{5.17}{0.60} = 8.62, \text{ or } 9 \quad \left[\frac{3312}{387} = 8.56\right]$$

$$\text{For No. 8 bars, } \frac{5.17}{0.79} = 6.54, \text{ or } 7 \quad \left[\frac{3312}{510} = 6.49\right]$$

$$\text{For No. 9 bars, } \frac{5.17}{1.00} = 5.17, \text{ or } 6 \quad \left[\frac{3312}{645} = 5.13\right]$$

$$\text{For No. 10 bars, } \frac{5.17}{1.27} = 4.07, \text{ or } 5 \quad \left[\frac{3312}{819} = 4.04\right]$$

$$\text{For No. 11 bars, } \frac{5.17}{1.56} = 3.31, \text{ or } 4 \quad \left[\frac{3312}{1006} = 3.29\right]$$

In real design situations there are always various additional considerations that influence the choice of the reinforcing bars. One general desire is that of having the bars in a single layer, as this keeps the centroid of the steel as close as possible to the edge (bottom in this case) of the member, giving the greatest value for d with a given height of concrete section. With the section as shown in Fig. 13.5, a beam width of 15 in. will yield a net width of

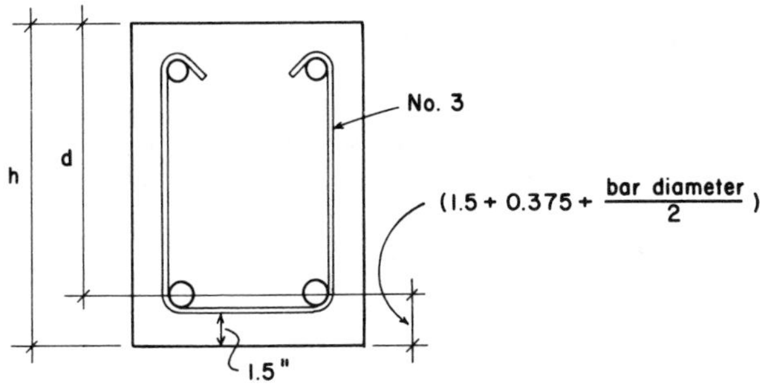

FIGURE 13.5

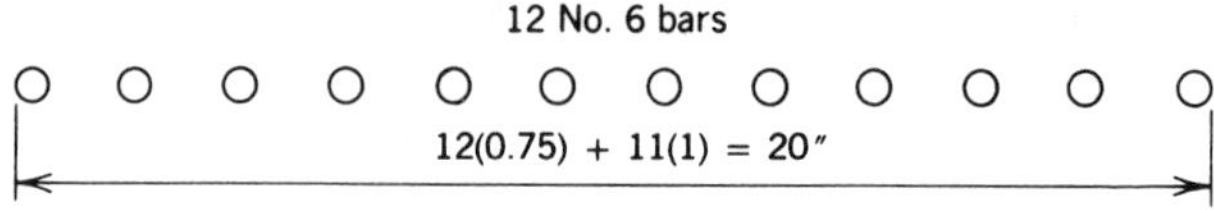

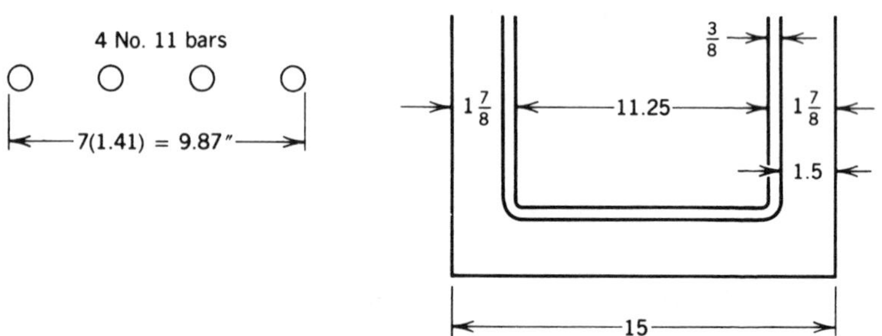

FIGURE 13.6 Beam width considerations for a single layer of bars.

11.25 in. inside the No. 3 stirrups. (Outside width of 15 − 2 × 1.5 cover and 2 × 0.375 stirrup diameter.) Applying the criteria for minimum spacing as described in Sec. 13.5, the required width for the various bar combinations can be determined. Two examples for this are shown in Fig. 13.6. It will be found that the four No. 11 bars are the only choice that will fit this beam width.

Example 2. A rectangular concrete beam of concrete with f'_c of 3000 psi [20.7 MPa] and steel with f_s of 20 ksi [138 MPa] has dimensions of $b = 15$ in. [380 mm] and $h = 36$ in. [910 mm]. Find the area required for the steel reinforcing for a moment of 200 kip-ft [271 kN-m].

Solution: The first step in this case is to determine the balanced moment capacity of the beam with the given dimensions. If we assume the section to be as shown in Fig. 13.5, we may assume an approximate value for d to be h minus 2.5 in. [64 mm], or 33.5 in. [851 mm]. Then with the value for R from Table 13.3,

$$M = Rbd^2 = 0.226 \times 15 \times (33.5)^2 = 3804 \text{ kip-in.}$$

or $$M = \frac{3804}{12} = 317 \text{ kip-ft}$$

$$[M = 1554 \times 0.380 \times (0.850)^2 = 427 \text{ kN-m}]$$

Since this value is considerably larger than the required moment, it is thus established that the given section is larger than that required for a balanced stress condition. As a result, the concrete flexural stress will be lower than the limit of $0.45 f'_c$, and the section is qualified as being under-reinforced, which is to say that the reinforcing required will be less than that required to produce a balanced section (with moment capacity of 317 kip-ft). In order to find the required area of steel, we use Formula (4) just as we did in the preceding example. However, the true value for j in the formula will be something greater than that for the balanced section (0.872 from Table 13.3).

As the amount of reinforcing in the section decreases below the full amount required for a balanced section, the value of k decreases and the value of j increases. However, the range for j is small: from 0.872 up to something less than 1.0. A reasonable procedure is to assume a value for j, find the corresponding required area, and then perform an investigation to verify the assumed value for j, as follows. Assume $j = 0.90$. Then

$$A_s = \frac{M}{f_s jd} = \frac{200 \times 12}{20 \times 0.90 \times 33.5} = 3.98 \text{ in.}^2 \ [2567 \text{ mm}^2]$$

and

$$p = \frac{A_s}{bd} = \frac{3.98}{15 \times 33.5} = 0.00792$$

Using this value for p in Fig. 13.4, we find $k = 0.313$. Using Formula (6) we then determine j to be

$$j = 1 - \frac{k}{3} = 1 - \frac{0.313}{3} = 0.896$$

which is reasonably close to our assumption, so the computed area is adequate for design.

Problem 13.7.A. A rectangular concrete beam has concrete with f'_c = 3000 psi [20.7 MPa] and steel reinforcing with f_s = 20 ksi [138 MPa]. Select the beam dimensions and reinforcing for a balanced section if the beam sustains a bending moment of 240 kip-ft [325 kN-m].

Problem 13.7.B. Find the area of steel reinforcing required and select the bars for the beam in Problem 13.7.A if the section dimensions are b = 16 in. and d = 32 in.

13.8 Strength Method: General Considerations

Application of the working stress method consists of designing members to *work* in an adequate manner (without exceeding established stress limits) under actual service load conditions. The basic procedure in strength design is to design members to *fail;* thus the ultimate strength of the member at failure (called its design strength) is the only type of resistance considered. Safety in strength design is not provided by limiting stresses, as in the working stress method, but by using a factored design load (called the *required strength*) that is greater than the service load. The code establishes the value of the required strength, called U, as not less than

$$U = 1.4D + 1.7L$$

in which D = the effect of dead load,
L = the effect of live load.

Other adjustment factors are provided when design conditions involve consideration of the effects of wind, earth pressure, differential settlement, creep, shrinkage, or temperature change.

The design strength of structural members (i.e., their *usable* ultimate strength) is determined by the application of assumptions and requirements given in the code and is further modified by the use of a *strength reduction factor* ϕ as follows:

ϕ = 0.90 for flexure, axial tension, and combinations of flexure and tension

= 0.75 for columns with spirals

= 0.70 for columns with ties

= 0.85 for shear and torsion

= 0.70 for compressive bearing

= 0.65 for flexure in plain (not reinforced) concrete

Thus while the formula for U may imply a relatively low safety factor, an additional margin of safety is provided by the stress-reduction factors.

13.9 Beams: Strength Method

Figure 13.7 shows the rectangular "stress block" that was described in Sec. 13.7 and is used for analysis of the rectangular section with tension reinforcing only by the strength method. This is the basis for investigation and design as provided for in the ACI Code.

The rectangular stress block is based on the assumption that a concrete stress of $0.85f'_c$ is uniformly distributed over the compression zone, which has dimensions equal to the beam width b

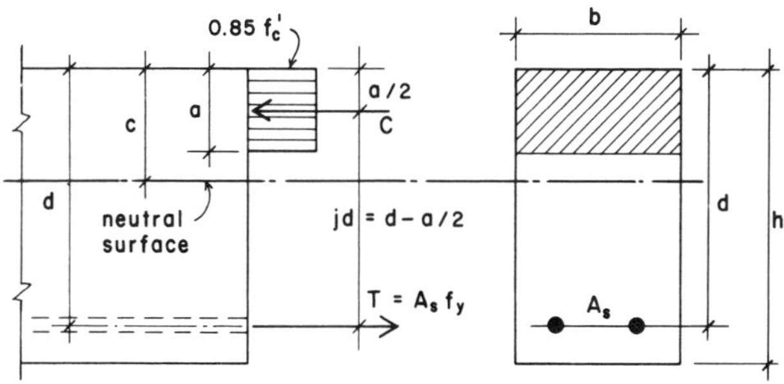

FIGURE 13.7

and the distance a which locates a line parallel to and above the neutral axis. The value of a is determined from the expression $a = \beta_1 \times c$, where β_1 (beta one) is a factor that varies with the compressive strength of the concrete, and c is the distance from the extreme fiber to the neutral axis. For concrete having f'_c equal to or less than 4000 psi [27.6 MPa], the Code gives $a = 0.85\ c$.

With the rectangular stress block, the magnitude of the compressive force in the concrete is expressed as

$$C = 0.85f'_c \times b \times a$$

and it acts at a distance of $a/2$ from the top of the beam. The arm of the resisting force couple then becomes $d - (a/2)$, and the developed resisting moment as governed by the concrete is

$$M_t = C\left(d - \frac{a}{2}\right) = 0.85f'_c ba \times \left(d - \frac{a}{2}\right)$$

With T expressed as $A_s \times f_y$, the developed moment as governed by the reinforcing is

$$M_t = T\left(d - \frac{a}{2}\right) = A_s f_y \left(d - \frac{a}{2}\right)$$

A formula for the dimension a of the stress block can be derived by equating the compression and tension forces; thus

$$0.85f'_c ba = A_s f_y \qquad \text{and} \qquad a = \frac{A_s f_y}{0.85f'_c b}$$

Expressing the area of steel in terms of a percentage ρ, the formula for a may be modified as follows:

$$\rho = \frac{A_s}{bd} \qquad \text{or} \qquad A_s = \rho bd$$

$$a = \frac{(\rho bd)f_y}{0.85f'_c b} = \frac{\rho d f_y}{0.85f'_c}$$

The balanced section for strength design is visualized in terms of strain rather than stress. The limit for a balanced section is expressed in the form of the percentage of steel required to produce balanced conditions. The formula for this percentage is

$$\rho_b = \frac{0.85 f'_c}{f_y} \times \frac{87}{87 + f_y}$$

in which f'_c and f_y are in units of ksi.

The ACI Code limits the percentage of steel to 75% of this balanced value in beams with tension reinforcing only.

Returning to the formula for the developed resisting moment, as expressed in terms of the steel, we see that a useful formula may be derived as follows:

$$M_t = A_s f_y \left(d - \frac{a}{2}\right) = (\rho b d)(f_y)\left(d - \frac{a}{2}\right)$$

$$= (\rho b d)(f_y)(d)\left(1 - \frac{1}{2}\frac{a}{d}\right)$$

$$= (bd^2)\left[\rho f_y \left(1 - \frac{1}{2}\frac{a}{d}\right)\right]$$

$$= Rbd^2$$

where $R = \rho f_y \left(1 - \frac{1}{2}\frac{a}{d}\right)$.

With the reduction factor applied, as discussed in Sec. 13.8, the design moment for a section is limited to nine-tenths of the theoretical resisting moment.

Values for the balanced section factors—ρ, R, and a/d—are given in Table 13.4 for various combinations of f'_c and f_y. The balanced section, as discussed in the preceding section, is not necessarily a practical one for design. In most cases economy will be achieved by using less than the balanced reinforcing for a given concrete section. In special circumstances it may also be possi-

TABLE 13.4 Balanced Section Factors—Strength Design

f'_c		f_y = 40 ksi [276 MPa]					f_y = 60 ksi [414 MPa]				
		Balanced	Usable a/d	Usable	Usable R		Balanced	Usable a/d	Usable	Usable R	
psi	MPa	a/d	(75% Balance)	ρ	k-in.	kN-m	a/d	(75% Balance)	ρ	k-in.	kN-m
2000	13.79	0.5823	0.4367	0.0186	0.580	4000	0.5031	0.3773	0.0107	0.520	3600
2500	17.24	0.5823	0.4367	0.0232	0.725	5000	0.5031	0.3773	0.0137	0.650	4500
3000	20.69	0.5823	0.4367	0.0278	0.870	6000	0.5031	0.3773	0.0160	0.781	5400
4000	27.58	0.5823	0.4367	0.0371	1.161	8000	0.5031	0.3773	0.0214	1.041	7200
5000	34.48	0.5480	0.4110	0.0437	1.388	9600	0.4735	0.3551	0.0252	1.241	8600

ble, or even desirable, to use compressive reinforcing in addition to tension reinforcing. Nevertheless, just as in the working stress method, the balanced section is often a useful reference when design is performed. The following example illustrates a procedure for the design of a simple rectangular beam section with tension reinforcing only.

Example. The service load bending moments on a beam are 58 kip-ft [78.6 kN-m] for dead load and 38 kip-ft [51.5 kN-m] for live load. The beam is 10 in. [254 mm] wide, f'_c is 4000 psi [27.6 MPa], and f_y is 60 ksi [414 MPa]. Determine the depth of the beam and the tensile reinforcing required.

Solution: (1) The first step is to determine the design moment, using the load factors, as discussed in Sec. 13.8. Thus

$$U = 1.4D + 1.7L$$

$$M_u = 1.4(M_{DL}) + 1.7(M_{LL})$$

$$= 1.4(58) + 1.7(38) = 145.8 \text{ kip-ft } [197.7 \text{ kN-m}]$$

(2) With the capacity reduction factor of 0.90 applied, as discussed in Sec. 13.8, the required moment capacity of the section is determined as

$$M_t = \frac{M_u}{0.90} = \frac{145.8}{0.90} = 162 \text{ kip-ft} \quad \text{or} \quad 1944 \text{ kip-in. } [220 \text{ kN-m}]$$

(3) From Table 13.4 we obtain the following factors for a balanced section: $a/d = 0.377$, $\rho = 0.0214$, $R = 1042$. To find the minimum required effective depth d, we use the formula for resisting moment,

$$M = Rbd^2 = (1042)(10)d^2 = 1{,}944{,}000$$

from which

$$d^2 = 186.6, \qquad d = \sqrt{186.6} = 13.7 \text{ in.}$$

If the limiting depth is used, the required steel area may be found using the balanced percentage factor, thus

$$A_s = \rho(bd) = 0.0214(10 \times 13.7) = 2.93 \text{ in.}^2$$

If a greater depth than that required for the balanced section is used, the balanced factors from Table 13.4 for a/d and ρ will not apply. Design aids are available from various handbooks to assist in the design of such a section. However, a relatively simple approach is to assume some value for a and to use it in a simple two-step procedure. To estimate a value for a, we first determine the balanced value for a as follows:

$$a/d = 0.377, \qquad a = 0.377(d) = 0.377(13.7) = 5.16 \text{ in.}$$

If a greater value is used for d, the value of a will be less than this, since the compression force is reduced as the internal moment arm $(d - a/2)$ increases. Let us assume a d of 20 in. and, for a first trial, assume a value of 4 in. for a. Then from the formula for the resisting moment based on the steel, we find A_s as

$$A_s = \frac{M_t}{f_y(d - a/2)} = \frac{1,944,000}{60,000(20 - 2)} = 1.80 \text{ in.}^2$$

With this steel area the actual percentage of steel is

$$\frac{A_s}{bd} = \frac{1.80}{10 \times 20} = 0.0090$$

and with the formula previously derived for a,

$$a = \frac{\rho d f_y}{0.85 f'_c} = \frac{(0.0090)(20)(60)}{(0.85)(4)} = 3.18 \text{ in.}$$

This indicates that our first guess for a was high and that the internal moment arm is even larger. Thus the required steel area is actually slightly less and will result in an a value less than 3.18. We therefore try a second time with an a value of 3.0 in. and determine the new value for A_s to be

$$\frac{1{,}944{,}000}{60{,}000(20 - 1.5)} = 1.75 \text{ in.}^2$$

Since this results in a change of only about 3% in the steel area, further correction is of no practical value, and the steel bars may be selected for the revised area of 1.75 in.^{2}

The ACI Code stipulates various other restrictions for beams, including a requirement for a minimum amount of reinforcing. Note that the procedure in the preceding example dealt only with consideration of bending moment. In a real design situation, consideration must be given also to shear and the required development length for the bars.

Problem 13.9.A,B,C. Using f'_c = 3 ksi [20.7 MPa] and f_y = 50 ksi [345 MPa], find the minimum depth required for a balanced section for the given data. Also find the area of reinforcement required if the depth chosen is 1.5 times that required for the balanced section. Use strength design methods.

	Moment Due To					
	Dead Load		Live Load		Beam Width	
	(kip-ft)	(kN-m)	(kip-ft)	(kN-m)	(in.)	(mm)
A	40	54.2	20	27.1	12	305
B	80	108.5	40	54.2	15	381
C	100	135.6	50	67.8	18	457

13.10 T-Beams

When a floor slab and its supporting beams are poured at the same time, the result is a monolithic construction described as *cast-in-place*. In this case, a portion of the slab on each side of the beam serves as a flange acting together with the beam stem to develop a T-beam action, as shown in Fig. 13.8*a*. For a simple beam the flange is in the compression zone, providing an extensive area for development of compression stress, as shown in Fig. 13.8*b*. However, in a continuous beam, for the negative moments at the supports, the flange area is in the tension zone, and compression must be developed in the beam stem, as shown in Fig.

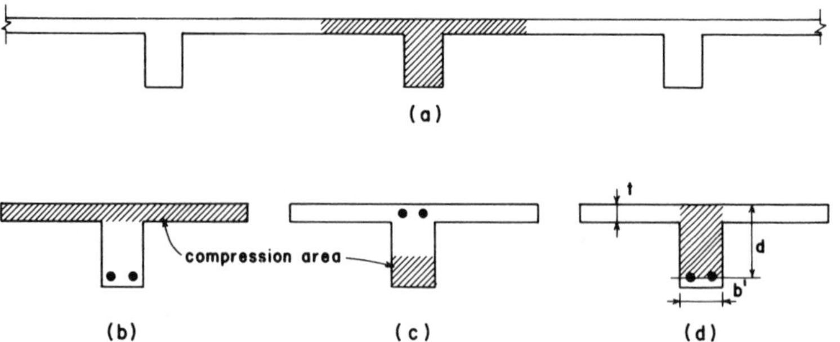

FIGURE 13.8 Beam actions in monolithic beam–slab construction.

13.8*c*. The stem must therefore be adequate for the negative bending moments, as well as for resistance to shear (as discussed in Sec. 13.13).

The ACI Code stipulates that the effective flange width (b in Fig. 13.9*c*) to be used in the design of symmetrical T-beams shall not exceed one-quarter of the beam span, and its overhanging width on either side of the beam stem shall not exceed eight times the slab thickness nor one-half the clear distance to the next beam. For T-beams with continuous spans, in which the negative moments are typically the greatest, the flange is seldom critical for compressive stress, and the stem size must be selected for shear and negative moment concerns.

Consider the T-beam shown in Fig. 13.9*a* in which the neutral axis for bending stress falls within the slab (the T-flange), whereas in Fig. 13.9*b* it falls below the slab. In cast-in-place construction either case may occur, but that shown in Fig. 13.9*a* is the more common occurrence. As both the magnitude of stress and the area in compression in the stem are quite small, the resistance to compression is typically assumed to be developed only by the flange.

The design of a T-beam is commonly limited to the determination of the required steel area for reinforcement. The slab thickness is usually established by the design of the slab itself, and the stem dimensions are derived from consideration of shear and negative bending moments. To find A_s for the positive bending

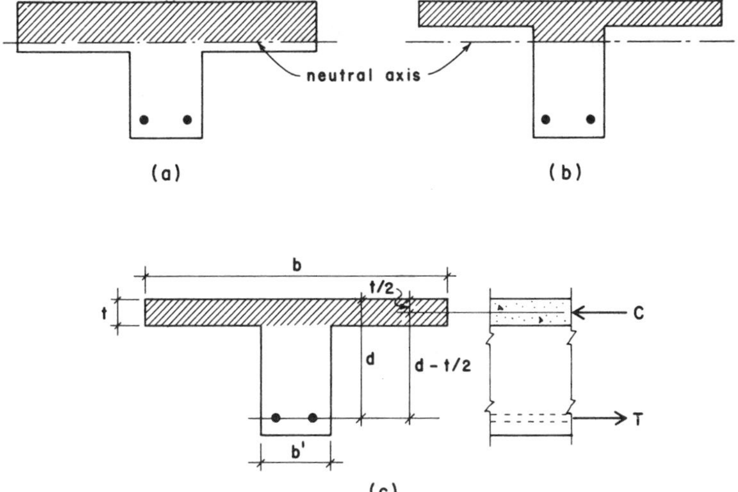

FIGURE 13.9 T-beam actions and design considerations.

moment between the supports, a formula commonly used is

$$A_s = \frac{M}{f_s[d - (t/2)]}$$

in which A_s = the required steel area for positive moment,
M = the maximum positive moment in the span,
f_s = allowable tensile stress in the steel,
d = effective depth for the T-beam,
t = thickness of the flange (slab) of the T-beam.

This is a slightly conservative formula based on the assumption that the compressive stress in the flange is uniform in magnitude. Answers obtained will be relatively approximate depending on the various dimensions of the section. For T-beams that occur in concrete beam and slab systems (as discussed in Chapter 14), the slab is typically quite thick and the formula produces quite reasonable results.

The following example demonstrates the use of the formula. An additional example is given in Part V.

Example. A concrete T-section has the following dimensions (see Fig. 13.9): d = 24 in. [610 mm], t = 5 in. [127 mm], and b' = 14 in. [356 mm]. Find the area of steel required if the allowable steel stress is 24 ksi [165 MPa] and the section sustains a bending moment of 160 kip-ft [217 kN-m].

Solution: Applying the formula with the data as given,

$$A_s = \frac{M}{f_s[d - (t/2)]} = \frac{160 \times 12}{24 \times 21.5} = 3.72 \text{ in.}^2 \ [2406 \text{ mm}^2]$$

Problem 13.10.A. Find the area of steel reinforcement required for a concrete T-beam for the following data: allowable f_s = 20 ksi [138 MPa], d = 28 in. [711 mm], t = 6 in. [152 mm], b' = 16 in. [406 mm], and the section sustains a bending moment of 240 kip-ft [325 kN-m].

13.11 Beams With Compressive Reinforcement

There are many situations in which steel reinforcing is used on both sides of the neutral axis in a beam. When this occurs, the steel on one side of the axis will be in tension and that on the other side in compression. Such a beam is referred to as a doubly reinforced beam or simply as a beam with compressive reinforcing (it being naturally assumed that there is also tensile reinforcing). Various situations involving such reinforcing have been discussed in the preceding sections. In summary, the most common occasions for such reinforcing include:

1. The desired resisting moment for the beam exceeds that for which the concrete alone is capable of developing the necessary compressive force.
2. Other functions of the section require the use of reinforcing on both sides of the beam. These include the need for bars to support U-stirrups and situations when torsion is a major concern.

3. It is desired to reduce deflections by increasing the stiffness of the compressive side of the beam. This is most significant for reduction of long-term creep deflections.
4. The combination of loading conditions on the structure result in reversal moments on the section; that is, the section must sometimes resist positive moment, and other times resist negative moment.
5. Anchorage requirements (for development of reinforcing) require that the bottom bars in a continuous beam be extended a significant distance into the supports.

The precise investigation and accurate design of doubly reinforced sections, whether performed by the working stress or by strength design methods, are quite complex and are beyond the scope of work in this book. The following discussion presents an approximation method that is adequate for preliminary design of a doubly reinforced section. For real design situations, this method may be used to establish a first trial design, which may then be more precisely investigated using more rigorous methods.

For the beam with double reinforcing, as shown in Fig. 13.10*a*, we consider the total resisting moment for the section to be the sum of the following two component moments.

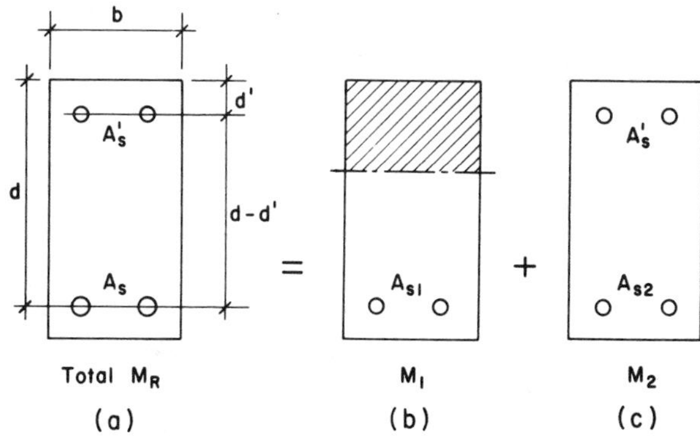

FIGURE 13.10 Basis for simplified analysis of a doubly reinforced section.

M_1 (Fig. 13.10*b*) is comprised of a section with tension reinforcing only (A_{s1}). This section is subject to the usual procedures for design and investigation, as discussed in Secs. 13.7 and 13.9.

M_2 (Fig. 13.10*c*) is comprised of two opposed steel areas (A_{s2} and A'_s) that function in simple moment couple action, similar to the flanges of a steel beam or the top and bottom chords of a truss.

Ordinarily, we expect that $A_{s2} = A'_s$, since the same grade of steel is usually used for both. However, there are two special considerations that must be made. The first involves the fact that A_{s2} is in tension, while A'_s is in compression. A'_s must therefore be dealt with in a manner similar to that for column reinforcing. This requires, among other things, that the compressive reinforcing be braced against buckling, using ties similar to those in a tied column.

The second consideration involves the distribution of stress and strain on the section. Referring to Fig. 13.11, it may be observed that, under normal circumstances (kd less than $0.5d$), A'_s will be closer to the neutral axis than A_{s2}. Thus the stress in A'_s will be lower than that in A_{s2} if pure elastic conditions are assumed. However, it is common practice to assume steel to be doubly stiff when sharing stress with concrete in compression, due to shrinkage and creep effects. Thus, in translating from linear strain conditions to stress distribution, we use the relation $f'_s/2n$ (where $n = E_s/E_c$, as discussed in Sec. 13.6). Utilization of this relationship is illustrated in the following examples.

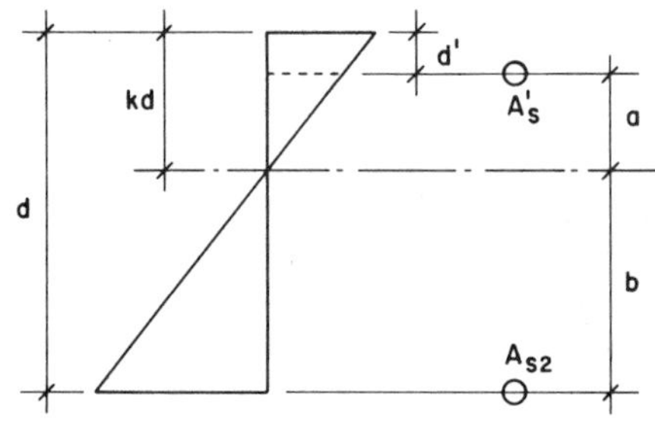

FIGURE 13.11

Example 1. A concrete section with $b = 18$ in. [0.457 m] and $d = 21.5$ in. [0.546 m] is required to resist service load moments as follows: dead load moment = 150 kip-ft [203.4 kN-m], live load moment = 150 kip-ft [203.4 kN-m]. Using working stress methods, find the required reinforcing. Use $f'_c = 4$ ksi [27.6 MPa] and $f_y = 60$ ksi [414 MPa].

Solution: For the Grade 60 reinforcing, we use an allowable stress of $f_s = 24$ ksi [165 MPa]. Then, using Table 13.3, find:

$$n = 8, \quad k = 0.375, \quad j = 0.875, \quad p = 0.0141$$

$$R = 0.295 \text{ in kip-in. units [2028 in kN-m units]}$$

Using the R value for the balanced section, the maximum resisting moment of the section is

$$M = Rbd^2 = \frac{0.295}{12} \times (18)(21.5)^2$$

$$= 205 \text{ kip-ft [278 kN-m]}$$

This is M_1, as shown in Fig. 13.10*b*. Thus

$$M_2 = \text{total } M - M_1 = 300 - 205 = 95 \text{ kip-ft}$$

$$[407 - 278 = 129 \text{ kN-m}]$$

For M_1 the required reinforcing (A_{s1} in Fig. 13.10*b*) may be found as

$$A_{s1} = pbd = 0.0141 \times 18 \times 21.5 = 5.46 \text{ in.}^2 \text{ [3523 mm}^2\text{]}$$

And, assuming $f'_s = f_s$, we find A'_s and A_{s2} as follows.

$$M_2 = A'_s(d - d') = A_{s2}(d - d')$$

$$A'_s = A_{s2} = \frac{M_2}{f_s(d - d')} = \frac{95 \times 12}{24 \times 19} = 2.50 \text{ in.}^2 \text{ [1613 mm}^2\text{]}$$

The total tension reinforcing is thus

$$A_s = A_{s1} + A_{s2} = 5.46 + 2.50 = 7.96 \text{ in.}^2 \text{ [5136 mm}^2\text{]}$$

For the compressive reinforcing, we must find the proper limit for f'_s. To do this, we assume the neutral axis of the section to be that for the balanced section, producing the situation that is shown in Fig. 13.12. Based on this assumption, the limit for f'_s is found as follows:

$$\frac{f'_s}{2n} = \frac{5.56}{8.06}(0.45 \times 4) = 1.24 \text{ ksi}$$

$$f'_s = 2n \times 1.24 = 2 \times 8 \times 1.24 = 19.84 \text{ ksi [137 MPa]}$$

Since this is less than the limit of 24 ksi, we must use it to find A'_s; thus

$$A' = \frac{M_2}{f'_s(d - d')} = \frac{95 \times 12}{19.84 \times 19} = 3.02 \text{ in.}^2 \text{ [1948 mm}^2\text{]}$$

In practice, compressive reinforcement is often used even when the section is theoretically capable of developing the necessary resisting moment with tension reinforcement only. This calls for a somewhat different procedure, as is illustrated in the following example.

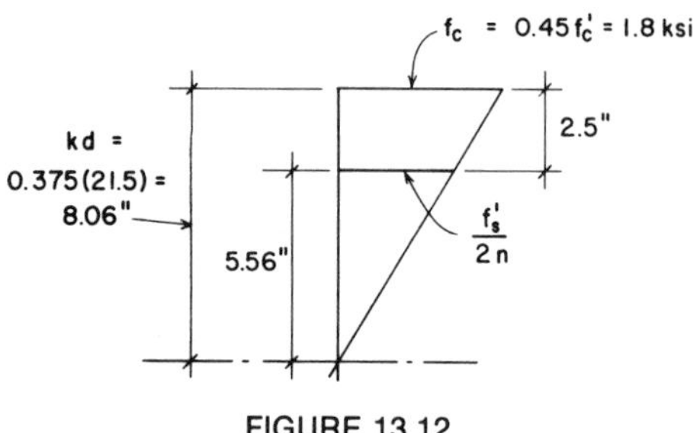

FIGURE 13.12

Example 2. Design a section by the working stress method for a moment of 180 kip-ft [244 kN-m]. Use the section dimensions and data given in Example 1.

Solution: The first step is to investigate the section for its balanced stress limiting moment, as was done in Example 1. This will show that the required moment is less than the balanced moment limit, and that the section could function without compressive reinforcing. Again, we assume that compressive reinforcing is desired, so we assume an arbitrary amount for A_s' and proceed as follows. We make a first guess for the total tension reinforcing as

$$A_s = \frac{M}{f_s(0.9d)} = \frac{180 \times 12}{24(0.9 \times 21.5)} = 4.65 \text{ in.}^2 \text{ [3000 mm}^2\text{]}$$

Try

$$A_s' = \tfrac{1}{3}A_s = \tfrac{1}{3}(4.65) = 1.55 \text{ in.}^2 \text{ [1000 mm}^2\text{]}$$

Choose two No. 8 bars,

$$\text{Actual } A_s' = 1.58 \text{ in.}^2 \text{ [1019 mm}^2\text{]}$$

Thus

$$A_{s1} = A_s - A_s' = 4.65 - 1.58 = 3.07 \text{ in.}^2 \text{ [1981 mm}^2\text{]}$$

Using A_{s1} for a rectangular section with tension reinforcing only (see Sec. 13.7):

$$p = \frac{3.07}{18 \times 21.5} = 0.0079$$

Then, from Fig. 13.4, we find $k = 0.30$ and $j = 0.90$.

Using these values for the section, and the formula involving the concrete stress in compression from Sec. 13.7, we find

$$f_c = \frac{2M_1}{kjbd^2} = \frac{2 \times 120 \times 12}{0.3 \times 0.9 \times 18 \times (21.5)^2}$$

$$= 1.28 \text{ ksi } [8.83 \text{ MPa}]$$

With this value for the maximum concrete stress and the value of 0.30 for k, the distribution of compressive stress will be as shown in Fig. 13.13. From this, we determine the limiting value for f'_s as follows:

$$\frac{f'_s}{2n} = \frac{3.95}{6.45}(1.28) = 0.784 \text{ ksi}$$

$$f'_s = 2n(0.784) = 2 \times 8 \times 0.784 = 12.5 \text{ ksi } [86.2 \text{ MPa}]$$

Since this is lower than f_s, we use it to find the limiting value for M_2. Thus

$$M_2 = A'_s f'_s(d - d')$$

$$= (1.58)(12.5)(19)\frac{1}{12} = 31 \text{ kip-ft } [42 \text{ kN-m}]$$

To find A_{s2} we use this moment with the full value of $f_s = 24$ ksi. Thus

$$A_{s2} = \frac{M_2}{f_s(d - d')} = \frac{31 \times 12}{24 \times 19.0} = 0.82 \text{ in.}^2 \ [529 \text{ mm}^2]$$

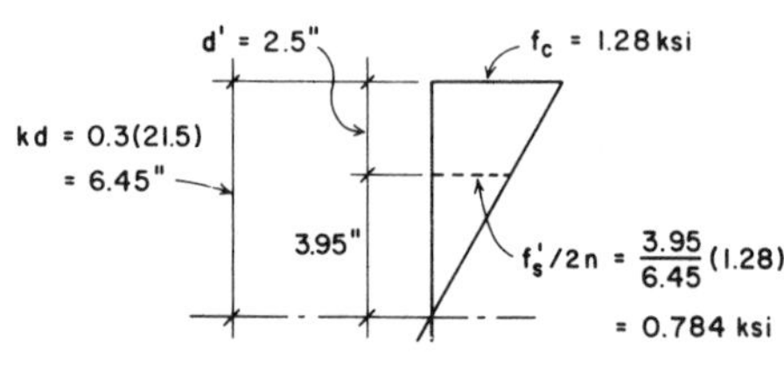

FIGURE 13.13

To find A_{s1}, we determine that

$$M_1 = \text{total } M - M_2 = 180 - 31 = 149 \text{ kip-ft [202 kN-m]}$$

$$A_{s1} = \frac{M_1}{f_s jd} = \frac{149 \times 12}{24 \times 0.9 \times 21.5} = 3.85 \text{ in.}^2 \text{ [2484 MPa]}$$

Then the total tension reinforcing is found as

$$A_s = A_{s1} + A_{s2} = 3.85 + 0.82 = 4.67 \text{ in.}^2 \text{ [3013 mm}^2\text{]}$$

Problem 13.11.A. A concrete section with b = 16 in. [0.406 m] and d = 19.5 in. [0.495 m] is required to resist service load moments as follows: dead load moment = 120 kip-ft [163 kN-m], live load moment = 110 kip-ft [136 kN-m]. Using working stress methods, find the required reinforcing. Use f'_c = 4 ksi [27.6 MPa] and Grade 60 bars with f_y = 60 ksi [414 MPa] and f_s = 24 ksi [165 MPa].

13.12 Structural Slabs

One of the most commonly used concrete floor systems consists of a solid slab that is continuous over parallel supports. The supports may consist of bearing walls of masonry or concrete but most often consist of sets of evenly spaced concrete beams. The beams are usually supported by girders, which in turn are supported by columns. In this type of slab the principal reinforcement runs in one direction, parallel to the slab span and perpendicular to the supports. For this reason it is called a *one-way solid slab*. The number and spacing of supporting beams depend on their span, the column spacing, and the magnitude of the loads. Most often the beams are spaced uniformly and frame into the girders at the center, third, or quarter points. The formwork for this type of floor is readily constructed, and the one-way slab is most economical for medium and heavy floor loads for relatively short spans, 6 to 12 ft. For long spans the slab thickness must be increased, resulting in considerable dead weight of the construction, which increases the cost of the slab and its reinforcing as well as the cost of supporting beams, girders, columns, and foundations.

To design a one-way slab, we consider a strip 12 in. wide (see

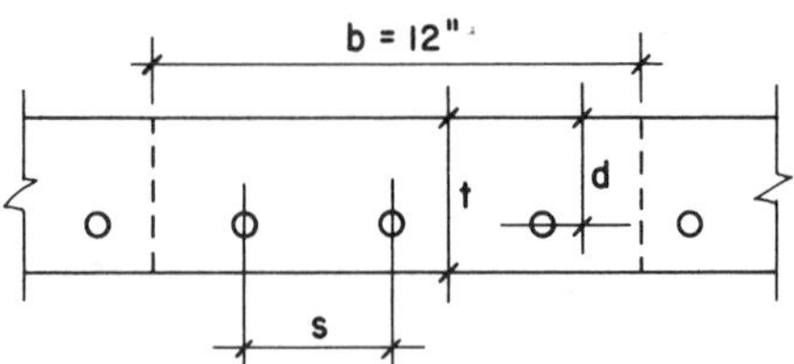

FIGURE 13.14 Reference for slab design.

Fig. 13.14). This strip is designed as a beam whose width is 12 in. and on which is a uniformly distributed load. As with any rectangular beam, the effective depth and tensile reinforcement are computed as explained in Sec. 13.7. A minimum slab thickness is often determined on the basis of the fire-rating requirements of the applicable building code. A minimum thickness is also required to prevent excessive deflection. Based on deflection limitations, slab thicknesses should not be less than those given in Table 13.9.

While flexural reinforcement is required in only one direction in the one-way slab, reinforcement at right angles to the flexural reinforcement is also provided for stresses due to shrinkage of the concrete and temperature fluctuations. The amount of this reinforcement is specified as a percentage p of the gross cross-sectional area of the concrete, as follows:

For slabs reinforced with Grade 40 or Grade 50 deformed bars,

$$p = \frac{A_s}{b \times t} = 0.0020$$

and for slabs reinforced with Grade 60 deformed bars,

$$p = 0.0018$$

Center-to-center bar spacing must not be greater than five times the slab thickness or 18 in.

The one-way slab is designed by assuming the slab to consist of a series of 12-in. [305-mm] wide segments. The bending moment for this 12-in.-wide rectangular element is determined, and the required effective depth and area of tensile reinforcement A_s is computed using the procedures for a rectangular beam as discussed in Sec. 13.7. The A_s thus determined is the average amount of steel per 1-ft width of slab that is required. The maximum spacing (s in Fig. 13.14) for this reinforcement is three times the slab thickness or a maximum of 18 in. [457 mm]. The size and spacing of bars may be selected by use of Table 13.5.

It is not practicable to use shear reinforcement in one-way slabs, and consequently the maximum unit shear stress must be kept within the limit for the concrete alone. The usual procedure is to check the shear stress with the effective depth determined for bending before proceeding to find A_s. Except for very short span slabs with excessively heavy loadings, shear stress is seldom critical.

The following example illustrates the design procedure for a one-way solid slab.

TABLE 13.5 Areas of Bars in Reinforced Concrete Slabs per Foot of Width

Bar Spacing (in.)	Areas of Bars (in.^{2}/ft)									
	No. 2	No. 3	No. 4	No. 5	No. 6	No. 7	No. 8	No. 9	No. 10	No. 11
3	0.20	0.44	0.80	1.24	1.76	2.40	3.16	4.00		
3.5	0.17	0.38	0.69	1.06	1.51	2.06	2.71	3.43	4.35	
4	0.15	0.33	0.60	0.93	1.32	1.80	2.37	3.00	3.81	4.68
4.5	0.13	0.29	0.53	0.83	1.17	1.60	2.11	2.67	3.39	4.16
5	0.12	0.26	0.48	0.74	1.06	1.44	1.89	2.40	3.05	3.74
5.5	0.11	0.24	0.44	0.68	0.96	1.31	1.72	2.18	2.77	3.40
6	0.10	0.22	0.40	0.62	0.88	1.20	1.58	2.00	2.54	3.12
7	0.08	0.19	0.34	0.53	0.75	1.03	1.35	1.71	2.18	2.67
8	0.07	0.16	0.30	0.46	0.66	0.90	1.18	1.50	1.90	2.34
9	0.07	0.15	0.27	0.41	0.59	0.80	1.05	1.33	1.69	2.08
10	0.06	0.13	0.24	0.37	0.53	0.72	0.95	1.20	1.52	1.87
11	0.05	0.12	0.22	0.34	0.48	0.65	0.86	1.09	1.38	1.70
12	0.05	0.11	0.20	0.31	0.44	0.60	0.79	1.00	1.27	1.56
13	0.05	0.10	0.18	0.29	0.40	0.55	0.73	0.92	1.17	1.44
14	0.04	0.09	0.17	0.27	0.38	0.51	0.68	0.86	1.09	1.34
15	0.04	0.09	0.16	0.25	0.35	0.48	0.63	0.80	1.01	1.25
16	0.04	0.08	0.15	0.23	0.33	0.45	0.59	0.75	0.95	1.17
18	0.03	0.07	0.13	0.21	0.29	0.40	0.53	0.67	0.85	1.04
24	0.02	0.05	0.10	0.15	0.22	0.30	0.39	0.50	0.63	0.78

Example. A one-way solid concrete slab is to be used for a simple span of 14 ft [4.27 m]. In addition to its own weight, the slab carries a superimposed dead load of 30 psf [1.44 kN/m^2] and a live load of 100 psf [4.79 kN/m^2]. Using $f'_c = 3$ ksi [20.7 MPa], $f_y = 40$ ksi [276 MPa], and $f_s = 20$ ksi [138 MPa], design the slab for minimum overall thickness.

Solution:

Working Stress Method

Using the general procedure for design of a beam with rectangular section (Sec. 13.7), we first determine the required slab thickness. Thus

For deflection, from Table 13.9,

$$\text{Minimum } t = \frac{L}{25} = \frac{14 \times 12}{25} = 6.72 \text{ in. [171 mm]}$$

For flexure we first determine the maximum bending moment. The loading must include the weight of the slab, for which we use the thickness required for deflection as a first estimate. Assuming a 7-in. [178-mm] thick slab, then slab weight is $\frac{7}{12}$ (150 pcf) = 87.5 psf, say 88 psf, and total load is 100 psf LL + 118 psf DL = 218 psf.

The maximum bending moment for a 12-in.-wide design strip of the slab thus becomes

$$M = \frac{w\,L^2}{8} = \frac{218(14)^2}{8} = 5341 \text{ ft-lb [7.24 kN-m]}$$

For minimum slab thickness, we consider the use of a balanced section, for which Table 13.3 yields the following properties:

$$j = 0.872,\ p = 0.0129,\ R = 0.226$$

Then

$$bd^2 = \frac{M}{R} = \frac{5.341 \times 12}{0.226} = 284 \text{ in.}^3$$

And, since b is the 12-in. design strip width,

$$d = \sqrt{\frac{284}{12}} = \sqrt{23.7} = 4.86 \text{ in. [123 mm]}$$

Assuming an average bar size of a No. 6 ($\frac{3}{4}$-in. nominal diameter) and cover of $\frac{3}{4}$ in., the minimum required slab thickness based on flexure becomes

$$t = 4.86 + \frac{0.75}{2} + 0.75 = 5.985 \text{ in. [152 mm]}$$

We thus observe that the deflection limitation controls in this situation, and the minimum overall thickness is the 6.72-in. dimension. If we continue to use the 7-in. overall thickness, the actual effective depth with a No. 6 bar will be

$$d = 7.0 - 1.125 = 5.875 \text{ in.}$$

Since this d is larger than that required for a balanced section, the value for j will be slightly larger than 0.872, as found from Table 13.3. Let us assume a value of 0.9 for j and determine the required area of reinforcement as

$$A_s = \frac{M}{f_s jd} = \frac{5.341 \times 12}{20(0.9)5.875} = 0.606 \text{ in.}^2 \text{ [391 mm}^2\text{]}$$

From Table 13.5, we find that the following bar combinations will satisfy this requirement:

Bar Size	Spacing From Center to Center (in.)	Average A_s in a 12-in. Width
No. 5	6	0.62
No. 6	8.5	0.62
No. 7	12	0.60
No. 8	15	0.63

The ACI Code requires a maximum spacing of three times the slab thickness (21 in. in this case). Minimum spacing is largely a matter of the designer's judgment. Many designers consider a minimum practical spacing to be one approximately equal to the slab thickness. Within these limits, any of the bar size and spacing combinations listed are adequate.

As previously described, the ACI Code requires a minimum reinforcement for shrinkage and temperature effects to be placed in the direction perpendicular to the flexural reinforcement. With the Grade 40 bars in this example, the minimum percentage of this steel is 0.0020, and the steel area required for a 12-in. strip thus becomes

$$A_s = p(b \times t) = 0.0020(12 \times 7) = 0.168 \text{ in.}^2$$

From Table 13.5, we find that this requirement can be satisfied with No. 3 bars at 8-in. centers or No. 4 bars at 14-in. centers. Both of these spacings are well below the maximum of five times the slab thickness.

Although simply supported single slabs are sometimes encountered, the majority of slabs used in building construction are continuous through multiple spans. An example of the design of such a slab is given in the next chapter.

Strength Design Method

Strength design procedures for the slab are essentially the same as for the rectangular beam, as described in Sec. 13.9. In most cases, slab sections will be reinforced with steel areas well below those for a balanced section, so the procedure for a so-called under-reinforced section should be used. If the procedure illustrated in Sec. 13.9 is used for this example, it will be found that the required steel area is approximately 15% less than that required from the working stress method computations.

Problem 13.12.A. A one-way solid concrete slab is to be used for a simple span of 16 ft [4.88 m]. In addition to its own weight, the slab carries a superimposed dead load of 60 psf [2.87 kN/m²] and a live load of 75 psf [3.59 kN/m²]. Using the working stress method with f'_c = 3 ksi [20.7 MPa], f_y = 40 ksi [276 MPa], and f_s = 20 ksi [138 MPa], design the slab for minimum overall thickness.

Problem 13.12.B. Using the data from Problem 13.12.A, design the slab using strength design methods.

13.13 Shear in Beams

Let us consider the case of a simple beam with uniformly distributed load and end supports that provide only vertical resistance (no moment restraint). For flexural resistance, it is necessary to provide longitudinal reinforcing bars near the bottom of the beam. These bars are oriented for primary effectiveness in resistance to tension stresses that develop on a vertical (90°) plane (which is the case at the center of the span, where the bending moment is maximum and the shear approaches zero).

Under the combined effects of shear and bending, the beam tends to develop tension cracks, as shown in Fig. 13.15*a*. Near the center of the span, where the bending is predominant and the shear approaches zero, these cracks approach 90°. Near the support, however, where the shear predominates and bending approaches zero, the critical tension stress plane approaches 45°, and the horizontal bars are only partly effective in resisting the cracking.

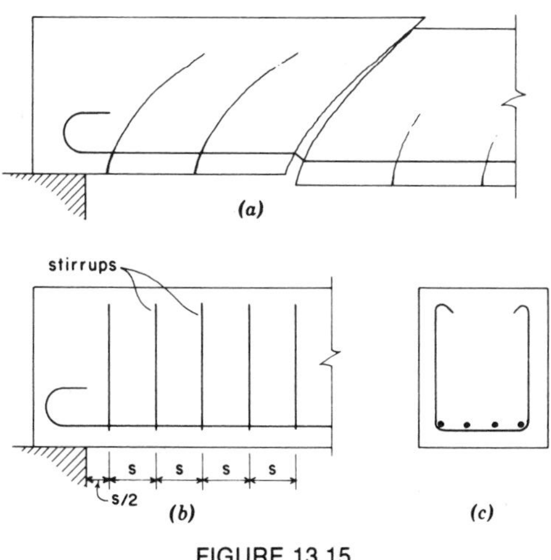

FIGURE 13.15

For beams, the most common form of shear reinforcement consists of a series of U-shaped bent bars called *stirrups* (Fig. 13.15*c*), placed vertically and spaced along the beam span, as shown in Fig. 13.15*b*. These bars are intended to provide a vertical component of resistance, working in conjunction with the horizontal resistance provided by the flexural reinforcement. In order to develop tension near the support face, the horizontal bars must be bonded to the concrete beyond the point where the stress is developed. Where the beam ends extend only a short distance over the support (a common situation), it is often necessary to bend or hook the bars, as shown in Fig. 13.15.

The rectangular section shown in Fig. 13.15 occurs only infrequently in building structures. The most common case is that of the beam section shown in Fig. 13.16*a*, which occurs when a beam is poured monolithically with a supported concrete slab. In addition, these beams normally occur in continuous spans with negative moments at the supports. Thus the stress in the beam near the support is as shown in Fig. 13.16*a*, with the negative moment producing compressive flexural stress in the bottom of the beam stem. This is substantially different from the case of the simple beam, where the moment approaches zero near the support.

For purpose of shear resistance, the continuous, T-shaped beam is considered to consist of the section indicated in Fig. 13.16*b*. The effect of the slab is ignored, and the section is considered to be a simple rectangular one. Thus for shear design, there is little difference between the simple span beam and the continu-

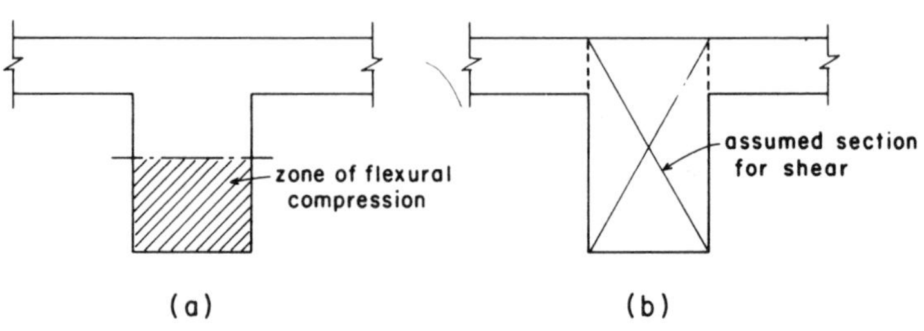

FIGURE 13.16

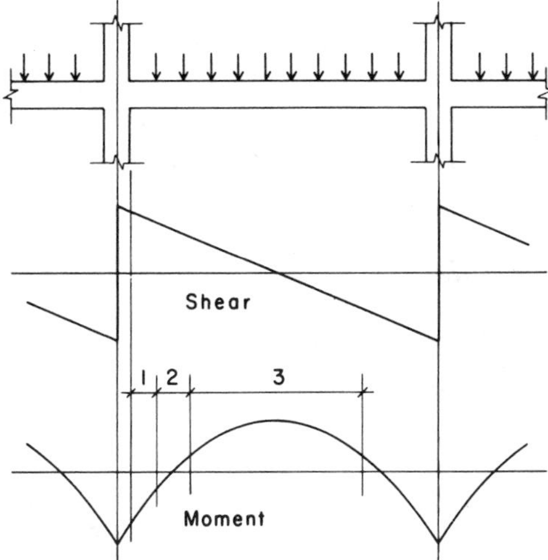

FIGURE 13.17

ous beam, except for the effect of the continuity on the distribution of shear along the beam span. It is important, however, to understand the relationships between shear and moment in the continuous beam.

Figure 13.17 illustrates the typical condition for an interior span of a continuous beam with uniformly distributed load. Referring to the portions of the beam span numbered 1, 2, and 3, we note:

1. In this zone the high negative moment requires major flexural reinforcing consisting of horizontal bars near the top of the beam.
2. In this zone, the moment reverses sign, moment magnitudes are low, and, if shear stress is high, the design for shear is a predominant concern.
3. In this zone, shear consideration is minor and the predominant concern is for positive moment requiring major flexural reinforcing in the bottom of the beam.

Vertical U-shaped stirrups, similar to those shown in Fig. 13.18*a*, may be used in the T-shaped beam. An alternate detail for the U-shaped stirrup is shown in Fig. 13.18*b*, in which the top hooks are turned outward; this makes it possible to spread the negative moment reinforcing bars to make placing of the concrete somewhat easier. Figure 13.18*c* and *d* show possibilities for stirrups in beams that occur at the edges of large openings or at the outside edge of the structure. This form of stirrup is used to enhance the torsional resistance of the section and also assists in developing the negative moment resistance in the slab at the edge of the beam.

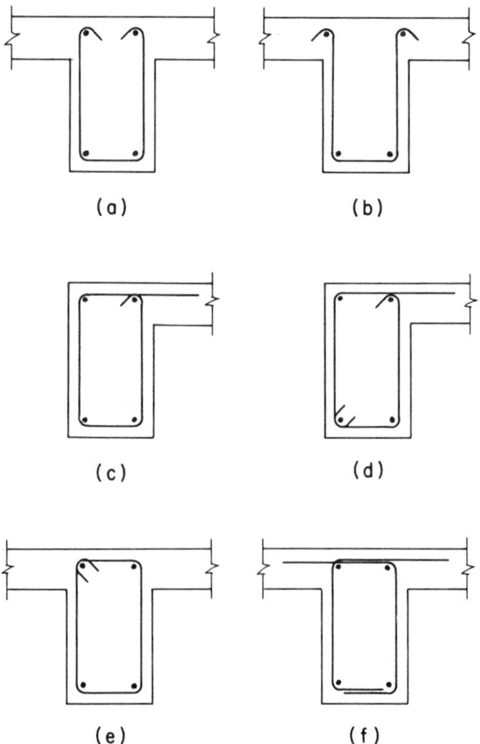

FIGURE 13.18 Various forms for vertical stirrups.

So-called closed stirrups, similar to ties in columns, are sometimes used for T-shaped beams, as shown in Fig. 13.18*e* and *f*. These are generally used to improve the torsional resistance of the beam section.

Stirrup forms are often modified by designers or by the reinforcing fabricator's detailers to simplify the fabrication and/or the field installation. The stirrups shown in Fig. 13.18*d* and *f* are two such modifications of the basic details in Fig. 13.18*c* and *e*, respectively.

The following are some of the general considerations and code requirements that apply to current practices of design for beam shear.

Concrete Capacity. Whereas the tensile strength of the concrete is ignored in design for flexure, the concrete is assumed to take some portion of the shear in beams. If the capacity of the concrete is not exceeded—as it sometimes is for lightly loaded beams—there may be no need for reinforcing. The typical case, however, is as shown in Fig. 13.19, where the maximum shear V exceeds the capacity of the concrete alone (V_c) and the steel reinforcing is required to absorb the excess, indicated as the shaded portion in the shear diagram.

Minimum Shear Reinforcing. Even when the maximum computed shear stress falls below the capacity of the concrete, the present code requires the use of some minimum amount of shear reinforcing. Exceptions are made in some situations, such as for slabs and very shallow beams. The objective is essentially to toughen the structure with a small investment in additional reinforcing.

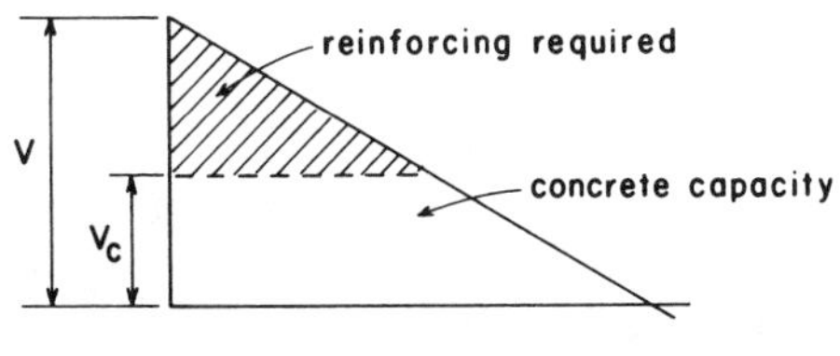

FIGURE 13.19

Type of Stirrup. The most common stirrups are the simple U-shaped or closed forms shown in Fig. 13.18, placed in a vertical position at intervals along the beam. It is also possible to place stirrups at an incline (usually 45°), which makes them somewhat more effective in direct resistance to the potential shear cracking near the beam ends. (See Fig. 13.15.) In large beams with excessively high unit shear stress, both vertical and inclined stirrups are sometimes used at the location of the greatest shear.

Size of Stirrups. For beams of moderate size, the most common size for U-stirrups is a No. 3 bar. These bars can be bent relatively tightly at the corners (small radius of bend) in order to fit within the beam section. For larger beams, a No. 4 bar is sometimes used, its strength (as a function of its cross-sectional area) being almost twice that of a No. 3 bar.

Spacing of Stirrups. Stirrup spacings are computed (as discussed in the following sections) on the basis of the amount of reinforcing required for the unit shear stress at the location of the stirrups. A maximum spacing of $d/2$ (i.e., one-half the effective beam depth d) is specified in order to assure that at least one stirrup occurs at the location of any potential diagonal crack. (See Fig. 13.15.) When shear stress is excessive, the maximum spacing is limited to $d/4$.

Critical Maximum Design Shear. Although the actual maximum shear value occurs at the end of the beam, the code permits the use of the shear stress at a distance of d (effective beam depth) from the beam end as the critical maximum for stirrup design. Thus, as shown in Fig. 13.20, the shear requiring reinforcing is slightly different from that shown in Fig. 13.19.

Total Length for Shear Reinforcing. On the basis of computed shear stresses, reinforcing must be provided along the beam length for the distance defined by the shaded portion of the shear stress diagram shown in Fig. 13.20. For the center portion of the span, the concrete is theoretically capable of the necessary shear resistance without the assistance of reinforcing. However, the code requires that some reinforcing be provided for a distance beyond this computed cutoff point. The 1963 ACI Code required

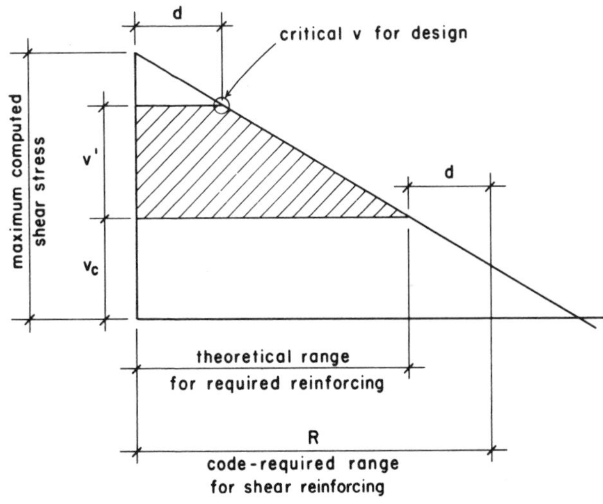

FIGURE 13.20

that stirrups be provided for a distance equal to the effective depth of the beam beyond the cutoff point. The 1983 ACI Code requires that minimum shear reinforcing be provided as long as the computed shear stress exceeds one-half of the capacity of the concrete. However it is established, the total extended range over which reinforcing must be provided is indicated as *R* on Fig. 13.20.

The following is a description of the procedure for design of shear reinforcing for beams that is in compliance with Appendix B of the 1983 ACI Code (Ref. 3).

Shear stress is computed as

$$v = \frac{V}{bd}$$

in which V = the total shear force at the section,
b = the beam width (of the stem for T shapes),
d = the effective depth of the section.

For beams of normal-weight concrete, subjected only to flexure and shear, shear stress in the concrete is limited to

$$v_c = 1.1\sqrt{f'_c}$$

When v exceeds the limit for v_c, reinforcing must be provided, complying with the general requirements previously discussed. Although the Code does not use the term, we coin the notation of v' for the excess unit shear for which reinforcing is required. Thus

$$v' = v - v_c$$

Required spacing of shear reinforcement is determined as follows. Referring to Fig. 13.21, we note that the capacity in tensile resistance of a single, two-legged stirrup is equal to the product of the total steel cross-sectional area times the allowable steel stress. Thus

$$T = (A_v) \times (f_s)$$

This resisting force opposes the development of shear stress on the area s times b, where b is the width of the beam and s is the

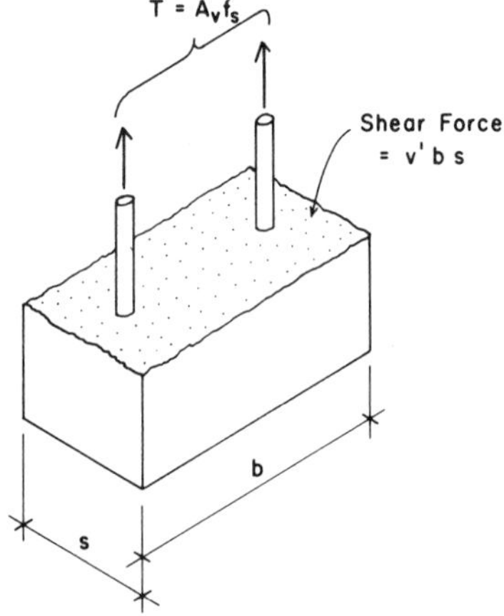

FIGURE 13.21

spacing (half the distance to the next stirrup on each side). Equating the stirrup tension to this force, we obtain the equilibrium equation

$$(A_v) \times (f_s) = (b) \times (s) \times (v')$$

From this equation, we can derive an expression for the required spacing; thus

$$s = \frac{A_v f_s}{v'b}$$

The following example illustrates the procedure for a simple beam.

Example 1. Using the working stress method, design the required shear reinforcing for the simple beam shown in Fig. 13.22. Use $f'_c = 3$ ksi [20.7 MPa] and $f_s = 20$ ksi [138 MPa] and single U-shaped stirrups.

Solution: The shear diagram for the beam will be of the form shown in case 2, Fig. 2.27. The maximum value for the shear is 40 kips [178 kN] and the maximum value for shear stress is computed as

$$v = \frac{V}{bd} = \frac{40{,}000}{12 \times 24} = 139 \text{ psi [957 kPa]}$$

We now construct the shear stress diagram for one-half of the beam, as shown in Fig. 13.22*c*. For the shear design, we determine the critical shear stress at 24 in. (the effective depth of the beam) from the support. Using proportionate triangles, this value is

$$\frac{72}{96}(139) = 104 \text{ psi [718 kPa]}$$

The capacity of the concrete without reinforcing is

$$v_c = 1.1\sqrt{f'_c} = 1.1\sqrt{3000} = 60 \text{ psi [414 kPa]}$$

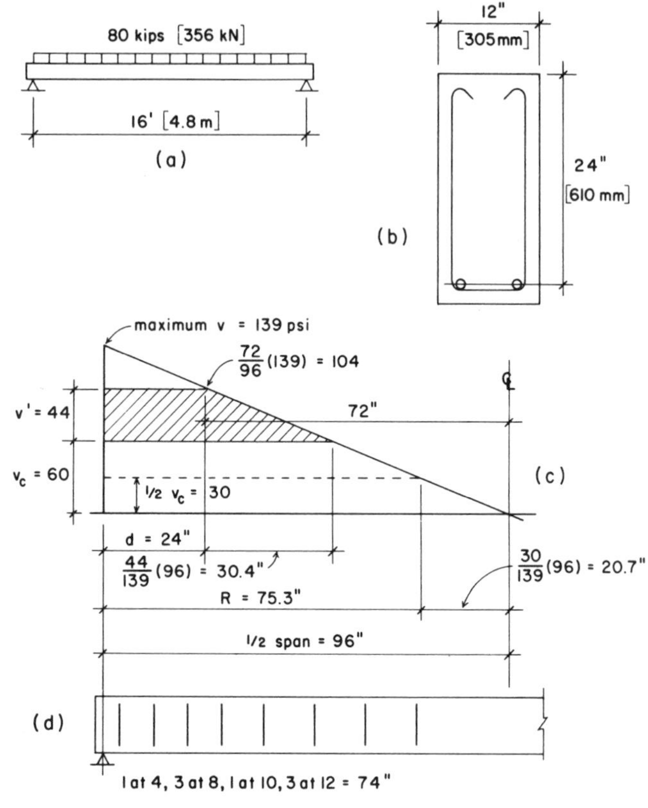

FIGURE 13.22

At the point of critical stress, therefore, there is an excess shear stress of 104 − 60 = 44 psi [718 − 414 = 304 kPa] that must be carried by reinforcing. We next complete the construction of the diagram in Fig. 13.22*c* to define the shaded portion, which indicates the extent of the required reinforcing. We thus observe that the excess shear condition extends to 54.4 in. [1.382 m] from the support.

In order to satisfy the requirements of the 1983 ACI Code, shear reinforcing must be used wherever the computed unit stress exceeds one-half of v_c. As shown in Fig. 13.22*c*, this is a distance of 75.3 in. from the support. The code further stipulates that the minimum cross-sectional area of this reinforcing be

$$A_v = 50 \frac{bs}{f_y}$$

If we assume an f_y value of 50 ksi [345 MPa] and use the maximum allowable spacing of one-half the effective depth, the required area is

$$A_v = 50 \frac{12 \times 12}{50{,}000} = 0.144 \text{ in.}^2$$

which is less than the area of 2 × 0.11 = 0.22 in.² provided by the two legs of the No. 3 stirrup.

For the maximum v' value of 44 ksi, the maximum spacing required is determined as

$$s = \frac{A_v f_s}{v'b} = \frac{0.22 \text{ in.}^2 \times 20{,}000 \text{ psi}}{44 \text{ psi} \times 12 \text{ in.}} = 8.3 \text{ in.}$$

Since this is less than the maximum allowable of 12 in., it is best to calculate at least one more spacing at a short distance beyond the critical point. We thus determine that the unit stress at 36 in. from the support is

$$v = \frac{60}{96} \times 139 = 87 \text{ psi}$$

and the value of v' at this point is 87 − 60 = 27 psi. The spacing required at this point is thus

$$s = \frac{0.22 \times 20{,}000}{27 \times 12} = 13.6 \text{ in.}$$

which indicates that the required spacing drops to the maximum allowed at less than 12 in. from the critical point. A possible choice for the stirrup spacings is shown in Fig. 13.22*d*, with a total of eight stirrups that extend over a range of 74 in. from the support. There is thus a total of 16 stirrups in the beam, 8 at each end.

Example 2. Determine the required number and spacings for No. 3 U-stirrups for the beam shown in Fig. 13.23. Use $f'_c = 3$ ksi [20.7 MPa] and $f_s = 20$ ksi [138 MPa].

Solution: As in the preceding example, the shear values and corresponding stresses are determined, and the diagram in Fig.

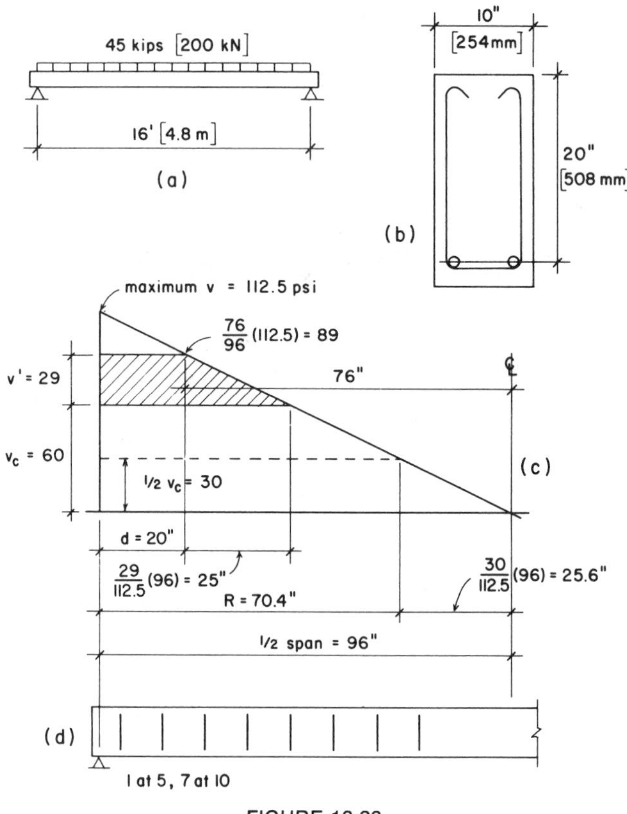

FIGURE 13.23

13.23*c* is constructed. In this case, the maximum critical shear stress of 89 psi results in a maximum v' value of 29 psi, for which the required spacing is

$$s = \frac{0.22 \times 20{,}000}{29 \times 10} = 15.2 \text{ in.}$$

Since this value exceeds the maximum limit of $d/2 = 10$ in., the stirrups may all be placed at the limiting spacing, and a possible arrangement is as shown in Fig. 13.23*d*.

Note that in both examples the first stirrup is placed at one-half the required distance from the support.

Example 3. Determine the required number and spacings for No. 3 U-stirrups for the beam shown in Fig. 13.24. Use $f'_c = 3$ ksi [20.7 MPa] and $f_s = 20$ ksi [138 MPa].

Solution: In this case, the maximum critical design shear stress is found to be less than v_c, which in theory indicates that reinforcing is not required. To comply with the code requirement for minimum reinforcing, however, we provide stirrups at the maximum permitted spacing out to the point where the shear stress drops to 30 psi (one-half of v_c). To verify that the No. 3 stirrup is adequate, we compute

$$A_v = 50 \frac{10 \times 10}{50{,}000} = 0.10 \text{ in.}^2 \quad \text{(See Example 1)}$$

which is less than the area of 0.22 in. provided, so the No. 3 stirrup at 10-in. spacing is adequate.

The preceding examples have illustrated what is generally the simplest case for beam shear design—that of a beam with uniformly distributed load and with sections subjected only to flexure and shear. When concentrated loads or unsymmetrical loadings produce other forms for the shear diagram, these must be used for design of the shear reinforcing. In addition, where axial forces of tension or compression exist in the concrete frame, consideration must be given to the combined effects when designing for shear.

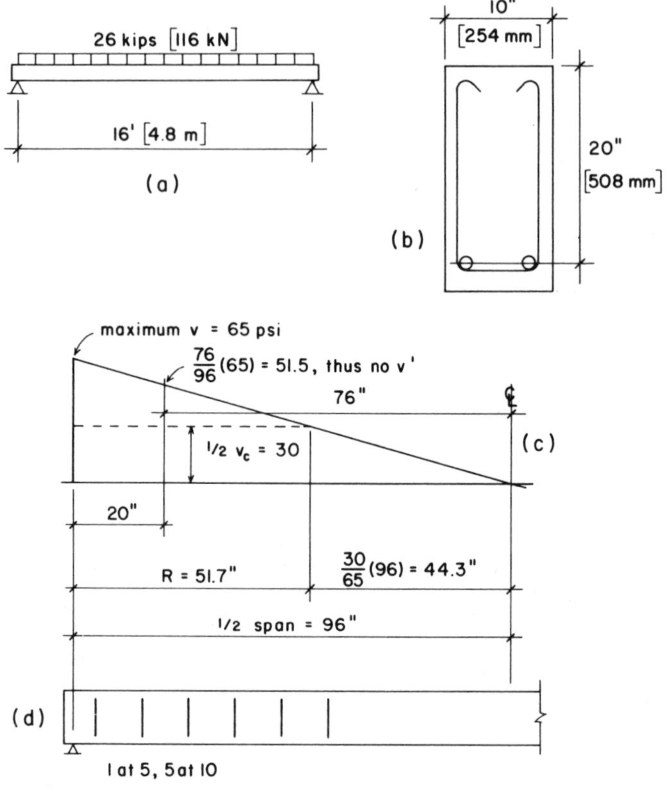

FIGURE 13.24

When torional moments exist (twisting moments at right angles to the beam), their effects must be combined with beam shear.

Problem 13.13.A. A concrete beam similar to that shown in Fig. 13.23 sustains a total load of 60 kips [267 kN] on a span of 24 ft [7.32 m]. Determine the layout for a set of No. 3 U-stirrups using f_s = 20 ksi [138 MPa] and f'_c = 3000 psi [20.7 MPa]. The section dimensions are b = 12 in. [305 mm] and d = 26 in. [660 mm].

Problem 13.13.B. Determine the layout for a set of No. 3 U-stirrups for a beam with the same data as Problem 13.13.A, except the total load on the beam is 30 kips [133 kN].

13.14 Development of Reinforcement

The 1983 ACI Code defines *development length* as the length of embedment required to develop the design strength of the reinforcing at a critical section. For beams, critical sections occur at points of maximum stress and at points within the span where some of the reinforcement terminates or is bent up or down. For a uniformly loaded simple span beam, one critical section is at midspan, where the bending moment is a maximum. The tensile reinforcing required for flexure at this point must extend on both sides a sufficient distance to develop the stress in the bars; however, except for very short spans with large bars, the bar lengths will ordinarily be more than sufficient.

In the simple beam, the bottom reinforcing required for the maximum moment at midspan is not entirely required as the moment decreases toward the end of the span. It is thus sometimes the practice to make only part of the midspan reinforcing continuous for the whole beam length. In this case it may be necessary to assure that the bars that are of partial length are extended sufficiently from the midspan point and that the bars remaining beyond the cutoff point can develop the stress required at that point.

When beams are continuous through the supports, top reinforcing is required for the negative moments at the supports. These top bars must be investigated for the development lengths in terms of the distance they extend from the supports.

For tension reinforcing consisting of bars of No. 11 size and smaller, the code specifies a minimum length for development (l_d) as follows:

$$l_d = 0.04 A_b \frac{f_y}{\sqrt{f'_c}}$$

but not less than $0.0004 d_b f_y$ or 12 in.

In these formulas A_b is the cross-sectional area of the bar and d_b is the bar diameter.

Modification factors for l_d are given for various situations, as follows:

For top bars in horizontal members with at least 12 in. of concrete below the bars	1.4
For sets of bars where the bars are 6 in. or more on center	0.8
For flexural reinforcement that is in excess of that required by computations	$\dfrac{A_s \text{ required}}{A_s \text{ provided}}$

Additional modification factors are given for lightweight concrete, for bars encased in spirals, and for bars with f_y in excess of 60 ksi.

Table 13.6 gives values for minimum development lengths for tensile reinforcing, based on the requirements of the 1983 ACI Code. The values listed under "Other Bars" are the unmodified length requirements; those listed under "Top Bars" are increased

TABLE 13.6 Minimum Development Length for Tensile Reinforcement (in.)[a]

	f_y = 40 ksi [276 MPa]				f_y = 60 ksi [414 MPa]			
	f'_c = 3 ksi [20.7 MPa]		f'_c = 4 ksi [27.6 MPa]		f'_c = 3 ksi [20.7 MPa]		f'_c = 4 ksi [27.6 MPa]	
Bar Size	Top Bars[b]	Other Bars	Top Bars[b]	Other Bars	Top Bars[b]	Other Bars	Top Bars[b]	Other Bars
3	12	12	12	12	13	12	13	12
4	12	12	12	12	17	12	17	12
5	14	12	12	12	21	15	21	15
6	18	13	16	12	27	19	15	18
7	25	18	21	15	37	26	32	23
8	32	23	28	20	48	35	42	30
9	41	29	36	25	61	44	53	38
10	52	37	45	32	78	56	68	48
11	64	46	55	40	96	68	83	59
14	87	62	75	54	130	93	113	81
18	113	80	98	70	169	120	146	104

[a] Lengths are based on requirements of the ACI Code (Ref. 3).
[b] Horizontal bars so placed that more than 12 in. [305 mm] of concrete is cast in the member below the reinforcement.

by the modification factor for this situation. Values are given for two concrete strengths and for the two most commonly used grades of tensile reinforcing.

When details of the construction restrict the ability to extend bars sufficiently to produce required development lengths, partial development can sometimes be achieved by use of a hooked end. Section 12.5 of the 1983 ACI Code provides a means by which a so-called standard hook may be evaluated in terms of an equivalent development length. Detailed requirements for standard hooks are given in Chapter 7 of the 1983 ACI Code. Bar ends may be bent at 90, 135, or 180° to produce a hook. The 135° bend is used only for ties and stirrups, which normally consist of relatively small diameter bars. (See Figs. 13.5 and 15.5.)

Table 13.7 gives values for standard hooks, using the same variables for f'_c and f_y that are used in Table 13.6. The table values given are in terms of the equivalent development length provided by the hook. Comparison of the values in Table 13.7 with those given for the unmodified lengths ("Other") in Table 13.6 will show that the hooks are mostly capable of only partial development. The development length provided by a hook may be added to whatever development length is provided by extension of the bar, so that the total development may provide for full utilization

TABLE 13.7 Equivalent Embedment Lengths of Standard Hooks (in.)

	f_y = 40 ksi [276 MPa]		f_y = 60 ksi [414 MPa]	
Bar Size	f'_c = 3 ksi [20.7 MPa]	f'_c = 4 ksi [27.6 MPa]	f'_c = 3 ksi [20.7 MPa]	f'_c = 4 ksi [27.6 MPa]
3	3.0	3.4	4.4	5.1
4	3.9	4.5	5.9	6.8
5	4.9	5.7	7.4	8.5
6	6.3	6.8	9.5	10.2
7	8.6	8.6	12.9	12.9
8	11.4	11.4	17.1	17.1
9	14.4	14.4	21.6	21.6
10	18.3	18.3	24.4	24.4
11	22.5	22.5	26.2	26.2

of the bar tension capacity (at f_y) in many cases. The following example illustrates the use of the data from Tables 13.6 and 13.7 for a simple situation.

Example. The negative moment in the short cantilever shown in Fig. 13.25 is resisted by the steel bar in the top of the beam. Determine whether the development of the reinforcing is adequate. Use $f'_c = 3$ ksi [20.7 MPa] and $f_y = 60$ ksi [414 MPa].

Solution: The maximum moment in the cantilever is produced at the face of the support; thus the full tensile capacity of the bar should be developed on both sides of this section. In the beam itself the condition is assumed to be that of a "top bar," for which Table 13.6 yields a required minimum development length of 27 in., indicating that the length of 46 in. provided is more than adequate. Within the support, the condition is unmodified, and the requirement is for a length of 19 in. The actual extended development length provided within the support is 14 in., which is measured as the distance to the end of the hooked bar end, as

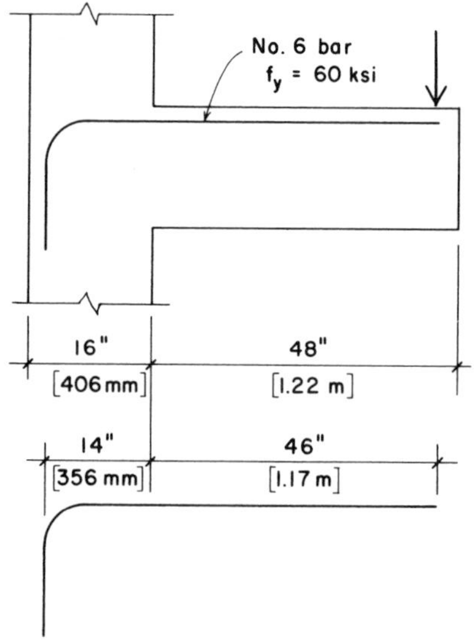

FIGURE 13.25

shown in the figure. If the hooked end qualifies as a *standard hook* (in accordance with the requirements of Chapter 7 of the 1983 ACI Code), the equivalent development length provided (from Table 13.7) is 9.5 in. Thus the total development provided by the combination of extension and hooking is 14 + 9.5 = 23.5 in., which exceeds the requirement of 19 in., so the development is adequate.

In a real situation, it is probably not necessary to achieve the full development lengths given in Table 13.6, since bar selection often results in some slight excess in the actual steel cross-sectional area provided. In such a case, the required development length can be reduced by the modification factor.

The discussion of development length so far has dealt with tension bars only. Development length in compression is, of course, a factor in column design and in the design of beams reinforced for compression.

The absence of flexural tension cracks in the portions of beams where compression reinforcement is employed, plus the beneficial effect of the end bearing of the bars on the concrete, permit shorter development lengths in compression than in tension. The ACI Code prescribes that l_d for bars in comrpession shall be computed by the formula

$$l_d = \frac{0.02 f_y d_b}{\sqrt{f'_c}}$$

but shall not be less than $0.0003 f_y d_b$ or 8 in., whichever is greater. Table 13.8 lists compression bar development lengths for a few combinations of specification data.

The ACI Code defines *development length* as the length of embedded reinforcement required to develop the design strength of the reinforcement at a critical section. Critical sections occur at points of maximum stress and at points within the span at which adjacent reinforcement terminates or is bent up into the top of the beam. For a uniformly loaded simple beam, one critical section is at midspan where the bending moment is maximum. This is a point of maximum tensile stress in the reinforcement

TABLE 13.8 Minimum Development Length for Compressive Reinforcement (in.)

Bar Size	f_y = 40 ksi [276 MPa]		f_y = 60 ksi [414 MPa]	
	f'_c = 3 ksi [20.7 MPa]	f'_c = 4 ksi [27.6 MPa]	f'_c = 3 ksi [20.7 MPa]	f'_c = 4 ksi [27.6 MPa]
3	8.0	8.0	8.0	8.0
4	8.0	8.0	11.0	9.5
5	9.2	8.0	13.7	11.9
6	10.9	9.5	16.4	14.2
7	12.8	11.1	19.2	16.6
8	14.6	12.7	21.9	19.0
9	16.5	14.3	24.8	21.5
10	18.5	16.1	27.8	24.1
11	20.6	17.9	31.0	26.8
14			37.1	32.1
18			49.5	42.8

(peak bar stress) and some length of bar is required over which the stress can be developed. Other critical sections occur between midspan and the reactions at points where some bars are cut off because they are no longer needed to resist the bending moment; such terminations create peak stress in the remaining bars that extend the full length of the beam.

When beams are continuous through their supports, the negative moments at the supports will require that bars be placed in the top of the beams. Within the span, bars will be required in the bottom of the beam for the positive moments. While the positive moment will go to zero at some distance from the supports, the codes require that some of the positive moment reinforcing be extended for the full length of the span and a short distance into the support.

Figure 13.26 shows a possible layout for reinforcing in a beam with continuous spans and a cantilevered end at the first support. Referring to the notation in the illustration, we make the following observations:

1. Bars a and b are provided for the maximum moment of positive sign that occurs somewhere near the beam mid-

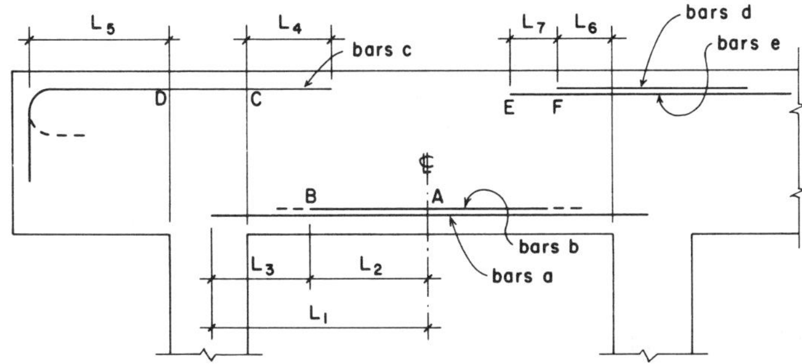

FIGURE 13.26 Various situations for consideration of development length.

span. If all these bars are made full length (as shown for bars a), the length L_1 must be sufficient for development (this situation is seldom critical). If bars b are partial length, as shown in the illustration, then length L_2 must be sufficient to develop bars b and length L_3 must be sufficient to develop bars a. As was discussed for the simple beam, the partial length bars must actually extend beyond the theoretical cutoff point (*B* in the illustration) and the true length must include the dashed portions indicated for bars b.

2. For the bars at the cantilevered end, the distances L_4 and L_5 must be sufficient for development of bars c. L_4 is required to extend beyond the actual cutoff point of the negative moment by the extra length described for the partial length bottom bars. If L_5 is not adequate, the bar ends may be bent into the 90° hook as shown or the 180° hook shown by the dashed line.
3. If the combination of bars shown in the illustration is used at the interior support, L_6 must be adequate for the development of bars d and L_7 adequate for the development of bars e.

For a single loading condition on a continuous beam it is possible to determine specific values of moment and their location along the span, including the locations of points of zero moment. In practice, however, most continuous beams are designed for

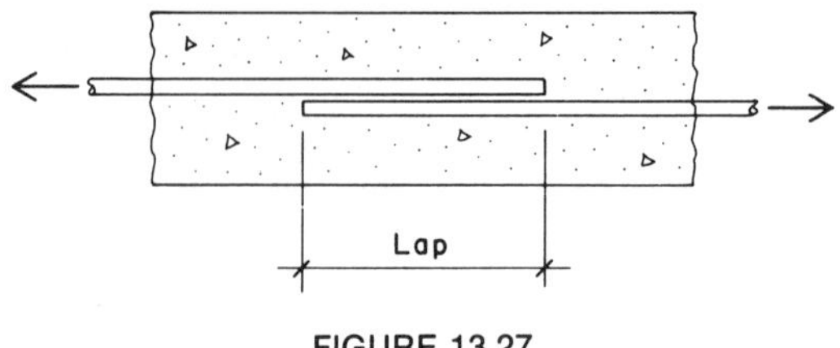

FIGURE 13.27

more than a single loading condition, which further complicates the problems of determining development lengths required.

In various situations in reinforced concrete structures it becomes necessary to transfer stress between steel bars in the same direction. Continuity of force in the bars is achieved by splicing, which may be effected by welding, by mechanical means, or by the so-called lapped splice. Figure 13.27 illustrates the concept of the lapped splice, which consists essentially of the development of both bars within the concrete. The length of the lap becomes the develoment length for both bars. Because a lapped splice is usually made with the two bars in contact, the lapped length must usually be somewhat greater than the simple development length required in Table 13.6.

Sections 12-14 to 13-19 of the 1983 ACI Code give requirements for various types of splices. For a simple tension lap splice, the full development of the bars requires a lap length of 1.7 times that required for simple development of the bars. Lap splices are generally limited to bars of No. 11 size and smaller.

For pure tension members, lapped splicing is not permitted, and splicing must be achieved by welding the bars or by some mechanical connection. End-to-end butt welding of bars is usually limited to compression splicing of large diameter bars with high f_y for which lapping is not feasible.

When members have several reinforcing bars that must be spliced, the splicing must be staggered. In general, splicing is not desirable, and is to be avoided where possible. Because bars are obtainable only in limited lengths, however, some situations unavoidably involve splicing. Horizontal reinforcing in long walls is one such case.

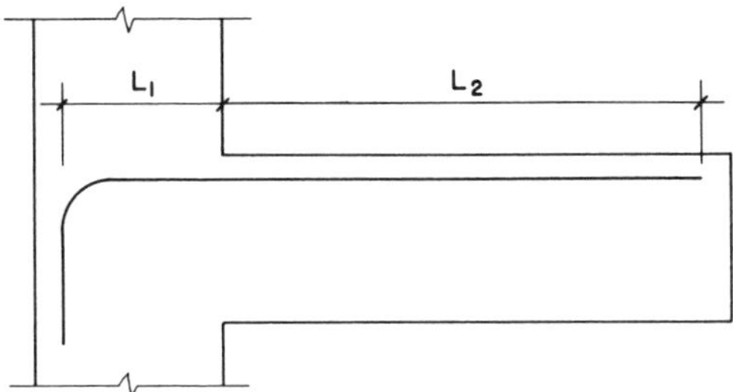

FIGURE 13.28

Problems 13.14.A,B,C,D. Determine whether the bar shown in Fig. 13.28 is adequately anchored for development of the full f_y capacity of the steel.

	f'_c (ksi)	f'_c (MPa)	f_y (ksi)	f_y (MPa)	L_2 (in.)	L_2 (mm.)	L_1 (in.)	L_1 (mm.)	Bar Size
A	3	20.7	60	414	36	914	12	305	8
B	4	27.6	60	414	42	1067	16	206	10
C	3	20.7	40	276	38	965	18	457	9
D	4	27.6	40	276	35	889	17	432	9

13.15 Deflection Control

Deflection of spanning slabs and beams of poured-in-place concrete is controlled primarily by using recommended minimum thicknesses (overall height) expressed as a percentage of the span. Table 13.9 is adapted from a similar table given in Section 9.5 of the 1983 ACI Code and yields minimum thicknesses as a fraction of the span. Table values apply only for concrete of normal weight (made with ordinary sand and gravel) and for reinforcing with f_y of 60 ksi [414 MPa]. The Code supplies correction factors for other concrete weights and reinforcing grades. It further stipulates that these recommendations apply only where beam deflections are not critical for other elements of the building construction, such as supported partitions subject to cracking caused by beam deflections.

Deflection of concrete structures presents a number of special problems. For concrete with ordinary reinforcing (not pre-

TABLE 13.9 Minimum Thickness of One-Way Slabs or Beams Unless Deflections Are Computed

Type of Member	End Conditions	Minimum Thickness of Slab or Height of Beam: f_y = 40 ksi [276 MPa]	Minimum Thickness of Slab or Height of Beam: f_y = 60 ksi [414 MPa]
Solid one-way slabs[a]	Simple support	$L/25$	$L/20$
	One end continuous	$L/30$	$L/24$
	Both ends continuous	$L/35$	$L/28$
	Cantilever	$L/12.5$	$L/10$
Beams or joists	Simple support	$L/20$	$L/16$
	One end continuous	$L/23$	$L/18.5$
	Both ends continuous	$L/26$	$L/21$
	Cantilever	$L/10$	$L/8$

Source: Data adapted from *Building Code Requirements for Reinforced Concrete* (ACI 318–83) (Ref. 3), 1983 ed., with permission of the publishers, American Concrete Institute.

[a] Valid only for members not supporting or attached to partitions or other construction likely to be damaged by large deflections.

stressed), flexural action normally results in tension cracking of the concrete at points of maximum bending. Thus the presence of cracks in the bottom of a beam at midspan points and in the top over supports is to be expected. In general, the size (and visibility) of these cracks will be proportionate to the amount of beam curvature produced by deflection. Crack size will also be greater for long spans and for deep beams. If visible cracking is considered objectionable, more conservative depth–span ratios should be used, especially for spans over 30 ft and beam depths over 30 in.

In heavily loaded concrete members there is a tendency for deformations to increase with lapse of time even under constant load conditions; this deformation is called *creep*. The effect of creep is equivalent to a decrease in the modulus of elasticity. The values for E_c, determined as described in Sec. 13.6, may therefore be used only for computation of deflections that occur immediately on application of the service loads. Long-time deflections of slabs, beams, and other floor and roof systems may be two or more times larger than the initial deflection. Control of deflections includes concerns for effects of creep. Creep deflections will be of greatest concern when any of the following exist:

1. Use of low percentages of tensile reinforcement and no compressive reinforcement.
2. Use of very low strength concrete (F'_c less than 3000 psi).
3. Total load that is predominantly dead load.
4. Use of lightweight aggregate to reduce the concrete density significantly.

In beams, deflections, especially creep deflections, may be reduced by the use of some compressive reinforcing. Where deflections are of concern, or where depth–span ratios are pushed to their limits, it is advisable to use some compressive reinforcing, consisting of continuous top bars.

When, for whatever reasons, deflections are deemed to be critical, computations of actual values of deflection may be necessary. Section 9.5 of the 1983 ACI Code provides directions for such computations; they are quite complex in most cases, and beyond the scope of this work. In actual design work, however, they are required very infrequently.

14

Concrete Framing Systems

There are many different reinforced concrete floor systems, both cast in place and precast. The cast-in-place systems are generally of one of the following types:

1. One-way solid slab and beam.
2. Two-way solid slab and beam.
3. One-way concrete joist construction.
4. Two-way flat slab or flat plate without beams.
5. Two-way joist construction, called waffle construction.

Each system has its distinct advantages and limitations, depending on the spacing of supports, magnitude of loads, required fire rating, and cost of construction. The floor plan of the building and the purpose for which the building is to be used determine loading conditions and the layout of supports.

14.1 Slab-and-Beam Systems

The most widely used and most adaptable poured-in-place concrete floor system is that which utilizes one-way solid slabs supported by one-way spanning beams. This system may be used for single spans, but occurs more frequently with multiple-span slabs and beams in a system such as that shown in Fig. 14.1. In the example shown, the continuous slabs are supported by a series of beams that are spaced at 10 ft center to center. The beams, in turn, are supported by a girder and column system with columns at 30-ft centers, every third beam being supported directly by the columns and the remaining beams being supported by the girders.

Because of the regularity and symmetry of the system shown in Fig. 14.1, there are relatively few different elements in the system, each being repeated several times. While special members must be designed for conditions that occur at the outside edge of the system and at the location of any openings for stairs, elevators, and so on, the general interior portions of the structure may be determined by designing only six basic elements: S1, S2,

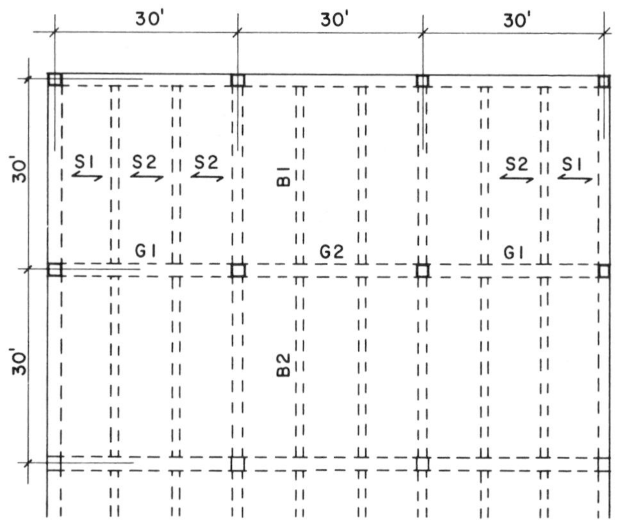

FIGURE 14.1 Plan of a typical slab–beam–girder framing system.

B1, B2, G1, and G2, as shown in the framing plan. The design of these typical elements is illustrated in Sec. 22.7 in Part V.

In computations for reinforced concrete, the span length of freely supported beams (simple beams) is generally taken as the distance between centers of supports or bearing areas; it should

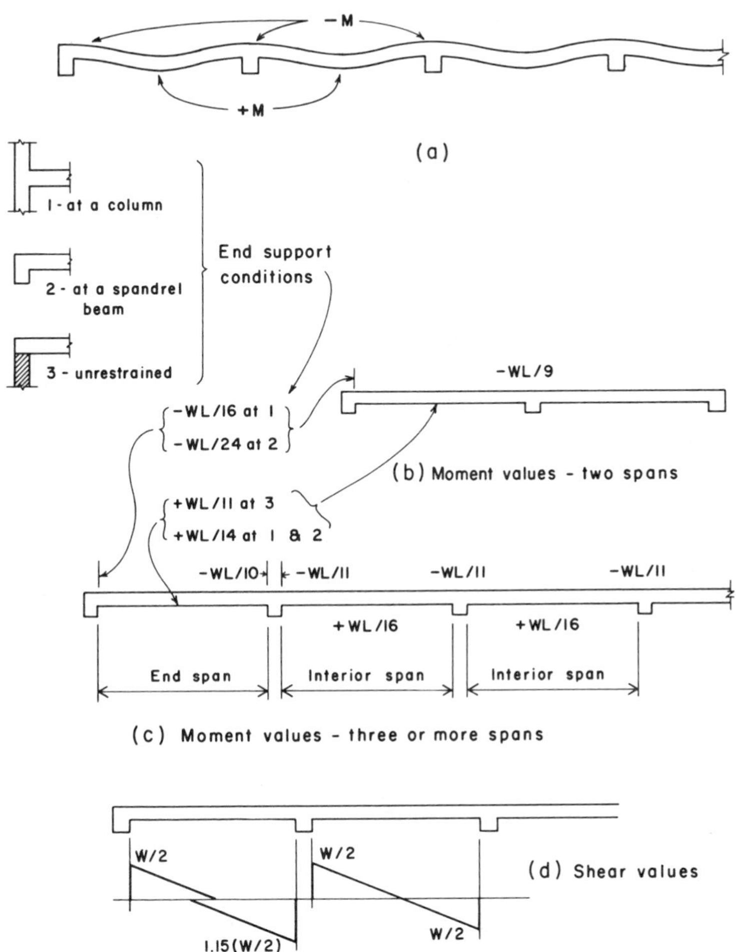

FIGURE 14.2 Approximate design factors for continuous structures.

not exceed the clear span plus the depth of beam or slab. The span length for continuous or restrained beams is taken as the clear distance between faces of supports. For a simple beam, that is, a single span having no restraint at the supports, the maximum bending moment for a uniformly distributed load is at the center of the span, and its magnitude is $M = WL/8$. The moment is zero at the supports and is positive over the entire span length. In continuous beams, however, negative bending moments are developed at the supports and positive moments at or near midspan. This may be readily observed from the exaggerated deformation curve of Fig. 14.2*a*. The exact values of the bending moments depend on several factors, but in the case of approximately equal spans supporting uniform loads, when the live load does not exceed three times the dead load, the bending moment values given in Fig. 14.2 may be used for design.

The values given in Fig. 14.2 are in general agreement with the recommendations of Chapter 8 of the 1983 ACI Code. These values have been adjusted to account for partial live loading of multiple-span beams. Note that these values apply only to uniformly loaded beams. Chapter 8 of the 1983 ACI Code also gives some factors for end-support conditions other than the simple supports shown in Fig. 14.2.

Design moments for continuous-span slabs are given in Fig. 14.3. Where beams are relatively large and the slab spans are small, the rotational (torsional) stiffness of the beam tends to minimize the effect of individual slab spans on the bending in adjacent spans. Thus most slab spans in the slab–beam systems tend to function much like individual spans with fixed ends.

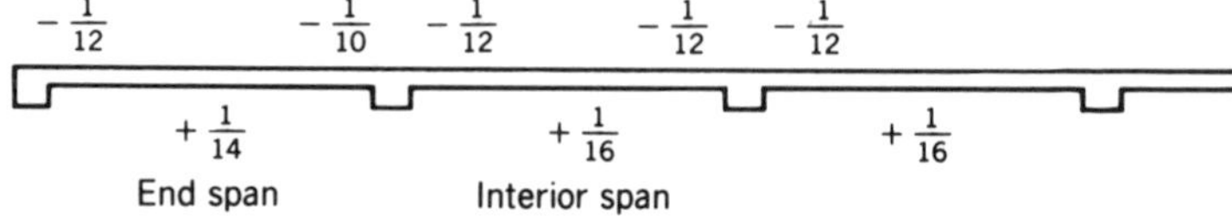

FIGURE 14.3 Approximate design factors for continuous slabs with spans of 10 ft or less.

14.2 Design of a One-Way Continuous Slab

The general design procedure for a one-way solid slab was illustrated in Sec. 13.12. The example given there is for a simple span slab. The following example illustrates the procedure for the design of a continuous solid one-way slab.

Example. A solid one-way slab is to be used for a framing system similar to that shown in Fig. 14.1. Column spacing is 30 ft, with evenly spaced beams occurring at 10 ft center to center. Superimposed loads on the structure (floor live load plus other construction dead load) are a total of 138 psf. Use $f'_c = 3$ ksi [20.7 MPa] and Grade 60 reinforcement with $f_y = 60$ ksi [414 MPa] and $f_s = 24$ ksi [165 MPa]. Determine the thickness for the slab and pick its reinforcement.

Solution: To find the slab thickness, we consider three factors: the minimum thickness for deflection, the minimum effective depth for the maximum moment, and the minimum effective depth for the maximum shear. For all of these we must first determine the span of the slab. For design purposes this is taken as the clear span, which is the dimension from face to face of the supporting beams. With the beams at 10-ft centers, this dimension is 10 ft, less the width of one beam. Since the beams are not given, we will assume a dimension for them. In practice we would proceed from the slab design to the beam design, after which the assumed dimension could be verified. For this example we will assume a beam width of 12 in., yielding a clear span of 9 ft.

We consider first the minimum thickness required for deflection. If the slabs in all spans have the same thickness (which is the most common practice), the critical slab is the end span, since there is no continuity of the slab beyond the end beam. While the beam will offer some restraint, it is best to consider this as a simple support; thus we use the factor of $L/30$ from Table 13.9.

$$\text{Minimum } t = \frac{L}{30} = \frac{9 \times 12}{30} = 3.6 \text{ in.}$$

We will assume here that fire-resistive requirements make it desirable to have a relatively thick slab and so will choose a 5-in. overall thickness, for which the dead weight of the slab will be

$$w = \frac{5}{12} \times 150 = 62 \text{ psf}$$

and the total design load is thus 62 + 138 = 200 psf.

We next consider the maximum bending moment. Inspection of the moment values given in Fig. 14.3 shows the maximum value to be $\frac{1}{10} wL^2$. With the clear span and the loading as determined, the maximum moment is thus

$$M = \tfrac{1}{10} wL^2 = (0.1)(200)(9)^2 = 1620 \text{ ft-lb}$$

This moment value should now be compared to the balanced moment capacity of the design section, using the relationships discussed for rectangular beams in Sec. 13.7. For this computation we must assume an effective depth for the design section. This dimension will be the slab thickness minus the concrete cover and one-half the bar diameter. With the bars not yet determined, we will assume an approximate effective depth of the slab thickness minus 1.0 in.; this will be exactly true with the usual minimum cover of 1 in. and a No. 4 bar. Then using the balanced moment R factor from Table 13.3, the maximum resisting moment for the 12-in.-wide design section is

$$M_R = Rbd^2 = (0.204)(12)(4)^2 = 39.17 \text{ kip-in.}$$

or

$$M_R = 39.17 \times \frac{1000}{12} = 3264 \text{ ft-lb}$$

As this value is in excess of the required maximum moment, the slab will be adequate for concrete flexural stress.

It is not practicable to use shear reinforcement in one-way slabs, and consequently the maximum unit shear stress must be kept within the limit for the concrete alone. The usual procedure is to check the shear stress with the effective depth determined for bending before proceeding to find A_s. Except for very short span slabs with excessively heavy loadings, shear stress is seldom critical.

Finally, before proceeding with the design of the reinforcing, we should verify our slab thickness for shear stress. For an interior span, the maximum shear will be $wL/2$, but for the end span it is the usual practice to consider some unbalanced condition for the shear due to the discontinuous end. We therefore use a maximum shear of $1.15wL/2$, or an increase of 15% over the simple beam shear value. Thus

$$\text{Maximum shear} = V = 1.15wL/2$$

$$= 1.15(200)(9/2) = 1035 \text{ lb}$$

and

$$\text{Maximum shear stress} = v = \frac{V}{bd} = \frac{1035}{12 \times 4} = 22 \text{ psi}$$

This is considerably less than the limit for the concrete alone ($v_c = 1.1\sqrt{f'_c} = 60$ psi), so the assumed slab thickness is not critical for shear stress.

Having thus verified our choice for the slab thickness, we may now proceed with the design of the reinforcing. For a balanced section, Table 13.3 yields a value of 0.886 for the j factor. However, since all of our reinforced sections will be classified as under-reinforced (actual moment less than the balanced limit), we will use a slightly higher value, say 0.90, for j in the design of the reinforcing.

Referring to Fig. 14.4, we note that there are five critical locations for which a moment must be determined and the required steel area computed. Reinforcing in the top of the slab must be computed for the negative moments at the end support, at the first interior beam, and at the typical interior beam. Reinforcing in the bottom of the slab must be computed for the positive moments at

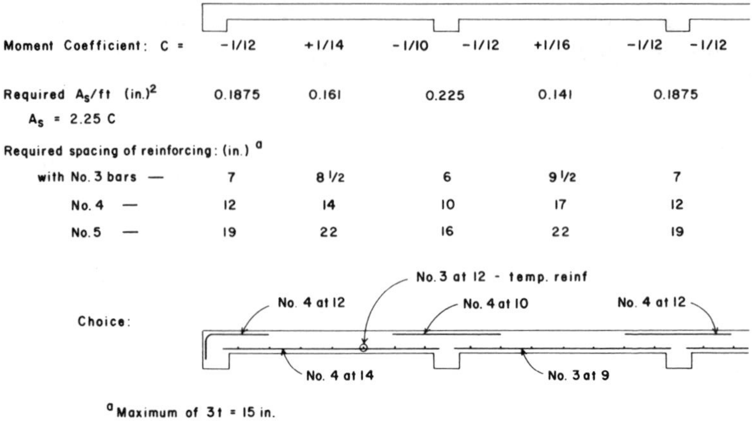

FIGURE 14.4 Design of the continuous slab.

midspan in the first span and the typical interior spans. The design for these conditions is summarized in Fig. 14.4. For the data displayed in the figure we note the following:

$$\text{Maximum spacing of reinforcing} = 3 \times t = 3 \times 5 = 15 \text{ in.}$$

$$\begin{aligned}\text{Maximum moment} = M &= (\text{moment factor } C)(wL^2)\\ &= C(200)(9)^2 \times 12\\ &= C(194{,}400) \qquad (\text{in in.-lb units})\end{aligned}$$

$$\text{Required } A_s = \frac{M}{f_s jd} = \frac{C(194{,}000)}{(24{,}000)(0.9)(4)} = 2.25C$$

Using data from Table 13.5, Fig. 14.4 shows required spacings for No. 3, 4, and 5 bars. A possible choice for the slab reinforcement, using all straight bars, is shown at the bottom of the figure.

Problem 14.2.A. A solid one-way slab is to be used for a framing system similar to that shown in Fig. 14.1. Column spacing is 36 ft [11 m], with regularly spaced beams occurring at 12 ft [3.66 m] center to center. Superimposed loads on the structure is a total of 180 psf [8.62 kN/m^2]. Use f'_c = 3 ksi [20.7 MPa] and Grade 40 reinforcing with f_y = 40 ksi [276 MPa] and f_s = 20 ksi [138 MPa]. Determine the thickness for the slab and select the size and spacing for the bars.

14.3 Continuous Concrete Beams

Multiple-span beams in poured-in-place concrete structures are ordinarily of a continuous-span nature. Special jointing is required to make them otherwise, and is only done in special situations, such as when separation is required for control of thermal expansion or seismic actions. By contrast, beams in structures of wood or steel are ordinarily of one piece per span and require special jointing to develop continuous actions. An exception to the latter is the case where beam joints are made off of the supports, as discussed in Sec. 2.14, which simulates some of the continuous-beam actions.

When the concrete construction is of the form shown in Fig. 14.1, beam cross sections will be of the T-form, and the beam will act as a T-beam for development of positive (midspan) moments. (See discussion in Sec. 13.10.) For negative moments, with tension in the top of the section, the T-form is not significant and the beam resisting moment must be developed by the beam stem alone, acting as a rectangular section, as discussed in Sec. 13.7. Since the negative moments at supports are ordinarily the maximum in magnitude, the stem size is frequently determined on the basis of this action, and it is quite common to use some compressive reinforcement to assist in development of the maximum bending resistance, as discussed in Sec. 13.11.

For floor construction in multistory buildings, the depth of the spanning structure is usually limited in order to reduce the floor-to-floor vertical distance. This often results in beam depths which are less than optimal, making necessary the use of higher percentages of reinforcement and the providing of compressive reinforcement for the critical maximum negative moments on the larger members of the framing systems.

Reinforcement must be designed and provided for the individual members of the framing system: slabs, beams, girders, and columns. Placement of reinforcing in the individual members must respond to requirements for cover and minimum or maximum bar spacings. A special problem is that which occurs where members intersect and the reinforcement for the various individ-

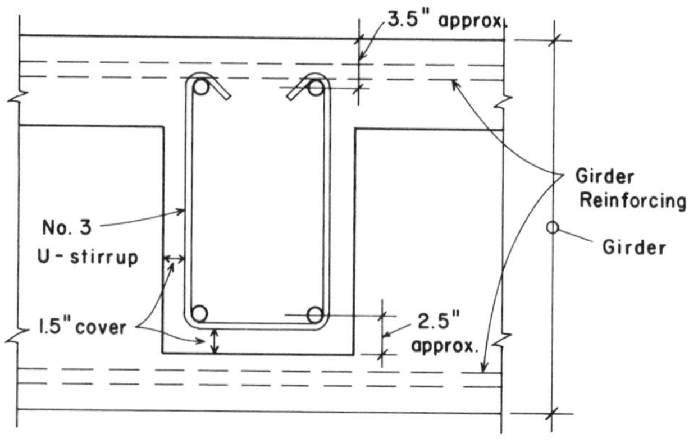

FIGURE 14.5

ual members must be merged; for example, where the slab meets a beam and the slab reinforcement must pass over the beam reinforcement. The latter is not usually a problem, as the cover required for slabs is considerably less than that for beams, so the top bars in the slab are closer to the top surface of the concrete. When beams intersect, however, and both have the same cover requirements, a potential conflict occurs, and it may be necessary to make a decision regarding which beam to give precedence so that its reinforcement is placed in the most desirable location. Figure 14.5 shows the situation for a beam in which the top bars have been lowered in position to allow the top bars of the girder to have the better position, probably a logical decision if the girder is more heavily loaded and stressed.

Another critical location is at the columns, where a three-dimensional traffic jam occurs with the intersection of reinforcement for the beam, girder, and column. This requires careful consideration of bar sizes and positions and may affect the choice of widths of the members to permit some leeway in bar locations in the individual members.

Examples of the design of continuous beams are given in the building design example in Chapter 22.

14.4 One-Way Joist Construction

Figure 14.6 shows a partial framing plan and some details for a type of construction that utilizes a series of very closely spaced beams and a relatively thin solid slab. Because of its resemblance to ordinary wood joist construction, this is called concrete joist construction. This system is generally the lightest (in dead weight) of any type of flat-spanning, poured-in-place concrete construction and is structurally well suited to the light loads and medium spans of office buildings and commercial retail buildings.

Slabs as thin as 2 in. and joists as narrow as 4 in. are used with this construction. Because of the thinness of the parts and the small amount of cover provided for reinforcement (typically $\frac{3}{4}$ to 1 in. for joists versus 1.5 in. for ordinary beams), the construction has very low resistance to fire, especially when exposed from the underside. It is therefore necessary to provide some form of fire protection, as for steel construction, or to restrict its use to situations where high fire ratings are not required.

The relatively thin, short-span slabs are typically reinforced with welded wire mesh rather than ordinary deformed bars. Joists are often tapered at their ends, as shown in the framing plan in Fig. 14.6. This is done to provide a larger cross section for increased resistance to shear and negative moment at the supports. Shear reinforcement in the form of single vertical bars may be provided, but is not frequently used.

Early joist construction was produced by using lightweight hollow clay tile blocks to form the voids between joists. These blocks were simply arranged in spaced rows on top of the forms, the joists being formed by the spaces between the rows. The resulting construction provided a flat underside to which a plastered ceiling surface could be directly applied. Hollow, lightweight concrete blocks later replaced the clay tile blocks. Other forming systems have utilized plastic-coated cardboard boxes, fiberglass-reinforced pans, and formed sheet-metal pans. The latter method was very widely used, the metal pans being pried off after the pouring of the concrete and reused for several additional pours. The tapered joist cross section shown in Fig. 14.6 is typical

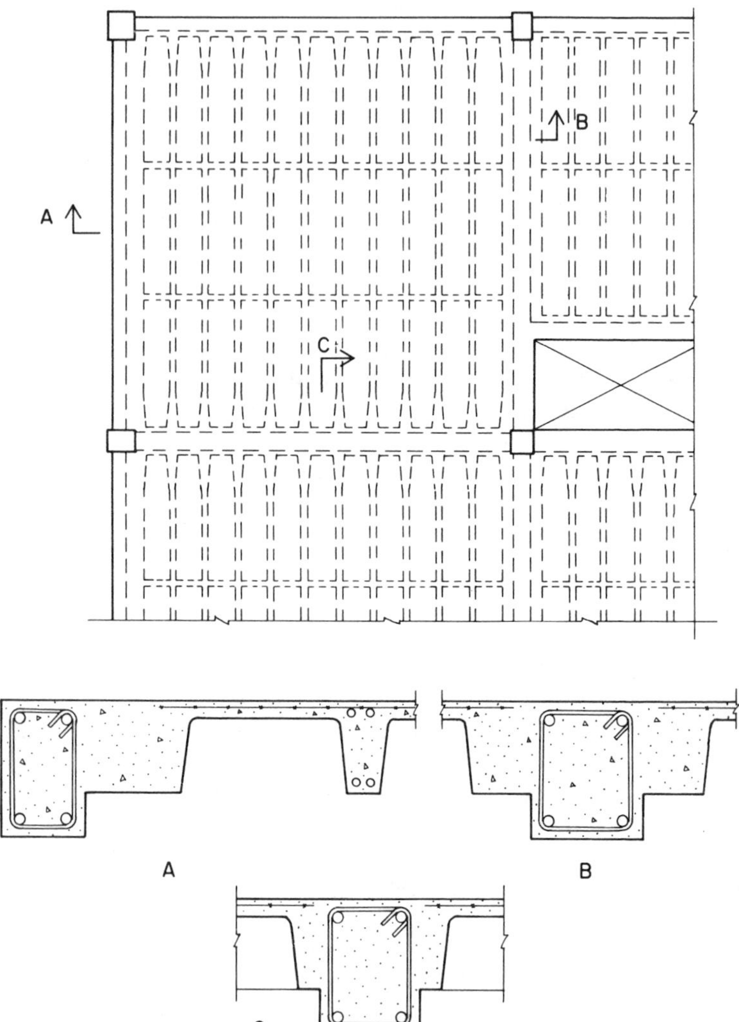

FIGURE 14.6 Typical concrete one-way joist construction.

of this construction, since the removal of the metal pans requires it.

Wider joists can be formed by simply increasing the space between forms, with large beams being formed in a similar manner or by the usual method of extending a beam stem below the construction, as shown for the beams in Fig. 14.6. Because of the narrow joist forms, cross-bridging is usually required, just as with wood joist construction. The framing plan in Fig. 14.6 shows the use of two bridging strips in the typical bay of the framing.

Design of joist construction is essentially the same as for ordinary slab-and-beam construction. Some special regulations are given in the ACI Code for this construction, such as the reduced cover mentioned previously. Because joists are so commonly formed with standard-sized metal forms, there are tabulated designs for typical systems in various handbooks. The *CRSI Handbook* (Ref. 6) has extensive tables offering complete designs for various spans, loadings, pan sizes, and so on. Whether for final design or simply for a quick preliminary design, the use of such tables is quite efficient.

One-way joist construction was highly popular in earlier times, but has become less utilized, due to its lack of fire resistance and the emergence of other systems. The popularity of lighter, less fire-resistive ceiling construction has been a contributing factor. In the right situation, however, it is still a highly efficient type of construction.

14.5 Waffle Construction

Waffle construction consists of two-way spanning joists that are formed in a manner similar to that for one-way spanning joists, using forming units of metal, plastic, or cardboard to produce the void spaces between the joists. The most widely used type of waffle construction is the waffle flat slab, in which solid portions around column supports are produced by omitting the void-making forms. An example of a portion of such a system is shown in Fig. 14.7. This type of system is analogous to the solid flat slab, which will be discussed in Sec. 14.7. At points of discontinuity in

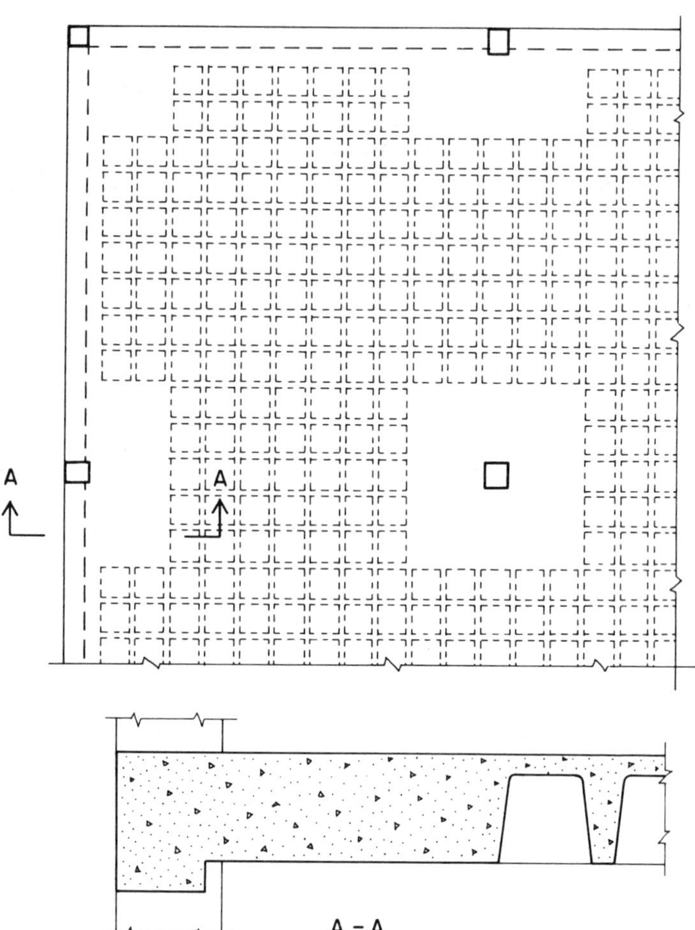

FIGURE 14.7 Typical concrete waffle construction.

the plan—such as at large openings or at edges of the building—it is usually necessary to form beams. These beams may be produced as projections below the waffle, as shown in Fig. 14.7, or may be created within the waffle depth by omitting a row of the void-making forms, as shown in Fig. 14.8.

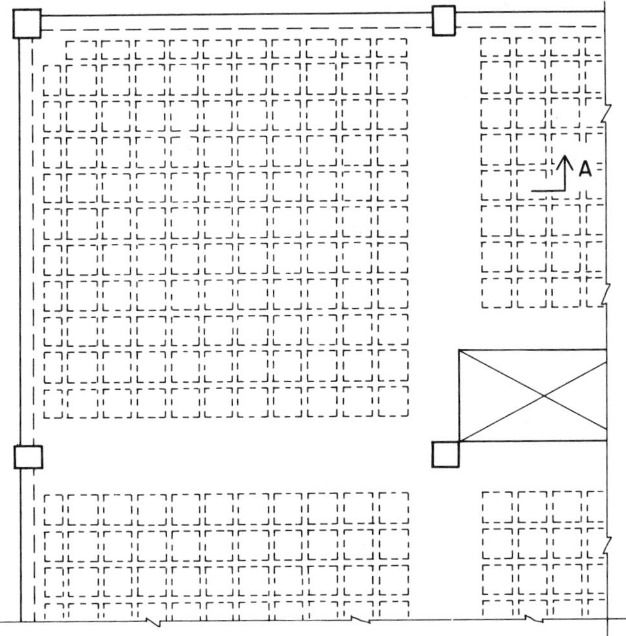

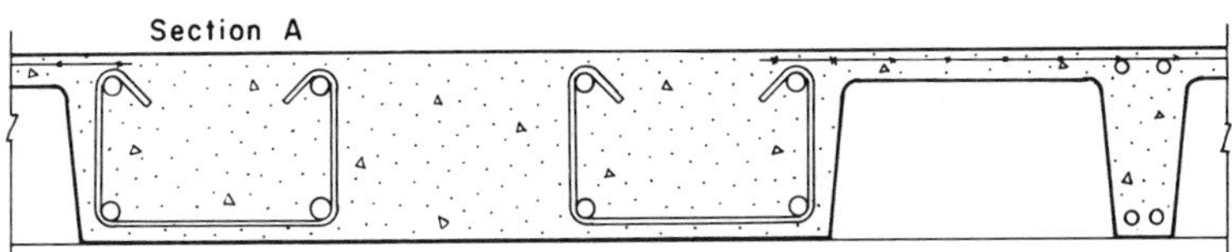

FIGURE 14.8 Typical concrete waffle construction with column line beams.

If beams are provided on all of the column lines, as shown in Fig. 14.8, the construction is analogous to the two-way solid slab with edge supports, as discussed in Sec. 14.7. With this system, the solid portions around the column are not required, since the waffle itself does not achieve the transfer of high shear or development of the high negative moments at the columns.

As with the one-way joist construction, fire ratings are low for ordinary waffle construction. The system is best suited for situations involving relatively light loads, medium-to-long spans, approximately square column bays, and a reasonable number of multiple bays in each direction.

For the waffle construction shown in Fig. 14.7, the edge of the structure represents a major discontinuity when the column supports occur immediately at the edge, as shown. Where planning permits, a more efficient use of the system is represented by the partial framing plan shown in Fig. 14.9, in which the edge occurs some distance past the columns. This projected edge provides a greater shear periphery around the column and helps to generate a negative moment, preserving the continuous character of the spanning structure. With the use of the projected edge, it may be possible to eliminate the edge beams shown in Fig. 14.7, thus preserving the waffle depth as a constant.

Another variation for the waffle is the blending of some one-way joist construction with the two-way waffle joists. This may be achieved by keeping the forming the same as for the rest of the

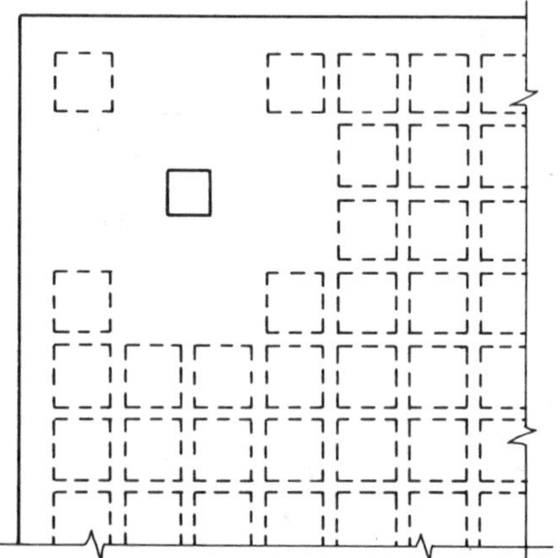

FIGURE 14.9 Plan of waffle construction with cantilevered edge and no edge beams.

waffle construction and merely using the ribs in one direction to create the spanning structure. One reason for doing this would be a situation similar to that shown in Fig. 14.8, where the large opening for a stair or elevator results in a portion of the waffle (the remainder of the bay containing the opening) being considerably out of square, that is, having one span considerably greater than the other. The joists in the short direction in this case will tend to carry most of the load due to their greater stiffness (less deflection than the longer spanning joists that intersect them). Thus the short joists would be designed as one-way spanning members and the longer joists would have only minimum reinforcing and serve as bridging elements.

The two-way spanning waffle systems are quite complex in structural behavior and their investigation and design are beyond the scope of this book. Some aspects of this work are discussed in the next article, since there are many similarities between the two-way spanning waffle systems and the two-way spanning solid slab systems. As with the one-way joist system, there are some tabulated designs in various handbooks that may be useful for either final or preliminary design. The *CRSI Handbook* (Ref. 6) mentioned previously has some such tables.

14.6 Two-Way Spanning Solid-Slab Construction

If reinforced in both directions, the solid concrete slab may span two ways as well as one. The widest use of such a slab is in flat-slab or flat-plate construction. In flat-slab construction, beams are used only at points of discontinuity, with the typical system consisting only of the slab and the strengthening elements used at column supports. Typical details for a flat-slab system are shown in Fig. 14.10. Drop panels consisting of thickened portions square in plan are used to give additional resistance to the high shear and negative moment that develops at the column supports. Enlarged portions are also sometimes provided at the tops of the columns (called column capitals) to reduce the stresses in the slab further.

Two-way slab construction consists of multiple bays of solid two-way spanning slabs with edge supports consisting of bearing

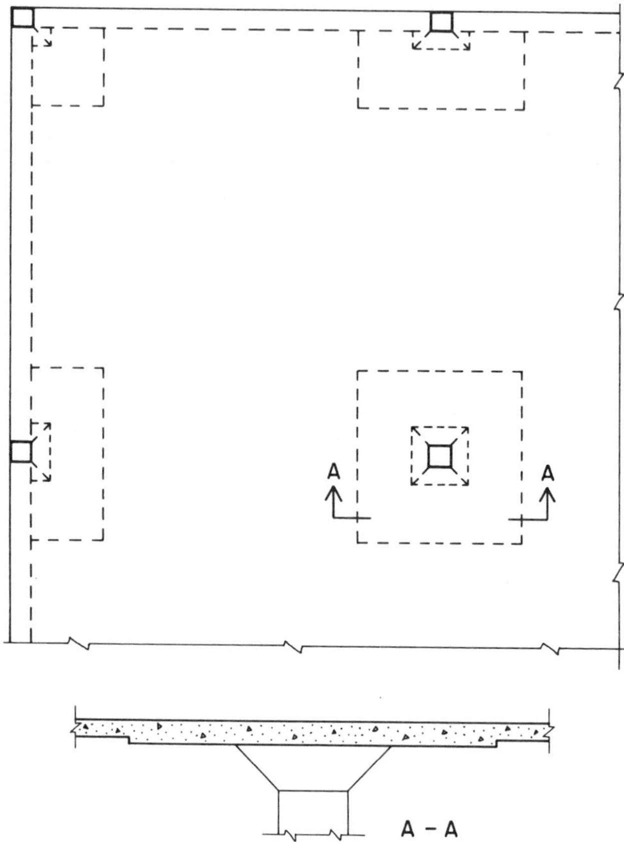

FIGURE 14.10 Typical concrete flat-slab construction with drop panels and column caps.

walls of concrete or masonry or of column-line beams formed in the usual manner. Typical details for such a system are shown in Fig. 14.11.

Two-way solid-slab construction is generally favored over waffle construction where higher fire rating is required for the unprotected structure or where spans are short and loadings high. As with all types of two-way spanning systems, they function most

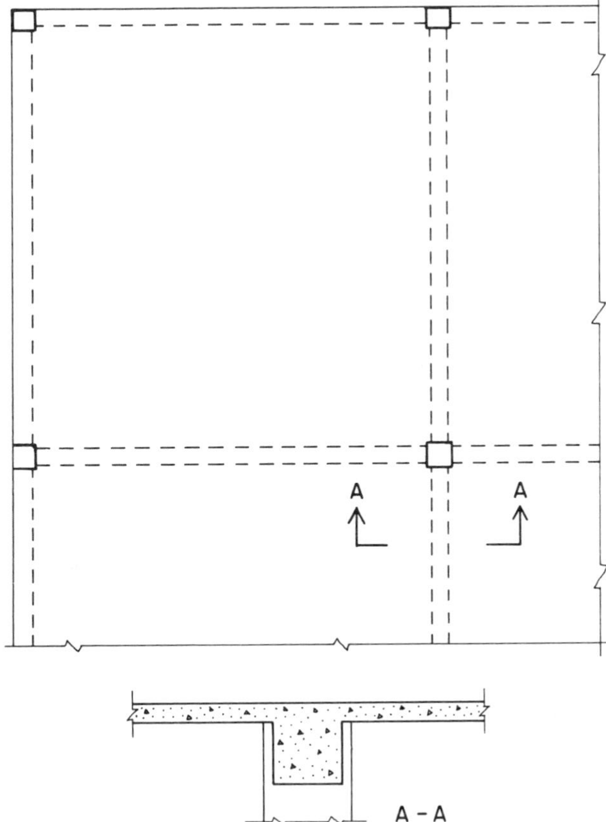

FIGURE 14.11 Typical two-way spanning concrete slab construction with edge supports.

efficiently where the spans in each direction are approximately the same.

For investigation and design, the flat slab (Fig. 14.10) is considered to consist of a series of one-way spanning solid-slab strips. Each of these strips spans through multiple bays in the manner of a continuous beam and is supported either by columns or by the strips that span in a direction perpendicular to it. The analogy for this is shown in Fig. 14.12*a*.

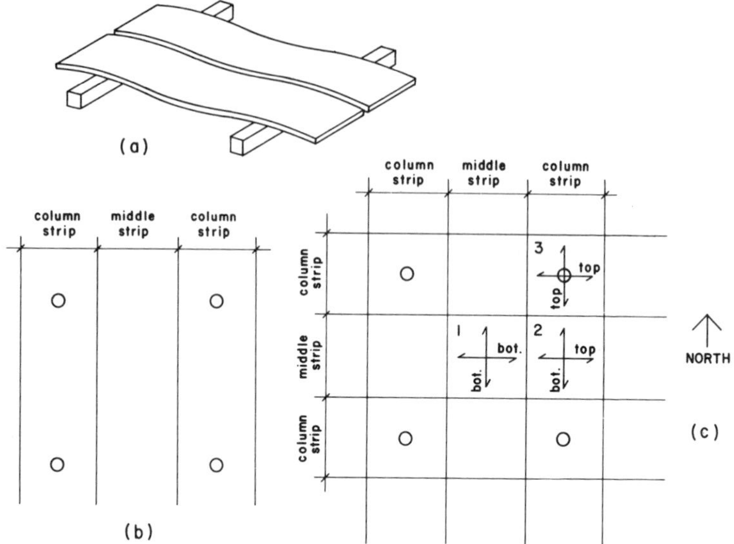

FIGURE 14.12 Development of the two-way concrete flat slab.

As shown in Fig. 14.12*b*, the slab strips are divided into two types: those passing over the columns, and those passing between columns, called middle strips. The complete structure consists of the intersecting series of these strips, as shown in Fig. 14.12*c*. For the flexural action of the system there is two-way reinforcing in the slab at each of the boxes defined by the intersections of the strips. In Box 1 in Fig. 14.12*c*, both sets of bars are in the bottom portion of the slab, due to the positive moment in both intersecting strips. In Box 2, the middle-strip bars are in the top (for negative moment), while the column-strip bars are in the bottom (for positive moment). And, in Box 3, the bars are in the top in both directions.

14.7 Use of Design Aids

The design of various elements of reinforced concrete can be aided—or in many cases totally achieved—by the use of various prepared materials. Handbooks (see Ref. 6) present complete data for various elements, such as footings, columns, one-way slabs, joist construction, waffle systems, and two-way slab systems. For the design of a single footing or a one-way slab, the handbook merely represents a convenience, or a shortcut, to a final design. For columns subjected to bending, for waffle construction, and for two-way slab systems, "longhand" design (without aid other than that from a pocket calculator) is really not feasible. In the latter cases, handbook data may be used to establish a reasonable preliminary design, which may then be custom-fit to the specific conditions by some investigation and computations. Even the largest of handbooks cannot present all possible combinations of values of f'_c, grade of reinforcing bars, value of superimposed loads, and so on. Thus only coincidentally will handbook data be exactly correct for any specific design job.

In the age of the computer, there is a considerable array of software available for the routine tasks of structural design. For many of the complex and laborious problems of design of reinforced concrete structures, these are a real boon for anyone able to utilize them.

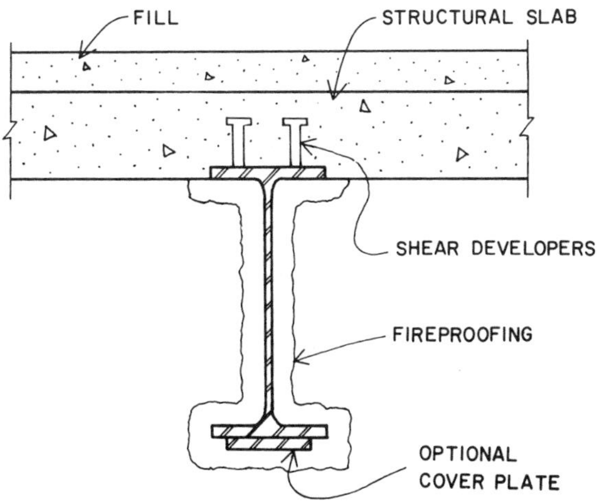

FIGURE 14.13 Steel frame with cast-in-place concrete slab; composite construction.

14.8 Composite Construction: Concrete Plus Structural Steel

Figure 14.13 shows a section detail of a type of construction generally referred to as *composite construction*. This consists of a poured-concrete spanning slab supported by structural steel beams, the two being made to interact by the use of shear developers welded to the top of the beams and embedded in the cast slab. The concrete slab may be formed by use of plywood sheets, resulting in the detail as shown in Fig. 14.13. However, a more popular form of construction is that in which a formed steel deck is used in the usual manner, welded to the top of the beams. The shear developers are then site-welded through the deck to the top of the beam. The steel deck may function essentially only to form the concrete, or may itself develop a composite action with the poured slab.

Part 2 of the AISC Manual (Ref. 2) contains data and design examples of this type of construction.

15

Concrete Columns

15.1 Introduction

The practicing structural designer customarily uses tables or a computer-aided procedure to determine the dimensions and reinforcing for concrete columns. The complexity of analytical formulas and the large number of variables make it impractical to perform design for a large number of columns solely by hand computation. The provisions relating to the design of columns in the 1983 ACI Code are quite different from those of the working stress design method in the 1963 Code. The current code does not permit design of columns by the working stress method, but rather requires that the service load capacity of columns be determined as 40% of that computed by strength design procedures. The strength design procedures are discussed briefly in Chapter 13. The discussions in this chapter are limited to the working stress design procedures, as provided for in the 1963 ACI Code.

Most concrete columns in building construction are relatively stout. Although the Code provides for reduction of axial compression on the basis of slenderness, the reductions do not become significant until the ratio of the column height to its least lateral dimension exceeds common practical limits.

15.2 Columns With Axial Load Plus Bending

Due to the nature of most concrete structures, current design practices generally do not consider the possibility of a concrete column with axial compression alone. That is to say, the existence of some bending moment is always considered together with the axial force. Figure 15.1 illustrates the nature of the so-called *interaction response* for a concrete column, with a range of combinations of axial load plus bending moment. In general, there are three basic ranges of this behavior, as follows:

1. *Large Axial Force, Minor Moment.* For this case the moment has little effect, and the resistance to pure axial force is only negligibly reduced.
2. *Significant Values for Both Axial Force and Moment.* For this case the analysis for design must include the full combined force effects, that is, the interaction of the axial force and the bending moment.
3. *Large Bending Moment, Minor Axial Force.* For this case the column behaves essentially as a doubly reinforced (tension and compression reinforced) member, with its capacity

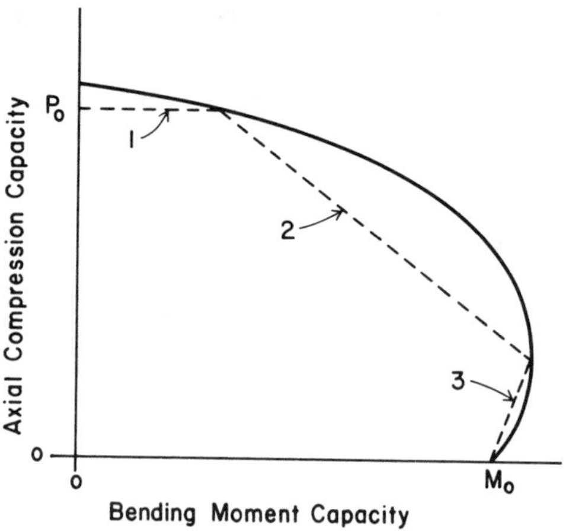

FIGURE 15.1 Interaction of axial compression and bending; idealized form of response for a reinforced concrete column.

for moment resistance affected only slightly by the axial force.

In Fig. 15.1 the solid line on the graph represents the true response of the column—a form of behavior verified by many load tests on laboratory specimens. The dashed line on the graph represents the generalization of the three types of response just described.

The terminal points of the interaction response—pure axial compression or pure bending moment—may be reasonably easily determined. The interaction responses between these two limits require complex analyses beyond the scope of this book.

15.3 Types of Reinforced Concrete Columns

Reinforced concrete columns for buildings generally fall into one of the following categories:

1. Square tied columns.
2. Round spiral columns.
3. Rectangular tied columns.
4. Columns of other geometries (hexagonal, L-shaped, T-shaped, etc.) with either ties or spirals.

In tied columns the longitudinal reinforcing is held in place by loop ties made of small-diameter reinforcing bars, commonly No. 2–No. 4. Such a column is represented by the square section shown in Fig. 15.2*a*. This type of reinforcing can quite readily accommodate other geometries as well as the square. The design of such a column is discussed in Sec. 15.5.

Spiral columns are those in which the longitudinal reinforcing is placed in a circle, with the whole group of bars enclosed by a continuous cylindrical spiral made from steel rod or large-diameter steel wire. Although this reinforcing system obviously works best with a round column section, it can be used also with other geometries. A round column of this type is shown in Fig. 15.2*b*.

Experience has shown the spiral column to be slightly stronger than an equivalent tied column with the same amount of concrete

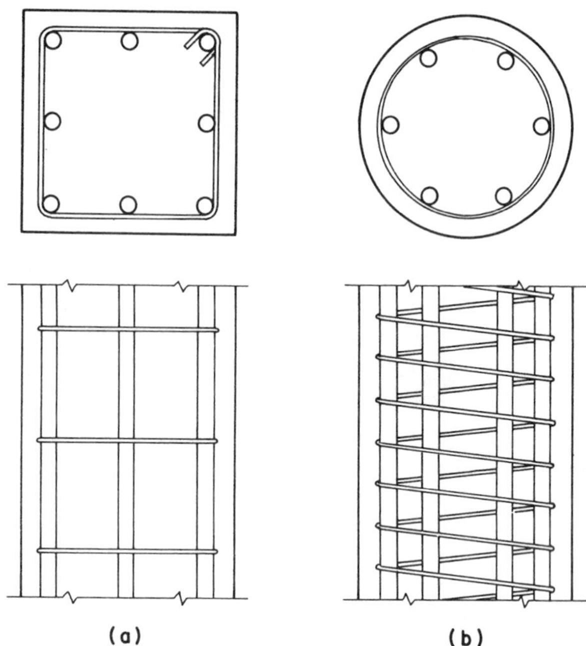

FIGURE 15.2 Typical reinforced concrete columns: (*a*) tied column with spaced, looped ties; (*b*) spiral column with continuous helical wrap of thick wire.

and reinforcing. For this reason code provisions allow slightly more load on spiral columns. Spiral reinforcing tends to be expensive, however, and the round bar pattern does not always mesh well with other construction details in buildings. Thus tied columns are often favored where restrictions on the outer dimensions of the sections are not severe.

15.4 General Requirements for Reinforced Concrete Columns

Code provisions and practical construction considerations place a number of restrictions on column dimensions and choice of reinforcing.

Column Size. In early codes rectangular tied columns were limited to a minimum area of 96 in.[2] and a side dimension of 10 in. if square and 8 in. if oblong rectangular. Spiral columns were limited to a minimum size of 12 in. if either round or square. These are still reasonable, practical limits.

Reinforcing. Minimum bar size is No. 5. The minimum number of bars is four for tied columns, six for spiral columns. The minimum amount of area of steel is 1% of the gross column area. A maximum area of steel 8% of the gross area is permitted, but bar spacing limitations makes this difficult to achieve; 4% is a more practical limit.

Ties. Ties shall be at least No. 3 for bars No. 10 and smaller. No. 4 ties should be used for bars that are No. 11 and larger. Vertical spacing of ties shall be not more than 16 times the bar diameter, 48 times the tie diameter, or the least dimension of the column. Ties shall be arranged so that every corner and alternate longitudinal bar is held by the corner of a tie with an included angle of not greater than 135°, and no bar shall be farther than 6 in. clear from such a supported bar. Complete circular ties may be used for bars placed in a circular pattern.

Concrete Cover. A minimum of 1.5 in. is needed when the column surface is not exposed to weather or in contact with the ground; 2 in. should be used for formed surfaces exposed to the weather or in contact with ground; 3 in. are necessary if the concrete is cast against earth.

Spacing of Bars. Clear distance between bars shall not be less than 1.5 times the bar diameter, 1.33 times the maximum specified size for the coarse aggregate, or 1.5 in.

15.5 Design of Tied Columns

The 1963 ACI Code limites the axial compression load on a tied column to

$$P = 0.85[A_g(0.25f'_c + f_s p_g)]$$

in which P = maximum permissible axial load,
A_g = gross area of the column,
f'_c = ultimate compressive strength of the concrete,
f_s = allowable compressive stress in the reinforcing, taken as 40% of the yield stress but not to exceed 30,000 psi,
p_g = percent of steel = A_s/A_g,
A_s = cross-sectional area of the reinforcing.

The following example illustrates the use of this formula for the determination of the allowable load on a given column.

Example 1. A 16-in. square tied column is reinforced with four No. 10 bars; $f'_c = 4000$ psi and $f_s = 20{,}000$ psi. Find the safe load for the column.

Solution: For use in the formula, we determine

$$A_g = 16 \times 16 = 256 \text{ in.}^2$$

$$p_g = \frac{A_s}{A_g} = \frac{(4 \times 1.27)}{256} = 0.0198$$

Then

$$P = 0.85[256(0.25 \times 4 + 20 \times 0.0198)] = 303.8 \text{ kips}$$

In most building structures, concrete columns will sustain some computed bending moment in addition to the axial compression load (see Fig. 15.3). Even when a computed moment is not present, however, it is well to consider some amount of accidental eccentricity or other source of moment. It is recommended, therefore, that the maximum safe load be limited to that given for a minimum eccentricity of 10% of the column dimension.

Figure 15.4 gives safe loads for a selected number of sizes of square tied columns. Loads are given for various degrees of eccentricity, which is a means for expressing axial load and bending moment combinations. The computed moment on the column is translated into an equivalent eccentric loading, as shown in Fig. 15.3. Data for the curves was computed by using 40% of the load

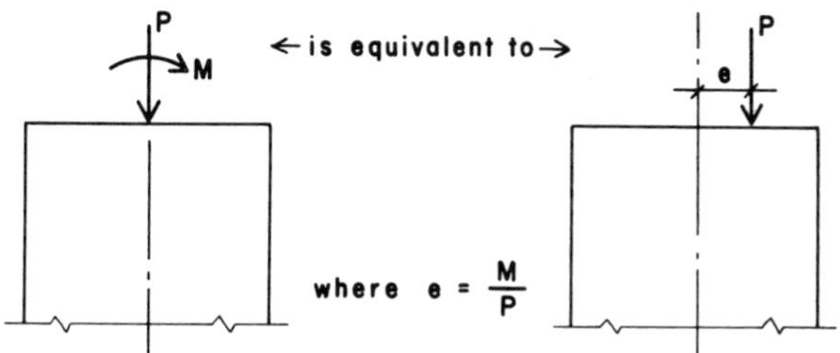

FIGURE 15.3

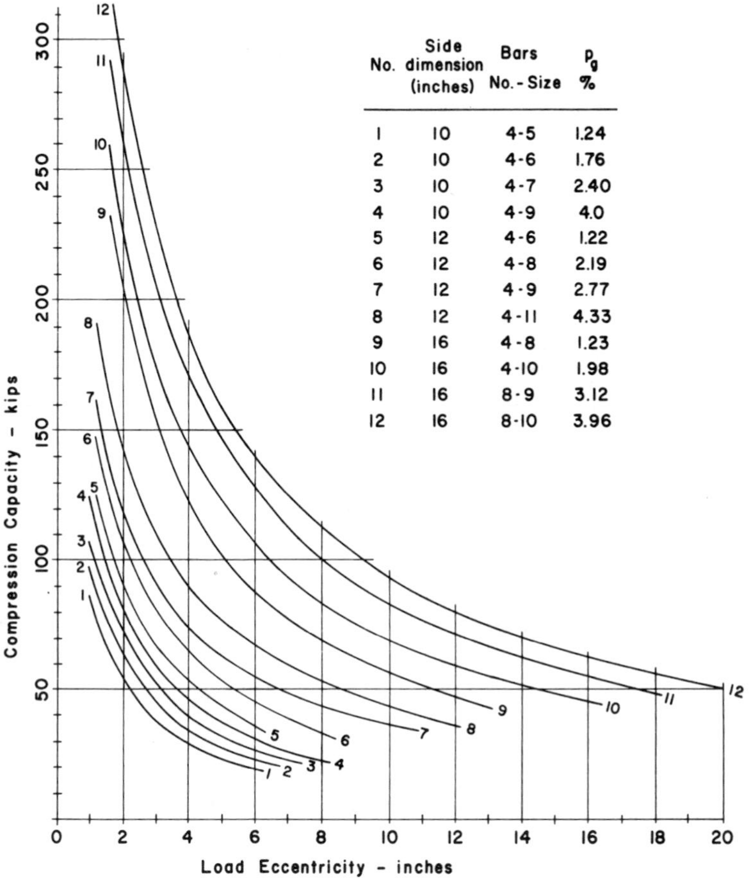

No.	Side dimension (inches)	Bars No.-Size	p_g %
1	10	4-5	1.24
2	10	4-6	1.76
3	10	4-7	2.40
4	10	4-9	4.0
5	12	4-6	1.22
6	12	4-8	2.19
7	12	4-9	2.77
8	12	4-11	4.33
9	16	4-8	1.23
10	16	4-10	1.98
11	16	8-9	3.12
12	16	8-10	3.96

Compression Capacity - kips

Load Eccentricity - inches

(a)

FIGURE 15.4 Safe service loads (40% of strength design load) for square tied columns with $f'_c = 4$ ksi and $f_y = 60$ ksi.

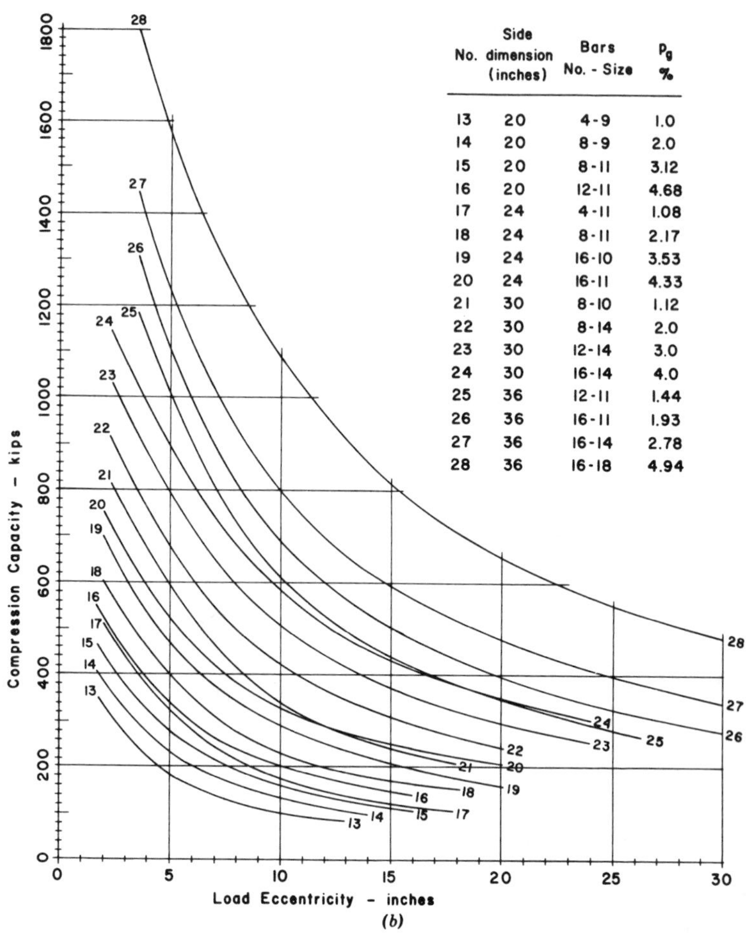

No.	Side dimension (inches)	Bars No. - Size	p_g %
13	20	4-9	1.0
14	20	8-9	2.0
15	20	8-11	3.12
16	20	12-11	4.68
17	24	4-11	1.08
18	24	8-11	2.17
19	24	16-10	3.53
20	24	16-11	4.33
21	30	8-10	1.12
22	30	8-14	2.0
23	30	12-14	3.0
24	30	16-14	4.0
25	36	12-11	1.44
26	36	16-11	1.93
27	36	16-14	2.78
28	36	16-18	4.94

FIGURE 15.4 *(Continued)*

determined by strength design methods, as required by the 1983 ACI Code.

For the column in Example 1, it may be noted that Fig. 15.4 yields a maximum value of approximately 260 kips, which is only about 85% of the value previously determined by the working stress formula. The discrepancy occurs because the Code requires the use of a minimum eccentricity for all columns. Thus the curves in Fig. 15.4 do not begin at zero eccentricity.

The following examples illustrate the use of Fig. 15.4 for the design of tied columns.

Example 2. A column with $f'_c = 4$ ksi and steel with $f_y = 60$ ksi sustains an axial compression load of 400 kips. Find the minimum practical column size if reinforcing is a maximum of 4% and the maximum size if reinforcing is a minimum of 1%.

Solution: Using Fig. 15.4*a*, we find from the sizes given:

Minimum column is 20 in. square with 8 No. 9 (Curve 14).
Maximum capacity is 410 kips, $p_g = 2.0\%$.
Maximum size is 24 in. square with 4 No. 11 (Curve 17).
Maximum capacity is 510 kips, $p_g = 1.08\%$.

It should be apparent that it is possible to use an 18- or 19-in. column as the minimum size and to use a 22- or 23-in. column as the maximum size. Since these sizes are not given in the figure, we cannot verify them for certain without using strength design procedures.

Example 3. A square tied column with $f'_c = 4$ ksi and steel with $f_y = 60$ ksi sustains an axial load of 400 kips and a bending moment of 200 kip-ft. Determine the minimum size column and its reinforcing.

Solution: We first determine the equivalent eccentricity, as shown in Fig. 15.3. Thus

$$e = \frac{M}{P} = \frac{200 \times 12}{400} = 6 \text{ in.}$$

Then, from Fig. 15.4, we find:

Minimum size is 24 in. square with 16 No. 10 bars.
Capacity at 6-in. eccentricity is 410 kips.

There is usually a number of possible combinations of reinforcing bars that may be assembled to satisfy the steel area require-

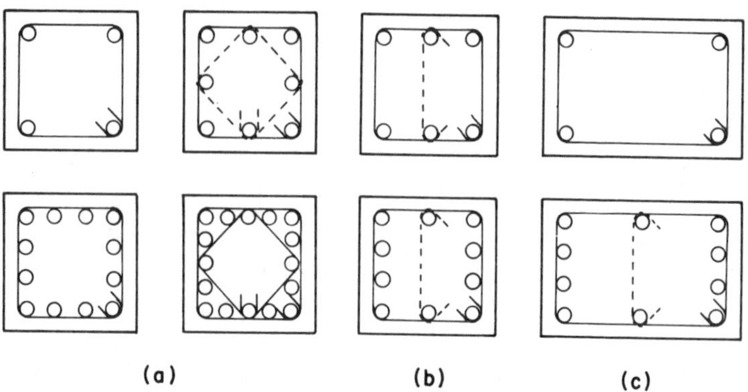

FIGURE 15.5 Typical bar placement and tie patterns for tied columns.

ment for a given column. Aside from providing for the area, the number of bars must also work reasonably in the layout of the column. Figure 15.5 shows a number of tied columns with various number of bars. When a column is small, the preferred choice is usually that of the simple four-bar layout, with one bar in each corner and a single peripheral tie. As the column gets larger, the distance between the corner bars gets larger, and it is best to use more bars so that the reinforcing is spread out around the column periphery. For a symmetrical layout and the simplest of tie layouts, the best choice is for numbers that are multiples of four, as shown in Fig. 15.5*a*. The number of additional ties required for these layouts depends on the size of the column and the considerations discussed in Sec. 15.4.

An unsymmetrical bar arrangement is not necessarily bad, even though the column and its construction details are otherwise not oriented differently on the two axes. In situations where moments may be greater on one axis, the unsymmetrical layout is actually preferred; in fact, the column shape will also be more effective if it is unsymmetrical, as shown for the oblong shapes in Fig. 15.5*c*.

Problems 15.5.A,B,C,D,E. Using Fig. 15.4, pick the minimum size square tied column and its reinforcing for the following combinations of axial load and bending moment:

	Axial Compressive Load (kips)	Bending Moment (kip-ft)
A	100	25
B	100	50
C	150	75
D	200	100
E	300	150

15.6 Column Form and Reinforcement

The simple rectangular column of square or oblong shape is most common in concrete-framed buildings. In many cases the columns will be of the tied classification, as this allows for the simplest construction. When other column section forms are required, the general form of the tied column reinforcement may be adapted but is not the only option.

Figure 15.6 shows a number of special column shapes developed as tied columns. Although spirals could be used in some cases for such shapes, the use of ties allows much greater flexibil-

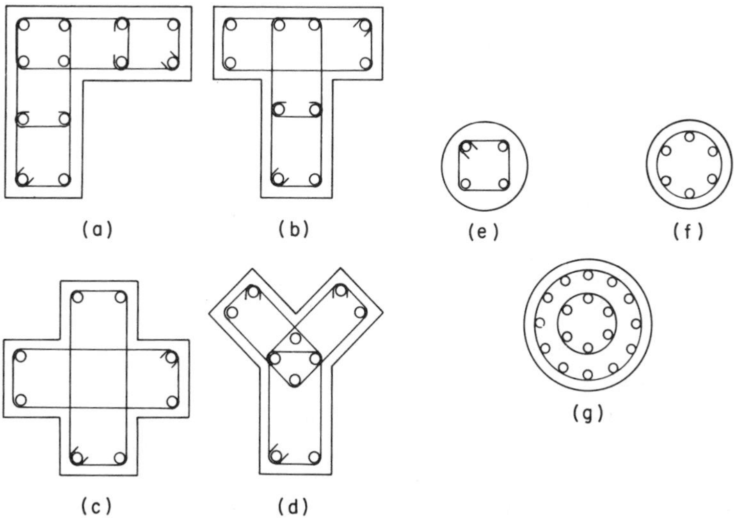

FIGURE 15.6 Typical reinforcement for columns of various shapes.

ity and simplicity of construction. One reason for using ties may be the column dimensions, there being a practical limit of about 12 in. in width for a spiral-bound column.

Round columns are most often formed as shown in Fig. 15.6*e* if built as tied columns. This allows for a minimum reinforcing with four bars. If a round pattern is used (as it must be for a spiral-bound column), the usual minimum number recommended is six bars. Spacing of bars is much more critical in spiral-bound circular arrangements, making it very difficult to use high percentages of steel in the column section. For very large diameter columns it is possible to use sets of concentric spirals, as shown in Fig. 15.6*g*.

For poured-in-place columns a concern that must be dealt with is that for vertical splicing of the steel bars. The two places where this commonly occurs are at the top of the foundation and at floors where a multistory column continues upward. At these points there are three ways to achieve the vertical continuity (splicing) of the steel bars, any of which may be appropriate for a given situation.

1. Bars may be lapped the required distance for development of the compression splice, as shown in Fig. 13.27. For bars of smaller dimension and lower yield strengths, this is usually the desired method.
2. Bars may have milled square-cut ends butted together with a grasping device to prevent separation in a horizontal direction.
3. Bars may be welded with full penetration butt welds or by welding of the grasping device described for method 2.

The choice of splicing methods is basically a matter of cost comparison, but is also affected by the size of the bars, the degree of concern for bar spacing in the column arrangement, and possibly for a need for some development of tension through the splice if uplift or high magnitudes of moments exist. If lapped splicing is used, a problem that must be considered is the bar layout at the location of the splice, at which point there will be twice the usual

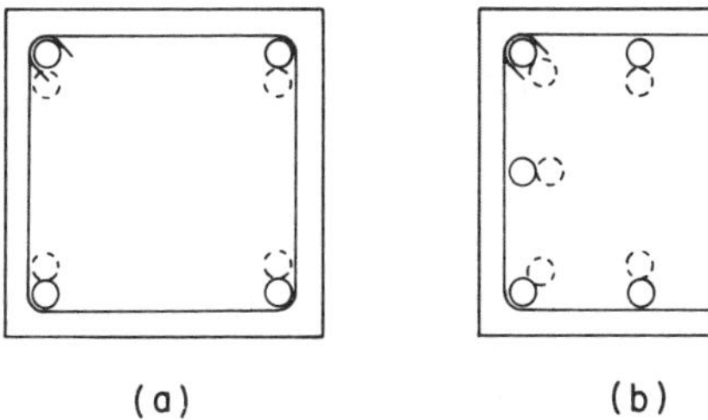

FIGURE 15.7 Options for placement of bars at lapped splices.

number of bars. The lapped bars may be adjacent to each other, but the usual considerations for space between bars must be made. If spacing is not critical, the arrangement shown in Fig. 15.7*a* is usually chosen, with the spliced sets of bars next to each other at the tie perimeter. If spacing does not permit the arrangement in Fig. 15.7*a*, that shown in Fig. 15.7*b* may be used, with the lapped sets in concentric patterns. The latter arrangement is commonly used for spiral-bound columns, where spacing is often critical.

15.7 Round Tied Columns

Round columns may be designed and built as spiral columns, as described in Sec. 15.3, or they may be developed as tied columns with the bars placed in a circle and held by a series of round circumferential ties. Because of the cost of spirals, it is usually more economical to use the tied column, so it is often used unless the additional strength or other behavioral characteristics of the spiral column are required.

It is also possible to use rectangular bar layouts and tie patterns as shown in Fig. 15.6*e* inside a round column form. In such cases, the column is usually designed as a square column using the square shape that can be included within the round form. It is thus possible to use a four-bar column for small-diameter, round column forms.

Figure 15.8 gives safe loads for round columns that are designed as tied columns. Load values have been adapted from

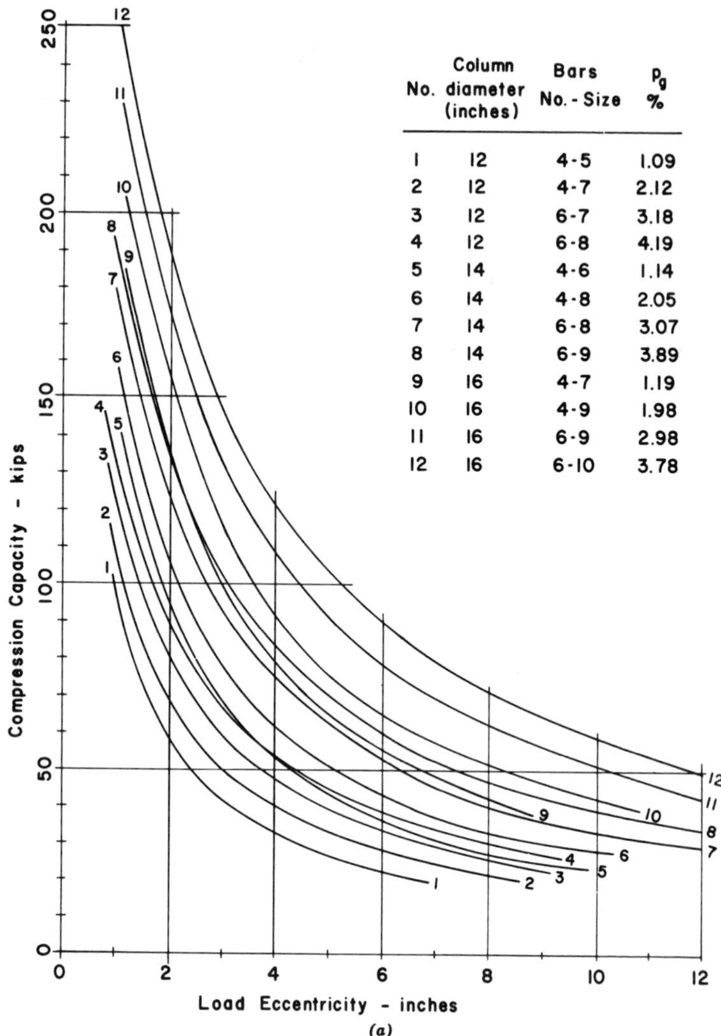

No.	Column diameter (inches)	Bars No. - Size	p_g %
1	12	4-5	1.09
2	12	4-7	2.12
3	12	6-7	3.18
4	12	6-8	4.19
5	14	4-6	1.14
6	14	4-8	2.05
7	14	6-8	3.07
8	14	6-9	3.89
9	16	4-7	1.19
10	16	4-9	1.98
11	16	6-9	2.98
12	16	6-10	3.78

FIGURE 15.8 Safe service loads (40% of strength design load) for round tied columns with $f'_c = 40$ ksi and $f_y = 60$ ksi.

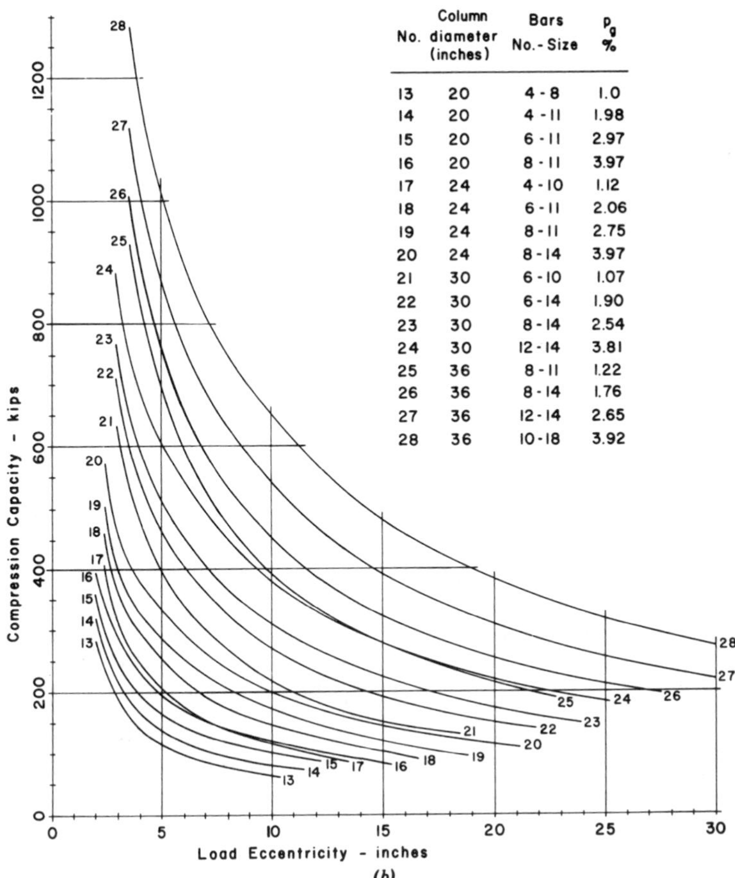

No.	Column diameter (inches)	Bars No.-Size	p_g %
13	20	4-8	1.0
14	20	4-11	1.98
15	20	6-11	2.97
16	20	8-11	3.97
17	24	4-10	1.12
18	24	6-11	2.06
19	24	8-11	2.75
20	24	8-14	3.97
21	30	6-10	1.07
22	30	6-14	1.90
23	30	8-14	2.54
24	30	12-14	3.81
25	36	8-11	1.22
26	36	8-14	1.76
27	36	12-14	2.65
28	36	10-18	3.92

(b)

FIGURE 15.8 *(Continued)*

values determined by strength design methods. The curves in Fig. 15.8 are similar to those for the square columns in Fig. 15.4, and their use is similar to that demonstrated in Examples 2 and 3.

Problems 15.7.A,B,C,D,E. Using Fig. 15.8, pick the minimum size round column and its reinforcing for the load and moment combinations in Problem 15.5.

15.8 Slenderness

Poured-in-place concrete columns tend to be quite stout in profile, so that slenderness is much less often a critical concern than with columns of wood or steel. The Code provides for consideration of slenderness, but permits the issue to be neglected when the L/r of the column falls below a controlled value. For rectangular columns this usually means that the effect is neglected when the ratio of unsupported height to side dimension is less than about 12. This is roughly analogous to the case for the wood column with L/D less than 11.

Slenderness effects must also be related to the conditions of bending for the column. Since bending is usually induced at the column ends, the two typical cases are those shown in Fig. 15.9. If two equal end moments, as shown in Fig. 15.9*a*, exist, the buckling effect is magnified, the *P*-delta effect is maximum, and the Code limits slenderness without reduction to L/d ratios of 6.6 or less. The condition in Fig. 15.9*a* is not the common case, however, the more typical condition in framed structures being that shown in Fig. 15.9*b*, for which the L/d limit for equal end moments jumps to 13.8 before reduction for slenderness is required.

When slenderness must be considered, the complex procedures required are simply built into your friendly neighborhood software program. One should be aware, however, that reduction for slenderness is not considered in the usual design aids, such as tables or graphs.

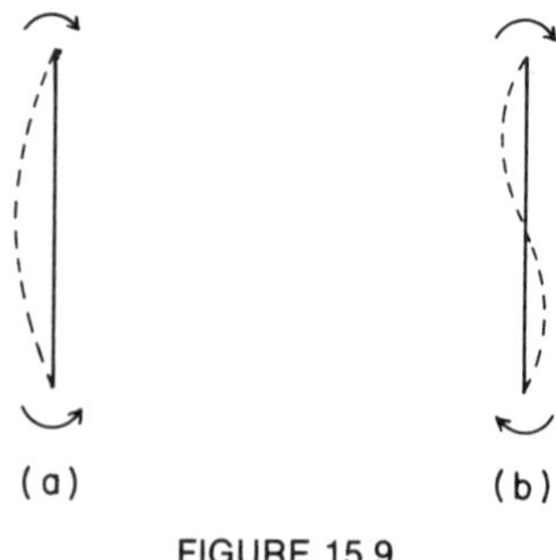

FIGURE 15.9

15.9 Bar Development in Columns

In reinforced concrete columns both the concrete and the steel bars share the compression force. Ordinary construction practices require the consideration of various situations for development of the stress in the reinforcing bars. Figure 15.10 shows a multistory concrete column with its base supported on a concrete footing. With reference to the illustration, we note the following.

1. The concrete construction is ordinarily produced in multiple, separate pours, with construction joints between the separate pours occurring as shown in the illustration.

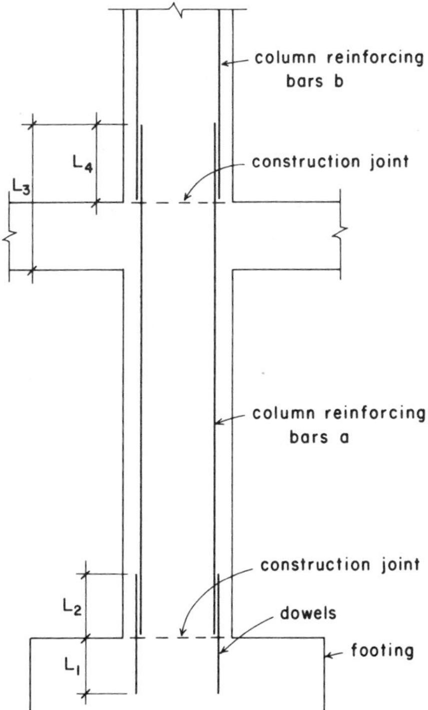

FIGURE 15.10 Bar development considerations for concrete columns.

2. In the lower column, the load from the concrete is transferred to the footing in direct compressive bearing at the joint between the column and footing. The load from the reinforcing must be developed by extension of the reinforcing into the footing: distance L_1 in the illustration. Although it may be possible to place the column bars in position during pouring of the footing to achieve this, the common practice is to use dowels, as shown in the illustration. These dowels must be developed on both sides of the joint: L_1 in the footing and L_2 in the column. If the f'_c value for both the footing and the column are the same, these two required lengths will be the same.
3. The lower column will ordinarily be cast together with the supported concrete framing above it, with a construction joint occurring at the top level of the framing (bottom of the upper column), as shown in the illustration. The distance L_3 is that required to develop the reinforcing in the lower column—bars a in the illustration. As for the condition at the top of the footing, the distance L_4 is required to develop the reinforcing in bars b in the upper column. L_4 is more likely to be the critical consideration for the determination of the extension required for bars a.

16

Footings

The primary purpose of a footing is to spread the loads so that the allowable bearing capacity of the foundation material is not exceeded. In cities where experience and tests have established the allowable strengths of various foundation soils, local building codes may be consulted to determine the bearing capacities used in design. In the absence of such information, borings or load tests should be made. For sizable structures, borings at the site should always be made and their results interpreted by a qualified geotechnical engineer.

Footings may be classified as wall footings and column footings. In the former type the load is brought to the foundation as a uniform load per linear ft of wall; in the latter it is concentrated at the base of the column. Column loads are sometimes combined on a single footing, especially where columns are very closely spaced. However, the independent footing supporting a single column is the most common type and the one to be considered in this chapter.

16.1 Independent Column Footings

The great majority of independent or isolated column footings are square in plan, with reinforcing consisting of two sets of bars at right angles to each other. This is known as two-way reinforcement. The column may be placed directly on the footing block, or it may be supported by a pedestal. A pedestal, or pier, is a short, wide compression block that serves to reduce the punching effect on the footing. For steel columns a pier may also serve to raise the bottom of the steel column above ground level.

The design of a column footing is usually based on the following considerations:

1. *Maximum Soil Pressure.* The sum of the superimposed load on the footing and the weight of the footing must not exceed the limit for bearing pressure on the supporting material. The required total plan area of the footing is determined on this basis.
2. *Control of Settlement.* Where buildings rest on highly compressible soil, it may be necessary to select footing areas that assure a uniform settlement of all the building columns rather than to strive for a maximum use of the allowable soil pressure.
3. *Size of the Column.* The larger the column, the less will be the shear, flexural, and bond stresses in the footing, since these are developed by the cantilever effect of the footing projection beyond the edges of the column.
4. *Shear Stress Limit for the Concrete.* For square-plan footings this is usually the only critical stress condition for the concrete. In order to reduce the required amount of reinforcing, the footing depth is usually established well above that required by the flexural stress limit for the concrete.
5. *Flexural Stress and Development Length Limits for the Bars.* These are considered on the basis of the moment developed in the cantilevered footing at the face of the column.
6. *Footing Thickness for Development of Column Reinforcing.* When a footing supports a reinforced concrete

column, the compressive force in the column bars must be transferred to the footing by bond stress—called *doweling* of the bars. The thickness of the footing must be sufficient for the necessary development length of the column bars. (See Sec. 15.9.)

The following example illustrates the design process for a simple, square column footing.

Example. A 16-in. [406-mm] square concrete column exerts a load of 240 kips [1068 kN] on a square column footing. Determine the footing dimensions and the necessary reinforcing using the following data: $f'_c = 3$ ksi [20.7 MPa], Grade 40 bars with $f_y = 40$ ksi [276 MPa] and $f_s = 20$ ksi [138 MPa], maximum permissible soil pressure = 4000 psf [192 kPa].

Solution: The first decision to be made is that of the height, or thickness, of the footing. This has to be a raw first guess unless the dimensions of similar footings are known. In practice this knowledge is generally available from previous design work or from handbook tables. In lieu of this, a reasonable guess is made, the design work is performed, and an adjustment is made if the assumed thickness proves inadequate. We will assume a footing thickness of 20 in. [508 mm] for a first try for this example.

The footing thickness establishes the weight of the footing on a per-square-ft basis. This weight is then subtracted from the maximum permissible soil pressure, and the net value is then usable for the superimposed load on the footing. Thus

$$\text{Footing weight} = \frac{20}{12}(150 \text{ psf}) = 250 \text{ psf [12 kPa]}$$

$$\text{Net usable soil pressure} = 4000 - 250 = 3750 \text{ psf [180 kPa]}$$

$$\text{Required footing plan area} = \frac{240{,}000}{3750} = 64 \text{ ft}^2 \text{ [5.93 m}^2\text{]}$$

$$\text{Length of the side of the square footing} = L = \sqrt{64} = 8 \text{ ft [2.44 m]}$$

Two shear stress situations must be considered for the concrete. The first occurs as ordinary beam shear in the cantilevered portion and is computed at a critical section at a distance d (effective depth of the beam) from the face of the column, as shown in Fig. 16.1a. The shear stress at this section is computed in the same manner as for a beam, as discussed in Sec. 13.13, and the stress limit is $v_c = 1.1\sqrt{f'_c}$. The second shear stress condition is that of peripheral shear, or so-called "punching" shear, and is investigated at a circumferential section around the column at a distance of $d/2$ from the column face, as shown in Fig. 16.1b. For this condition the allowable stress is $v_c = 2.0\sqrt{f'_c}$.

With two-way reinforcing, it is necessary to place the bars in one direction on top of the bars in the other direction. Thus, although the footing is supposed to be the same in both directions, there are actually two different d distances—one for each layer of bars. It is common practice to use the average of these two distances for the design value of d; that is, d = the footing thickness less the sum of the concrete cover and the bar diameter. With the bar diameter as yet undetermined, we will assume an approxi-

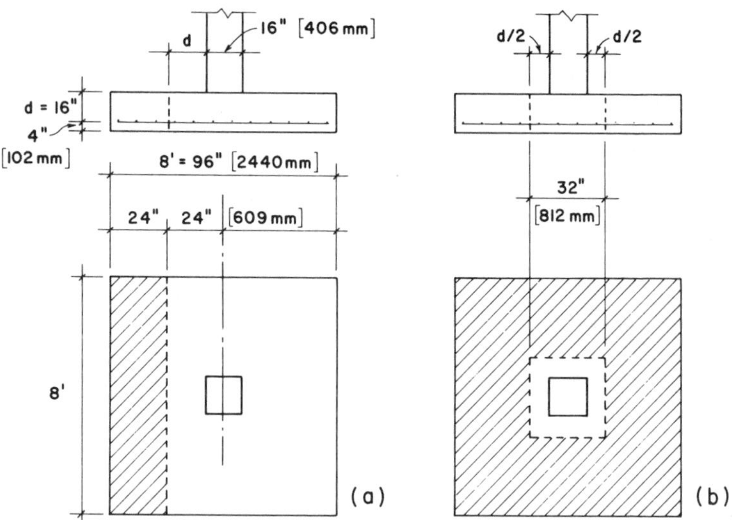

FIGURE 16.1 Shear considerations for the column footing.

mate d of the footing thickness less 4 in. [102 mm] (a concrete cover of 3 in. plus a No. 8 bar). For the example this becomes

$$d = t - 4 = 20 - 4 = 16 \text{ in. [406 mm]}$$

It should be noted that it is the *net* soil pressure that causes stresses in the footing, since there will be no bending or shear in the footing when it rests alone on the soil. We thus use the net soil pressure of 3750 psf [180 kPa] to determine the shear and bending effects for the footing.

For the beam shear investigation, we determine the shear force generated by the net soil pressure acting on the shaded portion of the footing plan area shown in Fig. 16.1*a*. Thus

$$V = 3750 \times 8 \times \frac{24}{12} = 60{,}000 \text{ lb [267.5 kN]}$$

and, using the formula for shear stress in a beam (Sec. 13.13),

$$v = \frac{V}{bd} = \frac{60{,}000}{96 \times 16} = 39.1 \text{ psi [0.270 MPa]}$$

which is compared to the allowable stress of

$$v_c = 1.1\sqrt{f'_c} = 1.1\sqrt{3000} = 60 \text{ psi [0.414 MPa]}$$

indicating that this condition is not critical.

For the peripheral shear investigation, we determine the shear force generated by the net soil pressure acting on the shaded portion of the footing area shown in Fig. 16.1*b*. Thus

$$V = 3750\left[8^2 - \left(\frac{32}{12}\right)^2\right] = 213{,}333 \text{ lb [953 kN]}$$

Shear stress for this case is determined with the same formula as for beam shear, with the dimension b being the total peripheral circumference. Thus

$$v = \frac{V}{bd} = \frac{213{,}333}{(4 \times 32) \times 16} = 104.2 \text{ psi [0.723 MPa]}$$

which is compared to the allowable stress of

$$v_c = 2\sqrt{f'_c} = 2\sqrt{3000} = 109.5 \text{ psi [0.755 MPa]}$$

This computation indicates that the peripheral shear stress is not critical, but since the actual stress is quite close to the limit, the assumed thickness of 20 in. is probably the least full-inch value that can be used. Flexural stress in the concrete should also be considered, although it is seldom critical for a square footing. One way to verify this is to compute the balanced moment capacity of the section with $b = 96$ in. and $d = 16$ in. Using the factor for a balanced section from Table 13.3, we find

$$M_R = Rbd^2 = 0.226(96)(16)^2 = 5554 \text{ kip-in. or } 463 \text{ kip-ft}$$

which may be compared with the actual moment computed in the next step.

For the reinforcing we consider the stresses developed at a section at the edge of the column, as shown in Fig. 16.2. The cantilever moment for the 40-in. [1016-mm] projection of the footing beyond the column is

$$M = 3750 \times 8 \times \frac{40}{12} \times \frac{1}{2}\left(\frac{40}{12}\right) = 166{,}667 \text{ lb-ft [227 kN-m]}$$

Using the formula for required steel area in a beam, with a conservative guess of 0.9 for j, we find (see Sec. 13.7)

$$A_s = \frac{M}{f_s jd} = \frac{166{,}667 \times 12}{20 \times 0.9 \times 16 \times 10^3} = 6.95 \text{ in.}^2 \text{ [4502 mm}^2\text{]}$$

This requirement may be met by various combinations of bars, such as those in Table 16.1. Data for consideration of the development length and the center-to-center bar spacing are also given

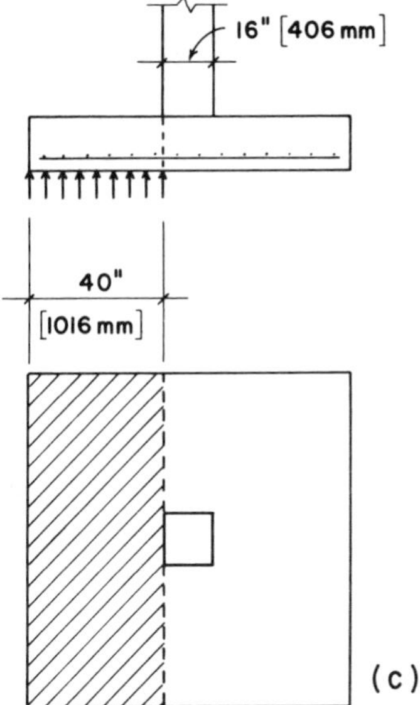

FIGURE 16.2 Bending and bar development length considerations for the column footing.

in the table. The flexural stress in the bars must be developed by the embedment length equal to the projection of the bars beyond the column edge, as discussed in Sec. 13.14. With a minimum of 2 in. [51 mm] of concrete cover at the edge of the footing, this length is 38 in. [965 mm]. The required development lengths indi-

TABLE 16.1 Reinforcing Alternatives for the Column Footing

Number and Size of Bars	Area of Steel Provided		Required Development Length[a]		Center-to-Center Spacing	
	(in.^{2})	(mm^2)	(in.)	(mm)	(in.)	(mm)
12 No. 7	7.20	4645	18	457	8.2	208
9 No. 8	7.11	4687	23	584	11.3	286
7 No. 9	7.00	4516	29	737	15	381
6 No. 10	7.62	4916	37	940	18	458

[a] From Table 13.6; values for "Other bars," $f_y = 40$ ksi, $f'_c = 3$ ksi.

cated in the table are taken from Table 13.6; it may be noted that all of the combinations in the table are adequate in this regard.

If the distance from the edge of the footing to the first bar at each side is approximately 3 in. [76 mm], the center-to-center distance for the two outside bars will be 96 − 2(3) = 90 in. [2286 mm], and with the rest of the bars evenly spaced, the spacing will be 90 divided by the number of total bars less one. This value is shown in the table for each set of bars. The maximum permitted spacing is 18 in. [457 mm], and the minimum should be a distance that is adequate to permit good flow of the wet concrete between the two-way grid of bars—say 4 in. [102 mm] or more.

All of the bar combinations in Table 16.1 are adequate for the footing. Many designers prefer to use the largest possible bar, as this reduces the number of bars that must be handled and supported during construction. On this basis, the footing will be the following:

8 ft square by 20 in. thick with 6 No. 10 bars each way.

Problem 16.2.A. Design a square footing for a 14-in. [356-mm] square concrete column with a load of 219 kips [974 kN]. The maximum permissible soil pressure is 3000 psf [144 kPa]. Use concrete with f'_c = 3 ksi [20.7 MPa] and reinforcing of Grade 40 bars with f_y = 40 ksi [276 MPa] and f_s = 20 ksi [138 MPa].

16.2 Load Tables for Column Footings

For ordinary situations we often design square column footings by using data from tables in various references. Even where special circumstances make it necessary to perform the type of design illustrated in Sec. 16.1, such tables will assist in making a first guess for the footing dimensions.

Table 16.2 gives the allowable superimposed load for a range of footings and soil pressures. This material has been adapted from a more extensive table in *Simplified Design of Building Foundations* (Ref. 14). Designs are given for concrete strengths of 2000 and 3000 psi. The low strength of 2000 psi is sometimes used for small buildings, since many building codes permit the omission of testing of the concrete if this value is used for design.

TABLE 16.2 Square Column Footings

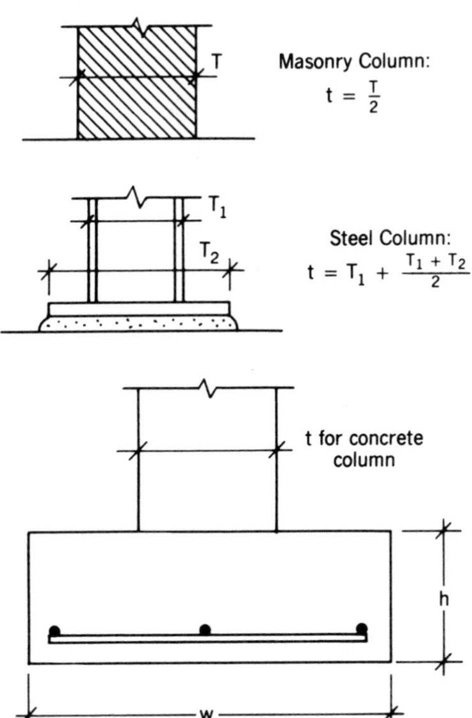

Maximum Soil Pressure (lb/ft^2)	Minimum Column Width t (in.)	f'_c = 2000 psi: Allowable Load[a] on Footing (k)	Footing Dimensions h (in.)	w (ft)	Reinforcing Each Way	f'_c = 3000 psi: Allowable Load[a] on Footing (k)	Footing Dimensions h (in.)	w (ft)	Reinforcing Each Way
1000	8	7.9	10	3.0	2 No. 3	7.9	10	3.0	2 No. 3
	8	10.7	10	3.5	3 No. 3	10.7	10	3.5	3 No. 3
	8	14.0	10	4.0	3 No. 4	14.0	10	4.0	3 No. 4
	8	17.7	10	4.5	4 No. 4	17.7	10	4.5	4 No. 4
	8	22	10	5.0	4 No. 5	22	10	5.0	4 No. 5
	8	31	10	6.0	5 No. 6	31	10	6.0	5 No. 6
	8	42	12	7.0	6 No. 6	42	11	7.0	7 No. 6
1500	8	12.4	10	3.0	3 No. 3	12.4	10	3.0	3 No. 3
	8	16.8	10	3.5	3 No. 4	16.8	10	3.5	3 No. 4
	8	22	10	4.0	4 No. 4	22	10	4.0	4 No. 4

TABLE 16.2 (*Continued*)

Maximum Soil Pressure (lb/ft²)	Minimum Column Width t (in.)	f'_c = 2000 psi: Allowable Load[a] on Footing (k)	Footing Dimensions h (in.)	Footing Dimensions w (ft)	Reinforcing Each Way	f'_c = 3000 psi: Allowable Load[a] on Footing (k)	Footing Dimensions h (in.)	Footing Dimensions w (ft)	Reinforcing Each Way
	8	28	10	4.5	4 No. 5	28	10	4.5	4 No. 5
	8	34	11	5.0	5 No. 5	34	10	5.0	6 No. 5
	8	48	12	6.0	6 No. 6	49	11	6.0	6 No. 6
	8	65	14	7.0	7 No. 6	65	13	7.0	6 No. 7
	8	83	16	8.0	7 No. 7	84	15	8.0	7 No. 7
	8	103	18	9.0	8 No. 7	105	16	9.0	10 No. 7
2000	8	17	10	3.0	4 No. 3	17	10	3.0	4 No. 3
	8	23	10	3.5	4 No. 4	23	10	3.5	4 No. 4
	8	30	10	4.0	6 No. 4	30	10	4.0	6 No. 4
	8	37	11	4.5	5 No. 5	38	10	4.5	6 No. 5
	8	46	12	5.0	6 No. 5	46	11	5.0	5 No. 6
	8	65	14	6.0	6 No. 6	66	13	6.0	7 No. 6
	8	88	16	7.0	8 No. 6	89	15	7.0	7 No. 7
	8	113	18	8.0	8 No. 7	114	17	8.0	9 No. 7
	8	142	20	9.0	8 No. 8	143	19	9.0	8 No. 8
	10	174	21	10.0	9 No. 8	175	20	10.0	10 No. 8
3000	8	26	10	3.0	3 No. 4	26	10	3.0	3 No. 4
	8	35	10	3.5	4 No. 5	35	10	3.5	4 No. 5
	8	45	12	4.0	4 No. 5	46	11	4.0	5 No. 5
	8	57	13	4.5	6 No. 5	57	12	4.5	6 No. 5
	8	70	14	5.0	5 No. 6	71	13	5.0	6 No. 6
	8	100	17	6.0	7 No. 6	101	15	6.0	8 No. 6
	10	135	19	7.0	7 No. 7	136	18	7.0	8 No. 7
	10	175	21	8.0	10 No. 7	177	19	8.0	8 No. 8
	12	219	23	9.0	9 No. 8	221	21	9.0	10 No. 8
	12	269	25	10.0	11 No. 8	271	23	10.0	10 No. 9
	12	320	28	11.0	11 No. 9	323	26	11.0	12 No. 9
	14	378	30	12.0	12 No. 9	381	28	12.0	11 No. 10
4000	8	35	10	3.0	4 No. 4	35	10	3.0	4 No. 4
	8	47	12	3.5	4 No. 5	47	11	3.5	4 No. 5
	8	61	13	4.0	5 No. 5	61	12	4.0	6 No. 5
	8	77	15	4.5	5 No. 6	77	13	4.5	6 No. 6
	8	95	16	5.0	6 No. 6	95	15	5.0	6 No. 6
	8	135	19	6.0	8 No. 6	136	18	6.0	7 No. 7
	10	182	22	7.0	8 No. 7	184	20	7.0	9 No. 7
	10	237	24	8.0	9 No. 8	238	22	8.0	9 No. 8
	12	297	26	9.0	10 No. 8	299	24	9.0	9 No. 9
	12	364	29	10.0	13 No. 8	366	27	10.0	11 No. 9
	14	435	32	11.0	12 No. 9	440	29	11.0	11 No. 10
	14	515	34	12.0	14 No. 9	520	31	12.0	13 No. 10
	16	600	36	13.0	17 No. 9	606	33	13.0	15 No. 10
	16	688	39	14.0	15 No. 10	696	36	14.0	14 No. 11
	18	784	41	15.0	17 No. 10	793	38	15.0	16 No. 11

[a] *Note:* Allowable loads do not include the weight of the footing, which has been deducted from the total bearing capacity. Criteria: f_s = 20 ksi, $v_c = 1.1\sqrt{f'_c}$ for beam shear, $v_c = 2\sqrt{f'_c}$ for peripheral shear.

TABLE 16.3 Wall Footings

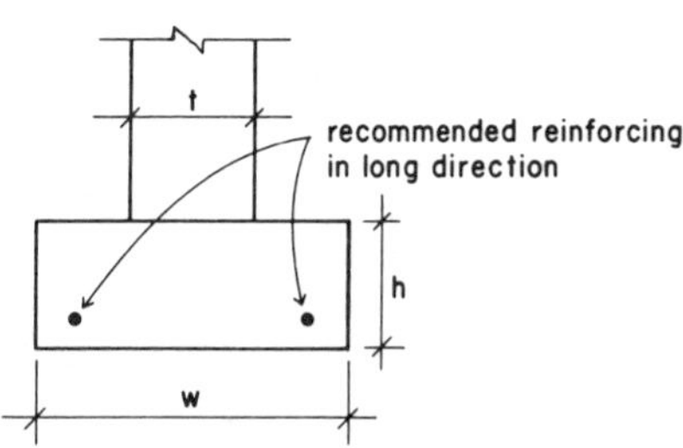

Maximum Soil Pressure (lb/ft²)	Minimum Wall Thickness: Concrete t (in.)	Minimum Wall Thickness: Masonry t (in.)	Allowable Load[a] on Footing (lb/ft)	Footing Dimensions: h (in.)	Footing Dimensions: w (in.)	Reinforcing: Long Direction	Reinforcing: Short Direction
1000	4	8	2,625	10	36	3 No. 4	No. 3 at 16
	4	8	3,062	10	42	2 No. 5	No. 3 at 12
	6	12	3,500	10	48	4 No. 4	No. 4 at 16
	6	12	3,938	10	54	3 No. 5	No. 4 at 13
	6	12	4,375	10	60	3 No. 5	No. 4 at 10
	6	12	5,250	10	72	4 No. 5	No. 5 at 11
1500	4	8	4,125	10	36	3 No. 4	No. 3 at 10
	4	8	4,812	10	42	2 No. 5	No. 4 at 13
	6	12	5,500	10	48	4 No. 4	No. 4 at 11
	6	12	6,131	11	54	3 No. 5	No. 5 at 15
	6	12	6,812	11	60	5 No. 4	No. 5 at 12
	8	16	8,100	12	72	5 No. 5	No. 5 at 10
2000	4	8	5,625	10	36	3 No. 4	No. 4 at 14
	6	12	6,562	10	42	2 No. 5	No. 4 at 11
	6	12	7,500	10	48	4 No. 4	No. 5 at 12
	6	12	8,381	11	54	3 No. 5	No. 5 at 11
	6	12	9,250	12	60	4 No. 5	No. 5 at 10
	8	16	10,875	15	72	6 No. 5	No. 5 at 9
3000	6	12	8,625	10	36	3 No. 4	No. 4 at 10
	6	12	10,019	11	42	4 No. 4	No. 5 at 13
	6	12	11,400	12	48	3 No. 5	No. 5 at 10
	6	12	12,712	14	54	6 No. 4	No. 5 at 10
	8	16	14,062	15	60	5 No. 5	No. 5 at 9
	8	16	16,725	17	72	6 No. 5	No. 6 at 10

Note: Allowable loads do not include the weight of the footing, which has been deducted from the total bearing capacity. Criteria: $f'_c = 2000$ psi, $f_s = 20$ ksi, $v_c = 1.1\sqrt{f'_c}$.

16.3 Load Tables for Wall Footings

Wall footings may be designed by the same process that was illustrated in Sec. 16.1, omitting the investigation for peripheral shear. The principal reinforcing in wall footings is placed in only one direction, perpendicular to the plane of the wall. Minimum reinforcing is commonly placed in the long direction (parallel to the wall) to provide for temperature and shrinkage stresses. Table 16.3 gives allowable loads for a range of wall footings and soil pressures. This material is adapted from a more extensive table in the reference cited in Sec. 16.2.

17

General Considerations for Masonry

Much of the masonry seen as finished surfaces in new construction these days is either nonstructural or of a few limited types of structural masonry. Most of what must be dealt with in using masonry falls in the general category of building construction rather than with strictly structural design considerations. The discussion here is limited to the most common uses of structural masonry and with the general design considerations relating to building structures.

17.1 Masonry Units

Masonry consists generally of a solid mass produced by bonding separate units. The traditional bonding material is mortar. The units include a range of materials, the common ones being the following:

Stone. These may be in essentially natural form (called rubble or field stone) or may be cut to specified shape.

Brick. These vary from unfired, dried mud (adobe) to fired clay (kiln-baked) products. Form, color, and structural properties vary considerably.

Concrete Blocks (CMUs). Called *concrete masonry units,* these are produced from a range of types of material in a large number of form variations.

Clay Tile Blocks. Used widely in the past, these are hollow units similar to concrete blocks in form. They were used for many of the functions now performed by concrete blocks.

Gypsum Blocks. These are precast units of gypsum concrete, used mostly for nonstructural partitions.

The potential structural character of masonry depends greatly on the material and form of the units. From a material point of view, the high-fired clay products (brick and tile) are the strongest, producing very strong construction with proper mortar, a good arrangement of the units, and good construction craft and work in general. This is particularly important if the general class of the masonry construction is the traditional, unreinforced variety. Although some joint reinforcing is typical in all structural masonry these days, the term "reinforced masonry" is reserved for a class of construction in which major vertical and horizontal reinforcing is used, quite analogous to reinforced concrete construction.

Unreinforced masonry is still used extensively, both in relatively crude form (rough, "native" construction with field stone or adobe bricks) and in highly controlled form with industrially produced elements. However, where building codes are sophisticated and strictly enforced, its use is limited in regions with critical wind conditions or high seismic risk. This makes for a somewhat regional division of use, reinforced masonry being most generally used in southern and western portions of the United States, for example, while unreinforced masonry is widely used in the east and midwest.

With reinforced masonry, the masonry unit takes a somewhat secondary role in determining the structural integrity of the construction. This is discussed more thoroughly in Sec. 17.4.

17.2 Mortar

Mortar is usually composed of water, cement, and sand, with some other materials added to make it stickier (so that it adheres to the units during laying up of the masonry), faster setting, and generally more workable during the construction. Building codes establish various requirements for the mortar, including the classification of the mortar and details of its use during construction. The quality of the mortar is obviously important to the structural integrity of the masonry, both as a structural material in its own right and as a bonding agent that holds the units together. While the integrity of the units is dependent primarily on the manufacturer, the quality of the finished mortar work is dependent primarily on the skill of the mason who lays up the units.

There are several classes of mortar established by codes, with the higher grades being required for uses involving major structural functions (bearing walls, shear walls, etc.) Specifications for the materials and required properties determined by tests are spelled out in detail. Still the major ingredient in producing good mortar is always the skill of the mason. This is a dependency that grows increasingly critical as the general level of craft in construction erodes with the passage of time.

17.3 Basic Construction and Terminology

Figure 17.1 shows some of the common elements of masonry construction. The terminology and details shown apply mostly to construction with bricks or concrete blocks.

Units are usually laid up in horizontal rows, called *courses,* and in vertical planes, called *wythes*. Very thick walls may have several wythes, but most often walls of brick have two wythes and walls of concrete block are single wythe. If wythes are connected directly, the construction is called *solid*. If a space is left between wythes, as shown in the illustration, the wall is called a *cavity wall*. If the cavity is filled with concrete, it is called a *grouted cavity wall.*

The multiple-wythe wall must have the separate wythes bonded together in some fashion. If this is done with the masonry

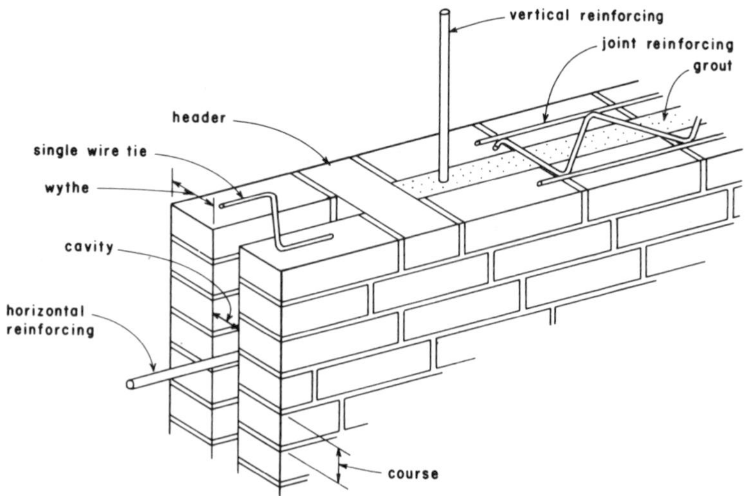

FIGURE 17.1 Elements of masonry construction.

units, the overlapping unit is called a *header*. Various patterns of headers have produced some classic forms of arrangement of bricks in traditional masonry construction. For cavity walls, bonding is often done with metal ties, using single ties at intervals, or a continuous wire trussed element that provides both the tying of the wythes and some minimal horizontal reinforcing.

The continuous element labeled *joint reinforcing* in Fig. 17.1 is now commonly used in both brick and concrete block construction that is code-classified as unreinforced. For seriously reinforced masonry, the reinforcement consists of steel rods (the same as those used for reinforced concrete) that are placed at intervals both vertically and horizontally, and are encased in concrete poured into the wall cavities.

17.4 Structural Masonry

Masonry intended for serious structural purpose includes that used for bearing walls, shear walls, retaining walls, and spanning walls of various type. The types of masonry most used for these purposes includes the following:

Solid Brick Masonry. This is unreinforced masonry, usually of two or more wythes, with the wythes directly connected (no cavities) and the whole consisting of a solid mass of bricks and mortar. This is one of the strongest forms of unreinforced masonry if the bricks and the mortar are of reasonably good quality.

Grouted Brick Masonry. This is usually a two-wythe wall with the cavity filled completely with lean concrete (grout). If unreinforced, it will usually have continuous joint reinforcing (see Fig. 17.1). If reinforced, the steel rods are placed in the cavity. For the reinforced wall, strength derives considerably from the two-way reinforced, concrete-filled cavity, so that considerable strength of the construction may be obtained, even with a relatively low-quality brick or mortar.

Unreinforced Concrete Block Masonry. This is usually of the single-wythe form shown in Fig. 17.2*a*. The faces of blocks as well as cross parts are usually quite thick, although thinner units of lightweight concrete are also produced for less serious structural uses. While it is possible to place vertical reinforcement and grout in cavities, this is not the form of block used generally for reinforced construction. Structural integrity of the construction derives basically from unit strength and quality of the mortar. Staggered vertical joints are used to increase the bonding of units.

Reinforced Concrete Block Masonry. This is usually produced with the type of unit shown in Fig. 17.2*b*. This unit has relatively large individual cavities, so that filled vertical cavities become small reinforced concrete columns. Horizontal reinforcing is placed in courses with the modified blocks shown in Fig. 17.2*c* or *d*. The block shown in Fig. 17.2*d* is also used to form lintels over openings.

The various requirements for all of these types of masonry are described in building codes and industry standards. Regional concerns for weather and critical design loads make for some variation in these requirements. Designers should be careful to determine the particular code requirements and general construction practices for any specific building location.

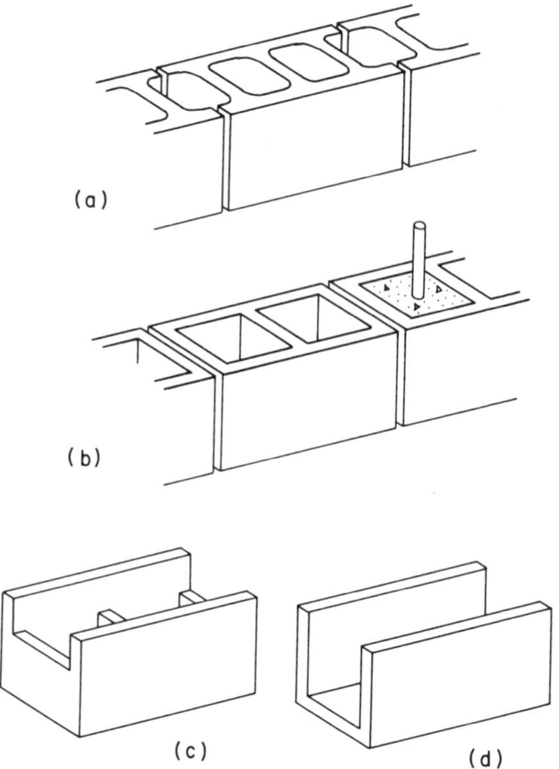

FIGURE 17.2 Construction with concrete masonry units.

Walls that serve structural purposes and have a finished surface of masonry may take various forms. All of the walls just described use the structural masonry unit as the finished wall face, although architecturally exposed surfaces may be specially treated in various ways. For the brick walls, the single brick face that is exposed may have a special treatment (textured, glazed, etc.). It is also possible to use the masonry strictly for structural purposes and to finish the wall surface with some other material, such as stucco or tile.

Another use of masonry consists of providing a finish of masonry on the surface of a structural wall. The finish masonry may merely be a single wythe bonded to other back-up masonry construction, or it may be a nonstructural veneer tied to a separate structure. For the veneered wall, the essential separation of the elements makes it possible for the structural wall to be something other than masonry construction. Indeed, many brick walls are really single-wythe brick veneers tied to wood or steel stud structural walls—still a structural wall, just not a masonry one.

Finally, a wall that appears to be masonry may not be masonry at all, but rather, a surface of thin tiles adhesively bonded to some structural wall surface. Entire "brick" buildings are now being produced with this construction.

17.5 General Design Considerations

Utilization of masonry construction for structural functions requires the consideration of a number of factors that relate to the structural design and to the proper details and specifications for construction. The following are some major concerns that must ordinarily be dealt with.

Units. The material, form, and specific dimensions of the units must be established. Where code classifications exist, the specific grade or type must be defined. Type and grade of unit, as well as usage conditions, usually set the requirements for type of mortar required.

Unit dimensions may be set by the designer, but the sizes of industrially produced products such as bricks and concrete blocks are often controlled by industry standard practices. As shown in Fig. 17.3*a*, the three dimensions of a brick are the height and length of the exposed face and the width that produces the thickness of a single wythe. There is no single standard-sized brick, but most fall in a range close to that shown in the illustration.

Concrete blocks are produced in families of modular sizes. The size of block shown in Fig. 17.3*b* is one that is equivalent to the 2 × 4 in wood—not the only size, but the most common. Concrete block have both nominal and actual dimensions. The nominal dimensions are used for designating the blocks and relate to

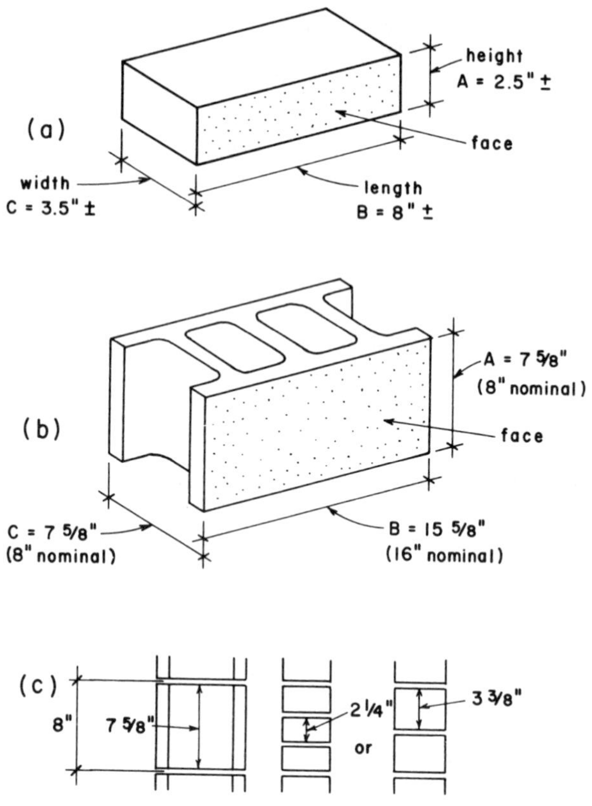

FIGURE 17.3 Dimensional considerations with masonry units.

modular layouts of building dimensions. Actual dimensions are based on the assumption of a mortar joint thickness of $\frac{3}{8}$–$\frac{1}{2}$ in. (the sizes shown in Fig. 17.3*b* reflect the use of $\frac{3}{8}$-in. joints).

A construction frequently used in unreinforced masonry is that of a single wythe of brick bonded to a single wythe of concrete block, as shown in Fig. 17.3*c*. In order to install the metal ties that effect the bonding, as well as to have the bricks and blocks come out even at the top of a wall, a brick of a special height is sometimes used, based on either two or three bricks to one block.

Since transportation of large quantities of bricks or concrete blocks is difficult and costly, units used for a building are usually those obtainable on a local basis. Although some industry standardization exists, the type of locally produced products should be investigated for any design work.

Unit Layout Pattern. When exposed to view, masonry units present two concerns: that for the face of the unit and that for the pattern of layout of the units. Patterns derive from unit shape and the need for unit bonding, if unit-bonded construction is used. Classic patterns were developed from these concerns, but other forms of construction, now more widely used, free the unit pattern somewhat. Nevertheless, classic patterns such as running bond, English bond, and so on, are still widely used.

Patterns also have some structural implications; indeed, the need for unit bonding was such a concern originally. For reinforced construction with concrete blocks, a major constraint is the need to align the voids in a vertical arrangement to facilitate installation of vertical bars. Generally, however, pattern as a structural issue is more critical for unreinforced masonry.

Structural Functions. Masonry walls vary from those that are essentially of a nonstructural character to those that serve major and often multiple structural tasks. The type of unit, grade of mortar, amount and details of reinforcement, and so on, may depend on the degree of structural demands. Wall thickness may relate to stress levels as well as to construction considerations. Most structural tasks involve force transfers: from supported structures, from other walls, and to supporting foundations. Need for brackets, pilasters, vertically tapered or stepped form, or other form variations may relate to force transfers, wall stability, or other structural concerns.

Reinforcement. In the broad sense, reinforcement means anything that is added to help. Structural reinforcement thus includes the use of pilasters, buttresses, tapered form, and other devices, as well as the usual added steel reinforcement. Reinforcement may be generally dispersed or may be provided at critical points, such as at wall ends, tops, edges of openings, and locations of concentrated loads. Both form variation and steel rods are used in both unreinforced masonry and in what is technically referred to as reinforced masonry.

Control Joints. Shrinkage of mortar, temperature variation, and movements due to seismic actions or settlement of foundations are all sources of concern for cracking failures in masonry. Stress concentrations and cracking can be controlled to some extent by reinforcement. However, it is also common to provide some control joints (literally, preestablished cracks) to alleviate these effects. Planning and detailing of control joints are complex problems and must be studied carefully as structural and architectural design issues. Code requirements, industry recommendations, and common construction practices on a local basis will provide guides for this work.

Attachment. Attachment of elements of the construction to masonry is somewhat similar to that required with concrete. Where the nature and exact location of attached items can be predicted, it is usually best to provide some built-in device, such as an anchor bolt, threaded sleeve, and so on. Adjustment of such attachments must be considered, as precision of the construction is limited. Attachment can also be effected with drilled-in anchors or adhesives. These tend to be less constrained by the problem of precise location, although the exact nature of the masonry at the point of attachment may be a concern. This is largely a matter of visualization of the complete building construction and of the general problem of integrating the structure into the whole building. It is simply somewhat more critical with masonry structures, since the simple use of nails, screws, and welding is not possible in the direct way that it is with structures of wood and steel.

17.6 Nonstructural Masonry

Masonry materials can be used for a variety of functions in building construction. Units of fired clay, precast concrete, cut stone, or field stone can be used to form floor surfaces, wall finishes, or nonload-bearing walls (partitions or curtain walls). Indeed, most walls that appear to be made of brick, cut stone, or field stone in present-day construction are likely to be of veneered construction, with the surface of masonry units attached to some back-up structure.

Although structural utilization of the masonry in these situations may be minor or nonexistent, it is still necessary to develop the construction with attention to many of the concerns given to the production of structural masonry for bearing walls, shear walls, and so on. The quality of the masonry units and the quality and workmanship of the mortar joints are often just as critical for these uses, although structural behavior or safety is not at issue. This also extends to concerns for shrinkage, thermal change, stress concentrations at discontinuities, and other aspects of the general behavior of the materials. From an appearance point of view, cracking is just as objectionable in nonstructural masonry facing or paving as it is in a masonry bearing wall.

Masonry veneer facings and nonload-bearing partitions must be provided with control joints and various forms of anchorage and support. Since nonstructural masonry is used extensively, there are many situations for these concerns and a large inventory of recommended construction details. Since these fall mostly in the category of general building construction issues, they are not treated generally in this book.

Masonry units used for nonstructural applications may be of the same structural character as those used for construction of serious masonry structures. Some of the strongest bricks are those that are high-fired to produce great hardness and color intensity for use in veneer construction. However, it is also possible to use some lower structural grades of material, or even some materials which are not usable for structural masonry. Gypsum tile, lightweight concrete blocks, and some very soft bricks may be limited to use in situations not involving structural demands—and probably also not involving exposure to weather or other hazardous situations.

A problem with some nonstructural masonry construction is that of unintended structural response. If the masonry is indeed real masonry (with real masonry units and joints of mortar or grout), it will usually have considerable stiffness. Thus, just as with a plaster surface on a light wood-framed wall structure, it may tend to absorb load due to deformation of the supporting structure. Much of the cracking of nonstructural masonry (and plaster) is due to this phenomenon. The whole construction must be carefully studied for potential problems of this kind. Flexible attachments, control joints, and possibly some reinforcement can be used to alleviate many of these situations.

A particular problem of the type just mentioned is that of the unintended bracing effect of nonstructural partitions during seismic activity. Actually, there are probably many buildings which have structures that are not adequate for significant lateral loads, but are being effectively braced by supposedly nonstructural walls. However, in some cases the general response of the building may be significantly altered by the stiffening effect of rigid partitions. This issue is discussed with regard to the design of rigid-frame structures for seismic effects in Part V.

17.7 Lintels

A lintel is a beam over an opening in a masonry wall. (It is called a header when it occurs in framed construction.) For a structural masonry wall the loading on a lintel is usually assumed to be developed only by the weight of the masonry occurring in a 45° isosceles triangle, as shown in Fig. 17.4*a*. It is assumed that the arching or corbeling action of the wall will carry the remaining wall above the opening, as shown in Fig. 17.4*b*. There are, however, many situations that can occur to modify this assumption.

If the wall height above the opening is short with respect to the width of the opening (Fig. 17.4*c*), it is best to design for the full weight of the wall. This may also be advisable when a supported load is carried by the wall, although the location of the loading and the height of the wall should be considered. If the loading is a considerable distance above the opening, and the opening is narrow, the lintel can probably be designed for only the usual triangular loading of the wall weight, as shown in Fig. 17.4*d*. However, if the loading is a short distance above the lintel (Fig. 17.4*e*), the lintel should be designed for the full applied loading, although the wall weight may still be considered in the usual triangular form.

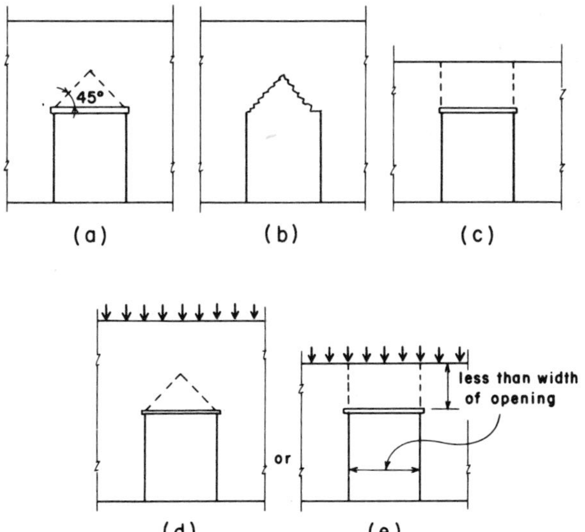

FIGURE 17.4 Loading conditions for lintels.

Lintels may be formed in a variety of ways. In times past lintels were made of large blocks of cut stone. In some situations today lintels are formed of reinforced concrete, either precast or formed and poured as the wall is built. With reinforced masonry construction, lintels are commonly formed as reinforced masonry beams, created with U-shaped blocks as shown in Fig. 17.5*a* when the construction is with hollow concrete units. For heavier

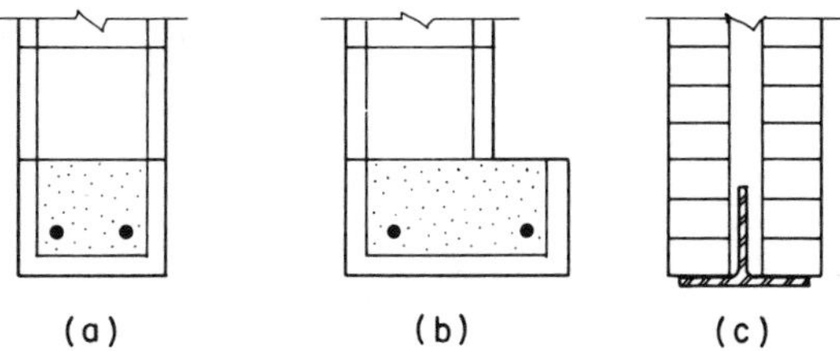

FIGURE 17.5 Typical details for lintels.

loadings, the size of the lintel may be increased by thickening the wall, as shown in Fig. 17.5*b*.

For unreinforced masonry walls a lintel commonly used is one consisting of a rolled steel section. A form that fits well with a two-wythe brick wall is the inverted tee shown in Fig. 17.5*c*. Single angles, double angles, and various built-up sections may also be used when the loading or the construction details are different.

18

Design of Structural Masonry

Masonry construction intended for structural purposes is subject to all the general concerns for the quality and correctness of the construction, as discussed in Chapter 17. In addition, if it is exposed to view—which it often is—the unit faces, mortar joints, and patterns of unit arrangements become architectural design concerns. Beyond these concerns are those for the structural actions which the construction must perform when major structural tasks are required. This chapter deals with some of the design considerations for various forms of structural masonry. Some design examples are also given in the work in Part V.

18.1 Unreinforced Structural Masonry

Many structures of unreinforced masonry have endured for centuries, and this form of construction is still widely used. Although it is generally held in low regard in regions that have frequent earthquakes, it is still approved by most building codes for use within code-defined limits. With good design and good-quality construction, it is possible to have structures that are more than adequate by present standards.

If masonry is essentially unreinforced, the character and structural integrity of the construction are highly dependent on the details and the quality of the masonry work. Strength and form of units, arrangements of units, general quality of the mortar, and the form and details of the general construction are all important. Thus the degree of attention paid to design, writing the specifications, detailing the construction, and careful inspection during the work must be adequate to ensure good finished construction.

There are a limited number of structural applications for unreinforced masonry, and structural computations for design of common elements are in general quite simple. We hesitate to show examples of such work, since the data and general procedures are considerably variable from one region to another, depending on local materials and construction practices as well as variations in building code requirements. In many instances forms of construction not subject to satisfactory structural investigation are tolerated simply because they have been used with success for many years on a local basis—a hard case to argue against. Nevertheless, there are some general principles and typical situations that produce some common problems and procedures. The following discussion deals with some major concerns of design of the unreinforced masonry structure.

Minimal Construction. As in other types of construction, there is a minimal form of construction that results from the satisfaction of various general requirements. Industry standards result in some standardization and classification of products, which is usually reflected in building code designations. Structural usage is usually tied to specified minimum grades of units, mortar, construction practices, and in some cases to need for reinforcement or other enhancement. This results in most cases in a basic minimum form of construction that is adequate for many ordinary functions, which is in fact usually the intent of the codes. Thus there are many instances in which buildings of a minor nature are built without benefit of structural computations, simply being produced in response to code-specified minimum requirements.

Design Strength of Masonry. As with concrete, the basic strength of the masonry is measured as its resistive compressive strength. This is established in the form of the *specified compressive strength,* designated f'_m. The value for f'_m is usually taken from code specifications, based on the strength of the units and the class of the mortar.

Allowable Stresses. Allowable stresses are directly specified for some cases (such as tension and shear) or are determined by code formulas that usually include the variable value of f'_m. There are usually two values for any situation: that to be used when special inspection of the work is provided (as specified by the code), and that to be used when it is not. For minor construction projects it is usually desirable to avoid the need for the special inspection. For construction with hollow units that are not fully grouted, stress computations are based on the net cross section of the hollow construction.

Avoiding Tension. While codes ordinarily permit some low stress values for flexural tension, many designers prefer to avoid tension in unreinforced masonry. An old engineering definition of mortar is "the material used to keep masonry units *apart,*" reflecting a lack of faith in the bonding action of mortar.

Reinforcement or Enhancement. The strength of a masonry structure can be improved by various means, including the insertion of steel reinforcing rods as is done to produce reinforced masonry. Vertical rods are sometimes used with construction that is essentially classified as unreinforced, usually to enhance bending resistance or to absorb localized stress conditions. Horizontal-wire-type reinforcing is commonly used to reduce stress effects due to shrinkage and thermal change.

Another type of structural reinforcement is achieved through the use of form variation, examples of which are shown in Fig. 18.1. Turning a corner at the end of a wall (Fig. 18.1*a*) adds stability and strength to the discontinuous edge of the structure. An enlargement in the form of a pilaster (Fig. 18.1*b*) can also be used to improve the end of a wall or to add bracing or concentrated strength at some intermediate point along the wall. Heavy

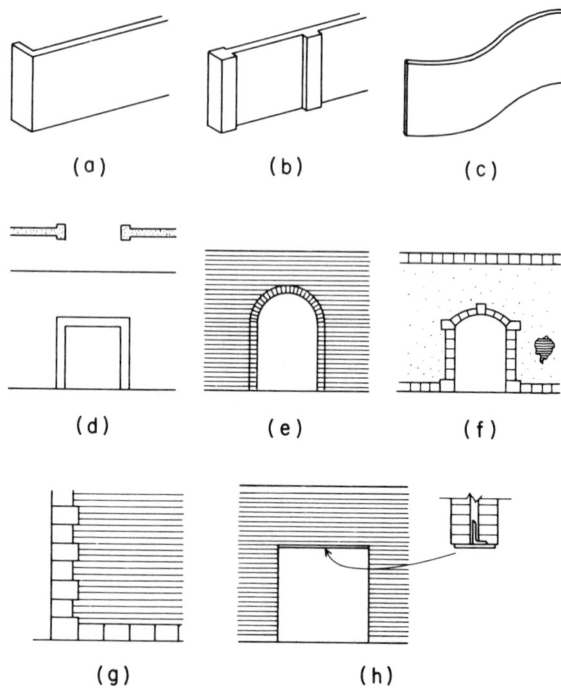

FIGURE 18.1 Techniques for reinforcement of masonry construction.

concentrated loads are ordinarily accommodated by using pilasters when wall thickness is otherwise minimal. Curving a wall in plan (Fig. 18.1*c*) is another means of improving the stability of the wall.

Building corners and openings for doors or windows are other locations where enhancement is often required. Figure 18.1*d* shows the use of an enlargement of the wall around the perimeter of a door opening. If the top of the opening is of arched form, the enlarged edge may continue as an arch to span the opening, as shown in Fig. 18.1*e*, or a change may be made to a stronger material to effect the edge and arch enhancement, as shown in Fig. 18.1*f*. Building corners in historic buildings were often strengthened by using large cut stones to form the corner, as shown in Fig. 18.1*g*.

While form variations or changes of the masonry units can be used to effect spans over openings, the more usual means of achieving this—especially for flat spans—is by using a steel lintel, as shown in Fig. 18.1*h*.

18.2 Reinforced Masonry

As we use it here, the term *reinforced masonry* designates a type of masonry construction specifically classified by building code definitions. Essential to this definition are the assumptions that the steel reinforcement is designed to carry forces and the masonry does not develop tensile stresses. This makes the design basically analogous to that for reinforced concrete, and indeed the present data and design procedures used for reinforced masonry are in general similar to those used for concrete structures. Until recently, the general methods of the working stress design procedure were used for masonry. However, recent industry standards and some building codes (including the 1988 edition of the UBC) have promoted the use of strength methods for investigation and design. As with concrete, the use of strength methods is quite complex and abstract. We have therefore chosen to use the simpler methods of the working stress design procedures for most of the example computations to permit a briefer treatment, our interest being more in demonstrating the problem and the basic concerns for investigation rather than particular means for investigation.

18.3 Reinforced Brick Masonry

Reinforced brick masonry typically consists of the type of construction shown in Fig. 17.1. The wall shown in the illustration consists of two wythes of bricks with a cavity space between them. The cavity space is filled completely with grout so that the construction qualifies firstly as *grouted masonry*, for which various requirements are stipulated by the code. One requirement is for the bonding of the wythes, which can be accomplished with the masonry units, but is most often done by using the steel wire joint reinforcement shown in Fig. 17.1. Added to this basic construction are the vertical and horizontal reinforcing rods in the grouted cavity space, making the resulting construction qualify as *reinforced grouted masonry*.

General requirements and design procedures for the reinforced brick masonry wall are similar to those for concrete walls. There are stipulations for minimum reinforcement and provisions for stress limits for the various structural actions of walls in vertical compression, bending, and shear wall functions. Structural investigation is essentially similar to that for the hollow unit masonry wall, which is discussed in the next section.

Despite the presence of the reinforcement, the type of construction shown in Fig. 17.1 is still essentially a masonry structure, highly dependent on the quality and structural integrity of the masonry itself—particularly the skill and care exercised in laying up the units and handling of the construction process in general. The grouted, reinforced cavity structure is in itself often considered to be a third wythe, constituted as a very thin reinforced concrete wall panel. The enhancement of the construction represented by the cavity wythe is considerable, but the major bulk of the construction is still basically just solid brick masonry.

18.4 Reinforced Hollow Unit Masonry

This type of construction most often consists of single-wythe walls formed as shown in Fig. 17.2*b–d*. Cavities are vertically aligned so that small reinforced concrete columns can be formed within them. At some interval, horizontal courses are also used to form reinforced concrete members. The intersecting vertical and horizontal concrete members thus constitute a rigid-frame bent inside the wall. This reinforced concrete frame is the major structural component of the construction. Besides providing forming, the concrete blocks serve to brace the frame, provide protection for the reinforcement, and interact in composite action with the rigid frame. Nevertheless, the structural character of the construction derives largely from the concrete frame created in the void spaces in the wall.

The code requires that reinforcement be a maximum of 48 in.

on center; thus the maximum spacing of the concrete members inside the wall is 48 in., both vertically and horizontally. With 16-in.-long blocks this means that every sixth vertical void space is grouted. With blocks having a net section of approximately 50% (half solid, half void), this means that the minimum construction is an average of approximately 60% solid. For computations based on the net cross section of the wall, the wall may therefore usually be considered to be a minimum of 60% solid.

If all void spaces are grouted, the construction is fully solid. This is usually required for structures such as retaining walls and basement walls, but may also be done simply to increase the wall section for the reduction of stress levels. Finally, if reinforcing is placed in all of the vertical voids (instead of every sixth one), the contained reinforced concrete structure is considerably increased and both vertical bearing capacity and lateral bending capacity are significantly increased. Heavily loaded shear walls are developed in this manner.

For shear wall actions, there are two conditions defined by the code. The first case involves a wall with minimum reinforcing, in which the shear is assumed to be taken by the masonry. The second case is one in which the reinforcing is designed to take all of the shear. Allowable stresses are given for both cases, even though the reinforcement must be designed for the full shear force in the second case.

18.5 Design of Structural Elements

Design of structural masonry is accomplished with the use of allowable stresses and analytical formulas from the applicable building code. Unreinforced elements are generally used for structural tasks that involve primary resistance by development of vertical compression: bearing walls, piers, and relatively stout (not slender) columns. Minor shear and flexural compression may also be developed, but tension stress is usually avoided or at least restricted to very low values. Strengths range from the relatively low capacity of voided, hollow unit masonry, up to the considerable capacity of solid brick masonry with the best mortar and high-strength bricks. In some cases usage is based largely on experience, and structural elements may be built to code specifications without recourse to structural computations.

A widely used form of reinforced masonry construction is that which is produced with hollow units of precast concrete—good old concrete blocks, although they are more formally referred to as CMUs (for concrete masonry unit). The relatively large voids in such units permit the insertion of steel reinforcing rods into vertically aligned voids and the pouring of concrete fill around the rods to produce the effect of reinforced concrete columns within the wall construction. These are combined with horizontal beams of reinforced concrete, created by filling some horizontal courses with concrete and reinforcement. Minimum code requirements ensure a dispersion of these enclosed bracing elements at some frequent intervals throughout the height and length of wall construction. Wall ends, tops, intersections, and the edges of all openings are also reinforced. The resulting composite masonry and reinforced concrete structure is quite resistive to various forms of structural action. This type of construction is required in the higher seismic-risk zones and is thus used quite extensively in regions where seismic design is a critical concern.

Reinforced masonry is generally designed by methods that are essentially similar to the working stress methods for reinforced concrete, as described in Chapter 13. The use of these methods with criteria from the 1988 edition of the *Uniform Building Code* (Ref. 1) is illustrated in some of the examples in Part V.

V

STRUCTURAL SYSTEMS FOR BUILDINGS

This part contains examples of the design of complete structural systems for buildings. The buildings selected for design are not intended as examples of good architectural design, but rather have been made to create a range of situations in order to be able to demonstrate the use of various structural components. Design of individual elements of the structural systems is largely based on the materials presented in the earlier chapters. The principal purpose here is to show the broader context of design work by dealing with whole structures and the building in general.

19

General Concerns for Structures

This chapter contains some discussion of general issues relating to design of building structures. These concerns have mostly not been addressed in the presentations in earlier chapters, but require some general consideration when dealing with whole building design situations.

19.1 Introduction

Materials, methods, and details of building construction vary considerably on a regional basis. There are many factors that affect this situation, including the real effects of response to climate and the availability of construction materials. Even in a single region, differences occur between individual buildings, based on individual styles of architectural design and personal techniques of builders. Nevertheless, at any given time there are usually a few predominant, popular methods of construction that are employed for most buildings of a given type and size. The construction methods and details shown here are reasonable, but in no way are they intended to illustrate a singular, superior style of building.

19.2 Dead Loads

Dead load consists of the weight of the materials of which the building is constructed such as walls, partitions, columns, framing, floors, roofs, and ceilings. In the design of a beam, the dead load must include an allowance for the weight of the beam itself. Table 19.1, which lists the weights of many construction materials, may be used in the computation of dead loads. Dead loads are due to gravity and they result in downward vertical forces.

Dead load is generally a permanent load, once the building construction is completed, unless frequent remodeling or rearrangement of the construction occurs. Because of this permanent, long-time, character, the dead load requires certain considerations in design, such as the following:

1. It is always included in design loading combinations, except for investigations of singular effects, such as deflections due to only live load.
2. Its long-time character has some special effects causing sag and requiring reduction of design stresses in wood structures, producing creep effects in concrete structures, and so on.
3. It contributes some unique responses, such as the stabilizing effects that resist uplift and overturn due to wind forces.

19.3 Building Code Requirements

Structural design of buildings is most directly controlled by building codes, which are the general basis for the granting of building permits—the legal permission required for construction. Building codes (and the permit-granting process) are administered by some unit of government: city, county, or state. Most building codes, however, are based on some model code, of which there are three widely used in the United States:

TABLE 19.1 Weights of Building Construction

	lb/ft²	kN/m²
Roofs		
3-ply ready roofing (roll, composition)	1	0.05
3-ply felt and gravel	5.5	0.26
5-ply felt and gravel	6.5	0.31
Shingles		
wood	2	0.10
asphalt	2–3	0.10–0.15
clay tile	9–12	0.43–0.58
concrete tile	8–12	0.38–0.58
slate, 1/4 in.	10	0.48
fiber glass	2–3	0.10–0.15
aluminum	1	0.05
steel	2	0.10
Insulation		
fiber glass batts	0.5	0.025
rigid foam plastic	1.5	0.075
foamed concrete, mineral aggregate	2.5/in.	0.0047/mm
Wood rafters		
2 × 6 at 24 in.	1.0	0.05
2 × 8 at 24 in.	1.4	0.07
2 × 10 at 24 in.	1.7	0.08
2 × 12 at 24 in.	2.1	0.10
Steel deck, painted		
22 ga	1.6	0.08
20 ga	2.0	0.10
18 ga	2.6	0.13
Skylight		
glass with steel frame	6–10	0.29–0.48
plastic with aluminum frame	3–6	0.15–0.29
Plywood or softwood board sheathing	3.0/in.	0.0057/mm
Ceilings		
Suspended steel channels	1	0.05
Lath		
steel mesh	0.5	0.025
gypsum board, 1/2 in.	2	0.10
Fiber tile	1	0.05
Dry wall, gypsum board, 1/2 in.	2.5	0.12
Plaster		
gypsum, acoustic	5	0.24
cement	8.5	0.41
Suspended lighting and air distribution systems, average	3	0.15

TABLE 19.1 (*Continued*)

Floors		
Hardwood, 1/2 in.	2.5	0.12
Vinyl tile, 1/8 in.	1.5	0.07
Asphalt mastic	12/in.	0.023/mm
Ceramic tile		
3/4 in.	10	0.48
thin set	5	0.24
Fiberboard underlay, 5/8 in.	3	0.15
Carpet and pad, average	3	0.15
Timber deck	2.5/in.	0.0047/mm
Steel deck, stone concrete fill, average	35–40	1.68–1.92
Concrete deck, stone aggregate	12.5/in.	0.024/mm
Wood joists		
2 × 8 at 16 in.	2.1	0.10
2 × 10 at 16 in.	2.6	0.13
2 × 12 at 16 in.	3.2	0.16
Lightweight concrete fill	8.0/in.	0.015/mm
Walls		
2 × 4 studs at 16 in., average	2	0.10
Steel studs at 16 in., average	4	0.20
Lath, plaster; see Ceilings		
Gypsum dry wall, 5/8 in. single	2.5	0.12
Stucco, 7/8 in., on wire and paper or felt	10	0.48
Windows, average, glazing + frame		
small pane, single glazing, wood or metal frame	5	0.24
large pane, single glazing, wood or metal frame	8	0.38
increase for double glazing	2–3	0.10–0.15
curtain walls, manufactured units	10–15	0.48–0.72
Brick veneer		
4 in., mortar joints	40	1.92
1/2 in., mastic	10	0.48
Concrete block		
lightweight, unreinforced—4 in.	20	0.96
6 in.	25	1.20
8 in.	30	1.44
heavy, reinforced, grouted—6 in.	45	2.15
8 in.	60	2.87
12 in.	85	4.07

1. The *Uniform Building Code* (Ref. 1), which is widely used in the West, as it has the most complete data for seismic design.
2. *The BOCA Basic National Building Code*, used widely in the East and Midwest.
3. *The Standard Building Code*, used in the Southeast.

These model codes are more similar than different, and are in turn largely derived from the same basic data and standard reference sources, including many industry standards. In the several model codes and many city, county, and state codes, however, there are some items that reflect particular regional concerns.

With respect to control of structures, all codes have materials (all essentially the same) that relate to the following issues:

1. *Minimum Required Live Loads.* This is addressed in Sec. 19.4; all codes have tables similar to those shown in Tables 19.2 and 19.3, which are reproduced from the *Uniform Building Code.*
2. *Wind Loads.* These are highly regional in character with respect to concern for local windstorm conditions. Model codes provide data with variability on the basis of geographic zones.
3. *Seismic (Earthquake) Effects.* These are also regional with predominant concerns in the western states. This data, including recommended investigations, is subject to quite frequent modification, as the area of study responds to ongoing research and experience.
4. *Load Duration.* Loads or design stresses are often modified on the basis of the time span of the load, varying from the life of the structure for dead load to a fraction of a second for a wind gust or a single major seismic shock. Safety factors are frequently adjusted on this basis. Some applications are illustrated in the work in the design examples in this part.
5. *Load Combinations.* These were formerly mostly left to the discretion of designers, but are now quite commonly

stipulated in codes, mostly because of the increasing use of ultimate strength design and the use of factored loads.

6. *Design Data for Types of Structures.* These deal with basic materials (wood, steel, concrete, masonry, etc.), specific structures (towers, balconies, pole structures, etc.), and special problems (foundations, retaining walls, stairs, etc.) Industry-wide standards and common practices are generally recognized, but local codes may reflect particular local experience or attitudes. Minimal structural safety is the general basis, and some specified limits may result in questionably adequate performances (bouncy floors, cracked plaster, etc.)
7. *Fire Resistance.* For the structure, there are two basic concerns, both of which produce limits for the construction. The first concern is for structural collapse or significant structural loss. The second concern is for containment of the fire to control its spread. These concerns produce limits on the choice of materials (e.g., combustible or noncombustible) and some details of the construction (cover on reinforcement in concrete, fire insulation for steel beams, etc.)

The work in the design examples in this part is based largely on criteria from the *Uniform Building Code* (Ref. 1). The choice of this model code reflects only the fact of the degree of familiarity of the author with specific codes in terms of his recent experience.

19.4 Live Loads

Live loads technically include all the nonpermanent loadings that can occur, in addition to the dead loads. However, the term as commonly used usually refers only to the vertical gravity loadings on roof and floor surfaces. These loads occur in combination with the dead loads, but are generally random in character and must be dealt with as potential contributors to various loading combinations, as discussed in Sec. 19.3.

Roof Loads. In addition to the dead loads they support, roofs are designed for a uniformly distributed live load that includes snow accumulation and the general loadings that occur during construction and maintenance of the roof. Snow loads are based on local snowfalls and are specified by local building codes.

Table 19.2 gives the minimum roof live load requirements specified by the 1988 edition of the *Uniform Building Code*. Note the adjustments for roof slope and for the total area of roof surface supported by a structural element. The latter accounts for the increase in probability of the lack of total surface loading as the size of the surface area increases.

Roof surfaces must also be designed for wind pressure, for which the magnitude and manner of application are specified by local building codes based on local wind histories. For very light roof construction, a critical problem is sometimes that of the upward (suction) effect of the wind, which may exceed the dead load and result in a net upward lifting force.

Although the term *flat roof* is often used, there is generally no such thing; all roofs must be designed for some water drainage. The minimum required pitch is usually $\frac{1}{4}$ in./ft, or a slope of approximately 1 : 50. With roof surfaces that are this close to flat, a potential problem is that of *ponding*, a phenomenon in which the weight of water on the surface causes deflection of the supporting structure, which in turn allows for more water accumulation (in a pond), causing more deflection, and so on, resulting in an accelerated collapse condition.

Floor Loads. The live load on a floor represents the probable effects created by the occupancy. It includes the weights of human occupants, furniture, equipment, stored materials, and so on. All building codes provide minimum live loads to be used in the design of buildings for various occupancies. Since there is a lack of uniformity among different codes in specifying live loads, the local code should always be used. Table 19.3 contains values for floor live loads as given by the 1988 edition of the *Uniform Building Code*.

Although expressed as uniform loads, code-required values are usually established large enough to account for ordinary concentrations that occur. For offices, parking garages, and some other occupancies, codes often require the consideration of a specified concentrated load as well as the distributed loading.

TABLE 19.2 Minimum Roof Live Loads

	Minimum Uniformly Distributed Load (lb/ft²)			Minimum Uniformly Distributed Load (kN/m²)		
	Tributary Loaded Area for Structural Member (ft^2)			Tributary Loaded Area for Structural Member (m^2)		
Roof Slope Conditions	0–200	201–600	Over 600	0–18.6	18.7–55.7	Over 55.7
1. Flat or rise less than 4 in./ft (1 : 3). Arch or dome with rise less than 1/8 span.	20	16	12	0.96	0.77	0.575
2. Rise 4 in./ft (1 : 3) to less than 12 in./ft (1 : 1). Arch or dome with rise 1/8 of span to less than 3/8 of span.	16	14	12	0.77	0.67	0.575
3. Rise 12 in./ft (1 : 1) or greater. Arch or dome with rise 3/8 of span or greater.	12	12	12	0.575	0.575	0.575
4. Awnings, except cloth covered.	5	5	5	0.24	0.24	0.24
5. Greenhouses, lath houses, and agricultural buildings.	10	10	10	0.48	0.48	0.48

Source: Adapted from the *Uniform Building Code*, 1988 ed. (Ref. 1), with permission of the publishers, International Conference of Building Officials.

TABLE 19.3 Minimum Floor Loads

Use or Occupancy		Uniform Load		Concentrated Load	
Description	Description	(psf)	(kN/m²)	(lb)	(kN)
Armories		150	7.2		
Assembly areas and auditoriums and balconies therewith	Fixed seating areas	50	2.4		
	Movable seating and other areas	100	4.8		
	Stages and enclosed platforms	125	6.0		
Cornices, marquees, and residential balconies		60	2.9		
Exit facilities		100	4.8		
Garages	General storage, repair	100	4.8	*	
	Private pleasure car	50	2.4	*	
Hospitals	Wards and rooms	40	1.9	1000	4.5
Libraries	Reading rooms	60	2.9	1000	4.5
	Stack rooms	125	6.0	1500	6.7
Manufacturing	Light	75	3.6	2000	9.0
	Heavy	125	6.0	3000	13.3
Offices		50	2.4	2000	9.0
Printing plants	Press rooms	150	7.2	2500	11.1
	Composing rooms	100	4.8	2000	9.0
Residential		40	1.9		
Rest rooms		**			
Reviewing stands, grandstands, and bleachers		100	4.8		
Roof decks (occupied)	Same as area served				
Schools	Classrooms	40	1.9	1000	4.5
Sidewalks and driveways	Public access	250	12.0	*	
Storage	Light	125	6.0		
	Heavy	250	12.0		
Stores	Retail	75	3.6	2000	9.0
	Wholesale	100	4.8	3000	13.3

Source: Adapted from the *Uniform Building Code*, 1988 ed. (Ref. 1), with permission of the publishers, International Conference of Building Officials.

* Wheel loads related to size of vehicles that have access to the area.

** Same as the area served or minimum of 50 psf.

Where buildings are to contain heavy machinery, stored materials, or other contents of unusual weight, these must be provided for individually in the design of the structure.

When structural framing members support large areas, most codes allow some reduction in the total live load to be used for design. These reductions, in the case of roof loads, are incorporated into the data in Table 19.2. The following is the method given in the 1988 edition of the *Uniform Building Code* for determining the reduction permitted for beams, trusses, or columns that support large floor areas.

Except for floors in places of assembly (theaters, etc.), and except for live loads greater than 100 psf [4.79 kN/m^2], the design live load on a member may be reduced in accordance with the formula

$$R = 0.08\,(A - 150)$$

$$[R = 0.86\,(A - 14)]$$

The reduction shall not exceed 40% for horizontal members or for vertical members receiving load from one level only, 60% for other vertical members, nor R as determined by the formula

$$R = 23.1\left(1 + \frac{D}{L}\right)$$

In these formulas

R = reduction in percent,
A = area of floor supported by a member,
D = unit dead load/sq ft of supported area,
L = unit live load/sq ft of supported area.

In office buildings and certain other building types, partitions may not be permanently fixed in location but may be erected or moved from one position to another in accordance with the requirements of the occupants. In order to provide for this flexibility, it is customary to require an allowance of 15–20 psf [0.72–0.96 kN/m^2] which is usually added to other dead loads.

19.5 Lateral Loads

As used in building design, the term *lateral load* is usually applied to the effects of wind and earthquakes, as they induce horizontal forces on stationary structures. From experience and research, design criteria and methods in this area are continuously refined, with recommended practices being presented through the various model building codes, such as the *Uniform Building Code* (UBC) (Ref. 1).

Space limitations do not permit a complete discussion of the topic of lateral loads and design for their resistance. The following discussion summarizes some of the criteria for design in the latest edition of the UBC. Examples of application of these criteria are given in the chapters that follow containing examples of building structural design. For a more extensive discussion the reader is referred to *Design for Lateral Forces* (Ref. 16).

Wind. Where wind is a major local problem, local codes are usually more extensive with regard to design requirements for wind. However, many codes still contain relatively simple criteria for wind design. One of the most up-to-date and complex standards for wind design is contained in the *American National Standard Minimum Design Loads for Buildings and Other Structures*, ANSI A58.1-1982, published by the American National Standards Institute in 1982.

Complete design for wind effects on buildings includes a large number of both architectural and structural concerns. The following is a discussion of some of the requirements for wind as taken from the 1988 edition of the UBC (Ref. 1), which is in general conformance with the material presented in the ANSI Standard just mentioned.

Basic Wind Speed. This is the maximum wind speed (or velocity) to be used for specific locations. It is based on recorded wind histories and adjusted for some statistical likelihood of occurrence. For the continental United States the wind speeds are taken from UBC, Figure No. 4. As a reference point, the speeds are those recorded at the standard measuring position of 10 m (approximately 33 ft) above the ground surface.

Exposure. This refers to the conditions of the terrain surrounding the building site. The ANSI Standard describes four conditions (A, B, C, and D), although the UBC uses only two (B and C). Condition C refers to sites surrounded for a distance of one-half mile or more by flat, open terrain. Condition B has buildings, forests, or ground-surface irregularities 20 ft or more in height covering at least 20% of the area for a distance of 1 mile or more around the site.

Wind Stagnation Pressure (q_s). This is the basic reference equivalent static pressure based on the critical local wind speed. It is given in UBC Table No. 23-F and is based on the following formula as given in the ANSI Standard:

$$q_s = 0.00256V^2$$

Example: For a wind speed of 100 mph

$$\begin{aligned} q_s &= 0.00256V^2 = 0.00256(100)^2 \\ &= 25.6 \text{ psf } [1.23 \text{ kPa}] \end{aligned}$$

which is rounded off to 26 psf in the UBC table.

Design Wind Pressure. This is the equivalent static pressure to be applied normal to the exterior surfaces of the building and is determined from the formula

$$p = C_e C_q q_s I$$

(UBC Formula 11-1, Section 2311)

in which p = design wind pressure in psf,
C_e = combined height, exposure, and gust factor coefficient as given in UBC Table No. 23-G,

C_q = pressure coefficient for the structure or portion of structure under consideration as given in UBC Table No. 23-H,

q_s = wind stagnation pressure at 30 ft given in UBC Table No. 23-F,

I = importance factor.

The importance factor is 1.15 for facilities considered to be essential for public health and safety (such as hospitals and government buildings) and buildings with 300 or more occupants. For all other buildings the factor is 1.0.

The design wind pressure may be positive (inward) or negative (outward, suction) on any given surface. Both the sign and the value for the pressure are given in the UBC table. Individual building surfaces, or parts thereof, must be designed for these pressures.

Design Methods. Two methods are described in the Code for the application of the design wind pressures in the design of structures. For design of individual elements particular values are given in UBC Table 23-H for the C_q coefficient to be used in determining p. For the primary bracing system the C_q values and their use is to be as follows:

Method 1 (Normal Force Method). In this method wind pressures are assumed to act simultaneously normal to all exterior surfaces. This method is required to be used for gabled rigid frames and may be used for any structure.

Method 2 (Projected Area Method). In this method the total wind effect on the building is considered to be a combination of a single inward (positive) horizontal pressure acting on a vertical surface consisting of the projected building profile and an outward (negative, upward) pressure acting on the full projected area of the building in plan. This method may be used for any structure less than 200 ft in height, except for gabled rigid frames. This is the method generally employed by building codes in the past.

Uplift. Uplift may occur as a general effect, involving the entire roof or even the whole building. It may also occur as a local phenomenon such as that generated by the overturning moment on a single shear wall. In general, use of either design method will account for uplift concerns.

Overturning Moment. Most codes require that the ratio of the dead load resisting moment (called the restoring moment, stabilizing moment, etc.) to the overturning moment be 1.5 or greater. When this is not the case, uplift effects must be resisted by anchorage capable of developing the excess overturning moment. Overturning may be a critical problem for the whole building, as in the case of relatively tall and slender tower structures. For buildings braced by individual shear walls, trussed bents, and rigid-frame bents, overturning is investigated for the individual bracing units. Method 2 is usually used for this investigation, except for very tall buildings and gabled rigid frames.

Drift. Drift refers to the horizontal deflection of the structure due to lateral loads. Code criteria for drift are usually limited to requirements for the drift of a single story (horizontal movement of one level with respect to the next above or below). The UBC does not provide limits for wind drift. Other standards give various recommendations, a common one being a limit of story drift to 0.005 times the story height (which is the UBC limit for seismic drift). For masonry structures wind drift is sometimes limited to 0.0025 times the story height. As in other situations involving structural deformations, effects on the building construction must be considered; thus the detailing of curtain walls or interior partitions may affect limits on drift.

Combined Loads. Although wind effects are investigated as isolated phenomena, the actions of the structure must be considered simultaneously with other phenomena. The requirements for load combinations are given by most codes, although common sense will indicate the critical combinations in most cases. With the increasing use of load factors the combinations are further modified by applying different factors for the various types of loading, thus permitting individual control based on the reliability of data and investigation procedures and the relative significance to safety of the different load sources and effects. Required load combinations are described in Section 2303 of the UBC.

Special Problems. The general design criteria given in most codes are applicable to ordinary buildings. More thorough investigation is recommended (and sometimes required) for special circumstances such as the following:

Tall Buildings. These are critical with regard to their height dimension as well as the overall size and number of occupants inferred. Local wind speeds and unusual wind phenomena at upper elevations must be considered.

Flexible Structures. These may be affected in a variety of ways, including vibration or flutter as well as the simple magnitude of movements.

Unusual Shapes. Open structures, structures with large overhangs or other projections, and any building with a complex shape should be carefully studied for the special wind effects that may occur. Wind-tunnel testing may be advised or even required by some codes.

Use of code criteria for various ordinary buildings is illustrated in the design examples in the following chapters.

Earthquakes. During an earthquake a building is shaken up and down and back and forth. The back-and-forth (horizontal) movements are typically more violent and tend to produce major unstabilizing effects on buildings; thus structural design for earthquakes is mostly done in terms of considerations for horizontal (called lateral) forces. The lateral forces are actually generated by the weight of the building—or, more specifically, by the mass of the building that represents both an inertial resistance to movement and the source for kinetic energy once the building is actually in motion. In the simplified procedures of the equivalent static force method, the building structure is considered to be loaded by a set of horizontal forces consisting of some fraction of the building weight. An analogy would be to visualize the building as being rotated vertically 90° to form a cantilever beam, with the ground as the fixed end and with a load consisting of the building weight.

In general, design for the horizontal force effects of earthquakes is quite similar to design for the horizontal force effects of wind. Indeed, the same basic types of lateral bracing (shear walls, trussed bents, rigid frames, etc.) are used to resist both force effects. There are indeed some significant differences, but in the main a system of bracing that is developed for wind bracing will most likely serve reasonably well for earthquake resistance as well.

Because of its considerably more complex criteria and procedures, we have chosen not to illustrate the design for earthquake effects in the examples in this part. Nevertheless, the development of elements and systems for the lateral bracing of the buildings in the design examples here is quite applicable in general to situations where earthquakes are a predominant concern. For structural investigation, the principal difference is in the determination of the loads and their distribution in the building. Another major difference is in the true dynamic effects, critical wind force being usually represented by a single, major, one-direction punch from a gust, while earthquakes represent rapid back-and-forth, reversing-direction actions. However, once the dynamic effects are translated into equivalent static forces, design concerns for the bracing systems are very similar, involving considerations for shear, overturning, horizontal sliding, and so on.

For a detailed explanation of earthquake effects and illustrations of the investigation by the equivalent static force method the reader is referred to *Design for Lateral Forces* (Ref. 16).

20

Building One

The building in this chapter consists of a simple, single-story structure that will be designed primarily as a light wood frame. Lateral bracing will be developed in response to an assumed wind load. Some possible alternatives for the construction of a similar building are presented and discussed in the last section of the chapter.

20.1 General Requirements and Design Criteria

Figure 20.1 shows a one-story, box-shaped building intended for commercial occupancy. Assuming that light wood framing is adequate for fire-resistance requirements, we will illustrate the design of the major elements of the structure for this building. We assume the following data for design:

Roof live load = 20 psf (reducible).

Wind load critical at 20 psf on vertical exterior surface.

Wood framing of Douglas fir–larch.

The general profile of the building is indicated in Fig. 20.1*c*, which shows a flat roof and ceiling and a short parapet at the

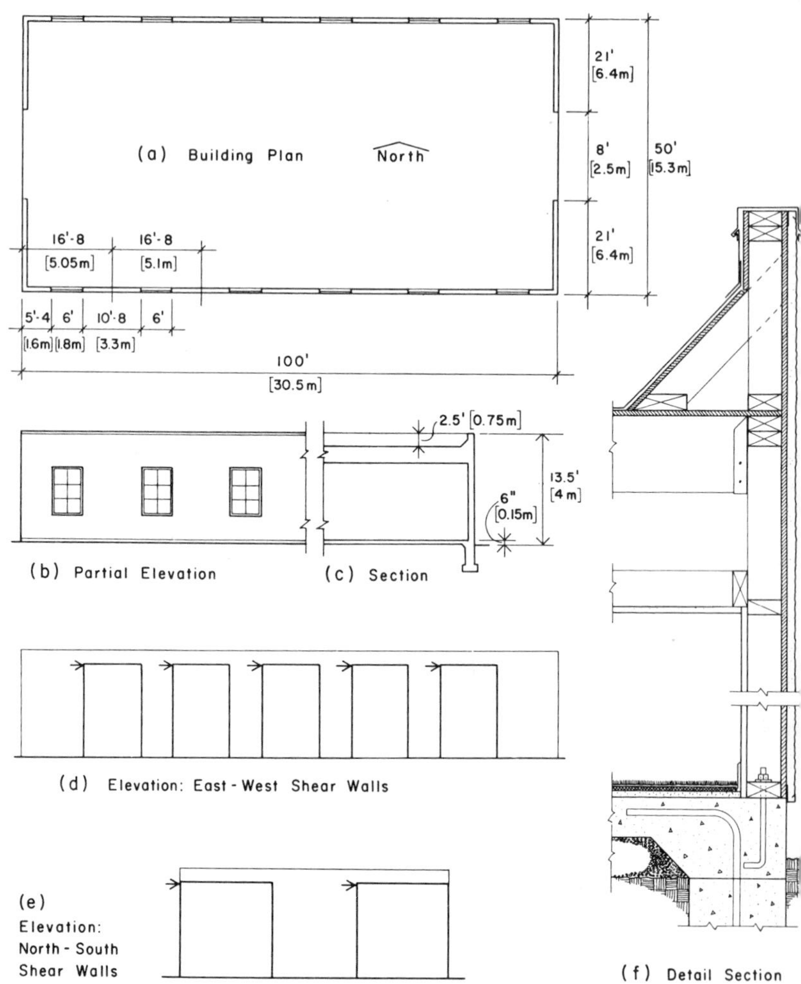

FIGURE 20.1 Building One: general form.

exterior walls. The general nature of the construction is shown in the detailed wall section in Fig. 20.1*f*. Specific details of the framing will depend on various decisions in the design of the structure as developed in the following discussion. The general form of the exterior shear walls is indicated in Fig. 20.1*d* and *e*. Design considerations for lateral loads are presented in Sec. 20.3. We first consider the design of the roof structural system for only gravity loads, although we keep in mind that the roof must eventually be developed as a horizontal diaphragm and the walls as shear walls.

20.2 Design for Gravity Loads

With the construction as shown in Fig. 20.1*f*, we determine the roof loads as follows:

Three-ply felt and gravel roofing	5.5 psf
Fiberglass insulation batts	0.5
$\frac{1}{2}$-in.-thick plywood roof deck	1.5
Rafters and blocking (estimate)	2.0
Ceiling joists	1.0
$\frac{1}{2}$-in. drywall ceiling	2.5
Ducts, lights, etc.	3.0
Total roof dead load	16.0 psf

Assuming a partitioning of the interior as shown in Fig. 20.2*a*, various possibilities exist for the development of the spanning roof and ceiling framing systems and their supports. Interior walls may be used for supports, but a more desirable situation in commercial uses is sometimes obtained by using interior columns that allow for rearrangement of the interior spaces. The roof framing system shown in Fig. 20.2*b* is developed with two rows of interior columns placed at the location of the corridor walls. If the partitioning shown in Fig. 20.2*a* is used, these columns will be architecturally out of view because they are incorporated within the wall space.

Figure 20.2*c* shows a second possibility for the roof framing using the same column layout as in Fig. 20.2*b*. There may be various reasons for favoring one of these framing schemes over the other. Problems of installation of ducts, lighting, wiring, roof drains, and fire sprinklers may influence the choice. We arbitrarily select the scheme shown in Fig. 20.2*b* to illustrate the design process for the elements of the structure.

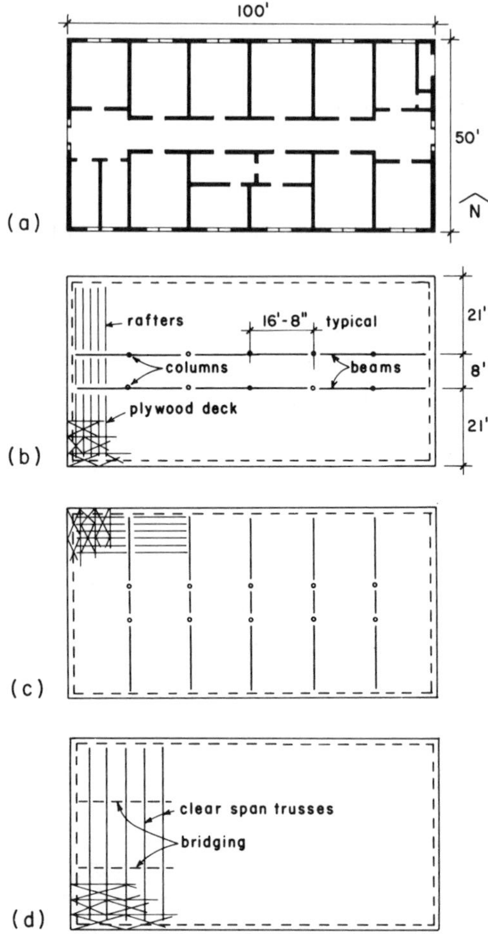

FIGURE 20.2 Alternatives for the roof framing.

Installation of a membrane-type roofing ordinarily requires at least a $\frac{1}{2}$-in.-thick roof deck. Such a deck is capable of up to 32-in. spans in a direction parallel to the face ply grain (the long dimension of ordinary 4 × 8 ft panels). If rafters are not over 24 in. on center—as they are likely to be for the schemes shown in Fig.

20.2*b* and *c*—the panels may be placed so that the plywood span is across the face grain. An advantage in the latter arrangement is the reduction in the amount of blocking required at edges not falling on the rafter. The reader is referred to the general discussions in Chapter 4 for uses of plywood structural decking.

For the rafters we assume No. 2 grade, an ordinary minimum structural usage. From Table 4.1 the allowable bending stress for repetitive use is 1450 psi and the modulus of elasticity is 1,700,000 psi. Since our loading falls approximately within the criteria of Table 4.6, we may use that table as a reference. Inspection of the table yields the possibilities of using either 2 × 10s at 16-in. centers or 2 × 12s at 24-in. centers. The reader can verify the adequacy of either of these choices using the procedures developed in Chapter 4. For stress investigation it is likely that an increase of the allowable stress of either 15% or 25% would be permitted. (See discussion of adjustment for load duration in Chapter 19.)

The ceiling joists may be made to span as separate framing, and a table similar to those shown here as Tables 4.5 and 4.6 can be used to select the joists. Ceiling framing is sometimes provided as a general structure—similar to the roof and floor framing—and is sometimes developed together with interior partitioning, which may be used to support it. Spacing of ceiling joists must, of course, relate to the surfacing material. It is also possible to suspend the ceiling joists from the rafters. The dead load tabulation made previously assumes the use of light ceiling framing—most likely 2 × 4 or 2 × 3 members—suspended from the rafters.

The beams as shown in Fig. 20.2*b* are continuous through two spans, with a total length of 33 ft 4 in. and a clear span of 16 ft 8 in. For the two-span beam the maximum bending moment is the same as that for a simple span, the principal advantage being a reduction of deflection. The total load area for one span is

$$16.67 \times \frac{21 + 8}{2} = 242 \text{ ft}^2$$

As indicated in Table 19.2, this permits the use of a live load of 16 psf. Thus the unit of the uniformly distributed load on the beam is found as

$$(16 + 16) \times \left(\frac{21 + 8}{2}\right) = 464 \text{ lb/ft}$$

Adding a bit for the beam weight, we will design for a load of 480 lb/ft. The maximum moment is thus found to be

$$M = \frac{wL^2}{8} = \frac{480(16.67)^2}{8} = 16{,}673 \text{ ft-lb}$$

The allowable bending stress depends on the beam size and the load duration assumed. We will assume a 15% increase for load duration and thus determine the following (see Table 4.1):

For a 4× member: $F_b = 1.15(1500) = 1725$ psi

For a 5× member or larger: $F_b = 1.15(1300) = 1495$ psi

Then

$$S = \frac{M}{F_b} = \frac{16{,}673 \times 12}{1725} = 116 \text{ in.}^3$$

thus indicating a 4 × 16 as a possible choice ($S = 135.7$ in.^{3}). Or

$$S = \frac{16{,}673 \times 12}{1495} = 134 \text{ in.}^3$$

thus indicating a 6 × 14 ($S = 167$ in.^{3}) or an 8 × 12 ($S = 165$ in.^{3}).

Although the 4 × 16 offers the least cross-sectional area and ostensibly the lower cost, various considerations of the development of the construction details may affect the beam selection. The beam can also be formed as a built-up member from a number of 2× elements for which the allowable stress is even higher. The latter may be preferred where good-quality heavy timber sections are not easily obtained.

Finally, the beam may consist of a glued laminated or rolled steel section. Advantages of these choices are a reduction in beam depth and a somewhat better long-term deflection response.

This is likely to be more critical for beams of longer span, for example, the beams in the framing scheme in Fig. 20.2*c*.

The beam should also be investigated for shear and deflection. Note that the maximum shear in the two-span beam is slightly larger than the simple beam shear of $wL^2/2$. This may be a reason for *not* using the two-span condition in wood beams where shear is often a critical concern for solid-sawn sections. For the example here, the span is quite short so that deflection is not a critical concern, especially if the two-span condition is used. Shear is also not critical if the stress increase for load duration is made and the load on the ends of the spans is eliminated, as discussed in Chapter 4.

We assume that a minimum roof slope of $\frac{1}{4}$ in. per ft is required to drain the roof surface. If the draining of the roof occurs at the outside walls, the building profile may thus be as shown in Fig. 20.3*a*. This requires a total elevation change of approximately $\frac{1}{4} \times 25 = 6.25$ in. from the center to the long side of the building. There are various ways to achieve this, including the tilting of the rafters.

Figure 20.3*b* shows some possibilities for the details of the construction at the center of the building. In this view the rafters are kept flat and the roof profile for drainage is achieved by attaching cut 2× elements on top of the rafters and using short profiled rafters at the corridor. Ceiling joists for the corridor are supported directly by the corridor walls. Other ceiling joists are supported at their ends by the walls, although long-span ceiling joists could be suspended in their spans by hangers from the rafters. Ceiling construction may be developed together with the roof structure or may be developed with the interior partitioning, depending on the occupancy needs.

The typical interior column supports a load approximately equal to the entire load on one beam or

$$P = 480 \times 16.67 = 8000 \text{ lb}$$

This is quite a light load, but the column height requires larger than a 4× size. (See Table 5.1.) If a 6 × 6 is not objectionable, it is adequate in the lower stress grades. However, it may be better to

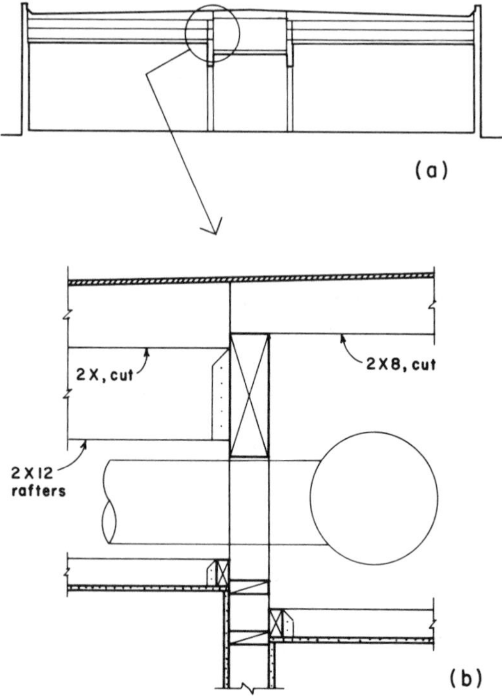

FIGURE 20.3 Building One: construction details.

consider use of a steel pipe or tubular section, either of which can be accommodated in a partition wall with 2 × 4 studs.

Structural design of the studs in the exterior walls is principally a matter of consideration for lateral bending due to wind. This is considered in the next section.

20.3 Design for Lateral Loads

Design of the building structure for wind includes the consideration for the following:

1. Inward and outward pressure on exterior walls, causing bending of the wall studs.

2. Total lateral force on the building, requiring bracing by the roof diaphragm and the shear walls.
3. Uplift on the roof, possibly requiring anchorage of the roof structure.
4. Total effect of lateral and uplift forces, possibly resulting in overturn of the entire building.

We will first investigate the wall studs on the north and south sides (see Fig. 20.1*a*), assuming them to be 2 × 6 members of Douglas fir–larch, Stud grade, and to be on 16-in. centers. These walls support the ends of the 21-ft span rafters as well as the parapet wall above. We will assume the total gravity dead load plus live load on the wall to be approximately 400 lb/ft of wall length. Thus a single stud supports a load of

$$P = 400 \times \frac{16}{12} = 533 \text{ lb}$$

Assuming the studs to span 10.5 ft vertically, each stud will carry a uniformly distributed wind load of w = 20 psf × (16/12) = 26.7 lb/ft and will thus sustain a maximum bending moment of

$$M = \frac{wL^2}{8} = \frac{26.7(10.5)^2}{8} = 368 \text{ ft-lb}$$

which results in a maximum bending stress in the stud of

$$f_b = \frac{M}{S} = \frac{368 \times 12}{7.563} = 584 \text{ psi}$$

The axial compression stress on a single stud is

$$f_c = \frac{P}{A} = \frac{533}{8.25} = 65 \text{ psi}$$

For Stud grade wood, Table 4.1 yields the following: F_b = 850 psi (repetitive member use), F_c = 675 psi, E = 1,500,000 psi. For the allowable column compression stress F'_c, the slenderness ratio is determined as $L/d = (10.5 \times 12)/5.5 = 22.9$. To establish the slenderness category, we determine

$$K = 0.671 \sqrt{\frac{E}{F_c}} = 0.671 \sqrt{\frac{1{,}500{,}000}{675}} = 31.6$$

Thus the allowable compression is determined as

$$F'_c = F_c \left[1 - \frac{1}{3}\left(\frac{L/d}{K}\right)^4\right] = 675 \left[1 - \frac{1}{3}\left(\frac{22.9}{31.6}\right)^4\right] = 631 \text{ psi}$$

Then the combined axial compression and lateral bending condition is investigated as follows:

$$\frac{f_c}{F'_c} + \frac{f_b}{F'_b - Jf_c} = \frac{65}{1.33(613)} + \frac{584}{1.33(850 - 65)}$$

$$= 0.080 + 0.560 = 0.640$$

As this total is less than one, the 2 × 6 stud is adequate in the Stud grade. An alternate solution would be to use a 2 × 4 in a higher stress grade such as No. 1. However, economy will probably be achieved by using the Stud grade if 2 × 6 size studs can be obtained in this grade.

Consideration for uplift force on the roof varies with different building codes. The maximum consideration usually consists of an uplift equal to the horizontal design pressure. If this is required, the uplift force exceeds the roof dead load by 4 psf in this example, and the roof must be anchored to its supports to resist this effect. With the use of metal framing anchors between the rafters and the beams or between the rafters and the stud walls, adequate anchorage will probably be provided.

Overturning of the entire building is not likely to be critical for a building with the profile in this example. Overturn is more critical in buildings of extremely light dead weight or those that are quite tall and narrow in profile. In any event, the investigation is similar to that for overturn of a single shear wall except for the inclusion of the uplift effect on the roof. If the total overturning moment is more than two-thirds of the available resisting moment due to building dead load, the anchorage of the building as a whole must be considered.

We proceed to consideration of the forces exerted on the principal bracing elements, that is, the roof diaphragm and the exterior shear walls. The building must be investigated for wind in the two directions: east–west and north–south. Consideration of the wall functions and determination of the forces exerted on the bracing system are illustrated in Fig. 20.4. The amount of the portion of the pressure on the exterior walls that is applied as load to the edge of the roof diaphragm depends partly on the spanning nature of the exterior walls. Figure 20.4*a* shows two common cases: with studs cantilevering to form the parapet and with simple span studs and a separate parapet structure. With the construction as shown in Fig. 20.1*f*, we assume the walls on the north and south sides to be case 2 in Fig. 20.4*a*. Thus the lateral wind load applied to the roof diaphragm in the north–south direction is

$$(20\text{ psf})\left(\frac{10.5}{2}\right) + (20\text{ psf})(2.5) = 155\text{ lb/ft}$$

In resisting this load, the roof functions as a spanning member supported by the shear walls at the east and west ends of the building. The investigation of this 100-ft simple span beam with uniformly distributed loading is shown in Fig. 20.5. The end reaction and maximum shear force is found as

$$(155)\left(\frac{100}{2}\right) = 7750\text{ lb}$$

which results in a maximum unit shear in the 50-ft-wide roof diaphragm of

$$v = \frac{\text{shear force}}{\text{roof width}} = \frac{7750}{50} = 155\text{ lb/ft}$$

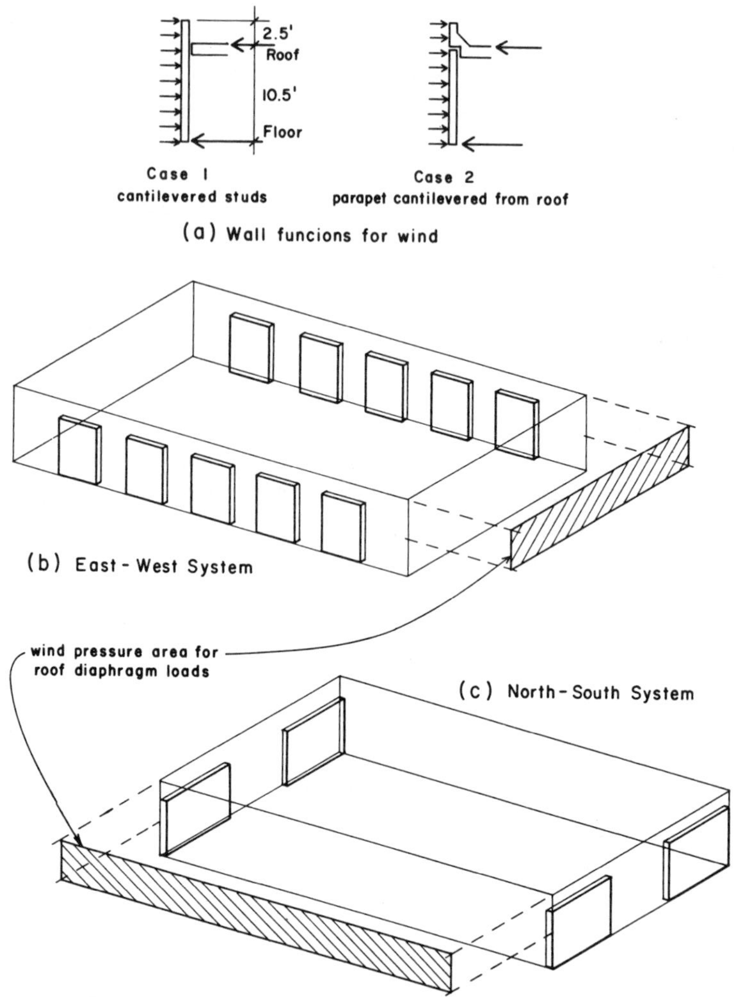

FIGURE 20.4 Building One: wall functions and wind pressure development for lateral load.

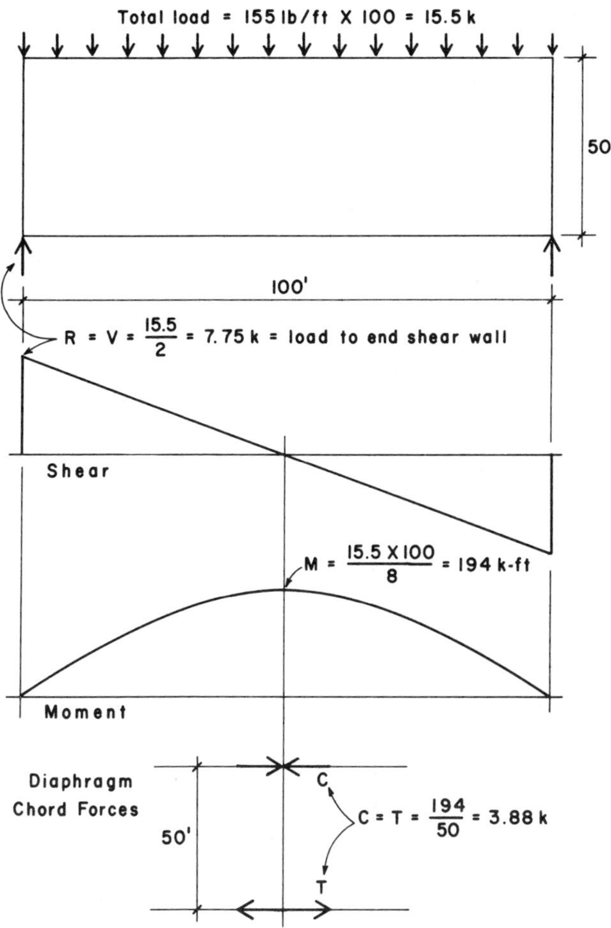

FIGURE 20.5 Functions of the roof diaphragm.

From Table 20.1 (UBC, Table 25-J-1) we may select a number of possible choices for the roof deck. Variables include the class of the plywood, the panel thickness, the width of supporting rafters, the nail size and spacing, and the use or omission of blocking (support for nailing at the plywood panel edges not falling on the

TABLE 20.1 Allowable Shear in Pounds per Foot for Horizontal Diaphragms With Framing of Douglas Fir–Larch or Southern Pine[1]

PLYWOOD GRADE	Common Nail Size	Minimum Nominal Penetration in Framing (In Inches)	Minimum Nominal Plywood Thickness (In Inches)	Minimum Nominal Width of Framing Member (In Inches)	BLOCKED DIAPHRAGMS: Nail spacing at diaphragm boundaries (all cases), at continuous panel edges parallel to load (Cases 3 and 4) and at all panel edges (Cases 5 and 6): 6; Nail spacing at other plywood panel edges: 6	4; 6	2½[2]; 4	2[2]; 3	UNBLOCKED DIAPHRAGM: Nails spaced 6" max. at supported end — Load perpendicular to unblocked edges and continuous panel joints (Case 1)	Other configurations (Cases 2, 3 & 4)
STRUCTURAL I	6d	1¼	5/16	2	185	250	375	420	165	125
				3	210	280	420	475	185	140
	8d	1½	3/8	2	270	360	530	600	240	180
				3	300	400	600	675	265	200
	10d	1⅝	15/32	2	320	425	640	730[2]	285	215
				3	360	480	720	820	320	240
C-D, C-C, STRUCTURAL II and other grades covered in U.B.C. Standard No. 25-9	6d	1¼	5/16	2	170	225	335	380	150	110
				3	190	250	380	430	170	125
			3/8	2	185	250	375	420	165	125
				3	210	280	420	475	185	140
	8d	1½	3/8	2	240	320	480	545	215	160
				3	270	360	540	610	240	180
			15/32	2	270	360	530	600	240	180
				3	300	400	600	675	265	200
	10d	1⅝	15/32	2	290	385	575	655[2]	255	190
				3	325	430	650	735	290	215
			19/32	2	320	425	640	730[2]	285	215
				3	360	480	720	820	320	240

[1]These values are for short-time loads due to wind or earthquake and must be reduced 25 percent for normal loading. Space nails 10 inches on center for floors and 12 inches on center for roofs along intermediate framing members.

Allowable shear values for nails in framing members of other species set forth in Table No. 25-17-J of U.B.C. Standards shall be calculated for all grades by multiplying the values for nails in STRUCTURAL I by the following factors: Group III, 0.82 and Group IV, 0.65.

[2]Framing shall be 3-inch nominal or wider and nails shall be staggered where nails are spaced 2 inches or 2½ inches on center, and where 10d nails having penetration into framing of more than 1⅝ inches are spaced 3 inches on center.

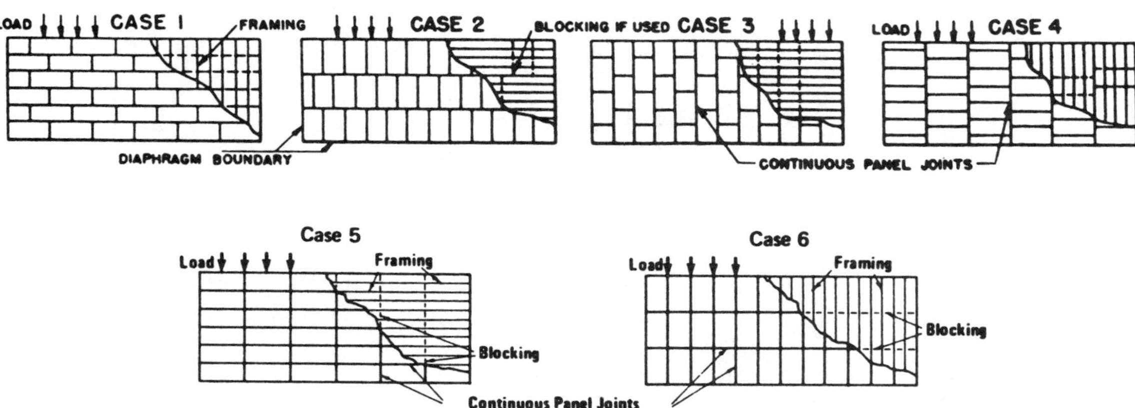

NOTE: Framing may be located in either direction for blocked diaphragms.

Source: This is Table 25-J-1 from the *Uniform Building Code*, 1988 ed. (Ref. 1), reproduced with permission of the publisher: International Conference of Building Officials.

rafters). In addition to satisfying the shear capacity requirement, there are many considerations for the choice of the construction that derive from the design for gravity loading, general construction details for the roof structure, and the problems of installation of roofing and insulation. For the flat roof with ordinary tar and felt membrane roofing, it is usually necessary to provide a minimum of $\frac{1}{2}$-in. (now $\frac{15}{32}$-in.) thick plywood. If this requirement is accepted, a possible choice from Table 20.1 is

> Structural II $\frac{15}{32}$-in. plywood with 2× framing and 8d nails at 6 in. at all panel edges and a blocked diaphragm.

For these criteria Table 20.1 yields a capacity of 270 lb/ft.

In this example, if the need for the minimum thickness of plywood is accepted, it turns out that the minimal construction is more than adequate for the required lateral force resistance. Had this not been the case, and the required capacity had resulted in considerable nailing beyond the minimal, it would be possible to graduate the nail spacings from the maximum required at the building ends to minimal nailing in the center portion of the roof.

The moment diagram shown in Fig. 20.5 indicates a maximum value of 194 kip-ft at the center of the span. This moment is used to determine the required chord force that must be developed in both tension and compression at the roof edges. With the construction as shown in Fig. 20.1*f*, the top plate of the stud wall is the most likely element to be used for this function. In this example the force is quite small and can be easily developed by the ordinary construction. The one problem that may require some special effort is the development of the member as a continuous tension element, since it is not possible to have a single, 100-ft-long piece for the plate. Splicing of the multipiece plates must therefore be developed to provide a continuity of the tension force.

The end reaction force for the roof diaphragm, as shown in Fig. 20.5, must be developed by the end shear walls. As shown in Fig. 20.1, there are two walls at each end, both 21 ft long in plan. Thus the total shear force is developed by a total of 42 ft of shear wall and the unit shear in the wall is

$$v = \frac{\text{total shear force}}{\text{total wall length}} = \frac{7750}{42} = 185 \text{ lb/ft}$$

As with the roof deck, there are various considerations for the selection of the wall construction. A common situation involves the use of a single facing of structural plywood on the exterior surface of the wall, which is considered as the resisting element for lateral force. Other elements of the wall construction are thus considered nonstructural in function. For this situation a possible choice from Table 20.2 (UBC, Table 25-K-1) is

Structural II, $\frac{3}{8}$-in.-thick plywood with 6d nails at 6-in. spacing at all panel edges.

Again, this is minimal construction. For situations that require much higher capacities of shear resistance it may be necessary to use thicker-than-normal plywood and nailing with larger, closer spaced nails. Unfortunately, the nailing cannot be graduated—as it may be for the roof—as the unit shear is a constant value throughout the height of the shear wall.

Figure 20.6*a* shows the loading condition for investigating the overturning effect on the shear wall. The loading shown includes only the lateral force (applied at the level of the roof deck) and the dead loads of the wall itself plus the portion of the roof carried by the wall. The overturning moment is determined as the product of the lateral force and its distance above the wall base. This value is then multiplied by a factor of 1.5, representing the usual minimum required safety factor for comparison with the dead load resisting moment (also called the restoring moment). If the minimum safety factor is not provided, an anchorage force called a *tiedown* (T in Fig. 20.6*a*) must be added to supplement the dead load resistance. The usual computations for this investigation are as follows:

$$\text{Overturning moment} = (3.875)(11)(1.5) = 64 \text{ kip-ft}$$

$$\text{Restoring moment} = (3 + 6)(21/2) = 94.5 \text{ kip-ft}$$

This indicates that no tiedown force is required. Actually, there is additional resistance at each end of the wall. At the build-

TABLE 20.2 Allowable Shear for Wind or Seismic Forces in Pounds per Foot for Plywood Shear Walls with Framing of Douglas Fir–Larch or Southern Pine[1,4]

PLYWOOD GRADE	MINIMUM NOMINAL PLYWOOD THICKNESS (Inches)	MINIMUM NAIL PENETRATION IN FRAMING (Inches)	NAIL SIZE (Common or Galvanized Box)	PLYWOOD APPLIED DIRECT TO FRAMING — Nail Spacing at Plywood Panel Edges				NAIL SIZE (Common or Galvanized Box)	PLYWOOD APPLIED OVER ½-INCH GYPSUM SHEATHING — Nail Spacing at Plywood Panel Edges			
				6	4	3	2[2]		6	4	3	2[2]
STRUCTURAL I	5/16	1¼	6d	200	300	390	510	8d	200	300	390	510
	3/8	1½	8d	230[3]	360[3]	460[3]	610[3]	10d	280	430	550[2]	730[2]
	15/32	1½	8d	280	430	550	730	10d	280	430	550[2]	730
	15/32	1⅝	10d	340	510	665[2]	870	—	—	—	—	—
C-D, C-C STRUCTURAL II and other grades covered in U.B.C. Standard No. 25-9.	5/16	1¼	6d	180	270	350	450	8d	180	270	350	450
	3/8	1¼	6d	200	300	390	510	8d	200	300	390	510
	3/8	1½	8d	220[3]	320[3]	410[3]	530[3]	10d	260	380	490[2]	640
	15/32	1½	8d	260	380	490	640	10d	260	380	490[2]	640
	15/32	1⅝	10d	310	460	600[2]	770	—	—	—	—	—
	19/32	1⅝	10d	340	510	665[2]	870	—	—	—	—	—
			NAIL SIZE (Galvanized Casing)					NAIL SIZE (Galvanized Casing)				
Plywood panel siding in grades covered in U.B.C. Standard No. 25-9.	5/16	1¼	6d	140	210	275	360	8d	140	210	275	360
	3/8	1½	8d	130[3]	200[3]	260[3]	340[3]	10d	160	240	310[2]	410

[1]All panel edges backed with 2-inch nominal or wider framing. Plywood installed either horizontally or vertically. Space nails at 6 inches on center along intermediate framing members for 3/8-inch plywood installed with face grain parallel to studs spaced 24 inches on center and 12 inches on center for other conditions and plywood thicknesses. These values are for short-time loads due to wind or earthquake and must be reduced 25 percent for normal loading.

Allowable shear values for nails in framing members of other species set forth in Table No. 25-17-J of U.B.C. Standards shall be calculated for all grades by multiplying the values for common and galvanized box nails in STRUCTURAL I and galvanized casing nails in other grades by the following factors: Group III, 0.82 and Group IV, 0.65.

[2]Framing shall be 3-inch nominal or wider and nails shall be staggered where nails are spaced 2 inches on center, and where 10d nails having penetration into framing of more than 1 5/8 inches are spaced 3 inches on center.

[3]The values for 3/8-inch-thick plywood applied direct to framing may be increased 20 percent, provided studs are spaced a maximum of 16 inches on center or plywood is applied with face grain across studs.

[4]Where plywood is applied on both faces of a wall and nail spacing is less than 6 inches on center on either side, panel joints shall be offset to fall on different framing members or framing shall be 3-inch nominal or thicker and nails on each side shall be staggered.

Source: This is Table 25-K-1 from the *Uniform Building Code*, 1988 ed. (Ref. 1), reproduced with permission of the publishers, International Conference of Building Officials.

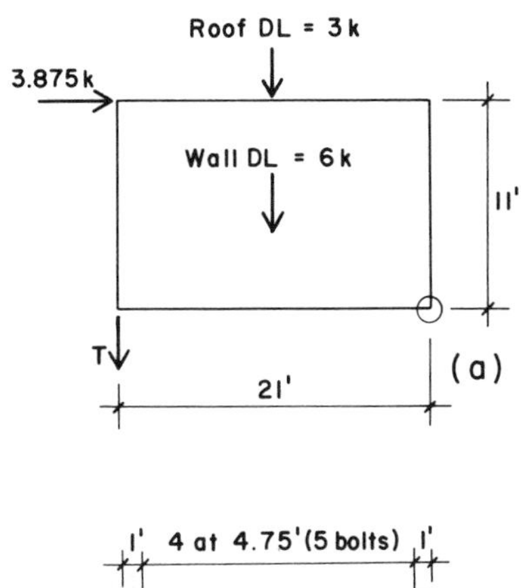

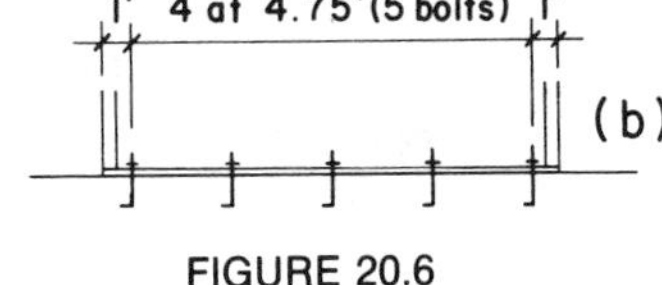

FIGURE 20.6

ing corner this wall is most likely quite well attached to the wall on the north or south of the building, which provides additional dead load resistance. At the end of the wall near the corridor a post would be provided for support of the beam. (See the framing plan in Fig. 20.2.) Thus the dead load portion of the beam reaction provides additional resistance. Finally, the wall sill will be bolted to the foundation providing some hold-down resistance to toppling of the wall. At present, most codes do not permit use of a computed value for the sill bolt resistance to uplift as this function involves cross-grain bending of the sill.

The sill bolts *will* be used for resistance to the horizontal sliding of the wall, however, and the bolting must satisfy this requirement. Code minimum bolting usually consists of $\frac{1}{2}$-in.-diameter bolts at 1 ft from the wall ends and at a maximum of 6 ft on center for the remainder of the wall length. A layout for this minimal bolting is shown in Fig. 20.6*b*. With a 2× sill member and $\frac{1}{2}$-in. bolts in single shear, Table 6.2 yields a value of 470 lb for one bolt. With a total of five bolts and an increase of one-third in the value for wind loading, the total sliding resistance of the minimum bolting is

$$(1.33)(470)(5) = 3125 \text{ lb}$$

As this is a bit short of the necessary resistance, it is necessary to increase the number and/or the size of the bolts. Within the limit of the sill width, it is probably best to increase the bolt size for economy. Setting of the bolts in the concrete is a major cost factor that is not much related to the bolt size. Thus a few large bolts are probably preferable to a lot of small bolts.

In some situations it may be necessary to consider the effects of the sliding and overturn forces on the wall foundations. In this example the forces are quite low and not likely to be critical. For buildings with shallow foundations, however, the overturning and sliding must be seriously considered in the foundation design when they are high in magnitude in relation to gravity forces.

A major consideration in the design for lateral force resistance is the development of the necessary force transfers between the various elements of the lateral bracing system. The sill bolting of the shear wall is one example of such a transfer. Another critical one is the transfer of force from the roof diaphragm to the shear wall. In this example the roof shear is vested in the roof plywood and the wall shear in the wall plywood. As these two elements are not directly attached, the details of the framing must be studied to determine how the transfer can be made. With the construction as shown in Fig. 20.1*f* a simple transfer can be made through the top plate of the wall to which both the roof and wall plywoods are directly attached. This is likely to occur normally for the roof plywood because this is the edge of the roof diaphragm (called the *boundary* in Table 20.1). For the wall plywood, however, this may not be a panel edge, and the necessary nailing for the required transfer should be indicated and specified on the wall section.

For other wall and roof constructions the force transfer may not be as simple and direct as that in Fig. 20.1*f*. In some cases there may not be any direct transfer through ordinary elements of the construction, thus requiring the addition of some framing or connecting devices.

20.4 Alternative Steel and Masonry Structure

Alternative construction for Building One is shown in Fig. 20.7. In this case the walls are made of concrete masonry units (concrete blocks) and the roof structure consists of a formed sheet steel deck on open-web joists. The masonry wall is made continu-

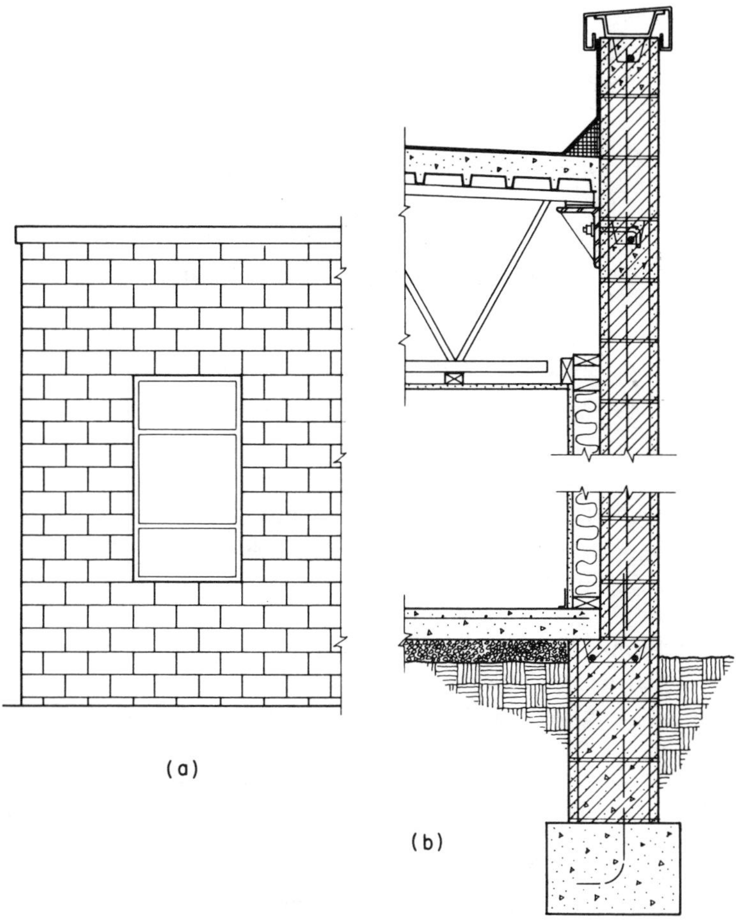

FIGURE 20.7 Building One: details for the alternative steel and masonry structure.

ous to the top of the parapet and the roof joists are supported by a continuous steel angle that is bolted to preset anchor bolts embedded in a grouted and reinforced course of the concrete units. While the ceiling could be supported in a number of ways, we show here that it is attached directly to the bottom of the steel joists. Although such open-web joists are standard manufactured products, they are individually specified in some detail for a particular installation. This permits some customizing, including the possibility of sloping the top chords to accommodate roof drainage, while maintaining a level bottom chord for a flat ceiling.

While the basic form of this construction is quite common, there is great variety in the various details of the construction, including choices for roofing, insulation, roof deck, specific type of joist, and ceiling construction. While not intended as a model for ideal construction, the specific details shown in Fig. 20.7 include the following:

K-Series open-web steel joists. (See Sec. 8.12.)

Reinforced hollow concrete unit masonry. (See Sec. 17.4.)

Formed sheet steel deck. (See Table 8.4.)

Deck surfaced with lightweight insulating concrete fill.

Multiple-ply, hot-mopped, felt and gravel roofing.

Gypsum drywall ceiling and interior wall surfacing on wood furring strips.

Deck units and open-web joists for this structure would be selected from available products, as marketed on a regional basis, although approximate designs may be achieved using industry standard references (illustrated here in Tables 8.4 and 8.5).

While the construction shown in Fig. 20.7 shows the use of reinforced masonry, the size and details of this building can be easily achieved with ordinary (unreinforced) construction, where such is permitted by building codes. Concerns for the structural design of the walls include the following:

Vertical Bearing Stress. This will be quite low in this case, even with the use of clear-span, 50-ft-long trusses. Minimum construction—reinforced or not—can easily achieve the necessary vertical compression resistance.

Bending Due to Wind—Direct Pressure. This involves the bending inward or outward, as discussed for the studs in Sec. 20.3. A reinforced wall will be much stronger in developing this resistance, but if the wall is not of excessive height and wind pressure is low, an ordinary wall will probably be adequate. Code requirements for limits on the ratio of wall height to thickness will generally ensure the use of blocks that will not result in a really thin wall.

Bending Due to Eccentric Vertical Load. If the detail for the support of the joists as shown in Fig. 20.7 is used, the roof load will induce a bending in the wall equal to the vertical end shear in the joists times the eccentric distance from the center of the wall to the center of bearing for the joist ends. This may be a serious stress condition on its own, but must also be considered in combination with the bending due to wind. If this results in overstress in the wall, or the need for undesirably heavy wall construction, there are various remedies. One solution is to revise the details at the roof-to-wall joint as shown in Fig. 20.8, with the joist ends bearing directly on the center of the wall. Again, this is less likely to be a critical problem with reinforced construction.

Stresses Induced by Shear Wall Actions. These walls will most likely be used for shear walls, and the walls themselves, as well as the roof diaphragm and roof-to-wall load transfers, must be investigated. For the wind conditions determined for the wood structure in Sec. 20.3, the masonry walls in this example would most likely be adequate in the form of "minimum construction," even if unreinforced. That is, the satisfying of minimum code requirements for structural masonry would ensure construction with adequate resistance to required loadings. For seismic loads or for high winds on larger buildings, this will possibly not be true, and the construction may require considerable enhancement for adequate resistance.

Attachment of Roof to Walls. This involves four possible concerns, any of which may require special detailing:

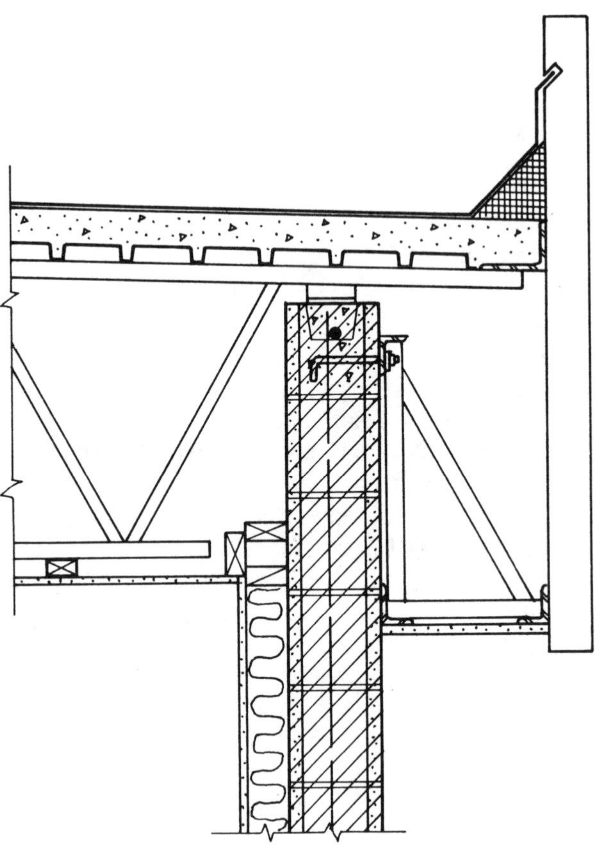

FIGURE 20.8 Building One: variation of the roof-to-wall detail. For comparison, see Fig. 20.7*b*.

1. Uplift due to wind, if the wind pressure is great and the total roof dead load is small. With foamed plastic insulation, single-ply roofing, and no ceiling, this is a real concern.
2. Vertical, downward gravity load, involving bearing pressure under the joist ends, if the detail in Fig. 20.8 is used, or the design of the anchor bolts if the detail in Fig. 20.7 is used.

3. Outward force on the wall, due to wind or seismic action. Codes usually require design for a minimum force for this, although actual forces may be great in some situations. This is particularly a concern for seismic action, although wind suction may be considerable on a large wall.
4. Shear wall load transfer from the roof diaphragm, involving force transfer from the roof deck to the wall.

20.5 Foundations

Foundations for Building One would be quite minimal. For the exterior bearing walls, the construction provided would generally depend on concerns for frost protection and the vertical location of good, undisturbed bearing materials. For warm climates, where good bearing is available close to the finished grade level, the detail shown in Fig. 20.9*a* may be used, consisting of a shallow grade beam, combining the functions of foundation wall and footing. For the wood structure, this beam may be as narrow as

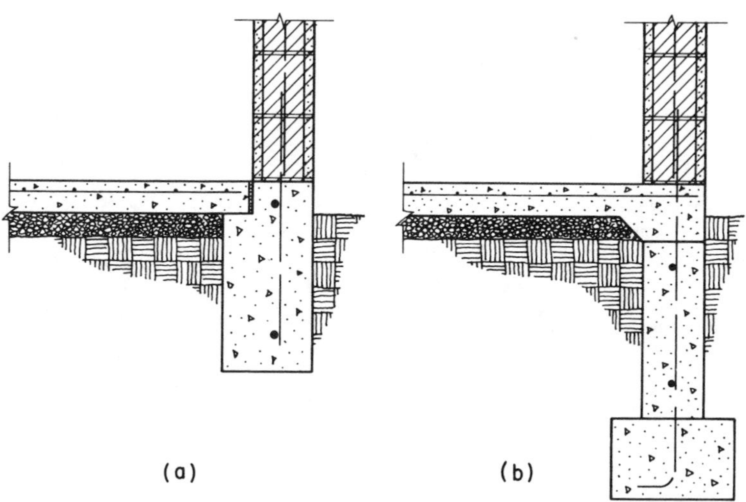

FIGURE 20.9 Building One: options for the wall foundation for the steel and masonry structure.

12 in. and provide adequate support with soil pressure as low as 1000 psf.

For cold climates, or where good bearing is some distance below grade, the construction shown in Fig. 20.9*b* may be used, employing separately poured footings and foundation walls. In this case the walls could also be made of concrete block, as shown in Fig. 20.7.

With the closely spaced columns, supporting only roof loads, column footings for the wood structure would be quite minimal in size. (See Table 16.2.) It is common practice to make small column footings (under about 4 ft square) without reinforcing, although we hesitate to recommend it.

21

Building Two

Figure 21.1 shows a building that consists essentially of stacking the plan for Building One to produce a two-story building. The profile section of the building shows that the structure for the second floor is developed essentially the same as the roof structure for Building One. For the roof, however, a clear span structure could be provided by 50-ft-span trusses—a relatively easy task for prefabricated, manufactured trusses as described in Sec. 8.12. This roof structure would also be an alternative for Building One and would provide for a greater degree of freedom for the arrangement of the partitioning in the building interior.

Although the second-floor structure for this building is similar to the roof structure for Building One, the principal difference is in the magnitude of live load that must be accommodated. The usual building code requirement for minimum load for offices is 50 psf (see Table 19.3). However, it is also usually required to add an allowance of 20 or 25 psf for movable partitioning. If we use a total live load of 75 psf, the design load will be several times greater than that for the roof. The difference in results is demonstrated in the next section.

The two-story building will also sustain a greater total wind force, although the shear walls in the second story will essentially

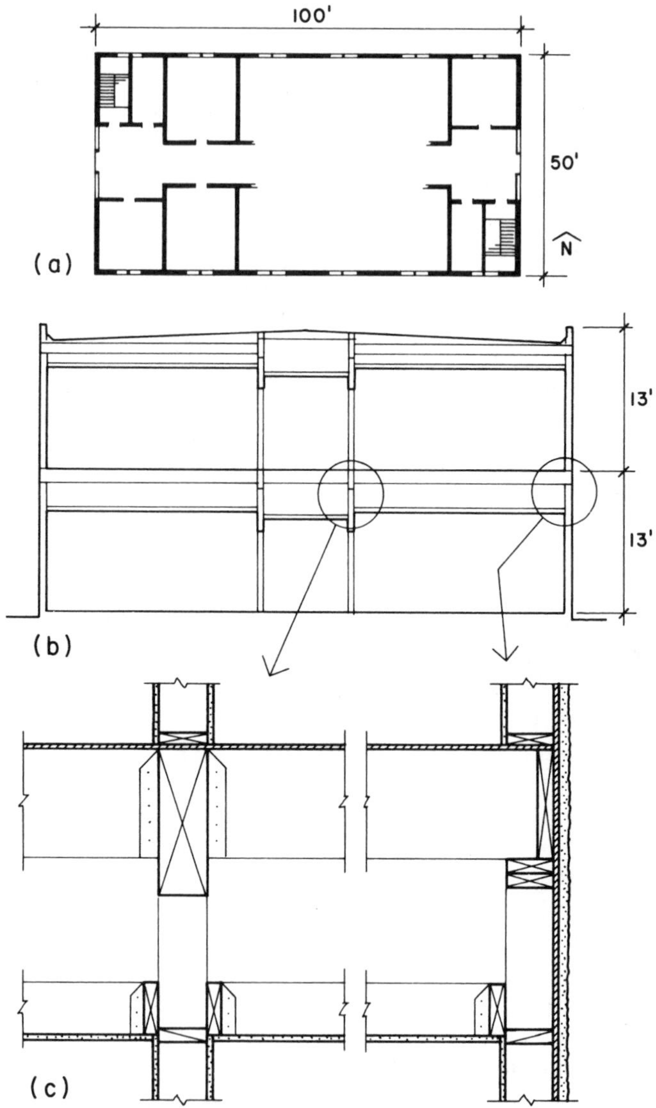

FIGURE 21.1 Building Two: general form.

be the same as those for Building One. The major effect in this building will be the forces in the first-story shear walls.

Another difference in the floor structure is that it will be dead flat, thereby eliminating the concerns for producing the sloped surface for draining of the roof. The details for the second-floor framing are shown in Fig. 21.1*c*, indicating the use of platform framing. Otherwise the details of the construction will be assumed to be similar to those for Building One, except for the possible use of the trusses for the roof. One way to economize with the trusses is by eliminating a general framing for the ceiling, and having the ceiling surface attached directly to the bottom chords of the trusses. Another economy is possible if the top chords of the trusses are sloped to provide the framing for the roof deck in the profile necessary for drainage. The planning of the building, the location of roof drains, and the framing plan for the trusses must be coordinated to bring these simplifications off effectively.

21.1 Design for Gravity Loads

Because of the increased loads, the member sizes for the framing of the floor with solid-sawn lumber will be considerably greater than those required for the roof for Building One. We now illustrate the design with structural lumber, but will discuss alternatives later.

We assume the following for the dead load of the floor construction, assuming the ceiling to be separately supported (not hung from the floor joists):

Carpet and pad	3.0 psf
Fiberboard underlay	3.0
Plywood deck, $\frac{1}{2}$ in.	1.5
Ducts, lights, wiring	3.5
Total without joists:	11.0 psf

With joists at 16-in. spacing, the loading for a single joist is

$$DL = \frac{16}{12}(11) = 14.6 \text{ lb/ft} + \text{the joist, say } 20 \text{ lb/ft}$$

$$LL = \frac{16}{12}(75) = 100 \text{ lb/ft}$$

$$\text{Total} = 120 \text{ lb/ft}$$

For the 21-ft span joists the maximum bending moment is

$$M = \frac{wL^2}{8} = \frac{(120)(21)^2}{8} = 6615 \text{ ft-lb}$$

For No. 1 Douglas fir–larch joists with 2-in. nominal thickness and 5-in. width or wider, F_b from Table 4.1 is 1750 psi for repetitive member use. Thus the required section modulus is

$$S = \frac{M}{S} = \frac{(6615)(12)}{1750} = 45.36 \text{ in.}^3$$

From Table 3.6 we find that this is just over the value for a 2 × 14, which is the largest member listed with a 2-in. nominal thickness. Alternatives are to increase the stress grade of the wood, use a thicker joist, or reduce the joist spacing. If the spacing is reduced to 12 in., the required section modulus drops by almost one-fourth, making the 2 × 14 adequate for flexure. Shear is unlikely to be critical with long-span joists, but deflection should be investigated.

The usual deflection limit for this situation is a maximum live load deflection of $\frac{1}{360}$ of the span, or (21)(12)/360 = 0.7 in. With the 2 × 14 at 12-in. spacing, the maximum deflection under live load only is

$$\Delta = \frac{5}{384}\frac{wL^4}{EI} \quad \text{or} \quad \frac{5}{384}\frac{WL^3}{EI}$$

$$= \frac{5}{384}\frac{(75 \times 21)(21 \times 12)^3}{(1{,}800{,}000)(291)} = 0.63 \text{ in.}$$

which indicates that deflection is not critical.

The beams support both the 21-ft joists and the shorter 8-ft joists at the corridor. The corridor live load is usually required to be 100 psf, but the short span will probably permit a small joist—usually a 2 × 6 minimum. Thus the dead load at the corridor is slightly less. Also the total area supported by a beam exceeds 150 ft^2, which allows for some minor reduction of the live load as described in Sec. 19.4. We will simplify the computations by using the dead load and the live load as determined for the 21 ft span joists as the design load for the beam for the total of 14.5 ft of joist span (half the distance to the other beam plus half the distance to the exterior wall). Thus the beam load is

$$
\begin{aligned}
DL = (16)(14.5) &= 232 \text{ lb/ft} \quad \text{(joist load)} \\
+ \text{ beam weight} &= 30 \quad \text{(assumed estimate)} \\
+ \text{ wall above} &= \underline{150} \quad \text{(2nd floor corridor)} \\
\text{Total } DL \quad &= 412 \text{ lb/ft} \\
LL = (75)(14.5) &= 1088 \text{ lb/ft} \\
\text{Total load} \quad &= 412 + 1088 = 1500 \text{ lb/ft}
\end{aligned}
$$

For a uniformly loaded simple span beam with a span of 16 ft 8 in. we determine

Total load $= W = (1.5)(16.67) = 25$ kips.
End reaction and maximum shear $= W/2 = 12.5$ kips.
Maximum moment $= wL^2/8 = (1.5)(16.67)^2/8 = 52.1$ kip-ft.

For a Douglas fir–larch, dense No. 1 grade beam, Table 4.1 yields values of $F_b = 1550$ psi, $F_v = 85$ psi, and $E = 1{,}700{,}000$ psi. To satisfy the flexural requirement, the required section modulus is

$$S = \frac{M}{F_b} = \frac{(52.1)(12)}{1.550} = 403 \text{ in.}^3$$

From Table 3.6 the least weight section that will satisfy this requirement is an 8 × 20 with $S = 475$ in.^{3}.

If the 20-in.-deep section is used, its effective bending resistance must be reduced, as discussed in Sec. 4.2. Thus the actual moment capacity of this section is reduced by use of the size factor from Table 4.3 and is determined as

$$M = C_F \times F_b \times S = (0.947)(1.550)(475)(1/12) = 58.1 \text{ kip-ft}$$

As this exceeds the requirement, the correction for size effect is not critical in the choice of the section.

If the actual beam depth is 19.5 in., the critical shear force may be reduced to that at a distance of the beam depth from the support. Thus we may subtract an amount of the load equal to the beam depth times the unit load. The critical shear force is thus

$$\begin{aligned} V &= \text{(actual end shear)} - \text{(unit load times beam depth)} \\ &= 12.5 \text{ kips} - (1.5)\left(\frac{19.5}{12}\right) = 12.5 - 2.44 = 10.06 \text{ kips} \end{aligned}$$

If the 8 × 20 is selected, the maximum shear stress is thus

$$f_v = \frac{3}{2}\frac{V}{A} = \frac{(3)(10{,}060)}{(2)(146.25)} = 103 \text{ psi}$$

Even with the reduction in the critical shear force, this exceeds the allowable stress of 85 psi. Thus the beam section must be increased to a 10 × 20 to satisfy the shear requirement. With this section, the flexural stress will be reduced and the use of the dense grade of wood may be unnecessary because the shear stress is a constant for all grades of the wood.

For beams of relatively short span and heavy loading, it is common for shear to be a controlling factor. This often rules against the practicality of using a solid-sawn timber section for which allowable shear stress is quite low. It is probably logical to modify the structure to reduce the beam span or to chose a steel beam or a glued laminated section in place of the solid timber.

Although deflection is often critical for long-span, lightly loaded joists, it is seldom critical for the short-span, heavily loaded beam. The reader should verify this by investigating the deflection of this beam, but we will dispense with the computation.

For the interior column at the first story the design load is approximately the same as the total load on a single beam, that is, 25 kips. For the approximately 10-ft-high column, Table 5.1 indicates a 6 × 6 column for a solid-sawn section. For various reasons it may be more practical to use a steel round pipe or a square tubular section whose size can fit into a 2 × 4 stud wall.

Columns must also be provided at the ends of the beams in the east and west walls. In the example these locations are the ends of the shear walls, and the normal use of a doubled stud at this point will probably result in an adequate column. If this location occurs in the center portion of a wall, a separate column, or special doubled stud, should be provided.

21.2 Design for Lateral Loads

The general design for wind includes the considerations enumerated at the beginning of Sec. 20.3 for Building One. Investigation of the second-story studs in the exterior walls would be similar to that made for Building One. At the first story in Building Two the studs carry considerably more axial compression, but the bending due to wind is approximately the same as at the second story. The 2 × 6 studs at 16-in. centers are probably adequate at the first story. If an investigation similar to that made in Sec. 20.3 shows an overstress condition, the stud spacing can be reduced to 12 in. or a higher-grade wood can be used.

For lateral load the roof deck in Building Two is basically the same as that in Building One. With the trusses it may be more practical to use an unblocked deck, and the investigation should consider this factor. The footnotes to Table 20.1 should be studied with regard to the pattern of the layout of the plywood panels, especially for unblocked decks. Various special deck panels with tongue-and-groove edges are available in thicknesses greater than $\frac{1}{2}$ in., thus permitting truss spacings up to 4 ft. Special data for lateral load resistance may be available from the manufacturers for these products, although local code approval must be determined.

The wind loading condition for the two-story building is shown in Fig. 21.2*a*. This indicates a loading to the second-floor diaphragm of 235 lb/ft. With a $\frac{15}{32}$-in.-thick deck as a minimum, the shear in the 50-ft-wide deck will not be critical. However, the stair wells at the east and west ends reduce the actual diaphragm width at the ends to only 35 ft. Figure 21.2*b* shows the loading for the second-floor deck and the critical shear and moment values for the diaphragm actions. At the ends the critical unit shear in the deck is

$$v = \frac{11{,}750}{35} = 336 \text{ lb/ft}$$

From Table 20.1 it may be determined that this requires a bit more than the minimum nailing for the deck. Options at this location include:

1. Using a $\frac{15}{32}$-in. Structural II deck with 8d nails at 4 in. at the diaphragm boundary and other critical edges.
2. Using $\frac{15}{32}$-in. Structural I deck with 10d nails at 6 in. throughout and 3× framing.
3. Using $\frac{19}{32}$-in. Structural II deck with 10d nails at 6 in. throughout and 3× framing.

At 8 ft from the building ends the deck resumes its full 50-ft width, and the unit shear at this point drops to

$$v = \frac{9870}{50} = 197 \text{ lb/ft}$$

Since this value is well below the capacity of the $\frac{15}{32}$-in. Structural II deck with minimum nailing, it may be the most practical to elect option 1, which involves only the use of 4-in. nailing in approximately 12% of the total second-floor deck.

The diaphragm chord force for the second-floor deck is approximately 6 kips, and must be developed in the framing at the wall, as shown in Fig. 21.1*c*. The most likely member to use for this is the continuous edge member at the face of the joists. The only real design consideration for this situation is developing the splicing of the member that will be made up of several pieces in the 100-ft length. Splicing may be achieved in a number of ways, the details must be developed to work within the construction of the wall and floor at this location. A joint using a steel strap with wood screws or one with bolts and steel plates will likely cause the least intrusion in the construction.

The loading for the two-story end shear wall is shown in Fig. 21.2*c* and the shear diagram for this load is shown in Fig. 21.2*d*. The second-story wall is essentially similar to the end wall in Building One for which an investigation was made in Sec. 20.3.

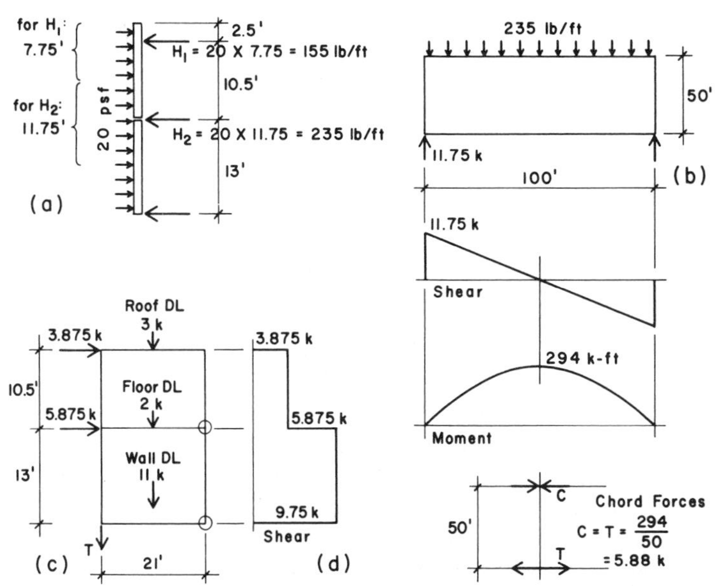

FIGURE 21.2 Building Two: development of lateral force: (*a*) generation of wind loads to the horizontal diaphragm, (*b*) functions of the second floor diaphragm, (*c*) loading of the end shear walls, (*d*) shear diagram for the two-story wall.

Because minimal construction is adequate here, and no anchorage for overturn is required, the only problem for concern is the development of sliding resistance to the lateral force of 3875 lb. Since this wall does not sit on a concrete foundation, other means of anchorage than steel anchor bolts must be considered.

The lateral force in the second-story wall must be transferred to the lower (first-story) wall. Essentially this occurs directly if the plywood is continuous past the construction at the level of the second floor, as it is indeed shown in Fig. 21.1*c*. A critical location for stress transfer in this location is at the top of the second-floor joists. At this point the lateral force from the second-floor deck is transferred to the wall through the continuous edge member. Therefore the nailing and plywood requirements for the first-story wall begin at this location. The last point for the nailing and plywood requirements for the second-story wall are at the location of the sill for the second-story wall (on top of the second floor deck, as shown in Fig. 21.1*c*). The plywood for the wall and its nailing from this point down must satisfy the requirements for the first-story wall.

In the first-story wall the total shear force is 9750 lb and the unit shear is

$$v = \frac{9750}{21} = 464 \text{ lb/ft}$$

If the $\frac{3}{8}$-in. Structural II plywood selected for Building One is used for the second floor (see Sec. 21.1), it may be practical to use the same plywood for the entire two-story wall and to simply increase the nail size and/or reduce the nail spacing at the first story. Table 20.2 yields a value of 410 lb/ft for $\frac{3}{8}$-in. Structural II plywood with 8d nails at 3 in. If the conditions of table footnote 3 are met, this value can be increased by 20% to 492 lb/ft. There are other options and many other design considerations for the choice of the wall construction, but this is an adequate choice for the lateral design criteria.

At the first-floor level, the investigation for overturn of the end shear wall is as follows (see Fig. 21.2*c*):

$$\begin{aligned}\text{Overturning moment} &= (3.875)(23.5)(1.5) = 136.6 \text{ kip-ft}\\ &+ (5.875)(13)(1.5) = \underline{114.6} \text{ kip-ft}\\ &\text{Total} = 251.2 \text{ kip-ft}\\ \text{Restoring moment} &= (3 + 2 + 11)(21/2) = 168 \text{ kip-ft}\\ \text{Net overturning moment} &= 83.2 \text{ kip-ft}\end{aligned}$$

This requires an anchorage force at the wall ends of

$$T = \frac{83.2}{21} = 3.96 \text{ kips}$$

Since the safety factor of 1.5 for the overturn has already been used in the computation, it is reasonable to consider reducing this anchorage requirement to 3.96/1.5 = 2.64 kips if it is used in the form of a service load. In addition, the wind loading permits an increase of one-third in allowable stress, which may also be used to reduce the requirement. Finally, there are added dead load resistances at both ends of the wall. At the corridor the beam sits on the end of the wall and at the building corner this wall is reasonably firmly attached to the wall around the corner. Thus the real need for an anchorage device is questionable. However, most structural designers would probably prefer the reassurance of such a device.

21.3 Alternative Steel and Masonry Structure

As with Building One, an alternative construction for this building would be one with masonry walls and interior steel framing. With ground-floor and roof construction essentially the same as that shown for Building One in Fig. 20.7, the second-floor construction may be achieved as shown in Fig. 21.3. Because of heavier loads, the floor structure here consists of a steel framing system with wide-flange beams supported by steel columns on the building interior and by masonry pilasters built into the exterior walls. The detail in Fig. 21.3*b* shows a plan section of a typical

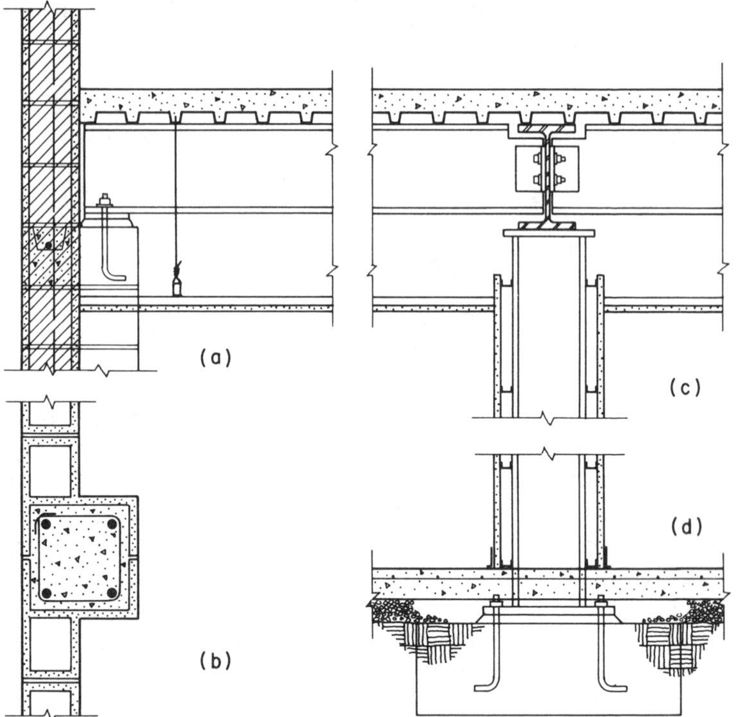

FIGURE 21.3 Building Two: details for the alternative steel and masonry structure.

pilaster as formed with reinforced concrete block construction (western style).

The steel floor deck could be the same type of units as used for the roof, although the fill in this case would consist of structural grade concrete. Fireproofing required for the steel framing and deck would depend on local fire zone and building code requirements.

Although the taller walls, carrying both roof and floor loads, would have greater vertical stress development, it is still possible that relatively minimal code-required construction may be adequate—in either reinforced or unreinforced masonry. Provisions for lateral loads would depend on the general form of the masonry walls. If the exterior walls are formed with individual segments of wall relatively short in plan length in proportion to their total height, overturning for shear wall effects may be critical, requiring considerable reinforcing. If the exterior walls are essentially continuous, with windows and doors "punched in" as holes, the critical concern would be for reinforcement around the wall discontinuities.

Another popular alternative for this building would be the use of interior construction of wood in combination with the masonry walls. Choices for the interior construction would be constrained largely by fire requirements, which are affected mostly by the size of the building. If wood is permitted, it may likely produce the most economical construction, steel components being used mostly where noncombustible construction is required.

21.4 Foundations

As with Building One, the primary elements of a bearing foundation system here would be the wall footings for the exterior walls and the individual square footings for any interior columns. As discussed for Building One, the details of the foundation construction would depend largely on concerns for frost and the vertical location of good bearing materials. For the walls, the details shown in Fig. 20.9 may be applicable, although the width of the bottom of the footings may be greater if design soil pressure is low. Both the wall footings and the column footings can be achieved with the simple elements described in Tables 16.2 and 16.3.

22

Building Three

Building Three is a three-story office building, designed for speculative rental. As with Buildings One and Two, there are many alternatives for the construction of such a building, although in any given place at any given time, the basic construction of such buildings will most likely vary little from a limited set of choices. In this chapter we will illustrate some of the work for the design of two schemes for the construction: an all-steel frame and a poured-in-place concrete frame.

22.1 General Considerations for the Building Design

Figure 22.1 presents a plan of the upper floor and a full building profile section. We assume that a fundamental requirement for the building is the provision of a significant amount of exterior window surface and the avoidance of long expanses of unbroken solid wall surface. Another assumption is that the building is freestanding on the site, with all sides having a clear view.

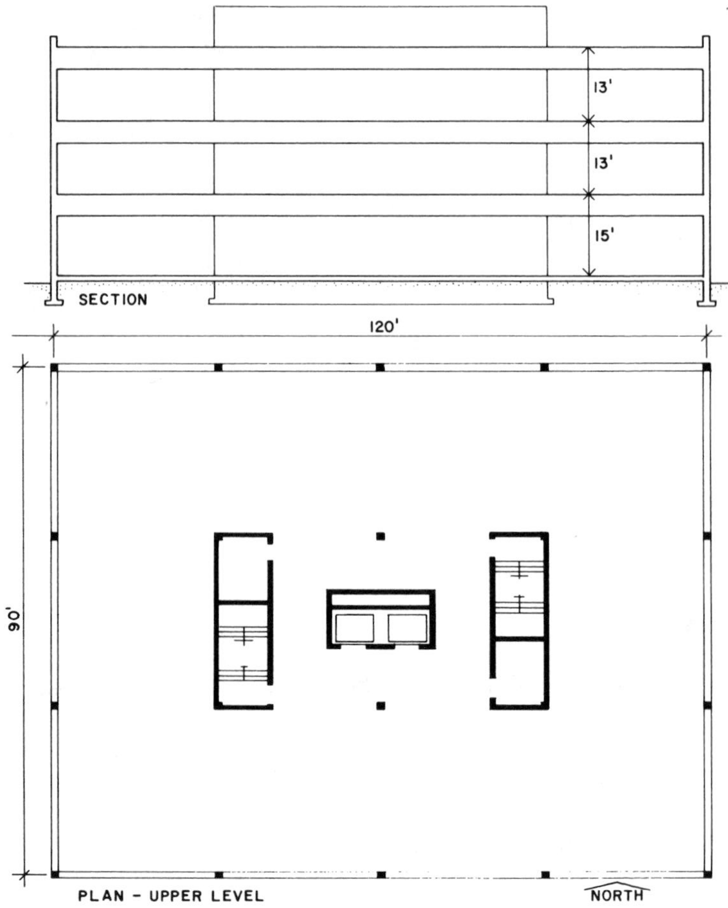

FIGURE 22.1 Building Three: general form.

The following will be assumed as criteria for the design work:

Building code: 1988 edition of Uniform Building Code (Ref. 1).
Live Loads:
  Roof: Table 19.2 (UBC Table 23-C).

Floors: Table 19.3 (UBC Table 23-A).
Office areas: 50 psf [2.39 kPa].
Corridor and lobby: 100 psf [4.79 kPa].
Partitions: 20 psf (UBC minimum per Sec. 2304) [0.96 kPa].

Wind: map speed, 80 mph [129 km/h]; exposure B.

Assumed construction loads:
Floor finish: 5 psf [0.24 kPa].
Ceilings, lights, ducts: 15 psf [0.72 kPa].
Walls (average surface weight):
Interior, permanent: 10 psf [0.48 kPa].
Exterior curtain wall: 15 psf [0.72 kPa].

22.2 Structural Alternatives

The plan as shown, with 30-ft square bays and a general open interior, is an ideal arrangement for a beam and column system in either steel or reinforced concrete. Other types of systems may be made more effective if some modifications of the basic plans are made. These changes may affect the planning of the building core, the plan dimensions for the column locations, the articulation of the exterior wall, or the vertical distances between the levels of the building.

The general form and basic type of the structural system must relate to both the gravity and lateral force problems. Considerations for gravity require the development of the horizontal spanning systems for the roof and floors and the arrangement of the vertical elements (walls and columns) that provide support for the spanning structure. Vertical elements should be stacked, thus requiring coordinating the plans of the various levels.

The most common choices for the lateral bracing system would be the following (see Fig. 22.2):

1. *Core Shear Wall System* (Fig. 22.2*a*). This consists of using solid walls to produce a very rigid central core. The rest of the structure leans on this rigid interior portion, and the roof and floor constructions outside the core, as well as the

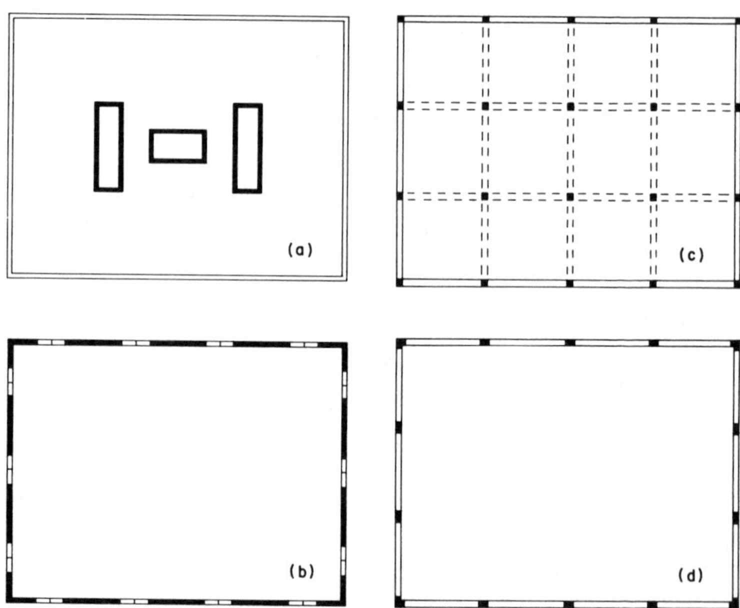

FIGURE 22.2 Options for the lateral bracing.

exterior walls, are free of concerns for lateral forces as far as the structure as a whole is concerned.

2. *Truss-Braced Core.* This is similar in nature to the shear-wall-braced core, and the planning considerations would be essentially similar. The solid walls would be replaced by bays of trussed framing (in vertical bents) using various possible patterns for the truss elements.
3. *Peripheral Shear Walls* (Fig. 22.2*b*). This in essence makes the building into a tubelike structure. Because doors and windows must pierce the exterior, the peripheral shear walls usually consist of linked sets of individual walls (sometimes called piers).
4. *Mixed Exterior and Interior Shear Walls.* This is essentially a combination of the core and peripheral systems.

5. *Full Rigid-Frame System* (Fig. 22.2*c*). This is produced by using the vertical planes of columns and beams in each direction as a series of rigid bents. For this building there would thus be four bents for bracing in one direction and five for bracing in the other direction. This requires that the beam-to-column connections be moment resistive.
6. *Peripheral Rigid-Frame System* (Fig. 22.2*d*). This consists of using only the columns and beams in the exterior walls, resulting in only two bracing bents in each direction.

In the right circumstances any of these systems may be acceptable. Each has advantages and disadvantages from both structural design and architectural planning points of view. The core-braced schemes were popular in the past, especially for buildings in which wind was the major concern. The core system allows for the greatest freedom in planning the exterior walls, which are obviously of major concern to the architect. The peripheral system, however, produces the most torsionally stiff building—an advantage for seismic resistance.

The rigid-frame schemes permit the free planning of the interior and the greatest openness in the wall planes. The integrity of the bents must be maintained, however, which restricts column locations and planning of stairs, elevators, and duct shafts so as not to interrupt any of the column-line beams. If designed for lateral forces, columns are likely to be large, and thus offer more intrusion in the building plan.

22.3 Design of the Floor Framing System

There are a number of options for this system. The spans, the plan layout, the fire code requirements, the anticipated surfacing of floors and ceilings, the need for incorporating elements for wiring, piping, heat and cooling, fire sprinklers, and lighting are all influences on the choice of construction. We assume that the various considerations can be met with a system consisting of a plywood deck, wood joists, and steel beams.

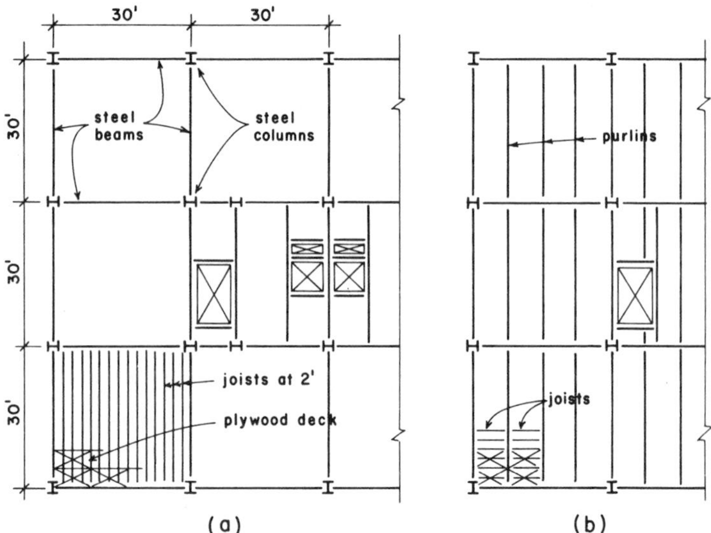

FIGURE 22.3 Building Three floor framing: (*a*) with 30-ft-span joists at 2-ft centers, (*b*) with purlins at 8-ft centers and joists at 2-ft centers.

Framing plans for the typical floor are shown in Fig. 22.3. Steel wide-flange members are used for the columns and the beams that occur on column lines. Steel members would also be used for the framing around large openings for the stairs and elevators. In Fig. 22.3*a* closely spaced joists are used to span between the column-line beams, making a short-span deck a possibility. In Fig. 22.3*b* a partial plan is shown for an alternative to the long-span joists, using a series of beams to reduce the joist spans. A similar system to that shown in Fig. 22.3*b* might be used with an all-steel framing system and a long-span deck. We will limit the illustration here to the design of the system shown in Fig. 22.3*a*. The extra columns shown at the center core area will be used to develop the core-braced system for lateral loads, which is discussed in Sec. 22.5.

The Deck. The plywood deck will be used for both spanning between the joists to resist gravity loads and as a horizontal diaphragm for distribution of wind and seismic forces to the shear walls. Choice of the deck material, the plan layout of the plywood panels, and the nailing must be done with both functions in mind. The spacing of the joists must also be coordinated with the panel layout. It is also necessary to anticipate floor surfacing in terms of both loading and construction accommodation.

We assume the use of the common size of plywood panel of 4 × 8 ft, and a joist spacing of 24 in., which is usually the maximum for floor systems of this type. Office floors will most likely be covered with carpet, thin tile, or wood flooring. The surface of the structural plywood deck, however, is too rough and uneven for these materials, thus resulting in needing some intermediate surfacing, which is most commonly either a second paneling with fiberboard or a thin coat of concrete.

Table 4.7 indicates that a $\frac{3}{4}$-in.-thick deck may be used with the joist spacing of 24 in. A footnote to the UBC table states that the unsupported edges must be blocked or have tongue-and-groove joints if underlayment is not provided. Paneling for floors is quite commonly available with tongue-and-groove joints, but this is not actually required as we will allow for a 2-in. concrete fill.

The panel layout, edge nailing, and the need for blocking must be investigated as part of the design for diaphragm action.

The Joists. It is unlikely that the joists for this span and loading would be solid timber members. As the calculations will show, the required section is quite large and would be very expensive for joists at 24-in. centers. In most regions it is likely that fabricated joists are available that are both lighter and less expensive. For purpose of comparison, however, we will demonstrate the joist design procedure for a solid member.

Economy dictates that a relatively low grade of wood be used for the joist. We will try Douglas fir–larch, No. 1 grade, which is about as high a grade as is feasible. For this grade, Table 4.1 yields the following data for sections 2- to 4-in. thick and 6 in. or more in width:

$$F_b = 1750 \text{ psi (repetitive use) [12.1 MPa]}$$

$$F_v = 95 \text{ psi [0.66 MPa]}$$

$$E = 1{,}800{,}000 \text{ psi [12.4 GPa]}$$

Using these data we will design a joist for the 100-psf live load at the corridor. The design loads for a single 30-ft [9.14-m] span joist are thus as follows:

Live load:
100 psf × 2 ft = 200 plf [2.92 kN/m]
Dead load:
Carpet + pad at 5 psf
2-in. concrete fill at 10 lb/in.
= 20 psf
$\frac{3}{4}$-in. plywood at 3 psf
Ceiling, lights, ducts at 15 psf
Total unit DL: 43 psf
Load on joist: 43 × 2
= 86 plf
Estimate joist weight: 20 plf
Design DL for joist: 106 plf [1.55 kN/m]

The total design load for a joist is thus

$$DL + LL = 106 + 200$$
$$= 306 \text{ plf } [4.46 \text{ kN/m}]$$

and the maximum bending moment is

$$M = \frac{wL^2}{8} = \frac{306(30)^2}{8}$$
$$= 34{,}425 \text{ ft-lb } [46.7 \text{ kN-m}]$$

for which the required section modulus is

$$S = \frac{M}{F_b} = \frac{34{,}425 \times 12}{1750}$$
$$= 236 \text{ in.}^3 \ [3.87 \times 10^6 \text{ mm}^3]$$

There is no member in the size range that was assumed that has an S value this high. We must therefore find a new value for S that corresponds to the proper size range. For the "Beams and stringers" category in Table 4.1 the allowable bending stress for grade dense No. 1 is 1550 psi [10.7 MPa]. The corresponding S is thus

$$S = \frac{34{,}425 \times 12}{1550}$$

$$= 267 \text{ in.}^3 \ [4.38 \times 10^6 \text{ mm}^3]$$

which may be satisfied with a 6 × 18 (S = 281 in.^{3}). This is a mammoth-sized member for joists at 24 in. on center. If the framing arrangement shown in Fig. 22.3*a* is to be retained, a fabricated joist will have to be used. One possibility is to use an open-web steel joist with provision for attachment of the plywood deck to its top flange. Using the joist loading determined previously, it will be found from Ref. 8 that a 24K4 joist at 8.4 lb/ft is adequate for both total load capacity and live load deflection of $L/360$. Other possibilities are a composite wood and steel truss, a vertically laminated wood joist, or a wood plus plywood member (wood flanges and plywood web).

Another possibility is to alter the framing system slightly to produce a system as shown in the partial framing plan in Fig. 22.3*b*. In this scheme the long-span joists are replaced by purlins at 8-ft centers that support short span joists at 2-ft centers. An advantage with this scheme is that it can be used to provide full edge support for the plywood panels, producing a blocked diaphragm without the need for added blocking.

The purlins for the system in Fig. 22.3*b* will carry approximately four times the joist load determined for the 2-ft-on-center joists. This would require the use of glued laminated members or steel wide-flange sections. We will not finish the design of this system, but will proceed with the design assuming the scheme in Fig. 22.3*a* with fabricated joists.

The Beams. The large interior beams may be of wood glued laminated construction, but are more likely to be steel wide-flange

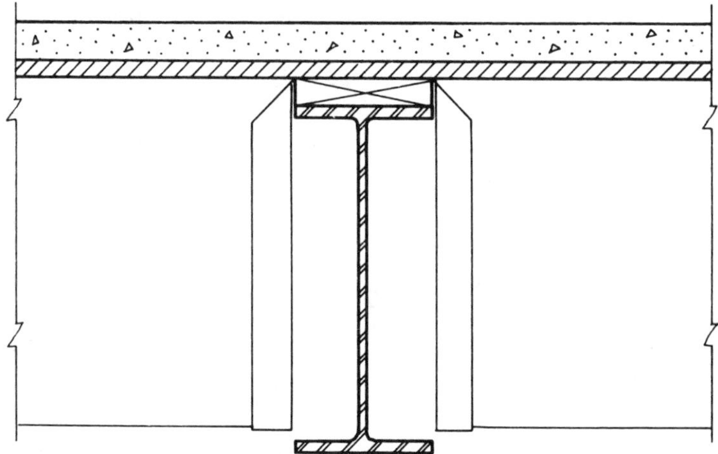

FIGURE 22.4 Detail at the steel beam.

sections. We assume the latter case, and assume the beam section to be as shown in Fig. 22.4, with a 2-in. nominal member bolted to the top of the steel section and the joists carried by saddle-type hangers, with the joist tops level with the top of the wood nailer. If open-web steel joists or composite wood and steel trusses are selected, their top chords would be supported on top of the beam, and the detail would be slightly different.

We will design the beam for a live load of 100 psf and a unit dead load of 55 psf, which includes allowances for the weight of the joists, the beams, and any bridging, blocking, and so on. The heaviest loaded beam is the 30-ft beam that occurs at the building ends. This beam has a total load periphery of 30 × 30 ft or 900 ft^2. The UBC allows a reduction of the live load equal to

$$R = 0.08(A - 150)$$
$$= 0.08(900 - 150) = 60\%$$

The reduction is further limited to a maximum of 40% or

$$R = 23.1\left(1 + \frac{\text{DL}}{\text{LL}}\right)$$

$$= 23.1\left(1 + \frac{55}{100}\right) = 35.8\%$$

We will therefore use a 35% reduction, or a design live load of 65 psf.

Using the reduced live load, the total load carried by the 30-ft-span beam is

$$W = (65 + 55) \times 30 \times 30$$

$$= 108{,}000 \text{ lb, or } 108 \text{ kips } [480 \text{ kN}]$$

and the maximum bending moment is

$$M = \frac{WL}{8} = \frac{108 \times 30}{8}$$

$$= 405 \text{ kip-ft } [549 \text{ kN-m}]$$

From Table 8.1, assuming full lateral support for the beam, possible choices are

W 27 × 84	$M_R = 426$ kip-ft
W 24 × 94	$M_R = 444$ kip-ft
W 18 × 106	$M_R = 408$ kip-ft

Deflection of the beam should be a minimum, for the total deflection of the joists must be added to that of the beam for the sag at the middle of the bay. An inspection of Fig. 8.2 shows that either the 27- or 24-in. members will have quite low values of deflection. If a member shallower than 24 in. is desired to allow for passage of ducts, the deflection should be carefully investigated.

The beam just designed is that which occurs on the interior column lines in the east–west direction. At the building edge the east–west spandrel beams will carry one-half of the joist loading, but must also carry the weight of one-story height of the exterior nonstructural wall. Depending on the weight of the wall construction, these beams may be more heavily loaded or less heavily loaded than the interior beams.

The column-line beams in the north–south direction are much less loaded by the floor construction, although the spandrels must also carry the wall at the exterior. The interior beams in this direction must carry a load equal to that for a single joist, but this is quite likely not the major consideration for their choice. In one way or another, these beams will assist in the collection and distribution of lateral loads, and are also necessary for the erection of the steel frame.

22.4 Design of the Columns for Gravity Load

There are several different cases for the columns, due to the framing arrangements and column locations. For a complete design it would be necessary to tabulate the loading for each different case, although simplification of the framing would probably result in some grouping of actual column sizes to produce the fewest different sections. The five likely typical columns are the following:

- The typical interior column.
- The typical exterior corner column.
- The intermediate column on the north and south sides.
- The intermediate column on the east and west sides.
- The four special columns at the building core, to be used to develop the lateral bracing system.

For an example design we will illustrate the computations for the intermediate exterior columns on the east and west sides. For

TABLE 22.1 Column Load and Design[a]

Level	Story	Unbraced Height (ft)	Loads (kips)				Section Choices[b]	
			Source	DL	LL	Total	8 in.	10 in.
Roof								
			Roof	25	9			
			Wall	7				
	Third	13	Column	2			W 8 × 24	W 10 × 33
			Total/story	34	9			
			LL reduction		0			
			Design/story	34	9	43		
Third floor								
			Floor	25	32			
			Wall	6				
	Second	13	Column	2			W 8 × 24	W 10 × 33
			Total/story	67	41			
			LL reduction		60%			
			Design/story	67	16	83		
Second floor								
			Floor	25	32			
			Wall	7				
	First	15	Column	2			W 8 × 31	W 10 × 33
			Total/story	101	73			
			LL reduction		60%			
			Design/story	101	29	130		
First floor								

[a] Assumptions: $K = 1$, roof and floor DL = 55 psf, wall DL = 15 psf, roof LL = 20 psf, floor LL = 70 psf (50 LL + 20 partition); LL reduction by UBC formulas.
[b] Lightest sections from Table 9.2.

gravity loading, Table 22.1 summarizes the design data and design load computations for this column. With the design loads for the three-story column and the unbraced heights for each story, choices for the column sections can be found from Table 9.2. Although the loads are quite modest, it is likely that the selection would be made from available 10-in.-nominal-depth wide-flange sections, as this provides for better access for beam framing connections. At this size, it is conceivable that the column might be made of a single piece—from footing to roof. This would make it redundantly strong in the upper stories, but the cost of splicing may likely be greater than cost of the extra weight of steel for the upper columns. We show the possibility for three column sizes in Table 22.1, but it is unlikely that more than a single size change—if any—would be made.

For the load tabulations in Table 22.1 we have assumed a dead weight for the roof construction approximately equal to that for the floor construction.

22.5 Design for Wind on the Steel Structure

Referring to Fig. 22.3, it may be noted that there are some extra columns in the framing plan at the location of the stair, rest room, and elevator walls. These columns will be used in conjunction with the regular columns and some of the horizontal framing to define vertical planes of framing for the development of the truss-bracing system shown in Fig. 22.5. These frames will be braced for resistance to lateral loads by the addition of diagonal X-braces. For a simplified design, we will consider the diagonal members to function only in tension, making the vertical frames consist of statically determinate trusses. There are thus four vertically cantilevered trusses in each direction, placed symmetrically at the building core.

With the symmetrical building exterior form and the symmetrically placed core bracing, this is a reasonable system to use in conjunction with the horizontal roof and floor structures to develop resistance to horizontal forces due to wind or seismic actions. We will illustrate the design of the trussed bents for wind.

The 1988 edition of the *Uniform Building Code* (Ref. 1) provides for the use of the projected profile method for wind using a pressure on vertical surfaces defined as

$$p = C_e C_q q_s I$$

where C_e is a combined factor including concerns for the height above grade, exposure conditions, and gust effects. From UBC Table 23-G, assuming exposure condition B,

$$C_e = 0.7 \text{ for surfaces from 0 to 20 ft above grade}$$

$$= 0.8 \text{ from 20 to 40 ft above grade}$$

$$= 1.0 \text{ from 40 to 60 ft above grade}$$

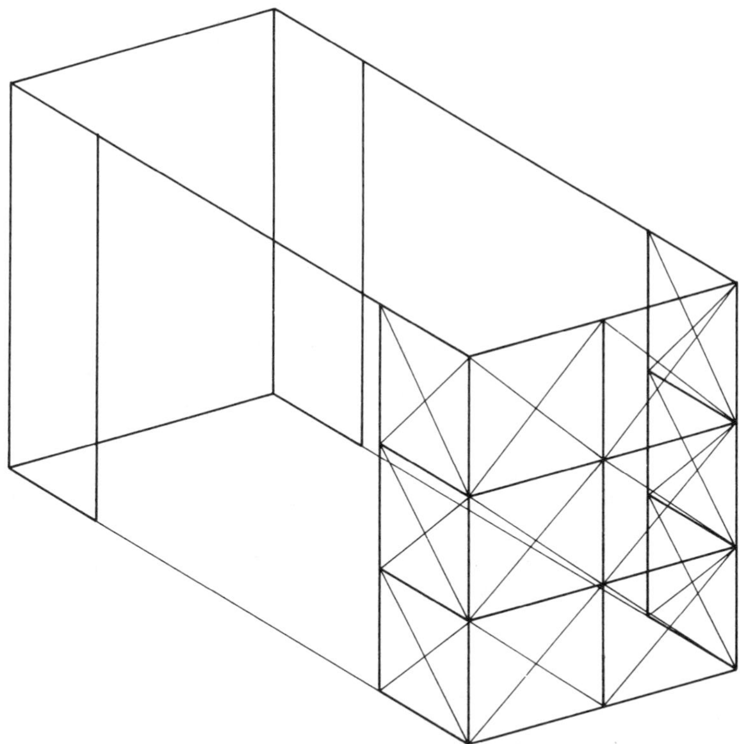

FIGURE 22.5 Development of the core-bracing system for lateral load resistance.

and C_q is the pressure coefficient, which the UBC defines as follows:

$$C_q = 1.3 \text{ for surfaces up to 40 ft above grade}$$

$$= 1.4 \text{ from 40 ft up}$$

The symbol q_s stands for the wind stagnation pressure as related to wind speed and measured at the standard height above ground of 10 m (approximately 30 ft). For the wind speed of 80 mph assumed earlier, the UBC yields a value for q_s of 17 psf.

TABLE 22.2 Design Wind Pressures for Building Three

Height above Average Level of Adjoining Ground (ft)	C_e	C_q	Pressure[a] p (psf)
0–20	0.7	1.3	15.47
20–40	0.8	1.3	17.68
40–60	1.0	1.4	23.80

[a] Horizontally directed pressure on vertical projected area: $p = C_e \times C_q \times 17$ psf.

Table 22.2 summarizes the forgoing data for the determination of the wind pressures at the various height zones on Building Three. For investigation of the wind effects on the lateral bracing system, the wind pressures on the exterior wall are translated into edge loadings for the horizontal roof and floor diaphragms, as shown in Fig. 22.6. Note that we have rounded off the wind pressures from Table 22.2 for use in Fig. 22.6.

The accumulated forces noted as H_1, H_2, and H_3 in Fig. 22.6 are shown applied to one of the vertical trussed bents in Fig.

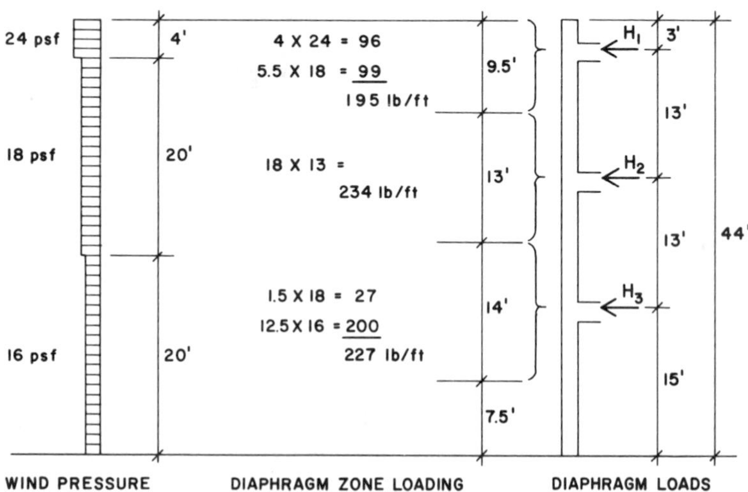

FIGURE 22.6 Building Three: wind loads.

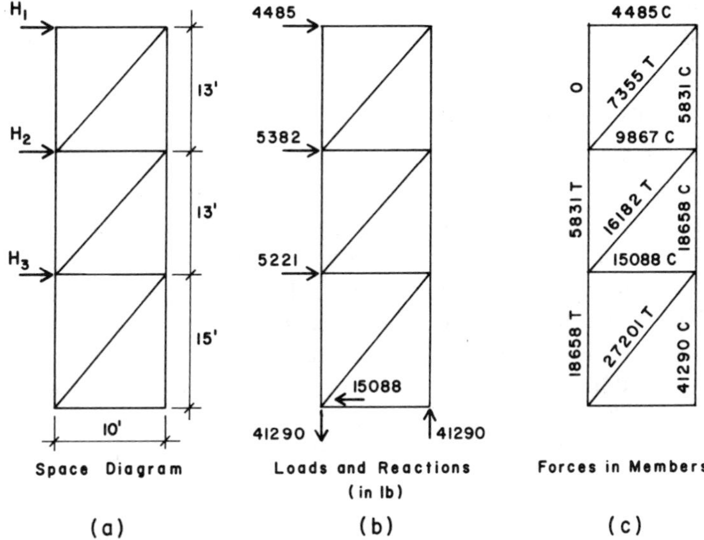

FIGURE 22.7 Investigation of the east–west trussed bents.

22.7*a*. For the east–west bents, the loads will be as shown in Fig. 22.7*b*. These loads are determined by multiplying the edge loadings for the diaphragms as shown in Fig. 22.6 by the 92-ft overall width of the building on the east and west sides. For a single bent, this total force is divided by four. Thus for H_1 the bent load is determined as

$$H_1 = 195 \times 92 \div 4 = 4485 \text{ lb}$$

Analyzed as a truss, ignoring the compression diagonals, the resulting internal forces in the bent are as shown in Fig. 22.7*c*. The forces in the diagonals may be used to design tension members, using the usual one-third increase in allowable stress. The forces in the vertical columns may be added to the gravity loads and checked for possible critical conditions for the columns previously designed for gravity load only. The anchorage force in tension (uplift) should be dealt with as for the shear walls in Chapters

20 and 21. This may reveal the need for tension-resistive anchor bolts and require some special considerations for the foundations.

22.6 The Concrete Structure

A structural framing plan for the upper floors in Building Three is presented in Fig. 22.8, where the use of a poured-in-place slab and beam system of reinforced concrete is indicated. Support for the spanning structure is provided by concrete columns. The system for lateral load resistance is that shown in Fig. 22.2*d*, which utilizes the exterior columns and spandrel beams as rigid-frame bents. This is a highly indeterminate structure for both gravity and lateral force design, and its precise engineering design would undoubtedly be done with a computer-aided system. We will discuss the major design considerations and illustrate the use of

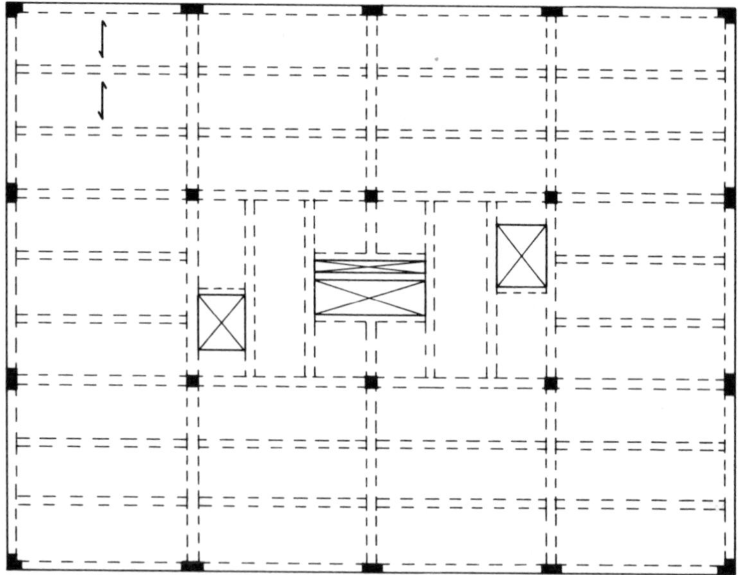

FIGURE 22.8 Building Three structural plan; upper floor with the concrete slab-and-beam system.

some simplified techniques for an approximate analysis and design of the structure.

22.7 Design of the Slab-and-Beam Floor Structure

As shown in Fig. 22.8, the basic floor-framing system consists of a series of beams at 10-ft centers that support a continuous, one-way spanning slab and are supported by column-line girders or directly by the columns. We will discuss the design of three elements of this system: the continuous slab, the four-span beam, and the three-span spandrel girder.

The design conditions for slab, beam, and girder are indicated in Fig. 22.9. Shown on the diagrams are the positive and negative moment coefficients as given in Chapter 8 of the ACI Code (Ref. 3). Use of these coefficients is quite reasonable for the design of the slab and beam. For the girder, however, the presence of the concentrated loads makes the use of the coefficients improper according to the ACI Code. But for an approximate design of the girder, their use will produce some reasonable results.

Figure 22.10 shows a section of the exterior wall that demonstrates the general nature of the construction. The exterior columns and spandrel beams are exposed to view and would

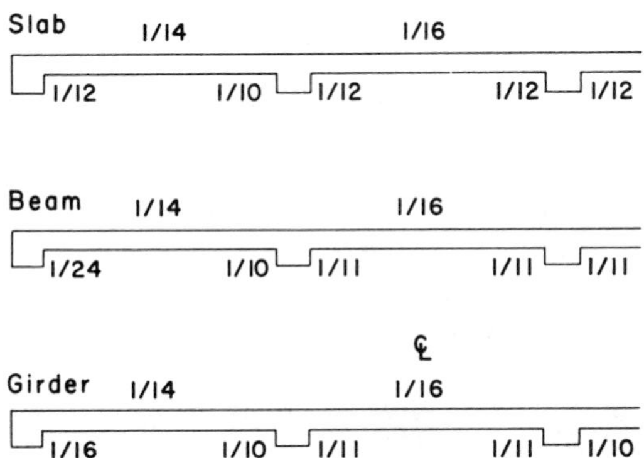

FIGURE 22.9 Approximate moment coefficients for the slab–beam–girder system.

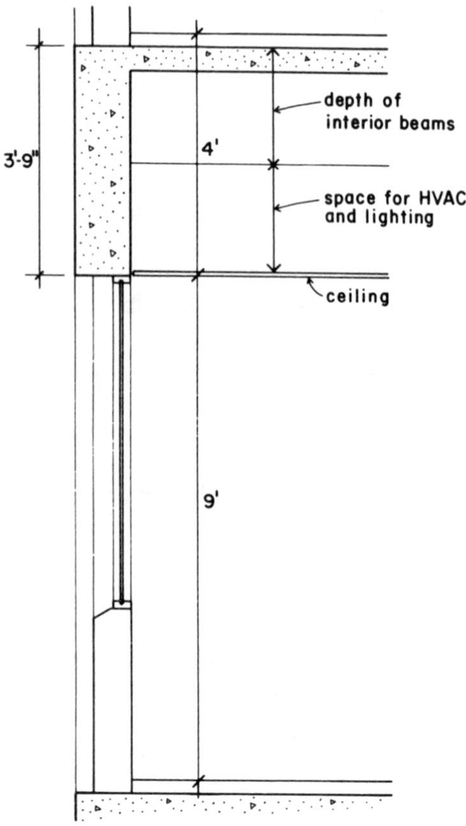

FIGURE 22.10 Section of the typical exterior wall at the upper floors.

receive some special treatment for a higher degree of control of the finished concrete. The use of the full available depth of the spandrel beams results in a much stiffened frame on the building exterior, which partly justifies the choice of the peripheral bent system for lateral bracing.

The design of the continuous slab is presented as the example in Sec. 14.2. The use of the 5-in. slab is based on assumed minimum requirements for fire protection. If a thinner slab is possible, the 9-ft clear span would not require this thickness based on limiting bending or shear conditions or recommendations for deflection control. If the 5-in. slab is used, however, the result will tend to be a slab with a relatively low percentage of steel bar weight per sq ft—a situation usually resulting in lower cost for the structure.

The unit loads used for the slab design in Sec. 14.2 are determined as follows:

Floor live load:
 100 psf (at the corridor) [4.79 kPa]
Floor dead load:
 Carpet and pad at 5 psf
 Ceiling, lights, and ducts at 15 psf
 2-in. lightweight concrete fill at 18 psf
 Assumed 5-in. thick slab at 62 psf
  Total dead load: 100 psf [4.79 kPa]

Inspection of the framing plan in Fig. 22.8 reveals that there are a large number of different beams in the structure for the floor with regard to individual loadings and span conditions. Two general types are the beams that carry only uniformly distributed loads as opposed to those that also provide some support for other beams; the latter produce a load condition consisting of a combination of concentrated and distributed loading. We now consider the design of one of the uniformly loaded beams.

The beam that occurs most often in the plan is the one that carries a 10-ft-wide strip of the slab as a uniformly distributed loading, spanning between columns or supporting beams that are 30 ft on center. Assuming the supports to be approximately 12 in. wide, the beam has a clear span of 29 ft and a total load periphery of $29 \times 10 = 290\ \text{ft}^2$. Using the UBC provisions for reduction of live load,

$$R = 0.08(A - 150)$$
$$= 0.08(290 - 150) = 11.2\%$$

We round this off to a 10% reduction, and, using the loads tabulated previously for the design of the slab, determine the beam loading as follows:

Live load per foot of beam span (with 10% reduction):

$$0.90 \times 100 \times 10 = 900 \text{ lb/ft} \quad \text{or} \quad 0.90 \text{ kip/ft [13.1 kN/m]}$$

Slab and superimposed dead load:

$$100 \times 10 = 1000 \text{ lb/ft} \quad \text{or} \quad 1.0 \text{ kip/ft [14.6 kN/m]}$$

The beam stem weight, estimating a size of 12 × 20 in. for the beam stem extending below the slab, is

$$\frac{12 \times 20}{144} \times 150 \text{ lb/ft}^3 = 250 \text{ lb} \quad \text{or} \quad 0.25 \text{ kip/ft [3.65 kN/m]}$$

The total uniformly distributed load is thus

$$0.90 + 1.0 + 0.25 = 2.15 \text{ kip/ft [31.35 kN/m]}$$

Let us now consider the design of the four-span continuous beam that occurs in the bays on the north and south sides of the building and is supported by the north–south spanning column-line beams that we will refer to as the girders. The approximation factors for design moments for this beam are given in Fig. 22.9, and a summary of the design is presented in Fig. 22.11. Note that

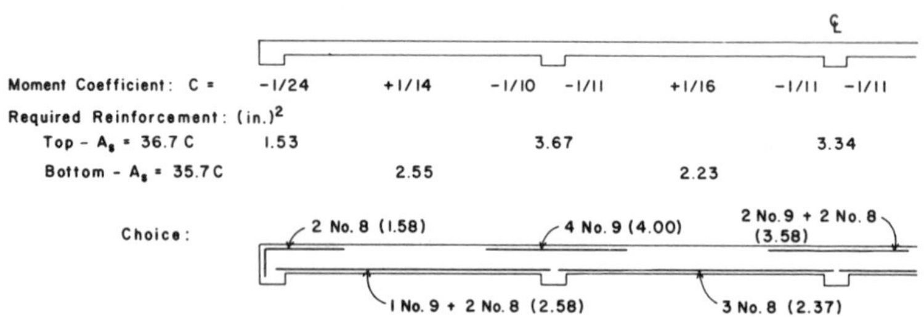

FIGURE 22.11 Design of the four-span beam.

the design provides for tension reinforcing only, thus indicating that the beam dimensions are adequate to prevent a critical condition with regard to flexural stress in the concrete. Using the working stress method, the basis for this is as follows.

Maximum bending moment in the beam is

$$M = \frac{wL^2}{10}$$

$$= \frac{(2.15)(29)^2}{10}$$

$$= 181 \text{ kip-ft } [245 \text{ kN-m}]$$

Then, for a balanced section, using factors from Table 13.3,

$$\text{Required } bd^2$$

$$= \frac{M}{R} = \frac{181 \times 12}{0.204}$$

$$= 10{,}647 \text{ in.}^3 \; [175 \times 10^6 \text{ mm}^3]$$

If $b = 12$ in.,

$$d = \sqrt{\frac{10{,}647}{12}} = 29.8 \text{ in. } [757 \text{ mm}]$$

With minimum concrete cover of 1.5 in. on the bars, No. 3 U-stirrups, and moderate-sized flexural reinforcing, this d can be approximately attained with an overall depth of 32 in. This produces a beam stem that extends 27 in. below the slab, and is thus slightly heavier than that assumed previously. Based on this size, we will increase the design load to 2.25 kip/ft for the subsequent work.

Before proceeding with the design of the flexural reinforcing, it is best to investigate the situation with regard to shear to make sure that the beam dimensions are adequate. Using the approximations given in Chapter 8 of the ACI Code, the maximum shear is considered to be 15% more than the simple span shear and to occur at the inside end of the exterior spans. We thus consider the following.

The maximum design shear force is

$$V = 1.15 \times \frac{wL}{2} = 1.15 \times \frac{2.25 \times 29}{2}$$

$$= 37.5 \text{ kips [167 kN]}$$

For the critical shear stress this may be reduced by the shear between the support and the distance of d from the support; thus

$$\text{Critical } V = 37.5 - \frac{29}{12} \times 2.25$$

$$= 32.1 \text{ kips [143 kN]}$$

Using a d of 29 in., the critical shear stress is

$$v = \frac{V}{bd} = \frac{32{,}100}{29 \times 12} = 92 \text{ psi [634 kPa]}$$

With the concrete strength of 3000 psi, this results in an excess shear stress of 32 psi that must be accounted for by the stirrups. The closest stirrup spacing would thus be

$$s = \frac{A_v f_s}{v'b} = \frac{0.22 \times 24{,}000}{32 \times 12}$$

$$= 13.75 \text{ in. [348 mm]}$$

Because this results in quite a modest amount of shear reinforcing, the section may be considered to be adequate.

For the approximate design shown in Fig. 22.11, the required area of tension reinforcing at each section is determined as

$$A_s = \frac{M}{f_s jd} = \frac{C \times 2.25 \times (29)^2 \times 12}{24 \times 0.89 \times 29}$$

$$= 36.7C$$

Based on the various assumptions and the computations we assume the beam section to be as shown in Fig. 14.5. For the beams the flexural reinforcing in the top that is required at the supports must pass either over or under the bars in the tops of the girders. Because the girders will carry heavier loadings, it is probably wise to give the girder bars the favored position (nearer the outside for greater value of d) and thus to assume the positions as indicated in Fig. 14.5.

At the beam midspans the maximum positive moments will be resisted by the combined beam and slab section acting as a T-section. For this condition we assume an approximate internal moment arm of $d - t/2$ and may approximate the required steel areas as

$$A_s = \frac{M}{f_s(d - t/2)}$$

$$= \frac{C \times 2.25 \times (29)^2 \times 12}{24 \times (29 - 2.5)} = 35.7C$$

The beams that occur on the column lines are involved in the lateral force resistance actions and are discussed in Sec. 22.9.

Inspection of the framing plan in Fig. 22.8 reveals that the girders on the north–south column lines carry the ends of the beams as concentrated loads at the third points of the girder spans. Let us consider the spandrel girder that occurs at the east and west sides of the building. This member carries the outer ends of the first beams in the four-span rows and in addition carries a uniformly distributed load consisting of its own weight and that of the supported exterior wall. The form of the girder and the wall was shown in Fig. 22.10. From the framing plan note that the exterior columns are widened in the plane of the wall. This is done to develop the peripheral bent system, as will be discussed later.

For the spandrel girder we determine the following:

Assumed clear span: 28 ft [8.53 m].

Floor load periphery, based on the carrying of two beams and half the beam span load, is

$$15 \times 20 = 300 \text{ ft}^2 \ [27.9 \text{ m}^2]$$

Note: This is approximately the same total load area as that carried by a single beam, so we will use the live load reduction of 10% as determined for the beam.

Loading from the beams:

Dead load: 1.35 kip/ft × 15 ft
= 20.35 kips

Live load: 0.90 kip/ft × 15 ft
= 13.50 kips

Total 33.85 kips, say 34 kips [151 kN]

Uniformly distributed load:

Spandrel beam weight: $\dfrac{12 \times 45}{144 \times 150}$
= 560 lb/ft

Wall assumed at 25 psf: 25 × 9
= 225 lb/ft

Total
= 785 lb/ft, say 0.8 kip/ft [11.7 kN/m]

For the uniformly distributed load approximate design moments may be found using the moment coefficients as was done for the slab and beam. Values for this procedure are given in Fig. 22.9. The ACI Code does not permit the use of this procedure for concentrated loads, but we may adapt some values for an approximate design using moments for a beam with the third-point loading. Values of positive and negative moments for the third-point loading may be obtained from various references, including Refs. 2, 5, and 6.

Figure 22.12 presents a summary of the work for determining the design moments for the spandrel girder under gravity loading. Moment values are determined separately for the two types of load and then added for the total design moment.

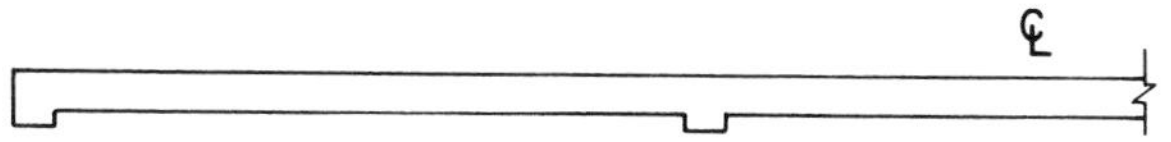

Moment due to distributed load: $M = C\,wL^2 = C \times 0.8 \times (28)^2 = 627\,C$

Coeff - C =	−1/16	+1/14	−1/10	+1/16
M (k-ft) =	−39.2	+44.8	−62.7	+39.2

Moment due to concentrated load: $M = CPL = C \times 34 \times 28 = 952\,C$

Coeff - C =	−1/6	+2/9	−1/3	+1/6
M =	−158.7	+211.6	−317.3	+158.7

Total gravity-induced moment:

M =	−197.9	+256.4	−380	+197.9

FIGURE 22.12 Gravity-load moments for the girder.

We will not proceed further with the girder design at this point, for the effects of lateral loading must also be considered. The moments determined here for the gravity loading will be combined with those from the lateral loading in the discussion in Sec. 22.9.

22.8 Design of the Concrete Columns

The four general cases for the columns are (Fig. 22.13):

The interior column carrying primarily only axial gravity loads.

The intermediate exterior columns on the north and south sides carrying the ends of the interior girders and functioning as members of the peripheral bents for lateral resistance.

FIGURE 22.13 Framing situations for the columns and beams.

The intermediate exterior columns on the east and west sides carrying the ends of the column-line beams and functioning as members of the peripheral bents.

The corner columns carrying the ends of the spandrel beams and functioning as the end members in both peripheral bents.

Summations of the design loads for the columns may be done as illustrated for the steel column in Sec. 22.4. As all columns will be subjected to combinations of axial compression and bending, these gravity loads represent only the axial compression action. For the interior columns, the bending moments will be relatively low in comparison to the compression loads, and it is reasonable for a preliminary design to ignore bending effects and design for axial compression only. The usual minimum required eccentricity, as described in Sec. 15.2, will most likely provide for suffi-

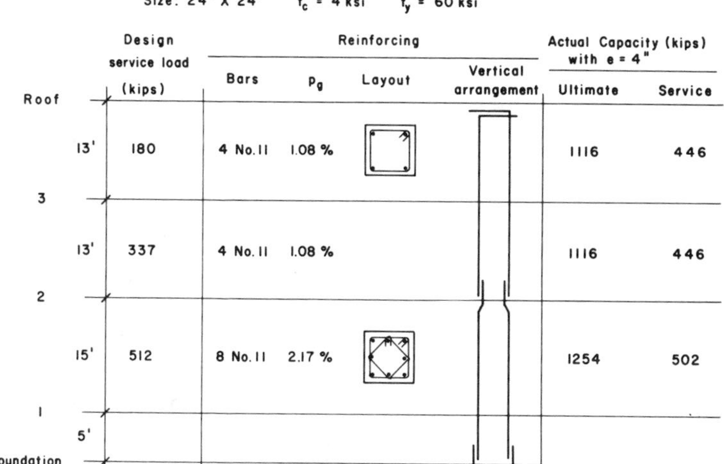

FIGURE 22.14 Design of the interior column.

cient bending. On this basis, a trial design for one of the interior columns is shown in Fig. 22.14. Design loads were obtained in a manner similar to that shown for the steel column in Sec. 22.4. A single size of 24 in. square is used for all three stories, a common practice permitting the reuse of column forming for cost savings. The service load capacities indicated may be compared with values obtained from the graphs in Fig. 15.4. Economy is also generally obtained with the use of low percentages of reinforcement when bending moments are not a critical concern; the percentages shown in Fig. 22.14 are minimal, but smaller column sizes could be used if floor space and planning problems are of major concern.

The interior column occurs at the location of the stairs and rest rooms, and it is possible that some form alteration may be made to allow the columns to fit more smoothly into the wall planning. This will add cost to the column construction, but is relatively easily achieved, as shown in Fig. 15.6.

For the intermediate exterior column there are four actions to consider:

1. The vertical compression induced by gravity.
2. Bending moment induced by the interior framing that intersects the wall column; the columns are what provides the end moments shown in Figs. 22.11 and 22.12.
3. Bending moments in the plane of the wall induced by unbalanced conditions in the spandrel beams and girders.
4. Bending moments induced by the actions of the peripheral bents in resisting lateral loads.

For the corner column the situation is similar to that for the intermediate exterior column, that is, bending on both axes. The forms of the exterior columns as shown on the plan in Fig. 22.8 have been established in anticipation of the major effects described. Further discussion of these columns will be deferred, however, until after we have investigated the situations of lateral loading.

22.9 Design for Wind on the Concrete Structure

The lateral force resisting systems for the concrete structure are shown in Fig. 22.15. For force in the east–west direction the resistive system consists of the horizontal roof and floor slabs and the exterior bents (columns and spandrel beams) on the north and south sides. For force in the north–south direction the system utilizes the bents on the east and west sides. Actually other elements of the structural frame will also resist lateral force, but by widening the columns and deepening the spandrel beams in these bents, an increased stiffness will be produced; the stiffer bents will then tend to offer the most resistance to lateral movements. We will design these stiffened bents for all the lateral force, ignoring the minor resistances offered by the other column and beam bents.

With the same building profile, the wind loads on this structure will be the same as those determined for the steel structure in

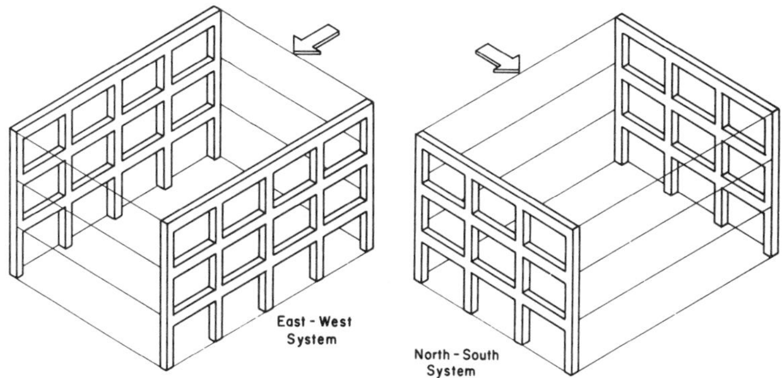

FIGURE 22.15 The peripheral bent bracing system for wind loads.

Sec. 22.5. As in the example in that section, we will illustrate the design of the bents in only one direction, in this case the bents on the east and west sides. From the data given in Fig. 22.6, the horizontal forces for the concrete bents are determined as follows:

$$H_1 = (195)(122)/2 = 11{,}895 \text{ lb}, \quad \text{say } 11.9 \text{ kips/bent}$$

$$H_2 = (234)(122)/2 = 14{,}274 \text{ lb}, \quad \text{say } 14.3 \text{ kips/bent}$$

$$H_3 = (227)(122)/2 = 13{,}847 \text{ lb}, \quad \text{say } 13.9 \text{ kips/bent}$$

Figure 22.16 shows a profile of one of the north–south bents with the bent loads.

For an approximate analysis we consider the individual stories of the bent to behave as shown in Fig. 22.17, with the columns developing an inflection point at their midheight points. Because the columns all move the same distance, the shear load in a single column may be assumed to be equal to the cantilever deflecting load and the individual shears to be proportionate to the stiffnesses of the columns. If the columns are all of equal stiffness in this case, the total load would be simply divided by four. However, the end columns are slightly less restrained as there is a beam on only one side. We will assume the net stiffness of the end

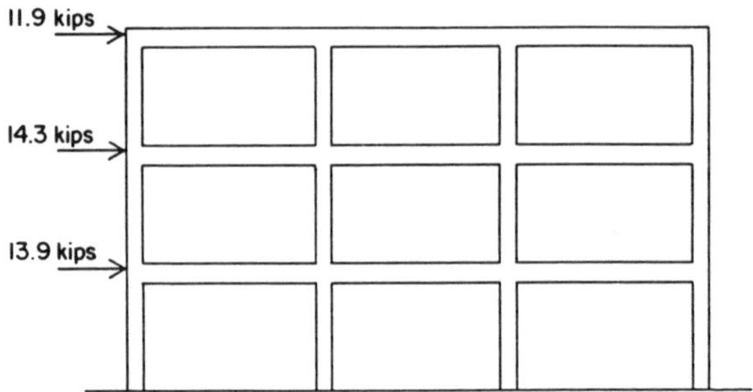

FIGURE 22.16 Wind loading for the north–south bent.

columns to be one-half that of the interior columns. Thus the shear force in the end columns will be one-sixth of the load and that in the interior columns one-third of the load. The column shears for each of the three stories is thus as shown in Fig. 22.18.

The column shear forces produce moments in the columns. With the column inflection points assumed at midheight, the moment produced by a single shear force is simply the product of the force and half the column height. These moments must be resisted by the end moments in the rigidly attached beams, and the actions are as shown in Fig. 22.19. These effects due to the lateral loads may now be combined with the previously determined effects of gravity loads for an approximate design of the columns and beams.

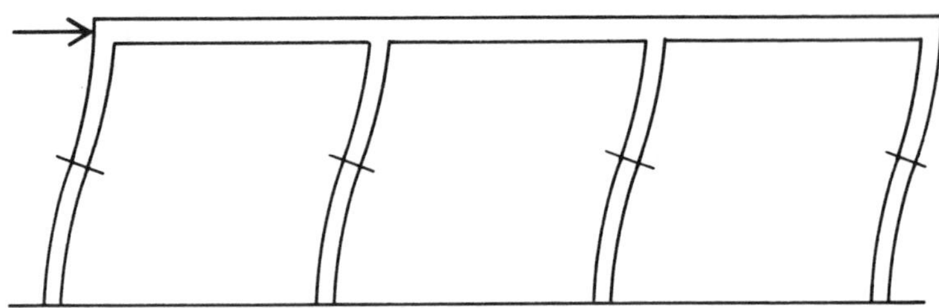

FIGURE 22.17 Assumed deformation of the bent columns.

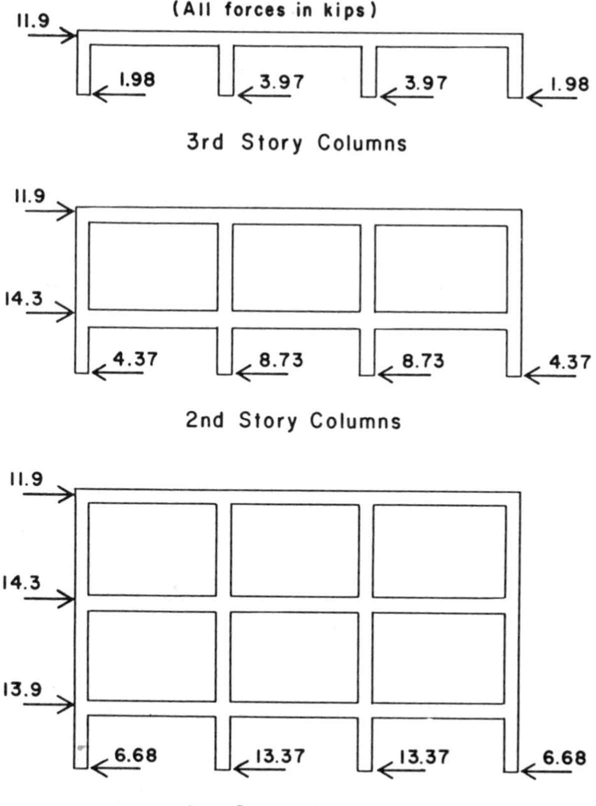

FIGURE 22.18 Investigation for column shear in the north–south bent.

For the columns, we combine the axial compression forces with any gravity-induced moments and first determine that the load condition without lateral effects is not critical. We may then add the effects of the moments caused by lateral loading and investigate the combined loading condition, for which we may use the one-third increase in allowable stress. Gravity-induced beam moments are taken from Fig. 22.12 and are assumed to

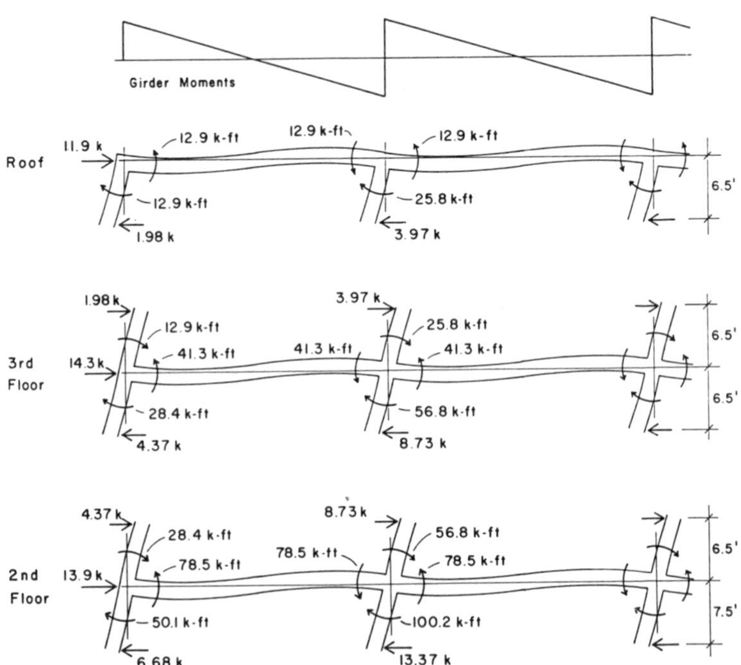

FIGURE 22.19 Investigation for column and girder moments in the north–south bent.

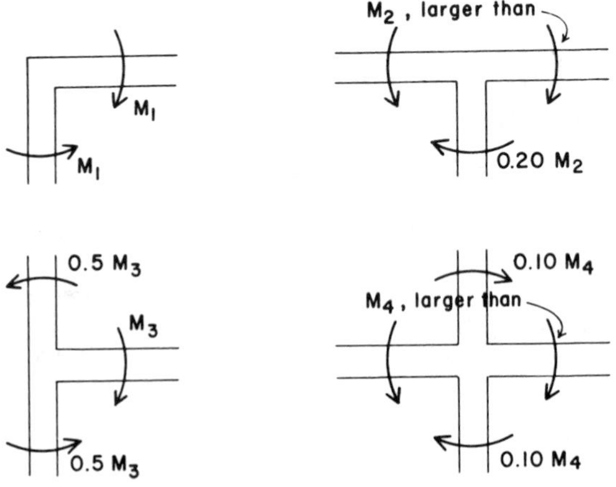

FIGURE 22.20 Assumptions for approximation of the distribution of bending moments in the columns due to gravity loads.

TABLE 22.3 Summary of Design Data for the Bent Columns

	Column	
	Intermediate	Corner
Axial gravity design load (kips)		
Third story	90	55
Second story	179	117
First story	277	176
Assumed gravity moment on bent axis (kip-ft) from Figs. 22.12 and 22.20		
Third story	60	120
Second story	39	100
First story	39	100
Moment from lateral force (kip-ft) from Fig. 22.19		
Third story	25.8	12.9
Second story	56.8	28.4
First story	100.2	50.1

induce column moments as shown in Fig. 22.20. The summary of design conditions for the corner and interior column is shown in Table 22.3. The design values for axial load and moment and approximate sizes and reinforcing are shown in Fig. 22.21. Column sizes and reinforcing were obtained from the tables in the

	Intermediate Column					Corner Column				
	Axial Load (kips)	Moment (kip-ft)	e (in.)	Column Dimensions (in.)	Reinforcement No. - Size	Axial Load (kips)	Moment (kip-ft)	e (in.)	Column Dimensions (in.)	Reinforcement No. - Size
Roof										
	90 X 3/4 = 68	85.8 X 3/4 = 64	11.3	20X28	6 - 9	55	120	35	20X24	6 - 10
3										
	179 X 3/4 = 134	95.8 X 3/4 = 72	6.5	20X28	6 - 9	117	100	13.6	20X24	6 - 10
2										
	277 X 3/4 = 208	139.2 X 3/4 = 105	6.1	20X28	6 - 10	176 X 3/4 = 132	150.1 X 3/4 = 113	10.3	20X24	6 - 10
1										

FIGURE 22.21 Design of the bent columns.

CRSI Handbook (Ref. 6) using concrete with $f'_c = 4$ ksi and Grade 60 reinforcing.

The spandrel beams (or girders) must be designed for the combined shears and moments due to gravity and lateral effects. Using the values for gravity-induced moments from Fig. 22.12 and the values for lateral load moments from Fig. 22.19, the combined moment conditions are shown in Fig. 22.22. For design we must consider both the gravity only moment and the combined effect. For the combined effect we use three-fourths of the total combined values to reflect the allowable stress increase of one-third.

Figure 22.23 presents a summary of the design of the reinforcing for the spandrel beam at the third floor. If the construction that was shown in Fig. 22.10 is retained with the exposed spandrel beams, the beam is quite deep. Its width should be approximately the same as that of the column, without producing too massive a section. The section shown is probably adequate, but

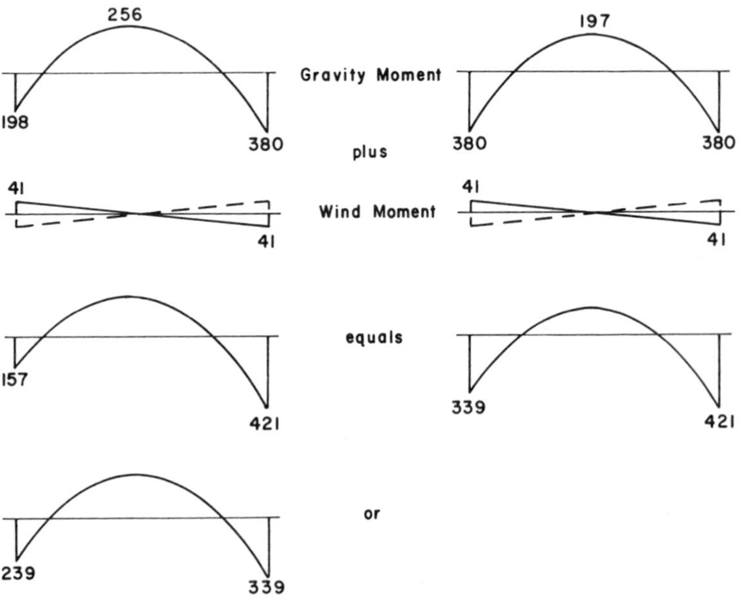

FIGURE 22.22 Combined moments for the bent girder.

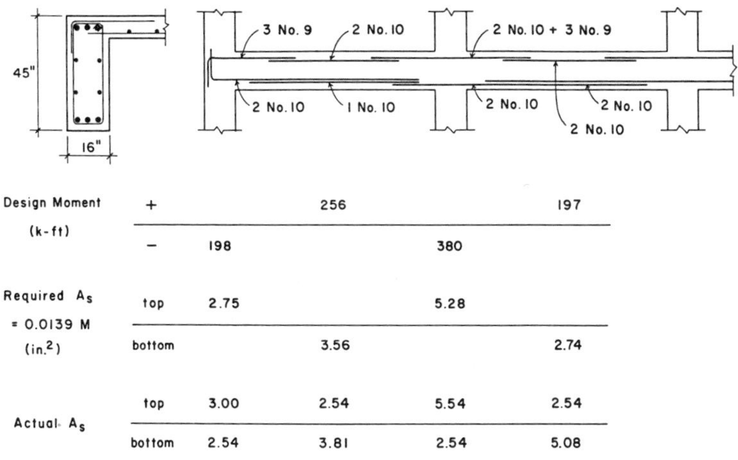

Design Moment (k-ft)	+		256		197
	−	198		380	
Required A_s = 0.0139 M (in.^{2})	top	2.75		5.28	
	bottom		3.56		2.74
Actual A_s	top	3.00	2.54	5.54	2.54
	bottom	2.54	3.81	2.54	5.08

FIGURE 22.23 Design of the bent girder for combined moments.

several additional considerations must be made as will be discussed later.

For computation of the required steel areas we assume an effective depth of approximately 40 in. and use

$$A_s = \frac{M}{f_s jd} = \frac{M(12)}{24(0.9)(40)} = 0.0139M$$

Because the beam is so deep, it is advisable to use some longitudinal reinforcing at an intermediate height in the section, especially on the exposed face.

Shear design for the beams should also be done for the combined loading effects. The closed form for the shear reinforcing, as shown in Fig. 22.23, is used for considerations of torsion as well as the necessity for tying the compressive reinforcing.

With all of the approximations made, this should still be considered to be a very preliminary design for the beam. It should, however, be adequate for use in preliminary architectural studies and for sizing the members for a dynamic seismic analysis and a general analysis of the actions of the indeterminate structure.

References

1. *Uniform Building Code*, 1988 ed., International Conference of Building Officials, 5360 South Workmanmill Road, Whittier, CA 90601. (Called simply the UBC.)
2. *Manual of Steel Construction*, 8th ed., American Institute of Steel Construction, Chicago, IL, 1980. (Called simply the AISC Manual.)
3. *Building Code Requirements for Reinforced Concrete*, ACI 318–83, American Concrete Institute, Detroit, MI, 1983. (Called simply the ACI Code.)
4. *National Design Specification for Wood Construction*, National Forest Products Association, Washington, D.C., 1986.
5. *Timber Construction Manual*, 3rd ed., American Institute of Timber Construction, Wiley, New York, 1985.
6. *CRSI Handbook*, 5th ed., Concrete Reinforcing Steel Institute, Schaumburg, IL, 1985.
7. *Masonry Design Manual*, 3rd ed., Masonry Institute of America, Los Angeles, CA, 1979.
8. *Standard Specifications, Load Tables, and Weight Tables for Steel Joists and Joist Girders*, Steel Joist Institute, Myrtle Beach, SC, 1986.
9. *Steel Deck Institute Design Manual for Composite Decks, Form Decks, and Roof Decks*, Steel Deck Institute, St. Louis, MO, 1981–1982.
10. C. G. Ramsey and H. R. Sleeper, *Architectural Graphic Standards*, 8th ed., Wiley, New York, 1988.
11. J. C. McCormac, *Structural Analysis*, 4th ed., Harper & Row, New York, 1984.

12. S. W. Crawley and R. M. Dillon, *Steel Buildings: Analysis and Design*, 3rd ed., Wiley, New York, 1984.
13. James Ambrose, *Simplified Design of Building Structures*, 2nd ed., Wiley, New York, 1986.
14. James Ambrose, *Simplified Design of Building Foundations*, 2nd ed., 1988.
15. H. Parker, *Simplified Design of Structural Steel*, 5th ed., Wiley, New York, 1983.
16. J. Ambrose and D. Vergun, *Design for Lateral Forces*, Wiley, New York, 1987.

Answers to Exercise Problems

Answers are given here for all of the exercise problems that are of a basic computational nature, resulting in general in a single correct answer. For the most part, numeric answers are carried to three or four significant figures except in cases where additional accuracy seemed desirable as an aid in interpreting the result. Illustrated solutions for all of these problems are presented in the Study Manual for this book, which is available from the publisher. Also contained in the Study Manual are some additional exercise problems and other study aids.

Chapter 1

1.1.A. 3.33 in.^{2} [2150 mm^2]
1.1.B. 99,275 lb [442 kN]
1.1.C. 0.874 in., or $\frac{7}{8}$ in. [22.2 mm]
1.1.D. 5.48 ft [1.67 m]
1.1.E. 18.4 kips [81.9 kN]
1.1.F. 196 kips [873 kN]
1.1.G. 2.5 in.^{2} [1613 mm^2]
1.1.H. 12 × 12

1.4.A. 19,333 lb [86 kN]
1.4.B. 0.072 in. [1.8 mm]
1.4.C. 29,550,000 psi [203 GPa]
1.4.D. 0.84 in. [21.4 mm] (assuming solid cable)
1.5.A. 1.18 in.^{2} [762 mm^2]
1.5.B. 8 × 8
1.5.C. 27,060 lb [120 kN]
1.5.D. 122,331 lb [544 kN]
1.5.E. f = 1212 psi [8.36 MPa], allowable f = 1150 psi; therefore the post is not adequate
1.5.F. f = 463 psi, allowable f = 900 psi; therefore the pier is OK
1.5.G. f = 14,136 psi [97.4 MPa], allowable f = 15,000 psi; therefore the bolt is OK
1.6.A. $s = 6t$, $y = 1.90t$; in 10 seconds y = 19 ft
1.6.B. $s = 1.5t^2$, $y = 0.474t^2$; in 10 seconds y = 47.4 ft
1.6.C. a = 10,800 miles/h^2, $v = 10{,}800t$, $s = 5400t^2$; in 20 seconds ($\frac{1}{180}$ h) $s = \frac{1}{6}$ mile

Chapter 2

2.3.A. R_1 = 3593.75 lb [15.98 kN], R_2 = 4406.25 lb [19.60 kN]
2.3.B. R_1 = 8375 lb [37.26 kN], R_2 = 10,625 lb [47.26 kN]
2.3.C. R_1 = 7667 lb [34.11 kN], R_2 = 9333 lb [41.53 kN]
2.3.D. R_1 = 4429 lb [19.71 kN], R_2 = 7571 lb [33.69 kN]
2.3.E. R_1 = 7143 lb [31.79 kN], R_2 = 11,857 lb [52.76 kN]
2.3.F. R_1 = 6750 lb [30.04 kN], R_2 = 5250 lb [23.36 kN]
2.4.A. Maximum shear = 10 kips [44.5 kN]
2.4.B. Maximum shear = 5250 lb [22.974 kN]
2.4.C. Maximum shear = 1114 lb [4.956 kN]
2.4.D. Maximum shear = 8.47 kips [37.55 kN]
2.4.E. Maximum shear = 9.375 kips [41.623 kN]
2.4.F. Maximum shear = 4333 lb [19.108 kN]
2.5.A. Maximum M = 60 kip-ft [80.1 kN-m]
2.5.B. Maximum M = 18,375 ft-lb [24.12 kN-m]
2.5.C. Maximum M = 4286 ft-lb [5.716 kN-m]
2.5.D. Maximum M = 61.3 kip-ft [81.7 kN-m]
2.5.E. Maximum M = 18.35 kip-ft [24.45 kN-m]
2.5.F. Maximum M = 20850 ft-lb [27.62 kN-m]
2.6.A. R_1 = 1860 lb [8.27 kN], maximum V = 1360 lb [6.05 kN], maximum $-M$ = 2000 ft-lb [2.66 kN-m], maximum $+M$ = 3200 ft-lb [4.27 kN-m]

2.6.B. R_1 = 10.32 kips [35.24 kN], maximum V = 7.32 kips [32.1 kN], maximum $-M$ = 4.5 kip-ft [5.9 kN-m], maximum $+M$ = 22.3 kip-ft [29.3 kN-m]

2.6.C. R_1 = 2760 lb [12.28 kN], maximum V = 2040 lb [9.07 kN], maximum $-M$ = 2000 ft-lb [2.67 kN-m], maximum $+M$ = 5520 ft-lb [7.37 kN-m]

2.6.D. $R_1 = R_2$ = 10 kips [43.8 kN], maximum V = 7 kips [30.66 kN], maximum $-M$ = 4.5 kip-ft [5.91 kN-m], maximum $+M$ = 20 kip-ft [26.28 kN-m]

2.7.A. Maximum V = 1500 lb [6.67 kN], maximum M = 12,800 ft-lb [17.1 kN-m]

2.7.B. Maximum V = 1500 lb [6.60 kN], maximum M = 9500 ft-lb [12.555 kN-m]

2.7.C. Maximum V = 1200 lb [5.27 kN], maximum M = 8600 ft-lb [11.33 kN-m]

2.7.D. Maximum V = 2700 lb [11.84 kN], maximum M = 12,750 ft-lb [16.8 kN-m]

2.9.A. M = 32 kip-ft [43.4 kN-m]

2.9.B. M = 90 kip-ft [122 kN-m]

2.11.A. At neutral axis f_v = 811.4 psi; at junction of web and flange f_v = 175 psi and 700 psi.

2.12.A. R_1 = 7.67 kips, R_2 = 35.58 kips, R_3 = 12.75 kips, $+M$ = 14.69 kip-ft and 40.64 kip-ft, $-M$ = 52 kip-ft

2.12.B. R_1 = 10.875 kips, R_2 = 38.25 kips, R_3 = 10.875 kips, maximum $+M$ = 59.133 kip-ft, maximum $-M$ = 99 kip-ft

2.13.A. Maximum V = 8 kips, maximum $+M$ = maximum $-M$ = 44 kip-ft, inflection at 5.5 ft from end

2.13.B. Maximum V = 5 kips, maximum $+M$ = 11.25 kip-ft, maximum $-M$ = 20 kip-ft

2.14.A. $A_x = C_x$ = 2.143 kips [9.643 kN], A_y = 2.587 kips [12.857 kN], C_y = 7.143 kips [32.143 kN]

2.14.B. $A_x = C_x$ = 1.33 kips [6 kN], $A_y = C_y$ = 2 kips [9 kN]

2.14.C. R_1 = 16 kips [72 kN], R_2 = 48 kips [216 kN], maximum $+M$ = 64 kip-ft [86.4 kN-m], maximum $-M$ = 80 kip-ft [108 kN-m]; inflection at pin location in both spans

2.14.D. R_1 = 5.5 kips [24.75 kN], R_2 = 15.6 kips [70.2 kN], R_3 = 2.9 kips [13.05 kN], $+M$ in left span = 15.125 kip-ft [20.4 kN-m], $-M$ = 21 kip-ft [28.4 kN-m]; inflection at 11 ft from left end and 5.8 ft from right end.

2.14.E. R_1 = 6.4 kips [28.8 kN], R_2 = 19.6 kips [88.2 kN], $+M$ = 20.48 kip-ft [27.7 kN-m] in end span and 24.5 kip-ft [33.1

kN-m] in center span, $-M$ = 25.5 kip-ft [34.4 kN-m]; inflection at 3.2 ft from R_2 in end span

2.15.A. R = 10 kips (up) and 110 kip-ft counterclockwise

2.15.B. R = 5 kips (up) and 24 kip-ft counterclockwise

2.15.C. R = 6 kips (to left) and 72 kip-ft counterclockwise

2.15.D. Left R = 4.5 kips (up), right R = 4.5 kips (down) and 12 kips to right

2.15.E. Left R = 4.5 kips (down) and 6 kips to left, right R = 4.5 kips (up) and 6 kips to left

Chapter 3

3.1.A. c_y = 2.6 in. [70 mm]

3.1.B. c_y = 1.75 in. [43.9 mm], c_x = 0.75 in. [18.9 mm]

3.1.C. c_y = 4.2895 in. [107.24 mm]

3.1.D. c_y = 3.4185 in. [85.185 mm], c_x = 1.293 in. [32.2 mm]

3.1.E. c_y = 4.4375 in. [110.9 mm], c_x = 1.0625 in. [26.6 mm]

3.1.F. c_y = 4.3095 in. [107.7 mm]

3.4.A. I = 535.86 in.^{4} [2.11×10^8 mm^4]

3.4.B. I = 205.33 in.^{4} [80.21×10^6 mm^4]

3.4.C. I = 447.33 in.^{4} [198.44×10^6 mm^4]

3.4.D. I = 5.0485 in.^{4} [2.036×10^6 mm^4]

3.4.E. I = 205.33 in.^{4} [80.21×10^6 mm^4]

3.4.F. I = 682.33 in.^{4} [2.670×10^8 mm^4]

3.4.G. I = 438 in.^{4}

3.4.H. I = 420.1 in.^{4}

3.4.I. I = 1672.49 in.^{4}

3.5.A. S = 51.5624 in.^{3}

3.5.B. S = 196.7 in.^{3}

3.5.C. S = 2.99 in.^{3}

3.5.D. S = 1.97 in.^{3}

3.6.A. r_y = 1.93 in.

3.6.B. r_x = 1.051 in.

3.6.C. r_x = 5.31 in.

Chapter 4

4.2.A. 3 × 12

4.2.B. 6 × 16

4.2.C. 2 × 10

4.3.A. v = 83.1 psi < allowable of 85 psi; beam is OK

4.3.B. v = 75.44 psi < allowable of 85 psi; beam is OK

4.3.C. $v = 68.65$ psi < allowable of 85 psi; beam is OK
4.3.D. Using shear at end: 10 × 12
4.4.A. Stress = 303 psi < allowable of 625 psi; beam is OK
4.4.B. Stress = 480 psi < allowable of 565 psi (without the permitted increase); beam is OK
4.5.A. $D = 0.31$ in. [7.6 mm] < allowable of 0.8 in.; beam is OK
4.5.B. $D = 0.19$ in. [4.9 mm] < allowable of 0.6 in.; beam is OK
4.5.C. $D = 0.23$ in. [6 mm] < allowable of 0.75 in.; beam is OK
4.5.D. $D = 0.30$ in. [8 mm] < allowable of 0.8 in.; beam is OK
4.5.E. Need $I = 911$ in.^{4}, lightest choice is 4 × 16
4.6.A. 2 × 10
4.6.B. 2 × 8
4.6.C. 2 × 12
4.6.D. 2 × 12
4.6.E. 2 × 6
4.6.F. 2 × 8
4.6.G. 2 × 10
4.6.H. 2 × 12

Chapter 5

5.1.A. 7.82 kips
5.1.B. 21.1 kips
5.1.C. 32.6 kips
5.1.D. 47.1 kips
5.1.E. 6 × 6
5.1.F. 8 × 8
5.1.G. 10 × 10
5.1.H. 12 × 12
5.2. See answers for Sec. 5.1

Chapter 6

6.1.A. Maximum T = 18,480 lb, limited by bolts
6.1.B. Maximum T = 1940 lb, limited by bolts
6.1.C. Maximum load = ±4800 lb
6.2.A. 64 6d nails (16 each side in each member) or 20 16d nails (10 in each member, driven through and clinched)

Chapter 8

8.2.A. M 14 × 18 is lightest shape, W 10 × 19 is lightest W
8.2.B. W 12 × 22

8.2.C. M 14 × 18
8.2.D. W 24 × 55
8.2.E. W 12 × 26 (U.S. data); W 14 × 22 (SI data)
8.2.F. W 16 × 31
8.2.G. M 14 × 18
8.2.H. W 12 × 22 or W 14 × 22
8.2.I. W 14 × 22
8.2.J. W 16 × 26
8.3.A. (a) M 12 × 11.8, (b) W 8 × 15 (W 5 × 19 lightest, but not in table)
8.3.B. (a) W 16 × 26, (b) W 10 × 45
8.3.C. (a) W 21 × 50, (b) W 16 × 57
8.3.D. (a) W 12 × 19, (b) W 10 × 26
8.3.E. (a) W 18 × 35, (b) W 14 × 48
8.3.F. (a) W 24 × 76, (b) W 21 × 101
8.5.A. W 14 × 34 is adequate
8.5.B. W 16 × 40 (for deflection)
8.6.A. 168.3 kips
8.6.B. 51.1 kips
8.6.C. 37.1 kips
8.6.D. 55.0 kips
8.7.A. $D = 0.80$ in.
8.7.B. $D = 0.69$ in.
8.7.C. $D = 0.83$ in.
8.9.A. $R = 65.2$ kips, stiffeners not required
8.9.B. $P = 93.9$ kips
8.10.A. $B = 15$ in., $t = 1$ in.
8.10.B. $B = 17$ in., $t = 1\frac{1}{8}$ in.
8.11.A. WR20
8.11.B. WR18
8.11.C. IR22 or WR22
8.12.A. 28K7
8.12.B. 28K7
8.12.C. (a) 22K4, (b) 18K7
8.12.D. 18K4

Chapter 9

9.3.A. 289 kips [1286 kN]
9.3.B. 558 kips [2484 kN]
9.3.C. 336 kips [1496 kN]

9.4.A. W 8 × 31
9.4.B. W 12 × 58
9.4.C. W 12 × 79
9.4.D. 4 in.
9.4.E. 5 in.
9.4.F. 6 in.
9.4.G. 8 in.
9.4.H. 3 × 3 × $\frac{1}{4}$ in.
9.4.I. 4 × 4 × $\frac{3}{16}$ in.
9.4.J. 4 × 4 × $\frac{3}{8}$ in.
9.4.K. 5 × $3\frac{1}{2}$ × $\frac{3}{8}$
9.4.L. 6 × 4 × $\frac{3}{8}$
9.4.M. 6 × 4 × $\frac{1}{2}$
9.5.A. W 12 × 58
9.5.B. W 14 × 120 (maybe W 14 × 109)
9.7.A. Possibility: 15 in. × 16 in. × $1\frac{1}{4}$ in.
9.7.B. Possibility: 12 in. × 14 in. × $1\frac{1}{8}$ in.

Chapter 10

10.5.A. 6 bolts in 2 rows, $\frac{5}{8}$-in. thickness for 8-in.-wide plates, 1-in. thickness for 10-in.-wide plate
10.5.B. 7 bolts required, use 9 in 3 rows for symmetry, $\frac{5}{8}$-in. thickness for outer plates, $1\frac{1}{8}$-in. thickness for middle plate
10.7.A. Could use 8 $\frac{3}{4}$-in. bolts, 6 $\frac{7}{8}$-in. bolts, 5 1-in. bolts, most critical with $\frac{3}{4}$-in. bolts, but OK for shear, bearing, and tearing
10.7.B. Could use 7 $\frac{3}{4}$-in. bolts, 5 $\frac{7}{8}$-in. bolts, 4 1-in. bolts; OK with $\frac{3}{4}$-in. bolts
10.7.C. Could use 4 $\frac{3}{4}$-in. bolts or 3 $\frac{7}{8}$-in. bolts; either OK

Chapter 11

11.5.A. Rounding off, use L_1 = 11 in., L_2 = 5 in.
11.5.B. Minimum: 4 $\frac{1}{4}$-in.-long welds on each side

Chapter 12

12.4.A. 114.6 in.^{3} [1884 × 10^3 mm^3]
12.6.A. (a) From Table 8.2: W 14 × 22; (b) need Z = 27.2 in.^{3}, could use W 12 × 22, but deflection on 32-ft span not good, so probably would also use W 14 × 22

Chapter 13

13.7.A. Width required to get bars in single layer is critical concern; least width: 16 in. with h = 31 in. and 5 No. 10 bars

13.7.B. From 13.7.A, this is under-reinforced section; find actual k = 0.886±, required A_s = 5.08 in.^{2}, use 4 No. 10 bars.

13.9.A. Use d = 11 in., A_s = 2.73 in.^{2}
If d = 16.5 in., A_s = 1.58 in.^{2}

13.9.B. Use d = 14 in., A_s = 4.35 in.^{2}
If d = 21 in., A_s = 2.74 in.^{2}

13.9.C. Use d = 14.5 in., A_s = 5.40 in.^{2}
If d = 22 in., A_s = 3.22 in.^{2}

13.10.A. 5.76 in.^{2} [3.71 × 10^3 mm^3]

13.11.A. M_R = 150 kip-ft; use tension reinforcement of 2 No. 11 plus 3 No. 10, compressive reinforcement of 3 No. 9

13.12.A. Use 8-in.-thick slab, reinforce with No. 4 at 3 in., No. 5 at 5 in., No. 6 at 7 in. or No. 7 at 10 in., use temperature reinforcement of No. 4 at 12 in.

13.12.B. Use 8-in. slab (for deflection), No. 4 at 4, No. 5 at 6, No 6 at 9, No. 7 at 12 (approximately 20° less than by working stress)

13.13.A. Possible choice: 1 at 6 in., 8 at 13 in.

13.13.B. Possible choice: 1 at 6 in., 4 at 13 in.

13.14.A. L_2 = 36 in., less than 48 in. required, L_1 plus hook equals 29.1 in., less than 35 in. required.

13.14.B. L_2 = 42 in., less than 68 in. required L_1 plus hook = 40.4 in., less than 48 in. required

13.14.C. L_2 = 38 in., less than 41 in. required, L_1 plus hook = 32.4 in., greater than required of 29 in.

13.14.D. L_2 = 35 in., just short of 36 in. required, L_1 plus hook = 31.4 in., greater than required of 25 in.

Chapter 14

14.2.A. Need 5-in. thickness for maximum bending moment, with all No. 5 bars spacings 9 in. at outside end and second interior support, 7 in. at first interior support, 10 in. at first span center, 12 in. at second span center

Chapter 15

Note: Two answers are given for Problems and 15.7A and C when the point is very close to a line in the graph.

15.5.A. Curve No. 8: 12-in. column, 4 No. 11
15.5.B. Curve No. 10: 16-in. column, 4 No. 10
15.5.C. Curve No. 13: 20-in. column, 4 No. 9
15.5.D. Curve No. 14: 20-in. column, 8 No. 9
15.5.E. Curve No. 16: 20-in. column, 12 No. 11
15.7.A. Curve No. 8: 14-in. column, 6 No. 9, or Curve No. 9: 16-in. column, 4 No. 7
15.7.B. Curve No. 14: 20-in. column, 4 No. 11
15.7.C. Curve No. 16: 20-in. column, 8 No. 11, or Curve No. 17: 24-in. column, 4 No. 10
15.7.D. Curve No. 18: 24-in. column, 6 No. 11
15.7.E. Curve No. 21: 30-in. column, 6 No. 10

Chapter 16

16.2.A. Footing from Table 16.2 checks out: 9 ft square, $h = 21$ in., 10 No. 8 bars each way

Index